D0860596

Conservation of Mass and Energy

McGRAW-HILL
CHEMICAL ENGINEERING SERIES

THE SERIES

ANDERSEN AND WENZEL *Introduction to Chemical Engineering*
ARIES AND NEWTON *Chemical Engineering Cost Estimation*
BADGER AND BANCHERO *Introduction to Chemical Engineering*
BENNETT AND MYERS *Momentum, Heat, and Mass Transfer*
BEVERIDGE AND SCHECHTER *Optimization: Theory and Practice*
CLARKE AND DAVIDSON *Manual for Process Engineering Calculations*
COUGHANOWR AND KOPPEL *Process Systems Analysis and Control*
GROGGINS *Unit Processes in Organic Synthesis*
HARRIOT *Process Control*
JOHNSON *Automatic Process Control*
JOHNSTONE AND THRING *Pilot Plants, Models, and Scale-up Methods in Chemical Engineering*
KATZ, CORNELL, KOBAYASHI, POETTMANN, VARY, ELENBAAS, AND WEINAUG *Handbook of Natural Gas Engineering*
KING *Separation Processes*
KNUDSEN AND KATZ *Fluid Dynamics and Heat Transfer*
LAPIDUS *Digital Computation for Chemical Engineers*
LUYBEN *Process Modeling, Simulation, and Control for Chemical Engineers*
MCADAMS *Heat Transmission*
MCCABE AND SMITH, J. C. *Unit Operations of Chemical Engineering*
MANTELL *Electrochemical Engineering*
MICKLEY, SHERWOOD, AND REED *Applied Mathematics in Chemical Engineering*
NELSON *Petroleum Refinery Engineering*
PERRY (EDITOR) *Chemical Engineers' Handbook*
PETERS *Elementary Chemical Engineering*
PETERS AND TIMMERHAUS *Plant Design and Economics for Chemical Engineers*
REED AND GUBBINS *Applied Statistical Mechanics*
REID AND SHERWOOD *The Properties of Gases and Liquids*
SCHECHTER *The Variational Method in Engineering*
SCHMIDT AND MARLIES *Principles of High-polymer Theory and Practice*
SCHWEYER *Process Engineering Economics*
SHERWOOD AND PIGFORD *Absorption and Extraction*
SHREVE *Chemical Process Industries*
SLATTERY *Momentum, Energy, and Mass Transfer in Continua*
SMITH, B. D. *Design of Equilibrium Stage Processes*
SMITH, J. M. *Chemical Engineering Kinetics*
SMITH, J. M., AND VAN NESS *Introduction to Chemical Engineering Thermodynamics*
TREYBAL *Liquid Extraction*
TREYBAL *Mass-transfer Operations*
VAN WINKLE *Distillation*
VILBRANDT AND DRYDEN *Chemical Engineering Plant Design*
VOLK *Applied Statistics for Engineers*
WALAS *Reaction Kinetics for Chemical Engineers*
WHITWELL AND TONER *Conservation of Mass and Energy*
WILLIAMS AND JOHNSON *Stoichiometry for Chemical Engineers*

CONSERVATION OF MASS AND ENERGY

John C. Whitwell
Richard K. Toner

PRINCETON UNIVERSITY

McGraw-Hill Book Company

New York
St. Louis
San Francisco
Düsseldorf
Johannesburg
Kuala Lumpur
London
Mexico
Montreal
New Delhi
Panama
Rio de Janeiro
Singapore
Sydney
Toronto

Conservation of Mass and Energy

Republished in 1973 by McGraw-Hill Book Company.
Printed and bound by Kingsport Press, Inc.

34567890KPKP7987

Library of Congress Cataloging in Publication Data
Whitwell, John C 1909–
Conservation of mass and energy.

(McGraw-Hill chemical engineering series)
Original ed. issued in series: A Blaisdell book in the pure and applied sciences.
1. Chemical engineering. I. Toner, Richard Kenneth, 1913– joint author. II. Title.
TP155.W45 1973 660′.2 73–1058
ISBN 0–07–070080–X

Contents

List of Tables

Tables, containing properties of various materials pertinent to the subject matter of the text, appear in a number of places where the properties are first discussed. The locations of these tables are summarized below for the convenience of the reader.

Preface

In the years which have intervened since the preparation of the original manuscript for this text, many changes have occurred in the field of chemical engineering as well as in chemical engineering curricula and secondary school preparation. For example, we find students better prepared in mathematics, in physical chemistry, and in computer programming, and we suspect that many of our colleagues may be experiencing the same improvements. Primarily, these changes allow for a varied approach to the initial course dealing with the principles of mass and energy balances. Less time has to be devoted to fundamentals, such as contained in Chapters 3, 7, and 8. More time is available for the interpretation of the usefulness of mass and energy balances in the future courses and in the real world of design and performance. Although the options for structuring courses around the combinations of chapters suggested in the first preface still exist, it seems that at least one new and important sequence offers great potential for interesting and inspiring the student toward the goals of upper-class, and eventually professional, work. The emphasis which we placed on degrees of freedom in design and performance testing (Chapters S4 and S9B) assumes new and greater importance as processes become more complex, competition becomes more demanding of the maximum return, and challenging new environmental problems appear.

In view of these remarks we suggest the following changes, which we shall employ in our own course. Eventually we plan to incorporate the more successful portions of this experiment into a revision and expansion of the present text. We shall continue to assign material in Chapters 3, 7, and 8, but shall not discuss these subjects in class, although we shall hold our students responsible for the material. We shall devote the time saved to the coordination of processing elements (reactors, stills, separators, pumps, etc.) into processing units and plants. In this work we shall rely upon the simpler of the principles with which Christensen, King, Lapidus, Rudd, Westerberg, and others are engaged. For example, the number of variables and relationships between them can provide a variety of orderings of the calculations depending upon which specifications one chooses. Simultaneous solutions can be transformed into sequential calculations, but one must be careful to note the effects imposed on the constraints. Knowledge of the degrees of freedom which our text presents in uniquely integrated form for both chemical and physical processes is essential to a proper study of this subject. The ability to analyze both the processing elements and their combination into units, for which we include the summing techniques, are also vital. It is essential to realize and note to the student that once one has established the number of degrees of freedom for mass and energy balances, which deal with matters external to the processing units, this number will not change when one becomes involved in the internals of the process (i.e., the design problems). At that time the necessity of introducing new parameters, such as rate constants, areas, pressure-temperature equilibrium relationships, and the like, will not change the number of degrees of freedom. Consequently, as each new parameter is introduced, it is necessary to add one and only one restrictive relationship. Thus the designer is aided in deciding the number of new relationships required, and he may guarantee that this number does appear. If he were erroneously to introduce a greater number, he has overspecified the problem, and no unique solution will appear. Unfortunately, this concept is not emphasized in the literature so that an ideal time to present it is in the initial course, obviously with the treatment of degrees of freedom. Finally, there are the interesting problems associated with the simplification of process flow and information flow sheets by combinations of simply linked elements, ineligible edges, and so forth, by which it is possible to treat a process with a very large number of processing elements in a reasonably-sized, reduced, form.

Even for those teaching courses where time must be devoted to the fundamentals of physical chemical principles we should like to emphasize the increasing importance of the degrees of freedom concepts to all subsequent design and performance calculations. Consequently, we urge those who have not been treating this subject as an integral part of a mass and energy balance course to begin to include these principles and to pursue the details as far as time limitations allow.

JOHN C. WHITWELL
RICHARD K. TONER

Conservation of Mass and Energy

1 The Role of the Chemical Engineer

The chemical engineer is responsible for producing chemicals in marketable quantities at a price the consumer will pay; the chemist is responsible for their invention; and industry organizes and finances both of these operations. Human nature being variable, there is some overlap of duties among chemists and chemical engineers. It is often difficult to tell whether the chemist performs as a chemical engineer, or vice versa, due to the nature of a particular project, due to the structure of the organization of his company, or due to the changing nature of the roles of the two professions as engineers devote less attention to the art of their professions and more to its scientific associations. The gray areas between the professions need not worry us. We are currently concerned with the broad general areas served by the chemical engineer and chemical industry and the consequent duties which the chemical engineer should be prepared to assume.

1.1 The Domain of Chemical Industry

Chemical industry in general takes materials in their natural state and converts them into a more useful form as, for example, crude oil to gasoline, wood to paper, salt to chlorine, sulfur to sulfuric acid, and gases to plastics. Even in this abbreviated list we see some products (gasoline) that will be used by the ultimate consumer and others (chlorine, acid, plastics, and also paper) that must undergo some other manufacturing modification before being attractive to the general public, the ultimate customer. It is logical to classify the first type as *consumer* goods and the second as *producer* goods. Obviously chemical industry produces both. The producer field has traditionally been predominant, but recent growth in areas such as fiber, film, and polymer products has moved the industry more strongly toward the consumer. Despite such trends, its efforts in the field of producer goods are of the utmost importance, for the chemical industry provides essentially *all* the materials required by the other branches of engineering and by all other manufacturing categories. As justification for this broad generality, Table 1.1 and Figure 1.1 are presented. The material presented in the table is fragmentary; it only illustrates the vast array of materials and products with which the industry is concerned.

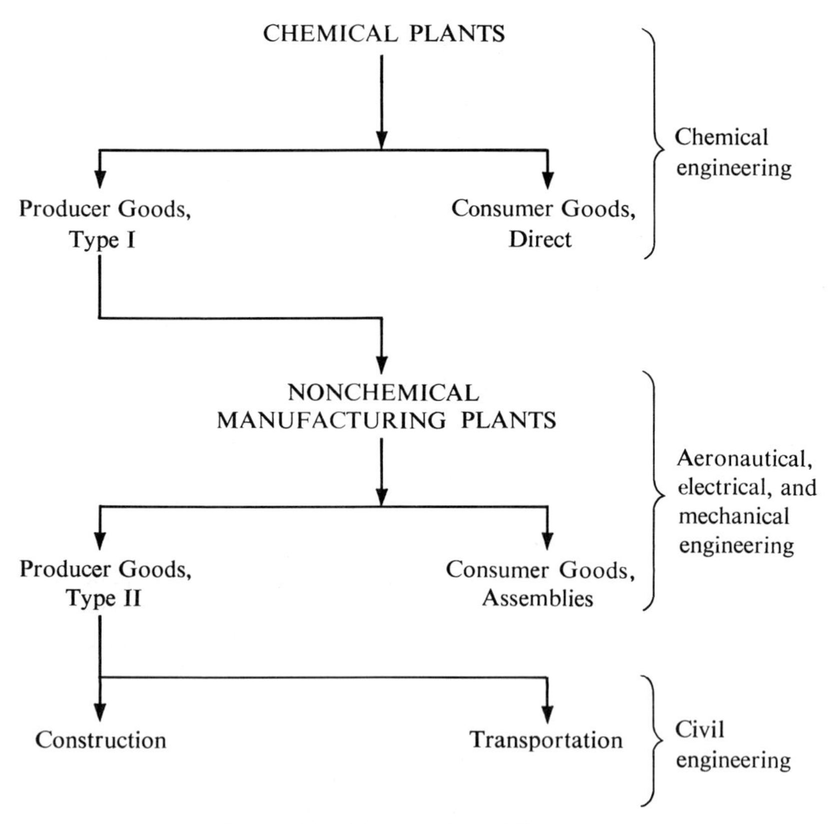

FIGURE 1.1 Organization of industry.

Chemical industry as the producer, and the chemical engineer as the artisan of the industry, are basic to every manufacturing field and to the standard of living that can be supported and supplied. The role of the chemical engineer is unique in this respect. The various responsibilities he must assume in his own field in order to discharge his duties effectively are of considerable interest, particularly with respect to the diversity of such duties and the consequent diversity of talents and personalities.

TABLE 1.1*

Raw Materials	Producer Goods, Type I	Producer Goods, Type II
Oil	Coke	Steel
Sugar	Petroleum Fuels	Plastics
Coal	Petrochemicals	Concrete
Limestone	Dyes	Piping
Gas	Chlorine	Electrical Components
Sea Water	Caustic	Motors
Salt	Alkalies	Chemical Plant Equipment
Metal Ores	Bromine	Petrochemicals
Silicates	Boron	Fabrics
Cotton	Glass	
Wood	Pulp	
Borax Ores	Rayon	
Sulfur	Acids	
Rubber	Plastics	
	Synthetic Yarns	
	Cement	
Consumer Goods, Assemblies	**Consumer Goods, Direct**	**Construction**
Automobiles	Gasoline	Highways
Airplanes	Fuels	Bridges
Rockets	Pharmaceuticals	Power Stations
Radios	Salt	Building Materials
Refrigerators	Sugar	Structures
Furniture	Calcium Chloride	Dams
Electrical Appliances	Glass	Drainage and Fills
Housing	Paper	
	Tires	
	Cement	

* Classification titles are identical to those appearing in Figure 1.1.

1.2 The Role of the Chemical Engineer

The duties of the chemical engineer can be summarized by division into staff and line functions. The latter follow the path of a product from concept to sale and are thereby the most suitable starting point.

A product is conceived in the laboratory, ordinarily by the chemist. In its invention, the author need have no basic interest in economics. His method of production will frequently involve very small fractions of the reactants being converted to the new product and may produce very large quantities of useless by-products. Parenthetically, we must note that "useless by-products" are frequently the source of entire new lines of products created to eliminate the problems of disposing of harmful wastes. The paper industry, with its lignin by-products, is an example of this situation. Petrochemicals is another enormous and profitable example. The first line job for the chemical engineer is the development of the process.

Development Functions

As a first step in the development, the chemical engineer must decide whether it is economical to devote any appreciable amount of the company's funds to this process. With the aid of the chemist he assumes maximum yields; with the aid of the sales force he assumes a market price and then calculates the potential net income, exclusive of manufacturing costs. If this figure is not favorable the project will be abandoned, granted no revision of prices and yields can be effected.

Once this first economic hurdle is successfully passed, the chemical engineer can propose equipment to do the job (alternate lines possibly being included) and reestimate potential profit. From here on we shall assume that each step yields favorable estimates, allowing progression to the next.

Equipment will be assembled for a test of the process, to see whether the optimistic estimates of yield will stand up in actual practice. An extensive experimental program may be undertaken at this point to decide upon reasonable estimates of the maximum that may be expected of the process. These programs may be governed by experimental designs in either, or both, of the categories of statistics or of mechanistic mathematical models.

The number of stages of equipment scale-up prior to the construction of the full-sized equipment will vary with the individual company and the individual process. Presumably, the first set of equipment does not simulate plant units but is similar to laboratory elements. In such a case, some further trial equipment, more closely resembling what is finally anticipated in the plant, would be a logical progression. Several such trial steps are not uncommon.

Pilot Plant Functions

A final development stage may involve a prototype plant, a small model of the intended final plant. It has the function of providing a moderately inexpensive testing ground for the final equipment design and operation, as well as being a source of products for customer testing.

All of these stages of process development involve knowledge of equipment and a constant assessment and reassessment of the economics of the process and product. The close association with chemical reactions requires a man with a strong chemistry background and one who is profit-oriented. The chemist fills the first qualification and is therefore a valuable member of the team. He cannot fill the second requirement. The chemical engineer fills both and is, therefore, generally in charge of the development work.

After successful completion of the development, or even in its final stages, the design, erection, and start-up of the full-sized plant constitute the next items in the program.

The Plant Functions

Design of chemical equipment is an obvious job of the chemical engineer. It might be well at this point to note a strong bifurcation of chemical equipment. On the one hand there are units in which the reactions take place, the chemical reactors. On the other there are units solely concerned with physical separations. The need for the latter group is due to the general rule that no material as produced is pure. The product is contaminated with by-products, with unreacted raw materials, or both. The process which will go to completion without by-products and without an excess of any reactant is so rare as to be essentially nonexistent.

The physical separation processes, being simpler, are more susceptible to analytical study than are those involving chemical reaction. It is not surprising that progress was more rapid in these areas than in the area of chemical reactors. Important advances in the formulation of realistic models around the separation processes was initiated in the 1920's and continued through the next three decades with the establishment of the *unit operations* dealing exclusively with this type of equipment. The unit operations became the basis for chemical engineering education in this period. Reactors were, in the same period, treated in a manner somewhat similar to "flying by the seat of one's pants." The trend toward engineering science has brought chemical reactor design into a more thoroughly studied and understood position in the last ten years. There is still, however, a clear demarcation between reactors and separation processes and a clear necessity for both types of equipment. This text deals with neither directly, but with the overall summations of materials and energies inherent in both. Calculations of these summations is an essential beginning in the design of either type, without which none of the design problems in their greater complexity can be even attacked.

The *erection* of a chemical plant is not fundamentally the aegis of the chemical engineer. His presence for the solution of problems pertaining to the location of lines, instruments, and auxiliaries is, however, an absolute necessity.

Once the plant is built, the start-up involves problems relating to chemical reaction, product separation, and equipment performance; the chemical engineer must take full responsibility for this operation. During continued operation, with full production and optimization of the process, the latter a continuing need throughout the life of the plant, the problems are again those of chemical reaction and separation in production equipment, and remain uniquely in the province of the chemical engineer.

Marketing

The problems of the chemical engineer do not end with the production of a good, or even the best, product in its competitive field. The fact that the field is competitive implies that others will be trying to get their share of the market; it is important to keep customers satisfied both with the product and with its ability to serve their needs as well as, or preferably better than, any other competing product. Technically trained personnel, who are familiar not only with the product and its properties, but also with the equipment in which it is subsequently processed, and who can help the customer solve his own production problems, are again an essential element in the chain of production and marketing. The chemical engineer is ideally educated for this position.

Staff Functions

It is often difficult to separate the development chemical engineer from the research chemical engineer in industry. When the researcher is not directly associated with any particular process or product, but is concerned with chemical engineering problems in general—the extension of the knowledge of the profession—he is truly in a staff position. Small companies rarely can afford this type of position. Large companies can, and do.

Other staff positions result from the necessity to attack problems not unique to any process or product but rather of general benefit to the operation of the company as a whole. In this category we find the statistician, the systems analyst (both mathematically strong and chemically oriented by educational background or by experience in the industry), the control expert, and the consultant. With the exception of the research man, those employed in these fields are not necessarily educated as chemical engineers, while those in line functions generally should be.

Administration

Administrators and executives usually have come through the ranks of positions described above. Their ability to assume and to hold these positions results from a combination of technical prowess and an appreciation of more than a single facet or process in the total company economics. This ability includes both personal and technical considerations. The assumption of duties in these areas usually limits their purely technical activities but results from their proficiency in this line. The day of the lawyer as the executive, with the engineer confined to the plant, is a thing of the past.

1.3 Qualifications of the Chemical Engineer

While the scientist may live in a world of his own, the engineer must live in the real world, solving real problems in real time. He must be motivated to attain specific objectives and to attain them in the most economical manner consistent with all aspects of each particular situation. He must realize that a reasonable solution of the problem in the allowable time is more important than a perfect understanding of all its facets.

To be successful, then, the chemical engineer must have certain characteristics, which have been summarized briefly in an article by E. E. Wilson.* First, he must have the ability to get the job done. This frequently involves defining the problem and scheduling the attack. An industrial assignment will not be specific, nor contain as neat a set of data as a textbook problem. To define the problem and accumulate data, it may be necessary to consult with many others in the organization; obtaining their cooperation is also an essential part of the task.

We thus encounter another qualification, communication through personal relations. Not only is it necessary to get help in attacking a problem, but it is generally necessary to persuade others that decisions reached are correct. Decisions calling for persuasive ability may run the gamut from the introduction of a new idea to the expenditure of thousands of dollars. The engineer frequently must make the latter decision without having available all the information that he would like. When time limitations prevent the accumulation of all the desired material, he must make sound approximations, estimate their effect upon his decisions, and be prepared to defend logically and clearly both methods and results. Personal communication, whether oral or written, becomes one of his basic tools.

Next, he must have initiative. It is not enough to perform well only on assigned tasks. The successful chemical engineer must be alert to find and suggest new assignments. He must be willing to "go out on a limb." He may well be told that he is out of his domain. He must then be ready to defend his ideas and suggestions until convinced that he has chosen the wrong course. The line between justified persistence and pure obstinacy is a difficult one. He must draw it well.

Finally, he should constantly be aware that he must grow in his work, keeping abreast of new developments and being ready to propose and

• • • • •

* *AIChE Student Members Bulletin*, Spring 1966.

support their application. He must search for new, useful techniques at their source, in the literature, at professional meetings, and among his younger, recently educated colleagues and subordinates. Since he cannot expect to be expert in every phase of the subject, he must guide the knowledge and judgment of others into channels that will serve useful purposes in dealing with his problems. Here again we see the tremendous importance of good personal relationships.

The occupation is demanding. If it were not, it would not be rewarding, mentally or financially. It offers unlimited opportunities for originality, perseverance, and cooperative effort. It is rarely humdrum and when it is, the chemical engineer ought to ask himself whether it really is dull or whether he, himself, is allowing it to be so. If the latter, he has a personal reorganization to accomplish. And if he asks the question, he has won half the battle.

1.4 Conservation of Mass and Energy in the Education of the Chemical Engineer

The most comprehensive view of industrial chemical change is contained in the application of the laws of conservation to balances around elements, units, or processes. This application provides information about the quantities of mass and energy which enter and leave the element, unit, or process, without concern for the internal workings which promoted the changes. In our application of these laws, then, we are not interested in the internal functioning. We are concerned with the usefulness of these laws in providing information about input or output streams and particularly with the possibility of providing information by calculation as a substitute for direct measurement. Chapters 4, 9, and 10 deal directly with this type of problem. The other chapters provide the fundamentals necessary for an understanding or performance of these applications.

2 Systems of Units

2.1 Existing Situation

All phases of engineering work are conducted on a dual system of units, the more scientific aspects using the metric system and the more applied dealing to a large extent in the English system. Undesirable and illogical as the English system is, there seems to be little chance that it will be discarded completely in the foreseeable future. There are so many practicing personnel who have grown up with the system and there is so much industrial information available in these units that a complete conversion to the metric system in a single operation would be most difficult, if not impossible.

2.2 Fundamental Systems

All systems deal with a group of fundamental units, defining measures of mass, length, and time. In metric units one encounters the c.g.s. or k.m.s. systems. The first is an abbreviation for centimeter (cm), gram (gm), second (sec); the second refers to the kilogram (kg), meter (m), second (sec). Either is equally appropriate, although the larger units of the k.m.s. system have some advantage for industrial applications. However, since the metric system is primarily used for scientific data, the c.g.s. system has been chosen for this text, due to its general use in handbooks and other sources of scientific data.

2.3 Procedure for Converting Between Systems

The procedure proposed here for converting between English and metric systems is simple. The units to be chosen are, in general, readily visualized. For example, the inch is a smaller unit than the foot, the centimeter is a smaller unit than the inch, the gram is a smaller unit than the pound.

The conversion factors are then reported as 12 in. per ft, 2.54 cm per in., and 454 gm per lb-mass. (The figure 454 is usually sufficiently accurate for engineering work; more precisely, it should be listed as 453.59.)

To use conversion factors one inserts both the numerical value and the units and uses the basic knowledge of the relative size of the units to double-check on the correct usage. For example, if one wishes to know the number of pounds-mass equivalent to 1500 gm:

$$1500 \text{ gm} \div 454 \text{ gm/lb-mass} = \left(\frac{1500}{454}\right) \text{lb-mass},$$

$$= 3.30 \text{ lb-mass}.$$

In this calculation the units of grams in both numerator and denominator have canceled, leaving pounds-mass in the numerator. The location of the numerical values is obvious, the 1500 remaining in the numerator and the 454 being the divisor. Note also that it is known that the gram is a small unit compared to the pound. Thus the number of pounds must be smaller than the number of grams, which gives a qualitative check on the direction of change from 1500 to 3.30.

To complicate the situation very slightly, suppose that one wishes to know the number of centimeters in 2 ft. Proceed as before but include both conversions, feet to inches and inches to centimeters: 2 ft × 12 in./ft × 2.54 cm/in. = (2)(12)(2.54) cm = 60.96 cm. Again the original dimension, feet, is canceled in the first factor, being replaced by inches; in turn, the dimension, inches, is replaced by centimeters in the second factor. Also, as before, the qualitative direction of change in numerical values is seen to be correct in each portion of the conversion.

In both these instances the magnitudes of the initial and final units were readily visualized. There are many less familiar quantities in which this visualization of the whole unit is impossible. For example, refer to the unit of viscosity, as detailed below. This unit has the advantage of illustrating conversions of mass, length, and time, all in the one property, viscosity. This type of combination is commonly encountered in greater or less complexity. If we take some number of poises (P), say 11.2, and wish to know the number of English viscosity units in pounds per (foot) (hour), each individual unit will be converted separately and an overall conversion factor need never be obtained. Such a single factor will be inherent in the combination of individual conversions and is available if desired. Obtaining this composite is not recommended in general, since it tends to make the user dependent upon conversion tables rather than on the simple elements from which it came. There are times when one will need such a factor time after time; under these circumstances the need for the composite provides the addition of this factor to one's general fund of knowledge; only under these circumstances is the use of the composite conversion factor recommended.

Example 2.1

11.2 P = 11.2 gm per (cm) (sec), by definition. From information already listed,

$$11.2 \frac{\text{gm}}{(\text{cm})(\text{sec})} \times \frac{1}{454} \frac{\text{lb}}{\text{gm}} \times (2.54)(12) \frac{\text{cm}}{\text{ft}} \times 3600 \frac{\text{sec}}{\text{hr}} =$$

$$\frac{(11.2)(2.54)(12)(3600)}{454} \frac{\text{lb}}{(\text{ft})(\text{hr})} = 2705 \text{ English hourly units.}$$

(The composite factor can readily be calculated to be 242.)

It is apparent that conversion between systems for any quantity can be reduced to conversions between the quantities of length, mass, time, and eventually temperature. These conversions, and some of their more common combinations, will be the subject of the remainder of this chapter. The interrelation between the pound-mass and the pound-force will also be considered.

Remember that conversion is accomplished by use of the numerical value *and* the dimensional ratio. Multiplication or division by the listed factor is determined by the units one wishes to replace; the numerical value enters automatically.

2.4 Useful Conversion Factors: Fundamental Units of Mass, Time, Length, Temperature

Units of Mass

The *necessary* conversion factors, the second of which is obvious, are as follows:

454 gm/lb,
1000 gm/kg.

Implied is 2.2 lb per kg, but this one is unnecessary after statement of the two previous factors.

Following no system, but having some general usage, the various tons are defined:

2000 lb/ton—the so-called short ton,
2240 lb/long ton,
1000 kg/metric ton.

A number of other measures exist, rarely useful in engineering work. However, to give a more comprehensive picture, the following less common measures are defined, with ounces and pounds referring to the avoirdupois, and not to the troy or the apothecary, system.

16 oz/lb,
7000 grains/lb,
16 drams/lb.

NOTE Only the first three of the eight listed factors will ordinarily be required.

Units of Time

There are no difficulties in this category. All of the necessary time units, seconds, minutes, hours, days, weeks, months, and years are well known and are common to both metric and English systems. The most commonly required conversions are:

60 sec/min,

60 min/hr.

The direct conversion, 3600 sec per hr, is clearly implied.

The only further time units which might be mentioned are the milli- and micro-second, usually abbreviated msec and μsec. They imply a subdivision of the smallest time unit, the second:

10^3 msec/sec,

10^6 μsec/sec.

Units of Length

The *necessary* conversion factors are as follows:

2.54 cm/in.—truly 2.539998,

100 cm/m,

1000 mm/m,

12 in./ft.

Other factors in less common demand, but well known, are:

3 ft/yard, and the implied 36 in./yard,

5280 ft/mile.

The rod has little use except in civil engineering. There are 16.5 ft per rod.

For measurement of extremely small lengths, the micron (μ), and the angstrom (Å), are very common and will be frequently encountered in scientific data. They are primarily concerned with further subdivision of the metric scale. The conversions are:

10^6 μ/m,

10^{10} Å/m.

NOTE Only the first four of the eight factors listed will ordinarily be required, and only the first of these might not be well known. The micron and the angstrom may, conceivably, be commonly encountered, depending upon the user's main fields of application. Certainly they will be needed in all phases of microscopy and in dealing with molecular spacings in compounds and crystals.

Units of Temperature

The only logical temperature scale is one based upon an absolute zero temperature. The *P-v-T* relationships for ideal gases (see Chapter 3) provide such a fixed temperature point and the Carnot heat engine cycle, working on an ideal gas, confirms that this zero temperature is also a thermodynamic zero temperature. (The Carnot cycle will not be considered in this text. The reader is referred to any standard introductory thermodynamics text.)

Unfortunately, a need for temperature scales preceded the knowledge of the thermodynamic zero and arbitrary scales were established with reference points at easily measurable and reproducible points. Thus the Centigrade scale used the freezing point of water as the zero point and the boiling point of water as 100°, splitting the region between into 100 equal parts. In the same manner, but somewhat less logically, the Fahrenheit scale used 32 and 212° for the freezing and boiling points of water, splitting the region between into 180 equal parts. These two arbitrary scales are the ones most commonly encountered. Their interrelation and their relation to absolute zero are then absolutely essential. The simplicity of the conversion factors and relationships leaves no excuse for reliance upon conversion tables, or even for any extensive use of such tables.

The relationships between the four temperature scales, two arbitrary and two absolute, is shown clearly in Figure 2.1. First, on the Centigrade scale the absolute zero is established at $-273.16°$, an experimentally determined figure obtained by extrapolation of very precise *PV*-data for ideal gases (see Figure 2.2 and Chapter 3). Taking this point as the zero on a new scale, termed the Kelvin scale, the relationship between Centigrade and Kelvin scales is, obviously,

$$°K = °C + 273.16$$

where °K and °C denote numbers of degrees on the Kelvin and Centigrade scales, respectively. This relationship is apparent from Figure 2.1.

Since there are 100 Centigrade degrees in the same range covered by 180 Fahrenheit degrees, the latter are smaller units, and we can write

$$\frac{1.8\ \Delta\ °F}{\Delta\ °C}; \quad \text{also} \quad \frac{1.8\ \Delta\ °R}{\Delta\ °K},$$

where °F stands for degrees Fahrenheit. As long as difference in degrees is all that is required, as with the rise in temperature as a material passes through a heater, this conversion is all that is necessary. Thus Δ °C and

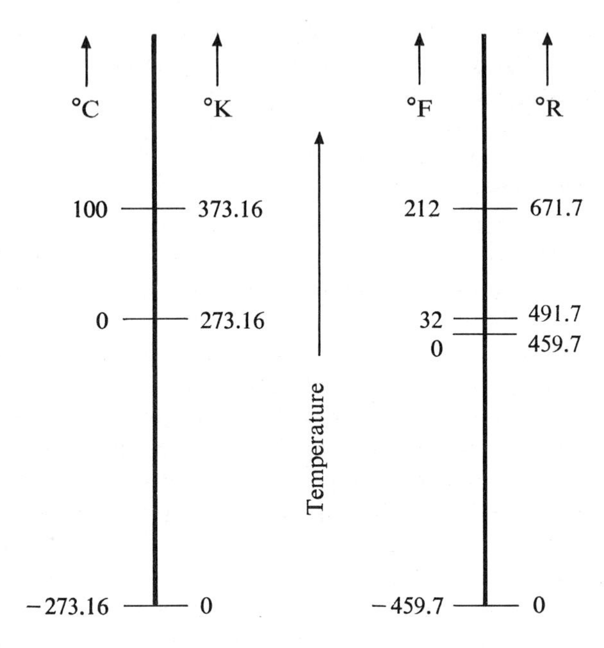

FIGURE 2.1 Interrelations between common temperature scales.

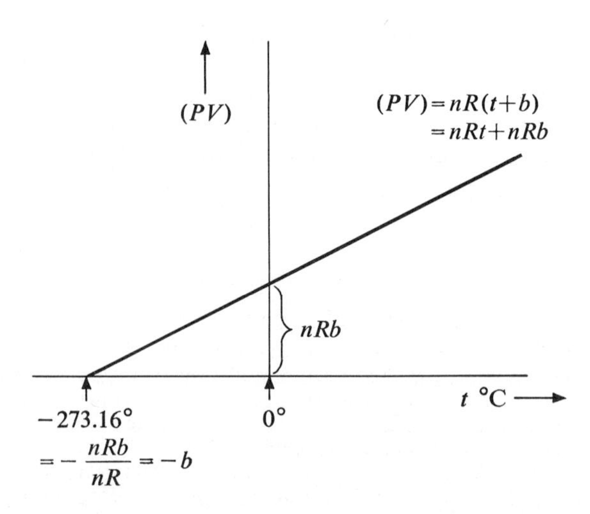

FIGURE 2.2 Experimental determination of absolute zero on Centigrade scale.

Δ °F have been written rather than simply °C and °F.* When a *point* on one scale is to be related to a point on another it is necessary to correct for the fact that the zero points do not represent the same "degree of hotness." Thus, a number of Centigrade degrees may be converted to another number of degrees sized according to the Fahrenheit scale; this number is the number of Fahrenheit degrees above 32 °F; 32 must then be added to the number of converted degrees to place the temperature on the true Fahrenheit scale. In reverse, to convert from Fahrenheit temperature to Centigrade temperature, 32 is first subtracted to obtain comparable scale ranges and the resulting number is divided by 1.8 Δ °F/Δ °C. Since the resulting temperature is based on the zero point of the Centigrade scale, the Δ °C so obtained is numerically equivalent to °C.

The absolute scale, in Fahrenheit-sized units, is known as the Rankine scale, with degrees on this scale designated °R. The zero point of this scale may be determined from that on the Centigrade scale, by multiplying 273.16 by 1.8. The resulting 491.7 is the number of degrees below 32 °F, so that

$$°\mathrm{R} = °\mathrm{F} + 459.7.$$

In engineering work, the two constants 273.16 and 459.7 are usually taken to be 273 and 460 respectively, with sufficient accuracy.

The previous remarks on temperature conversions may be *summarized* as follows:

$$°\mathrm{K} = °\mathrm{C} + 273,$$
$$°\mathrm{R} = °\mathrm{F} + 460,$$
$$°\mathrm{F} = 1.8\,°\mathrm{C} + 32,$$
$$°\mathrm{C} = \frac{°\mathrm{F} - 32}{1.8}.$$

These relationships are quite correct (within the limits of accuracy denoted by 273 and 460) but have the danger of being considered as formulas. If so treated, they are invariably memorized, an action which begs for subsequent improper application and which is entirely unnecessary in light of the simple visual interrelationships.

• • • • •

* The use of °F, °C, °K, °R, to indicate scale readings of temperature, and Δ °C, Δ °F, etc., when temperature differences are designated, is correct. The Δ-terminology becomes unwieldy, particularly in typing. Two recourses are available. One is simply to drop the Δ, relying upon the user to provide the distinction through instinctive recognition of the proper choice, according to connotation. The other is to reverse the symbols (i.e., F°, C°, etc.) when temperature difference is intended. The latter procedure is logical since one reads the symbol as Fahrenheit degrees, Centigrade degrees, etc., rather than degrees Fahrenheit, degrees Centigrade, etc. The implication is then clearly one of "degrees of a given size" rather than a temperature reading.

Because texts and handbooks rarely make the distinction, the authors have chosen not to reverse the symbol; thus the reader must learn to make the distinction for himself.

There have been other temperature scales (such as the Réamur scale which uses the freezing and boiling points of water to define the zero point and 80° point, respectively) but none has found wide scientific and industrial application. There was a time when science used the Centigrade scale exclusively and industry the Fahrenheit scale equally exclusively. Now the Centigrade scale is found extensively in industry so that conversion between scales is an everyday task for the engineer. The Centigrade scale is also known as the Celsius scale, and while there is a trend toward the use of the latter term, the former is more commonly encountered and will be used in this text.

2.5 Generality of Fundamental Units

When one understands the units of mass, time, length, and temperature, one should have all the information necessary to handle any quantity, for all other quantities are products and ratios of these fundamental units. A few quantities are listed in Table 2.1, with the fundamental units abbreviated as M, L, t (time); T (temperature) will not appear until one encounters some unit dependent upon a temperature difference. (See heat capacity, Chapter 7.) Thus, if it were not for complications caused by special definitions of arbitrary measures, a variety of empirically defined energy units, and the introduction of reference states to produce dimensionless quantities, there would be no need for any more remarks about units. As all these problems exist extensively in everyday usage, there is a need for considerable further discussion.

TABLE 2.1

Quantity	Definition	Fundamental Units
Velocity	Distance per unit time	L/t
Acceleration	Change in velocity per unit time	L/t^2
Force	Mass × acceleration (Newton's second law of motion)	ML/t^2
Pressure	Force per unit area perpendicular to it	M/Lt^2
Area	Product of two mutually perpendicular lengths	L^2
Volume	Product of three mutually perpendicular lengths	L^3
Density	Concentration of mass	M/L^3
Weight	Force with which body is attracted to earth	ML/t^2
Work (also energy)	Force acting over unit distance	ML^2/t^2
Power	Rate of performing work	ML^2/t^3

2.6 Force-Mass

From Newton's second law of motion it is apparent that the units of force and mass cannot be identical, for mass has fundamental unit, M, and force has fundamental units, ML/t^2. In the basic English system, there is, however, no clear distinction between force and mass, both being expressed in pounds. Then Newton's law can hold only through the introduction of a dimensional constant, k, so that the equation reads

$$F = kMa, \qquad \text{where} \qquad a = \text{acceleration.}$$

In any system where a clear and proper distinction is made between force and mass, $k = 1$ and is dimensionless, and in any system where such distinction is not made, k *must* have units and *may* differ from unity.

There are four systems in frequent use, the metric, the English gravitational, the English absolute, and the third English system, which does not distinguish clearly between force and mass units, and has earned the unfortunate designation of English engineering. The units used in each are summarized in Table 2.2.

TABLE 2.2

System	Force	Mass	Acceleration	k
Metric	Dyne	Gram	cm/sec^2	1
English absolute	Poundal	Pound	ft/sec^2	1
English gravitational	Pound	Slug	ft/sec^2	1
English "engineering"	Pound-force	Pound-mass	ft/sec^2	$1/g_c$

where g_c = (lb-mass/lb-force)(ft/sec^2)

Two problems remain. The first is the relationships between pounds, poundals, and slugs. The second is the proper notation in the English "engineering" system to avoid the mistakes which could result from failing to take into account the dimensional character of k.

From the Newton equation, it is apparent that 1 dyne will accelerate 1 gm at 1 cm per sec^2. Similarly, one poundal (or 1 lb) will accelerate 1 lb (or one slug) at 1 ft per sec^2. Now if g_c is chosen as the value of acceleration for a freely falling body in vacuum under the earth's gravitational field alone, 32.17 (lb-mass per lb-force)(ft per sec^2),* the relative size of the poundal and slug

• • • • •

* The acceleration due to gravity is dependent upon both the latitude and the elevation. The value adopted as a standard by the International Committee on Weights and Measures is 32.174 ft/sec^2 (equivalent to 980.665 cm/sec^2). This value is the acceleration due to gravity at sea level and latitude between 45 and 46°.

may be readily argued. The poundal moves a pound at 1 ft per sec^2; the pound-force moves 32.17 lb-mass at 1 ft per sec^2. Thus the poundal is a small unit, relative to the pound-force, with conversion factor

$$32.17 \text{ poundals/lb-force.}$$

Similarly, the slug is a large unit relative to the pound-mass, with conversion factor

$$32.17 \text{ lb-mass/slug.}$$

Predominantly, the poundal is used in physics, while the slug is used in the aeronautical and space sciences.

Now consider the problem of proper notation in the "English engineering" system. If the action on the body is entirely due to gravity, so that g, the acceleration due to gravity alone can be used in place of a in the Newton equation, we can write

$$\frac{g}{g_c} = \frac{F}{M}.$$

This ratio must have dimensions of pounds-force per pounds-mass in the "English engineering" system. Suppose that pounds-mass are taken as the basic use of the pound and pounds-force are required; pounds-force must then be represented as $(g/g_c)M$. For example, if pressure, which is defined as force per unit area, is expressed as pounds per unit area it must be written $(g/g_c)M$/unit area, since M has been reserved for pounds-mass. Several examples are given in Table 2.3, the definitions of the quantities being found in Table 2.2. In this table it is evident that whenever F appears in the listing of fundamental units, it is replaced by $(g/g_c)M$ in the English engineering. In a large number of engineering calculations, the geographical location of the plant, unit, or process is at, or near, sea level in temperate zones. Under these circumstances, g is equal to, or nearly equal to, g_c. Failure to include the factor, g/g_c, when technically required, would not result in a serious error. For instance, at 15,000 ft in the Montana Rockies, $g = 32.13$ ft per sec^2. Stationary equipment on the earth's surface will not, then, be greatly affected by the inclusion or omission of the ratio, g/g_c, since the ratio under these conditions is nearly 1.000.

TABLE 2.3

Quantity	"English Engineering" Units (M = pounds-mass)	Fundamental Units (M = mass)
Weight	$(g/g_c)M$	$ML/t^2 = F$
Density	M/L^3	M/L^3
Work (also energy)	$(g/g_c)ML$	$ML^2/t^2 = FL$
Pressure	$(g/g_c)M/L^2$	$M/Lt^2 = F/L^2$

In order to avoid as much as possible the confusions which can accrue in the use of the pound-mass and the pound-force in the English system, the following convention will be used in this book. The word *pound* and the abbreviation lb will be used for pound-mass only. The term *pound-force* will be written out or abbreviated as lb_f.

2.7 Combination of Fundamental Units

Area Units

Area units can generally be derived from length units. For example, if one wishes the number of square centimeters equivalent to 7.2 sq in.,

$$7.2 \text{ in.}^2 \times (2.54 \text{ cm/in.})^2 = 46.44 \text{ cm}^2.$$

Of the special, arbitrarily defined area units, the only one likely to be encountered (and that due to its use as a measure of real estate) is the acre. It can be defined in terms of square rods or square feet. Only the latter will be reported here, the rod having been excluded from interest in this text.

$$43{,}560 \text{ sq ft/acre.}$$

Volume Units

Volume units are also normally derived from length units. The liter and the gallon are useful arbitrary standards. The *liter* is defined as the volume of pure water, measured at 4 °C and barometric pressure of 760 mm of mercury, which has a mass of 1 kg. This liter is equivalent to slightly more than 1000 cu cm. The relationship, exactly stated, is

$$1000.027 \text{ cm}^3\text{/liter.}$$

For ordinary purposes there is no meaningful difference between the milliliter and the cubic centimeter, since the latter is only 0.0027% smaller.

The *U.S. gallon* of water at 15 °C contains 8.337 lb-mass. As a measure of capacity, it is better reported in volume units as noted below, but the specification is consistent with that for the liter, a volume containing a specified mass. The most convenient capacity factors are

$$7.48 \text{ U.S. gal/ft}^3,$$
$$231 \text{ in.}^3\text{/U.S. gal.}$$

Each of these factors is convenient at times. The individual user will probably prefer one to the other and use this preference almost exclusively.

Density

As defined in Table 2.1, density is a concentration of mass. It is commonly expressed in the metric system as grams per cubic centimeter or grams per milliliter, and in the English system as pounds-mass per cubic foot. The temperature must be designated, since volumetric dependence on temperature will affect the mass contained. The relationship between values of density in the two systems is easily obtainable by combination of fundamental conversion factors. For example, suppose that at a specified temperature, the density in grams per cubic centimeter is unity. The density in pounds-mass per cubic foot may then be calculated as follows:

$$1\,\frac{\text{gm}}{\text{cm}^3} \times \frac{1}{454}\,\frac{\text{lb}}{\text{gm}} \times \left(\frac{2.54\ \text{cm}}{\text{in.}}\right)^3 \times \left(\frac{12\ \text{in.}}{\text{ft}}\right)^3 = 62.37\ \text{lb/ft}^3.$$

Specific Gravity

In contrast to density, specific gravity has no units. The use of the word "specific" often implies that reference is made to some standard. In the case of specific gravity, the density of the substance in question is divided by the density of some standard substance, frequently water when liquids are being considered and air when gases are being considered. The division of one density by another cancels the units and leaves a number, i.e., a dimensionless quantity. The absolute density of water is a maximum at 4 °C, with a value of 1.0000 gm per ml. With this condition as the reference state, specific gravities, and densities, are numerically equal *in the metric system.* If the specific gravity is reported as $0.85^{25/4}$, the superscript is indicative of the two temperatures involved, the first being for the substance in question and the second for the reference material. In the case noted, the density of the material in question is 0.85 gm per ml since the density of water at 4 °C is 1.0000 gm per ml. At 15 °C the density of water is 0.99913 and at 20 °C, it is 0.99823, both in grams per milliliter. With ordinary engineering accuracy, either could be taken to be equal to 1.00.

The two temperatures may also be expressed in the Fahrenheit scale. In this instance the reference temperature is *usually* 60 °F (corresponding to 15.6 °C, not to the temperature for maximum density). The specific gravity might be recorded as $1.35^{60/60}$ in this system. Although the temperature scales are not usually identified, there should be no difficulty in distinguishing between the use of Centigrade and Fahrenheit.

It is apparent that a specific gravity must note not only the numerical value but also the two temperatures. In the case of air as a reference, the pressure condition must also be noted. (See Chapter 3.)

If one has a specific gravity and wishes to obtain the density in English units, there is no numerical equivalence as in the previous example where density is expressed in the metric system. The density of water at 4 °C has already been shown to be 62.37 lb per ft^3, since this is the equivalent of 1.000 gm per ml. Thus, for the substance whose specific gravity is reported as $0.850^{25/4}$, the density in pounds per cubic foot will be (0.850)(62.37) = 53.04

lb per ft^3. For a specific gravity of $1.35^{60/60}$, the density in English units is (1.35)(62.34) = 84.16 lb per ft^3.

CAUTION A word of caution with regard to liquid mixtures is worthwhile at this point, and the reader is referred to the chapter on material balances for a complete explanation. There is a great temptation for the beginner to believe that the knowledge of the density of a mixture and of its pure components also provides information regarding the composition of the mixture. This conclusion is invariably wrong; if it were to be true, the components would have to form a perfectly ideal mixture, a situation so rare or so limited in range of compositions as to be nonexistent for all practical purposes. Therefore, if one has density or specific gravity information and wishes to extract therefrom the composition of the mixture, such information can only be obtained by a table relating these two properties; the *relationship cannot be calculated.*

Special Specific Gravity Scales • A number of special specific gravity scales have had fairly common usage, generally as hydrometer scales, with readings as a number of degrees. The API, Baumé, and Twaddell relationships are summarized below. Little if any further attention will be paid to them here, since their use adds nothing to our basic knowledge of engineering calculations; one must first convert the scale readings to specific gravity and then proceed normally.

$$\text{API:} \quad \text{sp. gr.}^{60/60} = \frac{141.5}{°\text{API} + 131.5},$$

$$\text{Baumé: sp. gr.}^{60/60} = \frac{140}{°\text{Bé} + 130}$$

(for liquids lighter than water, with scale from 10 to 60°).

$$\text{sp. gr.}^{60/60} = \frac{145}{145 - °\text{Bé}}$$

(for liquids heavier than water, with scale from 0 to 66°).

$$\text{Twaddell: sp. gr.}^{60/60} = 1 + \frac{°\text{Tw}}{200}$$

(for liquids heavier than water only).

The reader will readily think of other special scales, such as that used for testing the charge on automobile batteries.

Pressure

Pressure is defined as force per unit area normal to the restraining surface. It should then be expressed as dynes per square centimeter in the metric system and in pounds-force per square foot in the English system. Since the dyne is a very small unit and since the pound-mass is the basic pound in the English system, neither of the two natural units is much used in engineering. The most common expression is pounds per square inch which must be expressed as:

$$\left(\frac{g}{g_c}\right)M/\text{sq in. in true force units.}$$

Suppose that a column of water, of density 62.4 lb-mass per cu ft, is 2 ft high and is supported on a horizontal surface of area equal to 1 sq ft. The mass of the column of water is 2(62.4) lb-mass. The force exerted by the water alone (independent of any force contributed by the atmosphere above the water column) is $(g/g_c)124.8$ pounds-force. This force acts on 144 sq in. and the pressure is then $124.8(g/g_c)/144 = 0.866(g/g_c)$ lb-force per sq in. In the usual case $g/g_c = 1.00$ and the pressure is simply 0.866 lb-force per sq in.

It is fairly obvious that, in this situation, the pressure is proportional to the height of the column of liquid. Such heights, designated as *heads* of liquid, then offer means of indicating pressures. Heads are converted to pressures in force per unit area by multiplying the head by density (both in equivalent units) and by g/g_c.

Example 2.2

▶ STATEMENT: what is the pressure in pounds per square inch corresponding to a column of mercury, at 25 °C, 760 cm in height?

▶ SOLUTION: The density of mercury at 25 °C is 13.534 gm/cm³.*

$$\text{gm/cm}^2 = 760 \text{ cm} \times 13.534 \text{ gm/cm}^3 = 10{,}285.8.$$

Converting to English units and multiplying by g/g_c:

$$10{,}285.8 \text{ gm/cm}^2 \times (g/g_c) \times \text{lb-mass}/454 \text{ gm} \times (2.54 \text{ cm/in.})^2 = 146.2(g/g_c) \text{ lb-force/in.}^2.$$

Note that the conversion of head to force per unit area cannot be performed accurately unless the temperature of the fluid head is known.

Pressure can be measured (and expressed) in absolute or relative terms. An absolute pressure is a record of the total pressure at the point in question. A relative pressure is the difference in pressure at the point in question from some arbitrary standard. Obviously the relative pressure can be above or below the standard.

• • • • •

* *Handbook of Chemistry and Physics*, 45th ed. (Cleveland, Ohio: The Chemical Rubber Publishing Company, 1964–65), p. F-5.

Relative pressures are common in two circumstances; the first denotes the excess (decrease) of pressure above (below) standard atmospheric pressure. All ordinary pressure gauges read in this fashion. They usually indicate pressures in pounds-force per square inch for pressures in excess of atmospheric and in inches of mercury below atmospheric pressure for pressures less than atmospheric. Thus the following types of statement may appear: "The pressure is 47 psig," or "The vacuum gauge reads 15 in. of mercury." The former means 47 lb per sq in., gauge, to which must be added barometric pressure to get the total (i.e., absolute) pressure. Assuming that barometric pressure is 14.7 lb per sq in., the absolute pressure is 61.7 lb per sq in., which would usually be represented as 61.7 psia. For the vacuum noted above (15 in.), if the barometer reading is 30.1 in the same units, the absolute pressure is $30.1 - 15 = 15.1$ in. of mercury, which might be abbreviated 15.1 in. Hg.

It should be noted that while data may frequently be recorded as gauge pressure, physical and mathematical relationships almost invariably require that absolute pressure be used.

Only where ΔP appears does it make no difference whether absolute or relative pressure is employed. Thus, a second use of relative pressures is in the recording of drop in pressure between two points as discussed in the appendix dealing with manometers.

Energy (and Work)

Energy (and work) have the units of force times distance, with resulting fundamental units of ML^2/t^2. The equivalence of energy and work is readily rationalized in terms of potential energy. If an object is moved through a distance, work is done in moving the object. On the other hand if an object exists at such a level that by removal of its support it will spontaneously move through a distance to a new level, it possesses the potential energy equivalent to the work that it would perform if it were to move to this new level. Its potential energy is then equal to the work which it can spontaneously perform. Further remarks in this section will be under the heading of energy but the reader will understand that they are equally applicable to work.

The *units of energy* may be in any system expressing force times distance. Thus one would expect units of dynes times centimeters and pounds-force times feet, in the metric and English systems, respectively. These units are, in fact, fundamental to the two systems, the dyne centimeter being given the new name, the erg, and the English unit commonly being called the foot-pound but the pound here is the pound-force. This latter unit is in general use, but the erg is so small that a larger unit, the absolute joule, is defined, equal in size to 10^7 ergs. The conversion factor is 10^7 ergs per J (joule).

To obtain the relationship between the erg and the foot-pound force, we convert the former, unit by unit.

$$1 \text{ erg} = 1 \text{ dyne-cm} = 1 \text{ gm-cm}^2/\text{sec}^2,$$

$$1 \text{ erg} = 1 \times \frac{\text{gm-cm}^2}{\text{sec}^2} \times \frac{\text{lb}_\text{m}}{454 \text{ gm}} \times \frac{\text{ft}^2}{(2.54 \times 12)^2 \text{ cm}^2},$$

$$= \frac{1}{42.12 \times 10^4} \frac{\text{ft}^2\text{-lb}_\text{m}}{\text{sec}^2}.$$

But

$$g_c = 32.2 \frac{\text{ft-lb}_\text{m}}{(\text{lb}_\text{f})(\text{sec}^2)},$$

so that, dividing by g_c we obtain

$$1 \text{ erg} = \frac{1}{1.356 \times 10^7} \text{ ft-lb}_\text{f}.$$

Thus, conversion factors of interest can be written:

$$1.356 \times 10^7 \frac{\text{ergs}}{\text{ft-lb}_\text{f}},$$

$$1.356 \text{ J (abs)/ft-lb}_\text{f}.$$

The conversion can, of course, be reversed by taking 1 ft-lb$_\text{f}$, first multiplying by g_c, and then converting unit by unit to the c.g.s. system.

Arbitrarily defined *heat units* often offer convenient forms for energy units. A number of such heat units exist due to the existence of different systems of units and temperature scales. All are defined as an amount of heat (heat being a form of energy) required to raise the temperature of some specific substance through a specific temperature range. The substance chosen as a standard is water. The temperature range must be closely defined since the amount of heat required per degree change in temperature varies with temperature. (See remarks in Chapter 7 on the variation of heat capacity with temperature.) Thus, a unit which shall be named the calorie (or more completely the gram-calorie) can be defined as amount of heat required to raise the temperature of 1 gm of water from 15 to 16 °C. It could also be defined as $\frac{1}{100}$ of the amount of heat required to raise the temperature of 1 gm of water from 0 to 100 °C. These two are not *exactly* equal, the former being designated the gram-calorie (15 °C) and the latter the gram-calorie (mean). Either may be related to the joule; and *mean* calorie equals 4.186 J (abs). To complicate the picture further, two other definitions of a calorie exist, one the thermochemical calorie and the other the steam calorie, the size of each being defined in terms of joules and standardized by electrical measurements. The thermochemical calorie is equal to 4.1833 international* joules

• • • • •

* The international and absolute joules are not quite equal, the relationship being 1.00032 J (abs) per J (int). Therefore 4.1846 J (abs) per thermochemical calorie.

and the steam calorie is essentially equal to the mean gram-calorie. It is important to know that these distinctions exist, but the use of one in preference to another is rarely important in engineering work. No further distinction will be made in this book. The calorie equal to 4.186 J(abs) will be used throughout.

The relationship between the calorie and the joule will, in fact, be infrequently required. More important relationships will be found to exist between the calorie, other defined heat units, and the foot-pound. The next steps are to define the other heat units, to relate all the heat units, and finally to relate one or all of them to the foot-pound.

The *British thermal unit* (Btu) is defined in the English system as the calorie was first defined, a quantity of heat required to raise the temperature of 1 lb-mass of water from 60 to 61 °F. According to strict international usage the Btu can be expressed as a number of international thermochemical calories. This number is 1 Btu/252.06 international thermal chemical calories.

The *Pound-centigrade unit* (Pcu)* is defined as the quantity of heat required to raise the temperature of 1 lb-mass of water from 15 to 16 °C. This unit is obviously a hybrid with roots partly in the English and partly in the Centigrade system, the latter being more generally associated with the metric than with the English system. The Pcu has importance due to the frequent use of the Centigrade scale in industry where the unit of mass remains the pound.

Conversions between heat units are simple if one disregards the slight differences between the heat capacity in the 15 to 16 °C range and that in the 60 to 61 °F range. One can then consider the units of heat to be those of mass × Δ-degrees. Interchange between units can be made by converting separately mass units and size of degrees in the two scales. For example suppose that one wishes to know the number of Pcu and the number of calories equivalent to 18 Btu.

$$
\begin{aligned}
18 \text{ Btu} &\doteq 18 \text{ lb-mass} \times \Delta\,{}^\circ\text{F} \times 454\,\frac{\text{gm}}{\text{lb}} \times \left(\frac{1}{1.8}\right) \Delta\,{}^\circ\text{C}/\Delta\,{}^\circ\text{F}, \\
&\doteq 4540 \text{ gm} \times \Delta\,{}^\circ\text{C} \doteq 4540 \text{ gm-cal.}
\end{aligned}
$$

And

$$
\begin{aligned}
18 \text{ Btu} &\doteq 18 \text{ lb-mass} \times \Delta\,{}^\circ\text{F} \times \left(\frac{1}{1.8}\right) \Delta\,{}^\circ\text{C}/\Delta\,{}^\circ\text{F}, \\
&\doteq 10 \text{ lb-mass} \times \Delta\,{}^\circ\text{C} = 10 \text{ Pcu.}
\end{aligned}
$$

No discussion of heat units would be complete without mention of the *kilogram-calorie* (or kcal). As the name implies, the unit of mass is the kilogram, with the unit being equal in size to 1000 gm-cal.

• • • • •

* Sometimes designated Chu, an abbreviation of Centigrade heat unit.

The most commonly required relationship between heat units and mechanical energy units is contained in the factor

$$778\,\frac{\text{ft-lb}_f}{\text{Btu}}.$$

The value of this factor may be derived from the relationship of each of the individual units to the joule.

The only remaining energy units which may be expected to have extensive use are the horsepower-hour (hph) and the electrical unit, the watt-hour (watt-hr). These may be related to the joule, according to arbitrary definition, as follows:

$$2.6845 \times 10^6 \text{ J (abs)/hph,}$$
$$3600 \text{ J/watt-hr,}$$
$$745.7 \text{ watt-hr/hph.}$$

Of the three factors, the last is the most commonly required. There are, of course, 1000 watt-hr/kilowatt-hour (kwhr).

Power

Power can be defined as the rate of performing work. Any of the following units (and, indeed, many more) are suitable power units: ergs per sec, ft-lb_f per min, Btu per sec, hp, watt, kw. When multiplied by the time period over which this power has been consumed, one obtains the total work done or energy expended. Thus horsepower (a power unit) when multiplied by hours duration gives horsepower-hours, the total energy expended.

The relationship between foot-pounds per unit time and horsepower is useful:

$$550 \text{ ft-lb}_f\text{/(sec)(hp);}$$

also,

$$33{,}000 \text{ ft-lb}_f\text{/(min)(hp).}$$

2.8 Systems of Composition and Concentrations

Of the many possibilities for denoting compositions, the most useful for the chemical engineer are *mass and mole percentages* (or fractions) and *mass and mole ratios*. These can be readily defined in terms of w_i, the mass of the ith component of a mixture, n_i, the number of moles of the ith component in the mixture, and M_i, the molecular weight of the ith component of the mixture. The molecular weight is defined as the number of units of mass per molecular weight (i.e., per mole). As such, it must be used consistently with the system of mass chosen for w_i. If this is done, we can write

$$n_i = w_i/M_i. \tag{a}$$

Then, n_i represents the number of *gram*-molecular weights in w_i *grams* of the substance when the unit of mass is the gram; it represents the number of *pound*-molecular weights in w_i *pounds* of the substance when the unit of mass is the pound. The latter is the more useful in most engineering calculations due to the larger size of unit. The student should make a concerted effort to perform as many calculations as possible in the English system and should, therefore, consistently attempt to use pounds and pound-moles as much as possible.

Mass percentages (or fractions) and *mass ratios* may be expressed as

$$\text{Fraction of } i\text{th component} = \frac{w_i}{\sum_{i=1}^{C} w_i}, \qquad (C\text{-components}) \tag{b}$$

Mass ratio of ith component referred to some other

$$j\text{th component} = \frac{w_i}{w_j} \qquad (i \neq j). \tag{c}$$

Mole percentages (or fractions) and *mole ratios* are identical in form and would read exactly the same with n's replacing all the w's in Equations (b) and (c).

Conversion of mole percentages to mass percentages can easily be represented by formulas. Since such formalization is unnecessary and tends to offer an undesirable crutch, formulas are omitted here and replaced by illustrations of the method.

Example 2.3

▶ Statement: A fuel gas is reported to analyze, on a molal basis, 20% methane (CH_4), 5% ethane (C_2H_6), and the remainder nitrogen. Calculate the analysis of the same fuel on a mass percentage basis.

▶ Solution: The molecular weights of methane, ethane, and nitrogen are 16, 30, and 28, respectively, to 0.2% accuracy or better. Using these values, Equation (a) and the molal counterpart of Equation (b), the conversion is accomplished on the basis of any chosen number of moles of gaseous mixture. Choose 100 moles.

100 moles of mixture contains

$$\begin{aligned} 20 \text{ moles } CH_4 \times 16 &= 320 \text{ lb } CH_4 \\ 5 \text{ moles } C_2H_6 \times 30 &= 150 \text{ lb } C_2H_6 \\ 75 \text{ moles } N_2 \times 28 &= 2100 \text{ lb } N_2 \\ &\overline{ 2570 \text{ lb Total}} \end{aligned}$$

$$\text{Mass \% } CH_4 = 320/2570 = 12.45\%.$$

Similarly,

$$\begin{aligned} \text{mass \% } C_2H_6 &= 5.84\% \\ \text{mass \% } N_2 &= 81.71\% \\ \text{Total} &= 100.00\% \end{aligned}$$

Example 2.4

▶ STATEMENT: An oil analyzes, on a mass basis, 11.1% hydrogen and 88.9% carbon. Calculate the analysis on a molal basis. What is the mole ratio of hydrogen to carbon? What is the simplest chemical formula which can be used to characterize the fuel?

▶ SOLUTION: Take as a basis 100 lb of oil. It contains then

$$11.1 \text{ lb } H_2 = \frac{11.1}{2} \text{ moles of } H_2 = 5.55 \text{ moles}$$

$$88.9 \text{ lb C} = \frac{88.9}{12} \text{ moles of C} = 7.408 \text{ moles}$$

$$12.958 \text{ total moles}$$

The mole percentage of C is

$$7.408/12.958 = 57.17.$$

The mole ratio, H_2/C, is

$$5.55/7.408 = 0.75.$$

The atomic ratio, H/C, is twice the mole ratio and, therefore, 1.5. Thus the simplest chemical formula for the oil is C_2H_3.

Concentrations

Concentration is defined as the mass per unit volume, with the mass units and the volume units to be specified as convenient. The most common units in these terms are the *normality* and the *molarity*. Normality is defined as the number of gram-equivalent weights (i.e., mass) per liter of solution. The gram-equivalent is, of course, the mass of the substance displacing, or otherwise reacting with, 1.008 gm of hydrogen, or combining with 8.000 gm of oxygen. The molarity is the same type of unit with gram-molecular weight replacing gram-equivalent weight as the unit of mass. Neither unit is particularly useful or commonly encountered in engineering work.

Molality is not really a concentration, but rather a mass ratio. It is defined as the number of gram-molecular weights per 1000 gm of solvent. It too is not common in engineering work.

Molal Gas Volumes

Although further remarks on molal volumes of gases will be found in Chapter 3, and appear logically there, the numerical values rightfully belong in a chapter on conversion factors. The molal volume (i.e., the volume occupied by 1 mole of an ideal gas) can be stated for either the gram-mole or the pound-mole, if one establishes some standard temperature and pressure conditions. The conditions normally chosen are 0 °C and 1 atm (760 mm of mercury at 0 °C) and are frequently designated STP (standard temperature and pressure). Under these conditions 1 gram-molecular weight of an ideal gas occupies 22.414 liters, and 1 pound-molecular weight of an ideal gas occupies 359 cu ft. As will be seen in Chapter 3, these figures are extremely convenient for converting from volume to mass, and vice versa.

One other abbreviation, equivalent to STP, is occasionally encountered. It is DSC, meaning dry standard conditions. Obviously, one might also encounter SC, meaning standard conditions for a gas that contains moisture (i.e., water in vapor form).

The condition of 0 °C and 760 mm of Hg is not the only standard used, but it is the commonest. The gas industry, for example, uses 60 °F and 30 in. of mercury. Under these conditions the molal volume is approximately equal to 371 cu ft. This standard will not be employed in this text.

Standard Composition Systems

Unless otherwise stated, it may always be assumed that the analysis of a solid or a liquid is reported on a mass basis. Analyses of gases are, however, reported on a molal basis. As noted in Chapter 3, this basis for gases is equivalent to a volume basis as well. These rules must be kept firmly in mind since the basis is not necessarily stated as long as these conditions are adhered to.

2.9 Summary of Important Conversion Factors

Table 2.4 summarizes the important conversion factors discussed in this chapter. The table is in two parts, the first dealing with factors related to fundamental units and the second listing the factors which are of frequent use and which involve more than one of the fundamental factors.

2.10 Consistency of Units in Equations

An important use of units, and one frequently overlooked by the student, is in the checking of the consistency of equations. Both the fundamental units and consistency within and between systems of units are important. It is, of course, obvious that both sides of an equation must reduce to the same units and that additive terms on any side of an equation must also reduce to the same units. Otherwise, one is in the position of adding apples and lemons, comparing forces to masses, or making similar errors.

TABLE 2.4

Fundamental Units

Mass

454 gm/lb-mass*
2.2 lb-mass/kg
2000 lb-mass/ton*
16 oz/lb-mass*
7000 grains/lb-mass

Time

60 sec/min*
60 min/hr*

Length

2.54 cm/in.*
12 in./ft*
5280 ft/mile
1000 microns/mm
10^{10} Å/m

Temperature

1.8 Δ °F/Δ °C*
°K ≐ °C + 273*
°R ≐ °F + 460*

Combinations of Fundamental Units

Area

(In general, obtain from the square of fundamental units.)

43,560 sq ft/acre

Volume

(In general, obtain from the cube of fundamental units.)

1000.027 cu cm/liter
7.48 gal/cu ft*
4 qt/U.S. gal

Pressure

760 mm of mercury, 0 °C/standard atmosphere*
29.92 in. of mercury, 0 °C/standard atmosphere*
14.696 (lb-force/sq in.)/standard atmosphere*

Energy or Work (Mechanical and Electrical)

1 dyne-cm/erg
10^7 erg/J (abs)
1.356×10^7 erg/ft-lb_f
2.6845×10^6 J (abs)/hph
3600 J (abs)/watt-hr
745.7 watt-hr/hph*
1000 watt-hr/kwhr*

Energy (Heat)

4.186 J (abs)/gm-cal (mean)
252 gm-cal/Btu*
454 gm-cal/Pcu*
1.8 Btu/Pcu*
(the three preceding: Derived from inherent mass × Δ° relationships)
778 ft-lb_f/Btu*

Power

550 (ft-lb_f/sec)/hp*
33,000 (ft-lb_f/min)/hp

Special Prefixes

Deca-	10 times
Hecto-	100 times
Kilo-	1000 times
Mega-	1,000,000 times
Milli-	1/1000th
Micro-	1/1,000,000th

* Absolutely essential in the everyday life of the chemical engineer.

Example 2.5

▶ STATEMENT: You have reason to believe that the loss in pressure (ΔP) incurred in forcing a fluid to flow, under restricted conditions,* through an ordinary cylindrical pipe is proportional to the fluid velocity (u), viscosity (μ), and length of pipe (L), and inversely proportional to the square of the pipe diameter (D). You wish to check a relationship of this sort for dimensional consistency.

▶ SOLUTION: The suggested form of equation is: $\Delta P = K\mu uL/D^2$, where K can be a numerical and/or dimensional constant. The pressure drop should be expressed as force per unit area, and the pound-force will be used. Viscosity has been noted earlier in this chapter to be expressed in the fundamental

• • • • •

* So stated to account for the limitation to the viscous flow region.

units M/Lt and will be in English units also. The other terms are familiar. Write then

$$\frac{\text{lb-force}}{L^2} = K\left(\frac{M}{Lt}\right)\left(\frac{L}{t}\right)L\left(\frac{1}{L^2}\right) = K\frac{M}{Lt^2}.$$

The M on the right represents pounds-mass. Obviously K *is* a dimensional constant. It must have the units of (lb-force per lb-mass)(t^2/L). But these dimensions are those of the constant $1/g_c$ used in the English engineering system. Thus we know that the equation should be written:

$$P = \frac{K'\mu uL}{g_c D^2}.$$

We cannot be certain whether or not a *numerical* constant, K', is needed. As may be shown by experimental data, K' is required and has the value 32.

Care must still be taken in the application of such an equation. The use of g_c (lb per lb-force)(ft per sec^2) has established the time and length dimensions. Velocity and viscosity must use these same dimensions. While feet per second would be practically automatic for velocity, great care must be taken in the case of viscosity; the units must be pounds per (second)(foot) rather than the other familiar English unit of pounds per (hour)(foot). Diameter of the pipe must now be expressed in feet. Since standard pipe sizes are usually reported in inches, the conversion to feet must not be omitted.

Equations containing sums (and/or differences) of terms can be checked in the same manner. Each term must, of course, have the same units. For this reason, the incorporation of integrals and differentials presents no difficulties. The differential terms are treated as differences and the integral terms are treated as products.

Example 2.6

▶ STATEMENT: Show that the equations for velocity and position of a body falling freely under the influence of gravity alone are dimensionally consistent.

▶ SOLUTION: Acceleration is defined as the change in velocity with time. Therefore, for a freely falling body $g = du/dt$. The units of acceleration, g, are L/t^2. Those of velocity are L/t. Those of difference in velocity are still L/t and those of difference in time are still units of time. Therefore, the units of du/dt are L/t^2.

Now if the variables are separated, and the integration is indicated,

$$\int du = g\int dt,$$

then the units of the left side are L/t while those on the right are $L/t^2 \times t = L/t$. Performing the indicated integration,

$$u = gt + \text{const.}$$

The constant must have the dimensions of velocity. Introducing as an initial condition the state of rest, the constant becomes zero, not a number, but rather zero velocity.

If this relationship is integrated once more,

$$\int u\,dt = g\int t\,dt.$$

The product on the left has dimensions $L/t \times t = L$ (a distance), and that on the right has dimensions $L/t^2 \times t \times t$, also equal to L. Integration and evaluation of the constant of integration for the same initial condition gives the well-known equation

$$s = \tfrac{1}{2}gt^2, \qquad \text{where } s \text{ is the integral of } u\,dt.$$

The arguments of exponentials and logarithms must, naturally, be dimensionless.

Example 2.7

▶ STATEMENT: The following equation represents the change in concentration in a perfectly stirred tank to which pure water is added at constant rate (see Chapter 5):

$$C = K\exp\left(\frac{-Rt}{V}\right).$$

Here, C is a concentration in mass per unit volume, R is the rate at which water is charged to the tank, and V is the (constant) volume of the tank. The value of the constant of integration, K, will be determined from initial conditions. You are to state the units in which K will be expressed and to state further the units which should be used to represent the rate of flow of the water.

▶ SOLUTION: Since the exponential term must be dimensionless, R must be expressed in units of V/t and the volumetric flow rate is required. Both units of volume and time are at the discretion of the engineer, as long as they are used consistently in R, V, and t. For example, volume could be in gallons and time in hours, in which case R is in gallons per hour, V is in gallons, and time is in hours.

Using the dimensionless character of the exponential term again, it is obvious that K must be in the units chosen for C. Both could be expressed in pounds per cubic foot, in pounds per gallon, in grams per liter, or any other choice, as long as it is used for both. Note also that the units (and even the system of units) chosen for C and K are independent of those used in the argument of the exponential.

2.11 Precision and Accuracy

During the course of this chapter there have been several remarks about engineering accuracy. Some clarifying statements seem essential at this point.

In the first place a clear distinction should be made between two words. *Precision* refers to the variation in the value of a measurement obtained from identically repeated experiments, with high precision indicating an ability to obtain very nearly the same values on repetitions. For example, consider a test on a piece of equipment known as a steam heater (see Figure 4.1). In this unit some fluid, maybe an oil, is heated, the heat coming from condensing steam, not in contact with the oil. A pressure gauge on the steam line may be read many times and may vary little, or even not at all, but this consistency does not mean that the recorded pressure is correct. The gauge may well be poorly calibrated, the needle may be bent or the bourdon tube may be distorted. In such a case, the reading is very precise but not accurate. On the other hand, the outlet temperature may fluctuate several degrees from reading to reading and yet the average of all readings may very closely represent the true average temperature of the heated fluid. This reading is very accurate but not very precise. The greater the number of readings taken when the precision is poor, the better the average value may be expected to be. The manner in which the values are distributed is the province of statistics. The subject will not be covered here, not due to any inherent difficulty of subject matter, but merely due to the lack of *direct* contribution to the study of the laws of conservation.

Introduction and Propagation of Error

Since any measurement involves some error, no experimental data can be expected to be free of error. If, in an engineering design, several items of experimental data are required, the design calculation must contain contributions from the errors of each item. It is true that the errors may cancel in some instances, but it is also true that all errors may contribute in the same direction in others. Both situations, and all intermediate ones, must be accountable in any system for evaluation of the potential error in the final design. All statistical predictions are based upon the potential additivity. It is then clear that the greater the number of experimental measurements entering into a calculation, the greater the potential error in the result. (This statement is obviously much too simple, ignoring the effect of errors of different sizes or, more important, of different magnitudes. It is, however, fundamentally true that with an increased number of sources of the same relative size, the error will increase.)

Any engineering design involves a fairly substantial number of properties, first for materials of which the equipment is composed and second for materials being processed. For example, take the relatively simple case of the steam heater. The diameter, length, and number of tubes may be measured. The number is the only one without possible error—neglecting the possibility of a gross inaccuracy. The diameter and length will, however, vary slightly from tube to tube and both are functions of temperature. Therefore a fixed value is not quite perfect. The magnitude of these errors may well be so small in relation to others as to be negligible, but it would be a mistake not to list them. The properties of the fluid being heated, which contribute to the design, are density, heat capacity, viscosity, and thermal conductivity, all determined experimentally and only as good as these measurements. Except in the case of well-known substances, some of the values for these properties may well be very poor. The properties of steam, the heating medium, must be known, but this substance is one on which a great deal of work has been done because of its general applicability and the precision here is good. It should, almost invariably, be an order of magnitude better than the properties of the less common industrial products. The properties of the substance to be heated and of steam are used along with flow rate, to obtain heat transfer coefficients which will, in turn, determine the size of the unit required to do the given job. The establishment of these coefficients from the fluid properties is again related through experimental measurements with new errors introduced by the correlation, and not necessarily small ones. The desired flow rates of the material being heated can be specified very closely, but the control will not necessarily be perfect. Since these rates contribute to the heat transfer coefficients, the failure in practice to maintain the exact velocity of the design introduces another error.

Heat transfer in the steam heater is a fairly simple process compared to situations where reactions occur, and heat transfer has been studied far more extensively than reactors. If the situation is so complicated for the steam heater, it will be vastly more so for more complex operations.

These remarks are the reasoning leading to the conclusion that in many engineering operations and pieces of equipment, there is no need for information to 0.1, 1, or even 10% precision. This sort of situation results in the extensive use of the term "engineering accuracy." Each engineer must size up his own situation and decide, for each calculation, the precision that he desires and that which he may hope to attain.

In this text we shall assume that engineering precision is generally limited and that an answer which is good to 1 or 2% is entirely satisfactory. For this reason we are unconcerned with the difference between cubic centimeters and milliliters; between 273 and the correct 273.16 in converting from Centigrade to Kelvin; between 2.54 cm per in. and the correct 2.539998; between the international joule and the absolute joule; between the differently defined calories, and so on. There may be some doubt in the reader's mind that it was even necessary to clutter up the presentation with these complications. When one uses an approximation, however, no matter how justifiably, he should be aware that he has done so and be prepared to use the exact value if the necessity ever arises.

In general, a number can be understood to be accurate to the last reported digit. Thus 2.54 does not mean 2.540. It may mean that nothing is known beyond the 4, or it may mean, as in this case, that the more precise figure has been rounded to this place. The number 254,000 is ambiguous since one is uncertain whether the zeros represent lack of knowledge beyond the 4 and are present only due to the magnitude of the number or whether they represent an exactness of the reported figure to even thousands. This difficulty can be surmounted by reporting numbers of this sort using digits to indicate the precision of the number and power of 10 to represent the magnitude of the number; e.g., 2.54×10^5 or 254×10^3. This system is recommended.

Some simple and reasonably satisfactory rules may be laid down for the operations of addition, subtraction, multiplication, and division when the numbers involved are not precise to the same orders of magnitude. All are based upon the number in the group with the minimum precision.

Addition

The precision of the sum is limited by that of the number with minimum precision. Thus to add 25×10^3, 268.9×10^3, 22.3, 16.1×10^2, the addition would appear as

$$\begin{array}{r} 25{,}000 \\ 268{,}900 \\ 22.3 \\ 1{,}610 \\ \hline 295{,}532.3 \end{array}$$

with precision no better than that of the 25×10^3. The sum is therefore rounded to, and reported as, 296,000, or 2.96×10^5.

In *rounding* one may encounter the rounding of a number ending in the digit 5. The decision must be made as to rounding the preceding digit up or down. (There was no question in the example above since the portion rounded as 532.2 was clearly greater than 500.0.) A reasonable procedure can readily be obtained by considering the procedure for rounding with *all* possible last digits. Thus a number ending in zero in the rounding position requires no rounding, the zero merely being dropped. Numbers ending in the digits 1, 2, 3, 4 are rounded by dropping these digits and numbers ending in 6, 7, 8, 9 are rounded by dropping these digits and increasing the preceding digit by one. Thus 4 digits round up and 4 down. If a single direction is chosen for rounding numbers ending in the digit 5, there will be a one-in-nine bias in the direction chosen. The usual rule to avoid this situation is to round the previous digit to an *even* digit. Thus, in cases where three significant digits are to be retained,

2526	rounds to	2530
2333	rounds to	2330
2345	rounds to	2340
2335	rounds to	2340
7689	rounds to	7690
8726	rounds to	8730
9075	rounds to	9080
3898	rounds to	3900

Subtraction

This operation follows the same procedure as addition, with precision of the difference determined by the minimum precision of the minuend and the subtrahend. Thus for the two cases 9.85293×10^6 minus 224×10^3, and 226.35 minus 2.6782, we have

$$\begin{array}{r} 9{,}852{,}930 \\ 224{,}000 \\ \hline 9{,}628{,}930 \end{array} \quad \text{and} \quad \begin{array}{r} 226.35 \\ 2.6782 \\ \hline 223.6718 \end{array}$$

and the differences, with precisions limited in the first case by the subtrahend and in the second by the minuend would be reported as 9.629×10^3 and 223.67.

The operation of subtraction results in considerable loss in precision when the numbers subtracted are of the same or closely similar orders of magnitude. Suppose that 11.83×10^3 is to be subtracted from 12.56042×10^3. Relatively, the latter number is without error and the former is good to 9 in the first unreported place. Effectively, we can say that the number 11.83×10^3 is good to approximately 10 in 12,000, or roughly 0.1%. Now perform the subtraction, obtaining 0.73×10^3, with precision limited by that of the less precise subtrahend. This number is good also to about 9 in the first unreported place, or roughly 10 in 800, or approximately 1%. The potential error is increased by roughly one order of magnitude due to the subtraction of numbers so closely similar.

If the minuend had been 12.56×10^3, this number would also have been good only to 0.009×10^3. The difference would then be good to approximately 20 in 800, or 2.5%.

The example taken here is deliberately chosen to emphasize the loss in precision resulting from subtractions. Situations of this severity can easily be encountered in dealing with heats of reaction, calculated from heats of formation and/or from heats of combustion (see Chapter 8).

Multiplication and Division

Limitation is again by the number with minimum precision but in this case one more significant figure is normally allowed in the product and the quotient than in the minimum precision number contributing to them. Thus $2.75 \div 4.000000 = 0.6875$. And 2.2×3.86293 is limited to three significant digits by the precision of the multiplicand and is, therefore, reported as 8.50.

2.12 Missing and Extra Data

Textbooks and courses in instruction often overlook the first problem that the practicing engineer will face. A real problem is often replete with missing items of data and, an opposite complication, an excess of data completely unnecessary for the problem at hand. The data supplied may be required for an exact solution but an approximation may be all that is justified at the moment. If so, the engineer who obtains the approximate answer is doing a much more efficient job than does his colleague who always gets the "best" answer possible.

No beginning text can provide a series of problems in which the student can be asked to pick the simplest solution. He does not have the technical judgment to decide when such a course is justifiable, nor does he usually have more than a small segment of a real problem. Knowledge of the whole problem is essential in deciding on the short cut solution. The student is therefore expected to use as exact a method as is available despite any doubts he may have as to the general desirability of such an action.

While the text must have the failing noted above, there is no reason why it must also fail to provide practice in the areas of too much and too little data. Both types of problems are to be expected. The second is the easier, since handbooks provide the necessary additional items of information. Use of additional sources should always be carefully noted, for the benefit of the checker, or for review by the original worker. It is often surprising to realize how quickly one can forget a source and second how often reference back to the source is needed. Another reason for delineation of the source is to provide a record of the conditions under which the data used were recorded. For example, heats of reaction are found at 18 and 25 °C. If one's main source is at 25 °C and a value for one material is obtained from a different source, it should also be at 25 °C, and one should be able to substantiate the fact that this was indeed the case. The source and the date of publication should then be noted.

The case of too much data cannot be discussed in any detail. It is clear, however, that as a problem is assigned to an engineer in industry he would expect that all pertinent information relating to the process would also be made available to him. In any subsection of his calculations he may expect to have an excess of data and he must sort out what he wants. A number of problems in this text will require this same discrimination.

LIST OF SYMBOLS AND ABBREVIATIONS

Latin letters

Å	Angstrom unit of length
a	Acceleration
abs	Absolute
API	American Petroleum Institute
Bé	Baumé
Btu	British thermal units
C	Concentration in moles per unit volume
C	Number of components
°C	Degrees Centigrade
cal	Gram-calories
c.g.s.	Centigrade-gram-second system of units
Chu	Centigrade heat unit
cm	Centimeters
DSC	Dry, standard conditions
F	Force
°F	Degrees Fahrenheit
ft	Feet
g	Acceleration due to gravity
g_c	Constant in Newton's second law, English engineering system
gal	Gallons
gm	Grams
hp	Horsepower
hph	Horsepower-hours
hr	Hours
in.	Inches
J	Joules
K	Constant
°K	Degrees Kelvin
k	General constant in Newton's second law
kcal	Kilogram calories
kg	Kilograms
k.m.s.	Kilogram-meter-second system of units
kw	Kilowatts
L	Length
lb	Pounds-mass
lb_f	Pounds-force
M	Mass
M_i	Molecular weight, component i
m	Meters
min	Minutes
ml	Milliliters
mm	Millimeters
n_i	Moles, component i

mph	Miles per hour
oz	Ounces
P	Poises
P	Pressure
Pcu	Pound centigrade units
qt	Quarts
R	Gas law constant
°R	Degrees Rankine
SC	Standard conditions
sec	Seconds
STP	Standard temperature and pressure
T	Temperature
t	Time, or temperature, ordinary scale
Tw	Twaddell
u	Velocity
v	Specific volume, volume/mole
V	Volume of tank
w_i	Mass, component i

Greek letters

Δ	Difference
μ	Viscosity, any consistent units

PROBLEMS

2.1 Convert the units of the following quantities as indicated:
   a. 5000 cal to Btu; to Pcu.
   b. 480 kcal to Btu; to Pcu.
   c. 20,000 cal per gm to Btu per pound; to Pcu per pound.
   d. 7.35 cal per (gm-mole) (°C) to Btu per (pound-mole) (°F); to Chu per (pound-mole) (°C).
   e. 0.29 cal per (gm) (°C) for $CaAl_2SiO_8$ to Btu per (pound-mole) (°F).
   f. 1500 Btu to foot-pounds force; to foot-pounds mass; to horsepower-hours.
   g. 2250 Btu to joules; to watt-hours; to ergs; to dyne-centimeters.
   h. 252 cal to horsepower-hours.
   i. 12,000 Btu per min to horsepower.

2.2 A diffusion coefficient has a value of 0.5 lb per (hr)(sq ft)(atm). Calculate the corresponding value in the metric units of gm per (sec)(cm^2)(mm Hg).

2.3 Calculate the number of:
   a. liters in a cubic foot.
   b. square centimeters in a square foot.
   c. cubic centimeters in a quart.
   d. cubic centimeters in a pound of water at 35 °C.

2.4 Aviation gasoline contains 3 cu cm lead tetraethyl per gal. Calculate concentration on a consistent volume basis.

2.5 A liter of beer is equivalent to how many quarts? Assume that a quart of water at 15 °C weighs 32 oz. Check the calculation without making this assumption.

2.6 The bar, a barometric measure, is defined at 10^6 dynes per cm^2. Standard atmospheric pressure is reported to be:

a. 760 mm Hg; density 13.5951 gm per cm^3.

b. 14.696 psia.

Convert each of these values to bars.

2.7 The acceleration of a body falling freely in a vacuum is 32.174 ft per sec^2. Express this quantity in c.g.s. units.

2.8 A pressure is recorded as 1 dyne per cm^2. Convert this value to one expressed in microns of mercury at 25 °C; to angstrom units.

2.9 A pressure gauge reads 26.5 in. of mercury, vacuum. The barometer stands at 763 mm Hg. What is the absolute pressure in the tank to which the pressure gauge is connected, expressed in pounds force per square inch, pounds force per square foot, bars, mm Hg at 25 °C, and inches Hg at 25 °C.

2.10 A submerged electrical heater converts perfectly the electrical input to heat. If 100 Btu per min are required, what is the proper heater size, in kilowatts.

2.11 A pump delivers 65% of the electrical input to the system as work. Its rating is 2 hp. How many foot-pounds force does it deliver in an hour?

2.12 A 4000-lb automobile is moving at a rate of 60 mph. What is the kinetic energy, expressed as foot-pounds force, joules, horsepower hours, and liter atmospheres?

2.13 A piston 5 in. in diameter moves through a distance of 9 in. against a pressure of 7 lb per sq in. Calculate the work done in foot-pounds force and in liter atmospheres.

2.14 Soil having an average specific gravity of 2.75 is eroded at a rate of 0.01 in. (depth) per yr from a square mile of farm land. It is carried away by a stream having an average flow rate of 200 gal per min. Calculate the concentration of suspended matter in the stream, by weight, in consistent units.

2.15 In the textile industry, filament and yarn sizes are reported in denier, defined as the weight in grams of a length of 9000 m. If a synthetic fiber has an average specific gravity of 1.32 and a filament of this material has a denier of 5.0, what is the weight per unit length in pounds per yard? What is the cross section of this fiber in square inches?

2.16 The density of wood varies widely with the species. Douglas fir averages 0.52 gm per cm^3. What is the weight of a log 50 ft long, 15 in. in diameter at the base and 9 in. in diameter at the top, in kilograms, and in pounds, if the taper from base to top is uniform?

2.17 A loss in energy may be expressed as $u^2/2g_c$ or as $u^2/2g$. What are the energy units in each case? How do you reconcile these units to an energy unit of force $\times$ distance?

2.18 One million cubic yards of earth are to be carted 10 miles, using dump trucks of 30-ton capacity each. If the average density of the earth is 120 lb per cu ft, how many ton-miles must be driven? What energy is expended in handling the earth alone if the dumping ground is an average of 30 ft below the level of the excavation?

2.19 One thousand pounds of rock are balanced at the edge of a precipice, 1000 ft above the canyon floor. What is the potential energy relative to the same mass on the canyon floor?

The rock falls to the floor of the canyon. Treating the fall as the equivalent of completely free fall in vacuum, calculate the kinetic energy when the rock reaches the floor of the canyon. Express the kinetic energy as a function of the linear distance, s, which the rock has fallen. Note that $ds/dt = gt$.

2.20 A gas from a petroleum refining process analyzes, by moles, 11% CH_4, 88% C_2H_4, and 1% C_2H_6. Report this analysis on a weight basis. What is the average molecular weight of the mixture? How many cubic feet at standard conditions are occupied by one ton of the mixture?

2.21 A flue gas analyzes, on a molal basis, 10% CO_2, 1% CO, 5% H_2O, 7% O_2, and the remainder nitrogen. What is the analysis on a weight basis? What is the molal analysis on a dry basis?

2.22 Pyrites, FeS_2, is a naturally occurring mineral which is a source of sulfur for the manufacture of sulfuric acid. If the pyrites is 93.2% pure, with the rest an inert clay, how many tons of sulfur are present per ton of pyrites? If recovery of sulfur in the process is 99.2%, how much sulfuric acid, 98%, can be produced?

2.23 A mixture of dolomite ($CaCO_3 \cdot MgCO_3$), alalite ($CaMg(SiO_3)_2$), and enstatite ($MgO \cdot SiO_2$) in proportions by weight of 95:3:2 is heated to drive off the CO_2 from the dolomite, leaving CaO and MgO as the solid residue. Assuming complete removal of the CO_2, how many tons of solid are produced from 100 tons of the mixture? How many tons of CO_2 are produced? How many moles?

2.24 The mineral apatite has the formula $CaF_2 \cdot 3\ Ca_3P_2O_8$. It occurs plentifully in nature and is a source of phosphorus. A source of apatite is 83% pure, with the remaining material being inert to the process for production of phosphorus. How much P_2O_5 is present in one ton of the impure apatite? How much phosphoric acid, 30%, can potentially be recovered from the ton? How many tons of sulfuric acid would be required to react with all the calcium in the rock, and how many pounds of hydrofluoric acid would be liberated? What is the phosphorus content of the impure rock, in mole percentage and in weight percentage?

2.25 Thirty pounds of NaOH is dissolved in 100 lb of water. What concentration of NaOH is formed in each of the following unit systems:

a. weight fraction.
b. mole fraction.
c. molality.
d. molarity.

The final solution is stored at 68 °F. What is its density? The specific gravity of solid NaOH is 2.130.

2.26 The Centigrade and Fahrenheit temperature scales are entirely arbitrary. Invent a new scale with the freezing and boiling points of ethyl alcohol as the zero and 100° points. Call degrees on this scale °A. Write relationships between °A and °F. Also between °A and °C. Required freezing and boiling points are −112.0 and 78.4 °C.

2.27 Calculate the weight per cent of $Na_2S_2O_3$ in a unimolar solution, measured at 68 °F.

2.28 Calculate the normality of 40% acetic acid solution if the specific gravity at 18 °C = 1.050.

2.29 The following definition is taken from the *Condensed Chemical Dictionary*:

"*Proof, degrees of*

"Number of gallons of 100-proof alcohol that can be made from 100 gal of the mixture.

"(100-) Proof spirit is defined by the regulation that 'it contains one half its volume of alcohol of sp. gr. 0.7939 (60 °F).' This latter is absolute alcohol and proof absolute alcohol is, therefore, 200 proof. The degree of proof is twice the percentage by volume of alcohol."

Calculate:

a. Percentage alcohol by weight in 100-proof alcohol.
b. Percentage alcohol by weight in 86-proof spirits.

2.30 One hundred pounds of a 20% zinc chloride solution at 20 °C is diluted by its addition to 100 lb of water. What is the volume in cubic feet before and after addition of the water?

2.31 To 100 ml of a saturated solution of silver chloride, at 25 °C, 100 ml of 0.1 N NaCl is added. How many milligrams of silver may be expected to precipitate? The solubility product, with ion concentrations expressed as moles per liter, is 1.56×10^{-10}, for silver chloride at this temperature.

2.32 A 10-ml sample of HCl solution is titrated with a standard NaOH solution, known to be 0.1004 normal. The end point is reached after 15.7 ml of this solution have been added. What is the concentration of the HCl solution, expressed as normality, as grams per liter, as pounds per cubic foot. What is the weight fraction of the HCl in the solution?

2.33 Sea water has a specific gravity of 1.03 (referred to water of density 1.0000...). Analysis indicates that sea water contains

2.8%	NaCl,
0.50%	$MgCl_2$,
0.0085%	NaBr.

Calculate the tons of sodium and magnesium in 1 cu mile of sea water; the pounds of each in 1 liter.

2.34 Water, containing acetic acid, is mixed with chloroform, well shaken, and allowed to separate into two layers. These layers analyze, at 18 °C:

1. 41% acetic in water, no $CHCl_3$; sp. gr. = 1.050.
2. 40% acetic in $CHCl_3$, no H_2O; sp. gr. = 1.52.

Calculate mole percentages of acetic in two layers.

2.35 The lifting power of a balloon is, of course, equal to the difference in weight of the gas with which it is filled and that of an equal volume of the surrounding atmosphere. A balloon which, when filled, has a volume of 1125 cu ft, is to lift a total load of 50 lb. What is the required density of the gas, providing that the outside air is at a temperature and pressure of 25 °C and 745 mm Hg, respectively. The density of air under these conditions may be found in handbooks, or may be calculated by use of the ideal gas law (see Chapter 3). Can this lifting power be obtained by use of hot air? If so, what is the temperature?

SUPPLEMENT TO CHAPTER 2

A. Manometers

Considerable advantage is taken of the relationship between head and pressure in measuring instruments known as *manometers*. Three basic forms are used, the open-ended manometer, the closed-end manometer, and the differential manometer. The first two are intended to indicate the pressure at the single point in a system at which it is attached; the last is intended to indicate only the pressure difference between the two points of attachment. The fundamental forms are illustrated in Figures S2.1A, S2.2A, and S2.6A. We shall now analyze the various heads in the instrument shown in Figure S2.1A so as to obtain the pressure or pressure difference as a function of the fluids which fill the arms of the manometer.

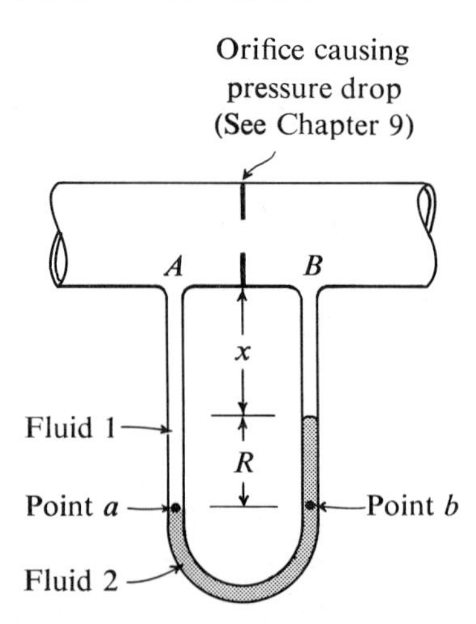

FIGURE S2.1A Ordinary two-fluid manometer.

S2.1A Ordinary Manometer

Consider first the ordinary manometer in Figure S2.1A. Analysis of this system will provide a generalized form of attack, applicable to the open-ended manometer in Figure S2.2A and to more complicated systems that will be considered later.

There are always at least two fluids in the ordinary manometer. They are, naturally, immiscible and they may be both liquids, or the top fluid may be a gas. Obviously Fluid 2 is always denser than Fluid 1. In the figure the manometer is attached to two points A and B in a line through which Fluid 1 is flowing as indicated. Between the points of attachment a constriction is indicated; this constriction accounts for the pressure drop between points A and B; as might well be concluded, such a constriction and incident pressure drop is a convenient means of measuring the flow rate in the main line (see Chapter 9 on energy balances).

The objective now shall be to determine the pressure drop between points A and B as a function of the reading, R, and of the fluid densities. A simple hydraulic principle is necessary. It states that, in a *single* motionless fluid, the pressure at any point in a horizontal plane is the same. This principle follows readily from previous remarks on the calculation of pressures from heads. Then, in Fluid 2, the pressures at points a and b must be equal. These pressures may be represented in terms of the pressures at points A and B and of the heads of the fluids between points a and A and between b and B. Letting ρ_i = the density of Fluid i,

$$P_a = P_A + \left(\frac{g}{g_c}\right)\rho_1(R + x),$$

and

$$P_b = P_B + \left(\frac{g}{g_c}\right)(\rho_1 x + \rho_2 R).$$

Equating and simplifying,

$$P_A - P_B = \left(\frac{g}{g_c}\right)R(\rho_2 - \rho_1). \tag{S2.1A}$$

Note that, one, the distance of the measuring fluid below the point of measurement is unimportant and that, two, the pressure drop is a function of the difference in density between the two fluids but *not* of the density of the heavier fluid alone. Only when the density of Fluid 1 is negligible with relation to the density of Fluid 2 can the pressure drop be expressed as a function of the manometer fluid only. This situation *usually* exists when Fluid 1 is a gas and Fluid 2 is a liquid. Exceptions would exist when the gas is very dense, usually due to high pressure.

S2.2A Open-Ended Manometers

The *open-ended manometer* offers a slightly different problem in that three fluids are involved. It will, however, be assumed that Fluid 3 is air and that its density may be considered negligible with relation to that of Fluid 2.

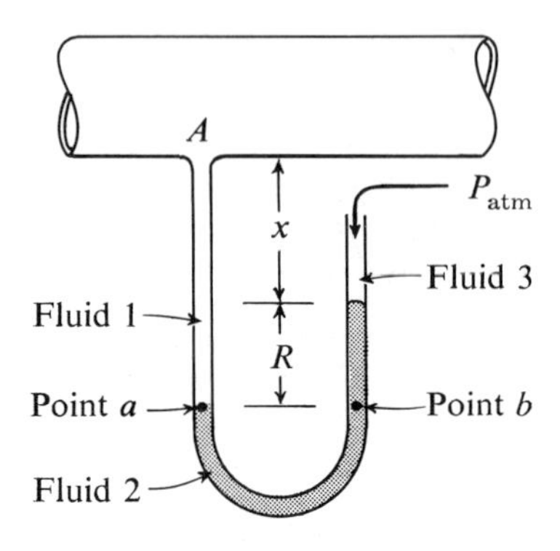

FIGURE S2.2A Open-ended manometer.

The analysis of the pressure at point A proceeds as in the case of the differential manometer. Writing the pressures at points a and b,

$$P_a = P_A + \left(\frac{g}{g_c}\right)\rho_1(R + x),$$

$$P_b = P_{atm} + \left(\frac{g}{g_c}\right)\rho_2 R.$$

Equating and simplifying,

$$P_A - P_{atm} = \left(\frac{g}{g_c}\right)[R(\rho_2 - \rho_1) - \rho_1 x].$$

In this case, due to the inequality of the two fluids above the manometer fluid, the location of the manometer with respect to the point at which pressure is to be indicated *is* important, as shown by the term containing the positioning distance, c. If the zero reading, R_o, is known, the distance, x, may be eliminated from the pressure drop expression, so that the latter reads

$$P_A - P_{atm} = \frac{g}{g_c}(R - R_o)(\rho_2 - \rho_1/2)$$

Many *other forms of manometers exist.* A popular device associated with these instruments is the expanded section at the junction of the two fluids, as illustrated in Figure S2.3A. If the cross section of this section is sufficiently large with relation to the cross section of the manometer tubing, the interface height is essentially constant and need not be read. The instrument becomes a single-armed instrument and is thereby much easier to read. If the motion in the expanded section cannot be neglected, a simple ratio of the cross sections of the expanded section and the manometer tubing can be incorporated into the equation and the instrument can still be a single-reading instrument. The scale attached to the instrument, on which R is read, is sometimes specially ruled to take this correction factor into account.

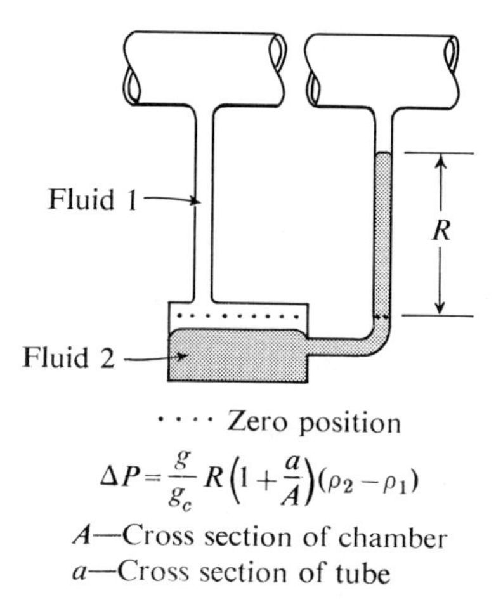

FIGURE S2.3A Manometer designed for reading one arm only.

For *magnifying very small pressure differences*, two devices are often seen. One is the three-fluid differential manometer shown in Figure S2.4A. Here the motion of the interface between Fluids 1 and 2 is minimized by the expanded section. Again, the ratio of the cross sections of tube and expanded section must be calculated to see whether this motion can truly be considered negligible. This form of manometer gives the greatest magnification when the densities of Fluids 2 and 3 most closely approach each other. The danger in this situation is that the densities may be brought so close that the interface is not clearly maintained; some mixing occurs when the fluids move in response to a change in pressure drop, and separation, to give a clear interface,

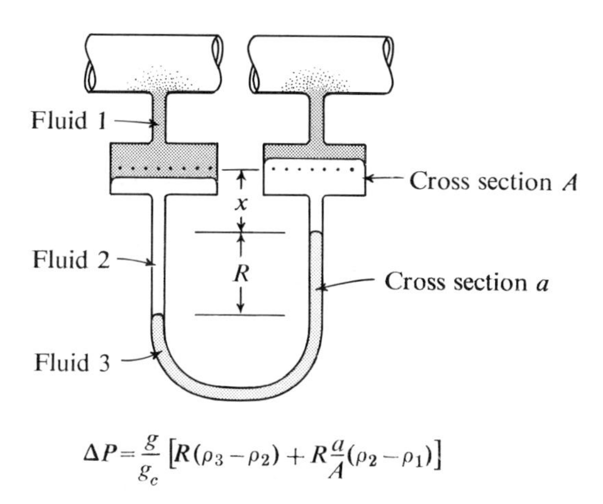

$$\Delta P = \frac{g}{g_c}\left[R(\rho_3 - \rho_2) + R\frac{a}{A}(\rho_2 - \rho_1)\right]$$

FIGURE S2.4A Differential manometer.

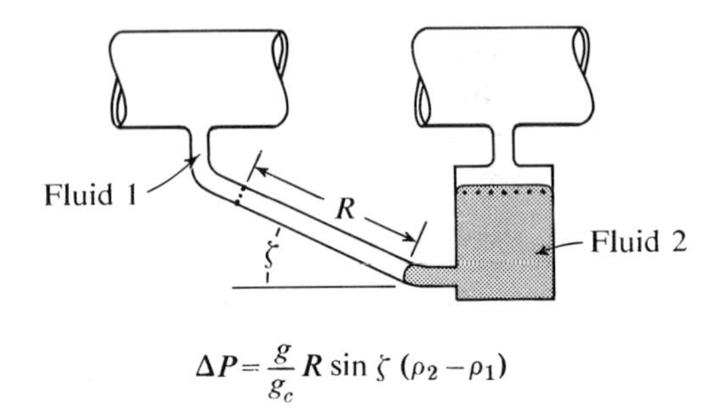

$$\Delta P = \frac{g}{g_c} R \sin \zeta \, (\rho_2 - \rho_1)$$

FIGURE S2.5A Inclined manometer; draft gauge.

is then slow. The second magnifying type is the *inclined manometer*, shown in Figure S2.5A. It is usually of the straight differential type, but it can incorporate any of the complications of expanded sections and multiple fluids. The angle at which it is inclined to the horizontal is of course of primary importance in converting the reading to the true pressure drop between the points of application. These systems are left to the student to analyze.

Two final head-indicating devices should be mentioned. Both have one end closed and there is zero pressure in this end above the liquid column (vapor pressure of the fluid makes this condition impossible but the pressure can be minimized by using a very low volatility fluid such as mercury). The first device is the standard, inverted-tube barometer, as shown in Figure S2.6A. The reading is direct. The other device, also shown in Figure S2.6A has a closed end, but in this instance, the tube is U-shaped and attached to the apparatus by the open end. The reading is direct (i.e., independent of x) if the density of the fluid at point A is negligible compared to that of the

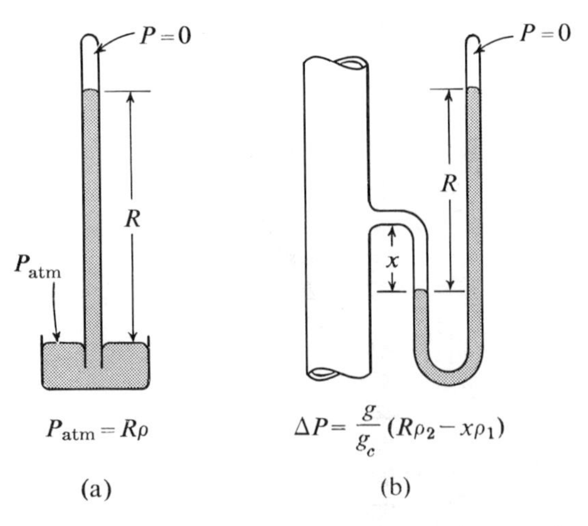

$P_{atm} = R\rho$ $\quad\quad$ $\Delta P = \frac{g}{g_c}(R\rho_2 - x\rho_1)$

(a) (b)

FIGURE S2.6A Manometers with one closed end.

manometer fluid. When this condition does not hold, there is a positional correction (dependent on x) that must be applied to the reading. The condition of negligible density at point A will always be approximately true for the situation where gases at ordinary pressures exist at point A and mercury is the manometer fluid.

LIST OF SYMBOLS

Latin letters

A	Cross-sectional area of expanded section of manometer arm
a	Cross-sectional area of manometer arm
g	Local acceleration due to gravity
g_c	Gravitational constant
P_A	Pressure in system at point of connection of high pressure manometer arm
P_B	Pressure in system at point of connection of low pressure manometer arm
P_a	Pressure at a point within heaviest fluid, high pressure arm
P_b	Pressure at a point within heaviest fluid, low pressure arm
P_{atm}	Pressure of atmosphere
R	Manometer reading
R_0	Manometer reading when no pressure difference is applied
x	Vertical distance

Greek letters

ζ	Angle with horizontal
ρ_1	Density, lighter (or lightest) fluid
ρ_2	Density, heavier fluid; intermediate density if three fluids are present
ρ_3	Density, heaviest fluid, when three are present.

PROBLEMS

S2.1A An ordinary manometer of the type shown in Figure S2.1A contains mercury as Fluid 2 and is at a temperature of 20 °C. The pipe, containing water, is of standard 3-in. brass. The reading is 10 in. of mercury. What is the pressure drop between points A and B, in psi; in feet of water?

S2.2A An ordinary manometer identical in design to that of Figure S2.1A is placed in a *vertical* pipe line, also 3-in. standard brass, all at 20 °C. The taps are 10 ft apart. What is the pressure drop represented by a reading of 10 in. of mercury, expressed as psi?

S2.3A An ordinary manometer (Figure S2.1A), with mercury as the sealing fluid, is placed across a restriction in a water line. Ten centimeters of carbon tetrachloride are placed on top of the mercury in the low pressure arm. The entire manometer is maintained at 20 °C. Calculate, based on difference in levels of the two mercury arms:

a. reading at zero pressure drop.
b. pressure drop in the pipe when mercury levels are equal.
c. pressure drop in the pipe when the difference in levels is 20 cm.

S2.4A The U-tube of an ordinary manometer may be inverted if the lighter fluid is placed in the upper end (i.e., in the top of the U). The open ends of the U-tube are then connected to the positions across which the pressure drop is to be measured. With such an instrument, in which air is the lighter fluid, a difference in level between the liquid in the two arms is found to be 35.4 cm. For the two cases: A. the liquid is water at 25 °C; B. the liquid is benzene at 25 °C, calculate the pressure drop across the two points of attachment:

a. in psi.
b. in feet of fluid flowing.

S2.5A Two manometers, one ordinary mercury-water and one inverted air-water (as described in Problem S2.4A), are connected in parallel across a common pressure differential position in a horizontal pipe. Valves are placed in the arms of each manometer so that either manometer may be used independently or both may be read simultaneously. By this arrangement, the mercury-water instrument may be used for large pressure drops in the line, while the inverted type will provide more sensitive readings for low pressure drops.

The maximum reading which may safely be obtained on the inverted manometer is 35 cm. Calculate the reading on the mercury-water manometer above which the inverted unit may not be used.

S2.6A A differential manometer (Figure S2.4A) is attached so as to indicate loss in pressure over a section of a line carrying a gas under moderate pressure. The manometer temperature is 25 °C. The heavier liquid is water and the lighter one is oil with specific gravity = $0.87(\frac{25}{4})$. Calculate:

a. pressure drop in the line in psi.
b. pressure drop in the line in inches of water.

S2.7A An ordinary manometer of the type represented in Figure S2.1A is used to register the pressure drop in a line carrying a gaseous mixture of average molecular weight = 26 lb per mole. With oil of density 0.81 gm per cm^3, the reading is only 0.7 mm. Since this reading is so small that no precision can be obtained, a differential manometer (Figure S2.4A) is being considered. If Fluid 2 is oil of the same type used in the ordinary manometer, and if Fluid 1 is water, what will be the reading for the same pressure drop in the line?

ADDITIONAL DATA The temperature of the room in which the manometer is housed is 75 °F.

The manometer tubing is of 1.5 cm diameter.

The expanded section of the manometer arms is 5.0 cm in diameter.

The temperature of the gas in the line averages 125 °F.

S2.8A A differential manometer (Figure S2.4A) has been in use with liquids whose specific gravities are 1.00 and 1.021, respectively. The fluid in the line is a gas. The instrument was constructed to give a maximum reading of 10 in., with $a/A = 0.01$. Difficulty has arisen because of the closeness of the densities, and their variation with temperature. Therefore, an inclined manometer (Figure S2.5A) is contemplated as replacement. If for the maximum pressure drop noted in connection with the first instrument, the new one is to read 5 in. and the manometer fluid is to be an oil of specific gravity = 0.81, what angle must be used?

S2.9A An open-ended manometer (Figure S2.2A) is attached to a pipe line carrying water. The temperature of the system is 77 °F. Mercury is the sealing fluid. With atmospheric pressure at the point of attachment, the reading is 1 in. Calculate the reading to be expected when the pressure in the pipe line is 10 psig.

S2.10A An open-ended manometer of the type represented in Figure S2.2A is placed so as to read the pressure at a specific point in a vertical iron pipe. What is the pressure in the pipe under the following conditions:

a. the reading is 5.3 cm of mercury; the fluid in the pipe is 20% caustic solution; both fluids are at 20 °C.

b. the reading is 5.3 cm of mercury; the fluid in the pipe is dry air; both fluids are at 45 °C.

c. At zero ΔP, the level of mercury in the manometer is 5 ft below the point of attachment.

S2.11A In Figure S2.2A the equation relating ΔP and R is reported to be:

$$\Delta P = \left(\frac{g}{g_c}\right)[R(\rho_2 - \rho_1) - x\rho_1].$$

Revise this equation so that x is eliminated in favor of R_o, the reading when $\Delta P = 0$. The answer is reported in the text.

HINT Draw diagram of zero reading. Write x in terms of x_o and R_o.

S2.12A A manometer with single reading arm (Figure S2.3A) is filled with mercury and used in a line carrying an oil of specific gravity $0.879^{25/4}$. You are to design the expanded section so that the correction factor, a/A, is not greater than 0.05. The two arms of the manometer, connecting to the line are constructed of glass, 10 mm in diameter. What is the cross section of the expanded section in square inches? If the mercury level is maintained 3 mm above the outlet tube when the maximum reading of 3 ft is reached, and if the reading arm enters 1 mm above the base of the section, and the tubing becomes vertical at a distance of 1 in. from the chamber, how much mercury, in milliliters, will be required to fill the instrument?

S2.13A A single-armed mercury manometer (Figure S2.3A) has an expanded section of 1.0 in. diameter in one arm; the glass arm is of 9 mm ID tubing. The manometer is placed across a restriction in a line carrying benzene. The temperature of the room in which the manometer is located is 25 °C. Calculate an expression for ΔP in psi, in terms of the reading, R, in inches, and in terms of the physical and dimensional parameters of the system.

S2.14A A draft gauge (i.e., inclined manometer with expanded section in the low pressure arm) is being designed. The angle of inclination is 15° from the horizontal. Oil of specific gravity = $0.87(\frac{25}{4})$ is to be the manometer fluid. The expanded section is large enough so that changes in level in this arm may be neglected. The scale is to be placed along the inclined arm, as in Figure S2.5A. This scale is to be marked so as to read directly in inches of water. Calculate the proper spacing for the "inch" divisions.

S2.15A A closed end, U-tube, vacuum gauge with mercury as the sealing fluid (Figure S2.6A) is to be designed so as to contain an expanded section in the arm adjacent to the point of attachment. Derive an expression for ΔP in terms of R.

S2.16A A closed-end manometer (Figure S2.6A) reads 4 cm of mercury. When no vacuum is on the system, the reading is 15 cm of mercury and mercury level is 10 cm below the point of attachment of side arm to tank. For the 4 cm reading calculate the absolute pressure in the tank to which the manometer is connected:

a. if the tank is filled with air? with hydrogen?
b. if the tank is filled with oil of sp. gr. = 0.92? what is the reading in this case if the pressure in the tank is zero?

SUPPLEMENT TO CHAPTER 2

B. Dimensional Analysis

Chemical engineers are frequently required to analyze systems involving many variables. Consider, for example, the relatively simple problem of a fluid flowing through a pipe. The pipe may be made of any of a number of materials, steel, iron, copper, etc. While pipes are normally circular in cross section, they may have other shapes for particular purposes. Regardless of cross-sectional shapes, pipes come in numerous sizes both in radial and length dimensions. The fluid flowing in the pipe may be a gas or liquid, highly fluid or very viscous, and its density may have any value within a wide range. Furthermore, it may be flowing slowly or very rapidly. On the surface, therefore, it would appear that innumerable experiments would be required to give data describing fluid flow and that the recording of such data in either tabular or graphical form would demand extensive space.

Dimensional analysis provides an approach that simplifies both the obtaining and the reporting of this sort of experimental data, as will be subsequently shown. Although the method has been known for many years, a vigorous mathematical substantiation of it did not appear until 1951. The reader is referred to Perry's *Chemical Engineers' Handbook* (4th ed.) for a more complete discussion of this topic with references to the original literature. We shall be content in this appendix with showing the application of the method through two examples, chosen to illustrate different ways in which the method may be used. Either method of attack could be applied to each example.

The first step in the method is to define the problem and list, with their fundamental dimensions, all the variables involved. These variables will be those that characterize the system, the material affected, and the response one desires. Suppose we are concerned with the power required to drive a specified mixer to be used in agitating a liquid in a cylindrical tank. Power is the response variable in which we are interested. Since the mixer is specified, the diameter of its impeller is not a variable, but since the tank in which the mixer is to be used may vary in size, some parameter to represent that size is required. Common sense tells us that tank diameter is likely to be the most useful dimension, since the distance between the tip of the impeller blades and the wall of the tank represents the principal channel through which the liquid will move. We naturally expect the power to increase as we increase the speed of agitation, so we shall list the rpm of the impeller as an important variable. Therefore, tank diameter and agitator speed are the system variables. The more viscous the liquid and the greater the density, the higher the power consumption is likely to be, so viscosity and density must also be listed; these are the material variables. We shall not include items like pressure and temperature, however, for two reasons. First of all, most agitations are usually carried out under a controlled pressure, very likely atmospheric. Secondly, variations in these variables will affect the density and viscosity of the liquid (pressure very little, temperature more so); thus density and viscosity will already reflect any changes in pressure or temperature.

We now are ready to tabulate our chosen variables, give each a symbol, and note the fundamental dimensions involved. Remember that power is the rate of doing work, that work is force acting through a distance, and that force is mass times acceleration, so

$$\text{Power} = \underbrace{\underbrace{(\text{mass})(\text{acceleration})}_{\text{force}}(\text{distance})}_{\text{work}}(\text{time})^{-1},$$

$$P = (M)\left(\frac{L}{t^2}\right)(L)\left(\frac{1}{t}\right) = \frac{ML^2}{t^3}.$$

Variable	Symbol	Dimensions
Power	P	ML^2/t^3
Tank diameter	D	L
Fluid density	ρ	M/L^3
Fluid viscosity	μ	M/Lt
Impeller speed (RPM)	N	$1/t$

The theorem upon which dimensional analysis is based states that π (a dimensionless quantity) results when the variables, each raised to an appropriate power, are multiplied, i.e.,

$$\pi = (P)^a(D)^b(\rho)^c(\mu)^d(N)^e.$$

Rewriting the variables in terms of their dimensions,

$$1 = \left(\frac{ML^2}{T^3}\right)^a (L)^b \left(\frac{M}{L^3}\right)^c \left(\frac{M}{Lt}\right)^d \left(\frac{1}{t}\right)^e.$$

For the latter equation to be valid, the exponents must have values so that all dimensions of a given kind will cancel out, thus,

for mass, $a + c + d = 0;$
for length, $2a + b - 3c - d = 0;$
for time, $-3a - d - e = 0.$

These relations give three equations in five unknowns, which means that two of the exponents must have known values or be specified before the equations can be solved. It also means, according to the theorem, that these two specifications imply that two dimensionless groups will describe the behavior of the system. How do we go about finding these groups?

First of all, we recognize that the relationship between two groups can readily be described graphically by plotting one against the other. Since we want to know how power is affected by the other variables, one group should contain P while the other group omits it (to avoid plotting power against itself). The second group should contain the variable to which power will be most sensitive, and the power group should not have this variable in it. In this problem, let us assume that we believe power is most sensitive to viscosity.

To get the group which constitutes the ordinate of our plot, we shall set $a = 1$ and $d = 0$. This procedure will assure a group in which P has an exponent of unity and in which μ is absent. The three equations now become:

$$1 + c + 0 = 0 \quad \text{or} \quad c = -1;$$
$$2 + b + 3 - 0 = 0 \quad \text{or} \quad b = -5;$$
$$-3 - 0 - e = 0 \quad \text{or} \quad e = -3;$$

from which

$$\pi_1 = P^{+1}D^{-5}\rho^{-1}\mu^0N^{-3} = \frac{P}{D^5\rho N^3}.$$

It is well to check the dimensions to see that no error has been made. Since,

$$\frac{ML^2}{t^3}\bigg/L^5\cdot\frac{M}{L^3}\cdot\frac{1}{t^3} = 1,$$

the group is dimensionless.

To get the other dimensionless group, the abscissa of the plot, we shall reverse the procedure and set $a = 0$ and $d = 1$. Then:

$$\begin{aligned} 0 + c + 1 &= 0 \quad \text{or} \quad c = -1; \\ 0 + b + 3 - 1 &= 0 \quad \text{or} \quad b = -2; \\ 0 - 1 - e &= 0 \quad \text{or} \quad e = -1; \end{aligned}$$

from which,

$$\pi_2 = P^0D^{-2}\rho^{-1}\mu^{+1}N^{-1} = \frac{\mu}{D^2\rho N}.$$

Checking,

$$\frac{M}{Lt}\bigg/L^2\cdot\frac{M}{L^3}\cdot\frac{1}{t} = 1.$$

Thus if $P/D^5\rho N^3$ is plotted against $\mu/D^2\rho N$ a single curve will result *providing* we have taken into account all the variables actually involved in the agitation process. All that dimensional analysis can do is to assemble the chosen variables into dimensionless groups. The choice of the variables is made by the engineer from his knowledge of the physical situation he is trying to describe. The test of accuracy comes when experimental data are plotted as described. If a single curve does result, regardless of the liquid being agitated or the tank size, or the rate of agitation, the essential variables have been properly chosen. And if this happens, note that a single graph is capable of answering the desired question of what power is required to drive the mixer no matter how varied the experimental conditions. Furthermore if the plot is (or approaches) a straight line, relatively few experimental determinations will be required to establish quite accurately the complete relationship.

The only functional form that can be predicted by dimensional analysis is the power function. There is little likelihood that such a function would be mechanistically correct. Dimensional analysis does not provide any basis for choice of mechanism but only a logical grouping of variables. The power function is most conveniently plotted on double logarithmic coordinates. Then, if all the variables have been accounted for, a single line will represent the data, as already noted. There is, however, no implication that the line will be straight.

Our analysis then has established that two dimensionless groups will completely describe the system being studied but, of course, we may not have chosen the best set of groups. In the above example P was paired with μ (i.e., alternately assigned to exponents a and d the values 1 and 0). If we had paired P with ρ or P with N, etc. different dimensionless groups would have resulted. Furthermore, if $\mu/D^2\rho N$ is dimensionless so is $D^2\rho N/\mu$. As an exercise, the reader might determine all the dimensionless groups which this method will generate.

It is also possible to generate new dimensionless groups by noting that dimensionless quantities may be multiplied or divided without changing their dimensionless characteristics. In the above example we have shown that

$$\underset{\pi_{\mathrm{I}}}{\frac{P}{D^5\rho N^3}} = \theta\Bigg(\underset{\pi_{\mathrm{II}}}{\frac{\mu}{D^2\rho N}}\Bigg).$$

If we would like to remove N from π_{I}, we may do so by dividing π_{I} by $\pi_{\mathrm{II}}{}^3$ to obtain a new group, π_{III}:

$$\pi_{\mathrm{III}} = \frac{P}{D^5\rho N^3}\,\frac{D^6\rho^3 N^3}{\mu^3} = \frac{P\rho^2 D}{\mu^3}.$$

Therefore,

$$\frac{P\rho^2 D}{\mu^3} = \phi'\left(\frac{\mu}{D^2\rho N}\right).$$

However, π_{III} would also have been obtained by letting $a = 1$ and $e = 0$, the procedure which would have been followed if we had assumed power most sensitive to the rate of agitation. When $e = 1$ and $a = 0$, the resulting dimensionless group is $D^2\rho N/\mu$, which is simply the reciprocal of π_{II}.

Only the actual plotting of data on various types of coordinates (cartesian, semi-log, or log-log) will determine which set of groups will be most useful.

The second example will illustrate a slightly different approach to the solution of a problem, and will also emphasize one or two points which did not arise in the first example. Consider a fluid flowing through round pipe. We are concerned with determining the energy lost in friction. The pressure drop from one end of the pipe to the other will be measured by the drop in pressure, $-\Delta P$*, the pertinent factors affecting friction, with their dimensions are thought to be:

• • • • •

* The minus sign indicates a *loss* in pressure from initial to final points in the direction of flow. Its use is optional but extremely useful when systems containing either increase or decrease are likely to be encountered.

Variable	Symbol	Dimensions
Pressure drop (response variable)	$-\Delta P$	M/Lt^2
Pipe length (system variable)	L	L
Pipe diameter (system variable)	D	L
Fluid velocity (material variable)	u	L/t
Fluid density (material variable)	ρ	M/L^3
Fluid viscosity (material variable)	μ	M/Lt

Thus,

$$\pi = (-\Delta P)^a (L)^b (D)^c (\rho)^d (\mu)^e (u)^f,$$

$$1 = \left(\frac{M}{Lt^2}\right)^a (L)^b (L)^c \left(\frac{M}{L^3}\right)^d \left(\frac{M}{Lt}\right)^e \left(\frac{L}{t}\right)^f.$$

In order that the various dimensions will cancel:

$$\begin{aligned} a + d + e &= 0, && \text{because of } M; \\ -a + b + c - 3d - e + f &= 0, && \text{because of } L; \\ -2a - e - f &= 0, && \text{because of } t. \end{aligned}$$

Again, there are three equations, but this time six unknowns. Thus, we shall expect three dimensionless groups since three of the exponents must be specified before a solution results. We could proceed as in the first example, by letting one exponent be unity and two others zero, the choice depending on what variables we wish to retain in a group and which ones we want to disappear. But we can also approach the problem by solving for three of the exponents in terms of the others. Choosing ΔP and μ as of primary importance,

$$\begin{aligned} d &= -a - e && \text{(from the first equation);} \\ f &= -2a - e && \text{(from the third equation);} \\ c &= -b - e && \text{(from the above and the second equation).} \end{aligned}$$

Then,

$$\pi = (-\Delta P)^a (L)^b (D)^{-b-e} (\rho)^{-a-e} (\mu)^e (u)^{-2a-e};$$

$$\pi = \left(\frac{-\Delta P}{\rho u^2}\right)^a \left(\frac{L}{D}\right)^b \left(\frac{\mu}{Du\rho}\right)^e.$$

Had we chosen to determine a different set of three exponents in terms of the others, a different result would have been obtained, although the reader will find if he does this, that some of the same groups (or their reciprocals) may reappear. But let us stay with the solution obtained. Although three dimensionless groups have been generated, inspection reveals that two of these may be combined. Obviously the pressure drop will be directly proportional to pipe length, so the exponents a and b must be the same (and will be arbitrarily unity, since we are interested in $-\Delta P$ and not $-\Delta P$ to a power).

To express all of this so that it is both mathematically and physically correct, write

$$-\frac{\Delta P}{\rho u^2} = \frac{L}{D}\,\boldsymbol{\theta}\left(\frac{Du\rho}{\mu}\right) \quad \text{or} \quad \frac{-\Delta PD}{\rho u^2 L} = \boldsymbol{\theta}\left(\frac{Du\rho}{\mu}\right).$$

Note that the value of π and $(\mu/Du\rho)^e$ are subsumed under $\boldsymbol{\theta}(Du\rho/\mu)$. It has already been pointed out that the reciprocal of a dimensionless group is also dimensionless. The ratio $Du\rho/\mu$ has been used because it turns out to be a very important one in fluid dynamics and is called the Reynolds' number (N_{Re}) after an early investigator into flow phenomena.

The possibility of combining groups in the above fashion simplifies the presentation of data. What has been shown is that if $-\Delta PD/\rho u^2 L$ is plotted against N_{Re}, a single curve should represent all the experimental data. At low N_{Re} this is what happens, but as N_{Re} increases, different lines are obtained for glass pipe, cast iron pipe, etc. (see Figure 9.10), which means we have omitted some variable from our analysis, for when three *independent* groups are present one is the ordinate, one the abscissa, and the third a parameter that identifies one of a family of curves. In this case, the third parameter is called a roughness factor, but further discussion of it will be omitted here. The reader should consider how to represent data graphically when four independent dimensionless groups are obtained. Fortunately, this situation is not common in practice.

These two illustrations will serve to exemplify how dimensionless groups are generated and why they are important. A further useful characteristic of such a group is that any consistent set of dimensions may be employed in its evaluation, e.g., metric or English, and, where convenient, different consistent units may be used for different groups. Many dimensionless groups appear repeatedly in the chemical engineering literature. Some of these are listed

TABLE S2.1B *Dimensionless Groups*

Name	Symbol	Definition
Fanning friction factor	f	$g_c D(-\Delta P_f)/2\rho u^2 L$
Reynolds' number	N_{Re}	$Du\rho/\mu$
Graetz number	N_{Gz}	wC_p/kL
Grashof number	N_{Gr}	$L^3\rho^2 g\beta T/\mu^2$
Nusselt number	N_{Nu}	hD/k
Peclet number	N_{Pe}	$Lu\rho C_p/k$
Prandtl number	N_{Pr}	$C_p\mu/k$
Stanton number	N_{St}	$h/C_p u\rho$
Schmidt number	N_{Sc}	$\mu/\rho D$
Arrhenius group	—	E/RT

above in Table S2.1B with their names and symbols. A longer list will be found in the September 1959 issue of *Chemical Engineering Progress* and in the March 1966 issue of *Industrial and Engineering Chemistry.* Note that the dimensional constant, g_c, is sometimes required to make a group dimensionless; such *dimensional* constants must be considered as one of the variables when the simultaneous equations are set up.

LIST OF SYMBOLS

Latin letters

A	Area, L^2
C_p	Heat capacity, (heat unit)/$(M)(\Delta°)$
D	Diameter, L
D_m	Molal diffusivity, M/Lt
E	Activation energy, (heat unit)/(M)
f	Fanning friction factor, dimensionless
G	Mass velocity, M/L^2t
g	Gravitational constant, L/t^2
g_c	Constant required in Newton's law of motion, English engineering units, $(\text{lb}/\text{lb}_f)L/t^2$
h	Heat transfer coefficient, (heat unit)/$(L^2)(t)(\Delta°)$
k	Thermal conductivity, (heat unit)/$(L)(t)(\Delta°)$
L	Length, L
N	R.p.m., l/t
ΔP_f	Pressure drop due to friction, M/Lt^2
R	Gas law constant, (heat unit)/$(M)(\Delta°)$
T	Absolute temperature, degrees
Δt	Temperature difference, $\Delta°$
u	Linear velocity, L/t
W_M	Average molecular weight, M

Greek letters

β	Coefficient of (fractional) volumetric expansion, $(\text{degrees})^{-1}$
Γ	Condensate loading, M/Lt
μ	Viscosity, M/Lt
ρ	Density, M/L^3

PROBLEMS

The following sets of variables represent situations in which dimensional analysis has been used effectively. Except where otherwise specified, obtain the minimum number of dimensionless groups in each case, expressing the group containing the independent variable as a function of the other groups unless otherwise instructed.

S2.1B Reynolds' number (N_{Re}) $= Du\rho/\mu$:

D—pipe diameter
u—average velocity
ρ—fluid density
μ—fluid viscosity

a. Prove that the number is dimensionless.
b. Eliminate u in favor of R, the mass rate of flow (fundamental units M/t).
c. The pipe diameter is suddenly halved; how is N_{Re} affected?

S2.2B A drag force is exerted on an object by a stream flowing past the object. The independent variable is the force, F. Other variables of interest in a general study of the force:

A—projection of area of object at right angles to axis of flow
u—stream velocity
ρ—stream density
μ—stream viscosity
D—diameter of sphere, when object is of that conformation
g_c—dimensional gravitational constant

Arrange the groups so that in one case the velocity, u, appears in each group; in another arrangement see that u appears in one group only.

S2.3B Power is required by six-bladed turbine mixers in baffled, or unbaffled tanks. The independent variable is power, P. Other variables of interest are:

N—r.p.m. of turbine
D_a—diameter of turbine
D_t—diameter of tank
ρ—fluid density
μ—fluid viscosity
g_c—dimensional gravitational constant
g—acceleration due to gravity

Arrange the grouping so that the r.p.m. appears in both groups in one instance and in only one group in another instance.

S2.4B A generalized correlation has been obtained of heat transmission. The independent variable is the heat transfer coefficient, h, expressed as Btu per (hour)(foot)^2(°F). Other variables of interest are:

C_p—heat capacity of the fluid being heated, Btu per (pound) (°F)
μ—viscosity of fluid being heated at average temperature
μ_w—viscosity of fluid being heated at wall temperature
ρ—density of fluid being heated
D—diameter of tube
G—mass velocity of fluid being heated, pounds per (hour)(foot)^2
k—thermal conductivity of fluid being heated

Arrange the groupings so that h and k do not appear in the same group.

S2.5B Heat is transferred to condensing films on vertical tubes. The independent variable is the heat transfer coefficient, h. Other variables of interest are:

- g—acceleration due to gravity
- Γ—condensate loading, expressed as weight rate per tube divided by the tube length
- μ—viscosity of condensing vapor film on tube wall
- ρ—density of condensing vapor on tube wall
- k—thermal conductivity of condensing vapor on tube wall, expressed as (Btu)(feet) per (hour)(foot)^2 (°F)

S2.6B In the case of unsteady state heating of a large slab, the independent variable is the temperature difference, ΔT, between the surface and the average slab temperature at any time. Other variables of interest are:

- Δt—temperature difference, heated surface, and initial slab temperature
- θ—time of heating
- k—thermal conductivity of slab (see S2.4B)
- C_p—heat capacity of slab (see S2.4B)
- L—slab thickness
- ρ—density of slab

S2.7B In mass transfer between fluids and beds of packed spheres, the independent variable is the mass transfer coefficient, k, expressed as pound-moles transferred per (square foot)(hour)(unit mole fraction difference). Other variables of interest are:

- W_M—average molecular weight of the gas
- G—mass velocity of the gas
- μ—viscosity of the gas
- D_m—molal diffusivity, expressed as lb-moles per (foot)(hour)
- D_p—diameter of the spheres

3 Behavior of Gases

Gases are an important subject of study for the chemical engineer, because many reactants and products in commercial syntheses are in this state of aggregation. In this chapter the elements of pressure-volume-temperature relationships of single gases as well as mixtures of gases will be discussed. More extensive treatment than is required here may be found in textbooks on thermodynamics.

3.1 The Ideal Gas Law

Imagine a cylinder, closed at the bottom, and fitted with a piston so well lubricated that it may be considered frictionless. At the beginning of an experiment, a quantity of a gas is contained within the cylinder. On the top of the piston there is a weight with known mass. Due to the force of gravity, a pressure is thereby exerted on the gas. Additional weights are now added slowly until the pressure is doubled. This process is carried out slowly so that any heat of compression can be dissipated through the cylinder walls. Thus, in effect, the process is an isothermal one; that is, there is no permanent change in the temperature of the gas.

At the end of the experiment the volume of the gas within the cylinder is one-half of what it was at the beginning of the test. If weights continue to be added slowly until the pressure is three times the initial value, the volume will be found to be one-third of what it was at the start. Similarly, increasing the mass by a factor of four, reduces the volume to one-fourth, and so forth.

Note that no specification was made for the gas within the cylinder. Oxygen, nitrogen, carbon dioxide, hydrogen, or any other gas would have served the purpose equally. Therefore, this experiment suggests that a universal law exists for gases treated at constant temperature, namely, that the volume of a specified quantity of gas is inversely proportional to the applied pressure. This pressure is total pressure, not gauge pressure.

The experiment is now repeated under different conditions. In this instance the cylinder is placed wholly within a chamber whose temperature can be accurately controlled, and the weight on the piston will not be changed; that is, the experiment will be conducted at constant pressure. As the temperature of the chamber (and thus the gas) is varied, it is noted that the higher the temperature, the greater the volume occupied by the gas; conversely, the lower the temperature, the less the volume of gas. The piston, under its constant weight, will move in or out of the cylinder to accommodate the changing volume. If a proper numerical scale is assigned to the temperature, a second general principle may be deduced: at constant pressure, the volume of a gas is directly proportional to the absolute temperature.

If these two principles are now combined, there results the more general statement that the volume of a gas is directly proportional to the absolute temperature and inversely proportional to the applied pressure. This statement suggests the existence of a proportionality constant that will be designated by the symbol, R. Thus, empirically, the following mathematical relationship has been obtained:

$$v = \frac{RT}{P} \quad \text{or} \quad Pv = RT. \tag{3.1}$$

This same relationship can be obtained from the kinetic theory of gases without the aid of experiment if one starts with an appropriate physical model and analyzes it mathematically. This analysis will not be discussed here, but it is important to realize that there is often more than one way to arrive at physical laws.

3.2 The Universal Constant, R

The constant, R, in Equation 3.1 is obviously dependent upon the dimensions of the other variables in the equation. If it is to be a truly universal constant, applicable to all gases, there are certain conventions that must be observed.

It has already been stated that P represents absolute pressure, not gauge pressure. The numerical value of R will vary with the units assigned to pressure, volume, and temperature, but given a particular set of units, R will not depend on the nature of the gas.

Gases differ in their density, some being composed of more massive molecules than others. Therefore, equal volumes of gases do not represent equal masses of these gases when mass is measured in such units as grams or pounds. A decision must therefore be reached as to whether a specified mass, e.g., 1 gm, should be used with the resultant variation in volume, or whether the volume should be kept the same regardless of mass. At one time the former of these two alternatives was chosen by engineers. Under these circumstances a different value of R is required for each gas. Tables of these values could be compiled, but it was a nuisance to have to consult such a table each time one wanted to work a problem.

The theory of gases presumes that a given volume of all gases at the same pressure and temperature contains the same number of molecules, i.e., that volume is inversely proportional to molecular weight. Thus it is reasonable to specify for v the volume occupied by one molecular weight of the gas, e.g., the number of cubic feet required to accommodate 32 lb of oxygen or the number of liters occupied by 32 gm of oxygen. Since both metric and English units are used by the chemical engineer, he will have to accustom himself to using R in either system.

The volume so defined is known as the molar specific volume. Thus, v has the units of L^3/n, where n = the number of moles. If the engineer wishes to know the total volume, he simply multiplies the specific volume by the number of moles under consideration, i.e., $V = nv$. Similarly $v = V/n$.

The remaining variable in the equation is the absolute temperature. Strictly speaking, to say that the absolute temperature (degrees Kelvin or degrees Rankine) must be used to make the equation valid, is to put the emphasis the wrong way around. Lord Kelvin used the relationship $Pv = RT$ to establish the absolute or thermodynamic temperature scale. Thus, knowledge of the temperature, T, really comes out of the ideal gas equation rather than being put into it. However, since most students are aware of the absolute temperature scale long before they have had a chance to study Lord Kelvin's development of it, perhaps all that needs to be emphasized here is that the numerical scale already familiar to us is the one to be used. Once again, there is a choice of the metric scale or its English equivalent. Normally, one uses degrees Kelvin if the other variables are expressed metrically and degrees Rankine if English units are employed elsewhere, but there is no physical or mathematical objection to mixing the system of units as long as the value of R is calculated for the same mixed system.

Since pressure, molar specific volume, and absolute temperature may be expressed in a variety of units, it would still seem necessary to have a table listing R in all of the possible variations. Tables of this sort do exist, but it is not necessary to use them. All that is required is to know one reference value. From this information all values of R may be computed.

At the freezing point of water (273.16 °K or 491.69 °R) and 1 atm, 1 lb-mole of an ideal gas occupies 359.046 cu ft, and 1 gm-mole of an ideal gas occupies 22.414 liters. These four numbers are usually rounded off to 273, 492, 359, and 22.4, respectively, as noted in Chapter 2. If these numbers are put into the equation $Pv = RT$, R can be readily calculated. Example 3.1 illustrates this procedure. By using the experience in unit conversion gained from Chapter 2, it may be shown that $R = 82.06$ (cu cm)(atm) per (gm-mole)(°K) $= 10.731$ (cu ft)(psi) per (lb-mole)(°R), etc.

There is one additional value of R which may be worth remembering. Regardless of the units assigned to P and v, the product Pv represents energy per unit mass. Energy can be expressed in thermal units as well as mechanical units. Since Btu per (lb-mole)(°R) = (numerically) gm-cal per (gm-mole)(°K), only one numerical value is needed. Thus, regardless of the gas in question, as long as consistent units are used $R = 1.987$ thermal (heat) units per (mole)(absolute degree), i.e., $Pv = 1.987T$.

3.3 Real Gases

Hitherto we have spoken as if the relationship $Pv = RT$ were an immutable law of nature. Unfortunately, if the experiments described in Section 3.1 are carried out very accurately, two things will be observed. First, the proportionality or inverse proportionality described is not exact. The more compressed the gas becomes, the less exact is the relationship proposed. Second, the amount of deviation varies with the gas used. These observed differences are not due to experimental error, because the more refined the work, the more clearly is the discrepancy between observed and expected results shown. Sometimes the deviation is small; at other times it is quite appreciable.

If the experiment is restricted to simple gases and is carried out at low pressures (considerably below atmospheric) and at relatively high temperatures, the equation $Pv = RT$ does hold almost exactly. For more complex gases, however, or at high pressures and low temperatures, the ideal gas law is a poor representation of actual behavior. Thus, the conclusion is reached that the ideal gas law proposed is truly valid only under limited conditions; that is, an ideal situation to which actual gases conform only more or less accurately. An ideal gas is, therefore, defined as one whose behavior can be described accurately by the equation $Pv = RT$.

We are led to inquire what there is about real gases that cause them to depart from the ideal. The empirical approach to the derived relationship does not assist us at this point, although it may suggest certain generalized conditions under which all gases tend to behave ideally. The value of a theoretical model is that it can be reexamined to see what assumptions have been made as an equation was derived, and to see if these assumptions are actually valid. If this procedure is followed with the theory of gases, it will be discovered that two basic assumptions are needed to get $Pv = RT$. It must, therefore, be concluded that these assumptions cannot be absolutely valid.

One of these assumptions is that the individual molecules of the gas do not affect one another. If the molecules are sufficiently far apart (as they would be at very low pressures), the magnitude of any interaction would be extremely slight. As the molecules approach each other when the gas is compressed, they exhibit the same type of attraction that the sun has for the earth. The combined effect of this "gravitational" force and the actual applied pressure causes the final volume to be less than that predicted by the applied pressure alone. The other assumption is that the molecules themselves occupy no volume. Again, when the total volume is very large compared to the molecular volume, the latter is insignificant and does not affect the results, but as the total volume is reduced, one may no longer ignore the volume of the molecules.

Therefore, the ideal gas law must be modified to allow for the complications present in real gases. Later in this chapter we shall consider some equations which have been proposed for this purpose; for the time being we shall concern ourselves only with those implications derivable from the ideal gas law.

3.4 Mixtures of Ideal Gases

Mixtures of ideal gases can be handled as readily as individual gases, because the relationship $Pv = RT$ implies that all molecules behave alike (under the same conditions of P, v, and T) regardless of type. Therefore, if we know the total number of moles of gas present and two of the three variables P, V, and T, the remaining variable may be determined from the basic equation $PV = nRT$. Similarly, a knowledge of P, V, and T permits one to determine n.

Although the meaning of V (the total volume occupied by the mixture of gases) is obvious, perhaps the significance of v in the equation $Pv = RT$ is not so clear when the gas consists of a mixture. It has already been pointed out that v refers to the specific molar volume, i.e., the volume occupied by 1 mole of gas. The inverse of this term, $1/v$, is the density. Thus v for a mixture of gases is the inverse of the density of that mixture. Given the pressure and temperature, then, one may readily calculate the density of a mixture of gases. The following example illustrates this point.

Example 3.1

▶ STATEMENT: A flue gas has the following analysis on a volume percentage basis:

CO_2	12.4%
CO	1.2%
O_2	5.4%
N_2	81.0%
	100.0%

Calculate the molar density (or, its reciprocal, the volume per mole) of this mixture at 600 °F and 740 mm Hg pressure in English units.

▶ SOLUTION: METHOD 1 The basis of calculation will obviously be 1 lb-mole.

$$\therefore\ v\,\frac{\text{(cu ft)}}{\text{(lb-mole)}} = \frac{RT}{P} = \frac{R(600 + 460)\ ^\circ\text{R}}{740\ \text{mm Hg}}.$$

Since 1 lb-mole of an ideal gas occupies 359 cu ft at 492 °R and 760 mm pressure,

$$R = \frac{760\ \text{mm}}{492\ ^\circ\text{R}}\left(359\,\frac{\text{cu ft}}{\text{lb-mole}}\right).$$

Using these two expressions, the one-line relationship for v results:

$$v = \frac{(760)(359)(1060)}{(492)(740)} = 794\ \text{cu ft/lb-mole}.$$

The density is the inverse of the number or 0.00126 lb mole/cu ft.

▶ SOLUTION: METHOD 2 An alternate, and generally faster, technique is to avoid the use of the gas constant R entirely by incorporating into the calculation a known volume at some standard temperature and pressure. In this problem the obvious choice is 359 cu ft per lb-mole at 760 mm Hg and 32 °F (492 °R). We now consider the changes in this specific volume, which would be occasioned by changing the pressure and temperature to the conditions specified in the problem, in this instance, 740 mm Hg and 600 °F.

The pressure is decreased so that the volume will expand by an amount represented by the ratio $\frac{760}{740}$. The temperature is increased so that the volume will expand by an amount represented by the ratio $(\frac{600+460}{492})$. Thus, the new volume per mole will be

$$359 \times \frac{760}{740} \times \frac{1060}{492} = 794\ \text{cu ft/lb-mole},$$

as before, but obtained directly in one step.

The simplicity of this type of calculation is enhanced by an easy check on the correct statement of the ratios. When properly used the standard temperature and pressure (i.e., 760 and 492) will appear in opposite parts of the fraction, one in the numerator and one in the denominator.

This method is recommended whenever practical.

One may still raise the question of what is meant by one mole of a mixture. For a single gas it is simply the molecular weight expressed in grams or pounds. A mixture, of course, does not really have a molecular weight in the same sense. However, it may be said to have an average molecular weight. If we know the analysis of the mixture on a molecular basis, the average molecular weight may be readily calculated by the formula:

$$\text{mol. wt.}_{\text{mix}} = N_A(\text{mol. wt.}_A) + N_B(\text{mol. wt.}_B) + N_C(\text{mol. wt.}_C) + \cdots,$$

where N is the mole fraction of component A, B, C, etc. (i.e., the number of moles of that component divided by the total number of moles of the mixture). $N \times 100$ = the mole percentage.

The next question is how the molecular analysis may be most readily expressed. A glance at the ideal gas equation ($Pv = RT$) shows that at a given pressure and temperature the same number of moles will occupy the same volume regardless of the gas involved. Therefore, the molecular analysis may be replaced with a volume analysis, and a volume percentage (or fraction) exactly equals a mole percentage (or fraction). Since gas analyses are usually reported as volume percentages, these figures may be used without conversion. However, it must be emphasized that the truth of these statements depends on the validity of the ideal gas law, i.e., for nonideal gases, volume percentages do not equal molar percentages.

To summarize what has been said to this point, if one is given the volume analysis of a gaseous mixture, one may calculate the average molecular weight, the molar density at any given pressure and temperature, and the density of the gas in pounds per cubic foot or grams per liter.

Example 3.2

▶ STATEMENT: For the flue gas described in Example 3.1, calculate (a) the average molecular weight, (b) the density in pounds per cubic foot, and (c) the specific gravity (referred to air at 60 °F and 1 atm).

▶ SOLUTION: a. Since volume analysis equals molar analysis for an ideal gas, 1 lb-mole of the flue gas will contain:

$$\begin{array}{lllll} CO_2 & 0.124 \text{ lb-moles} \times \dfrac{44 \text{ lb}}{\text{lb-mole}} & = & 5.456 \text{ lb} \\ CO & 0.012 \text{ lb-moles} \times \dfrac{28 \text{ lb}}{\text{lb-mole}} & = & 0.336 \text{ lb} \\ O_2 & 0.054 \text{ lb-moles} \times \dfrac{32 \text{ lb}}{\text{lb-mole}} & = & 1.728 \text{ lb} \\ N_2 & 0.810 \text{ lb-moles} \times \dfrac{28 \text{ lb}}{\text{lb-mole}} & = & 22.680 \text{ lb} \\ \hline & 1.000 \text{ lb-mole} & = & 30.200 \text{ lb} \end{array}$$

Therefore, the average molecular weight of the flue gas is 30.20 lb per lb-mole.

b. Example 3.1 has already given us the molar density of the flue gas. The density in pounds per cubic foot is obtained by multiplying the molar density by the average molecular weight:

$$\text{density} = 0.00126 \frac{\text{lb-mole}}{\text{cu ft}} \times 30.20 \frac{\text{lb}}{\text{lb-mole}} = 0.0381 \frac{\text{lb}}{\text{cu ft}} .$$

c. To get the specific gravity with reference to air at 60 °F and 1 atm, one must calculate the density of air at these conditions by the same method as above and then divide the density of the flue gas at its specified conditions by the density of the air at its condition. As a first step in the determination of the density of air, we must determine its average molecular weight and then determine its molar volume.

Although air contains traces of argon and other materials, its composition on a volume (and, therefore, molar) basis is closely approximated by 21% O_2 and 79% N_2. Thus, the average molecular weight is:

$$O_2 \qquad 0.21 \text{ lb-moles} \times 32 \frac{\text{lb}}{\text{lb-mole}} = 6.7 \text{ lb}$$

$$N_2 \qquad 0.79 \text{ lb-moles} \times 28 \frac{\text{lb}}{\text{lb-mole}} = 22.1 \text{ lb}$$

$$1.00 \text{ lb-mole} \qquad = 28.8 \text{ lb}$$

NOTE Because the above answer is an approximate one which will be increased by considering the rare gases present, the average molecular weight is usually taken as 29 lb per mole.

To illustrate a slightly different approach to the calculation of molar volume (without formally evaluating R) we can write the ideal gas law for both the given and standard conditions and divide them, noting that R in the two equations will cancel. This method is similar to Example 3.1, method 2.

Given conditions:

$$Pv = RT$$

$$(1 \text{ atm})\left(\frac{v \text{ cu ft}}{\text{lb-mole}}\right) \text{ at } 520\ °\text{R}.$$

Standard conditions:

$$P_s v_s = RT_s$$

$$(1 \text{ atm})\left(359 \frac{\text{cu ft}}{\text{lb-mole}}\right) \text{ at } 492\ °\text{R},$$

Dividing the first expression by the second,

$$v = \frac{(520)(359)}{(492)} = 379 \frac{\text{cu ft}}{\text{lb-mole}}.$$

The density of air is:

$$\frac{1 \text{ lb-mole}}{379 \text{ cu ft}} \times 28.8 \frac{\text{lb}}{\text{lb-mole}} = 0.0759 \frac{\text{lb}}{\text{cu ft}}.$$

The desired specific gravity is:

$$\frac{0.0381}{0.0759} = 0.502.$$

Suppose, however, that one is not concerned with the behavior of the mixture as a whole but with the behavior of the individual components that make up the mixture. How do the separate substances contribute to the overall mixture? There are two classical laws that answer this question: one is Amagat's law and the other is Dalton's law, both of which are exact and give the same results when applied to perfect gases.

Although there is no limit to the number of components that may be present in the mixture, the following discussion will be limited to two for the sake of simplicity. Refer to Section 3.1 and the piston-cylinder apparatus described there. For this analysis three identical units of this description will be required. Unit 1 will contain any given quantity of the mixture. Unit 2 will have only component A in it, the amount to be exactly equal to the amount of A in the mixture. If Unit 1 contains n moles of mixture, the amount of A in Unit 2 will be n_A. Similarly, Unit 3 will contain only gas B in exactly the same amount present in the mixture, thus n_B.

If more than two components were present in the original mixture, one additional piston-cylinder unit would be required for each additional component.

It will now be specified that the weights on the top of the pistons are all equal, i.e., the pressure is the same for each cylinder. It is also necessary to state that the temperature is the same for each unit. Under these circumstances, Amagat said that the volume of gas in Unit 2 when added to the volume of gas in Unit 3 would equal the volume of the mixture in Unit 1. Amagat's law may be stated as a proposition: the volume of a mixture of gases is equal to the sum of the volumes of the individual components when the latter are measured at the same pressure and temperature as the mixture.

The situation may be expressed mathematically by applying the ideal gas law three times.

$$PV = nRT, \tag{3.2}$$

$$PV_A = n_A RT, \tag{3.3}$$

$$PV_B = n_B RT. \tag{3.4}$$

Since P and T are the same in all equations, the second and third may be added to give the first, i.e., $P(V_A + V_B) = (n_A + n_B)RT$. Since $n_A + n_B = n$, $V_A + V_B = V$.

The individual volumes, V_A and V_B, are called partial volumes. When n_A and n_B are equal to unity, v_A and v_B would be specific partial volumes.

Division of Equation 3.3 by Equation 3.2 (and similar division of Equation 3.4 by Equation 3.2) shows that $V_A = N_A V$ and that $V_B = N_B V$; i.e., the partial volume is equal to the total volume times the mole fraction. This statement is strictly true only when the ideal gas law applies, although, as will be shown later, Amagat's law may often be used for mixtures of real gases.

It must be remembered that for the total volume to be equal to the sum of the partial volumes, the pressure and temperature used to calculate the partial volumes must be those of the mixture.

A discussion of Dalton's law requires us to return again to the cylinder-pistons previously described. Each of the three units still contains the same mass of gas as it did when we were considering Amagat's law, and the temperature is still maintained constant for all three. Now, however, the weights on the pistons are so adjusted that the volume of gas in each unit will be the same as that in the other two. According to Dalton's law, the pressure exerted by the mixture (which is, of course, the same as the pressure exerted on the mixture) will be equal to the sum of the pressures of the individual gases determined at the same total volume and temperature. As before, three equations may be written to show the interrelationship.

$$p_A V = n_A RT, \tag{3.5}$$

$$p_B V = n_B RT, \tag{3.6}$$

$$(p_A + p_B)V = (n_A + n_B)RT = nRT = PV. \tag{3.7}$$

The pressures p_A and p_B are called partial pressures. As long as the ideal gas law holds, the partial pressure of any component is equal to the total pressure times the mole fraction of that compound. Otherwise it does not. However, Dalton's law may also often be applied to mixtures of real gases as will be shown later.

As will be subsequently shown, if Amagat's law holds for a mixture of *real* gases, Dalton's law will not. Similarly, if Dalton's law holds for the nonideal gas, Amagat's law will not. Further, it is possible for neither of these laws to predict accurately the behavior of mixtures of real gases. However, for the ideal case now under consideration, both Amagat's law and Dalton's law hold equally well. The choice of which one to use depends on the data given and on the information required. When partial pressure data are given, or when partial pressure information is required, the problem is usually solved using Dalton's law (see Example 3.4). On the other hand, when data are in the form of partial volumes, or if that is the type of information sought, Amagat's law is indicated (see Example 3.3).

It may be well to add a warning at this point. When compositional changes occur in the process being analyzed, errors will be incurred if the volumes and partial volumes do not take these changes into account. Therefore, under these conditions particularly (though in most cases generally), it is better to convert volumetric quantities immediately to moles, with a reconversion to volumes or partial volumes as the final step in the calculations. (See Example 3.3, method 2, and Example 3.5.)

By way of summary and emphasis, Amagat's law applies to conditions of fixed total pressure and temperature. In this case, the total volume is the sum of the partial volumes. Dalton's law applies to conditions of fixed total volume and temperature, in which case the total pressure is the sum of the partial pressures. When $Pv = RT$ (but only under this condition), $V_i = N_i V$ and $p_i = N_i P$.

Example 3.3 (Amagat's law)

▶ STATEMENT: In the manufacture of hydrochloric acid by the Mannheim process, concentrated sulfuric acid and sodium chloride are heated in a mechanically agitated, indirectly heated retort. The following equation summarizes the reactions involved:

$$2NaCl + H_2SO_4 \rightarrow Na_2SO_4 + 2HCl.$$

To prevent the escape of noxious hydrogen chloride into the atmosphere, the furnace is operated under a small suction; therefore the hydrogen chloride is mixed with air when it leaves the reactor. Assume that in a particular plant, the product is 40% HCl and 60% air (by volume). Although the gas leaving the furnace is at a very high temperature, it is cooled before it enters the absorbers. In this plant the pressure and temperature at the entrance to the absorber are 740 mm Hg and 50 °C, respectively. It will be assumed that the final gases rising from the absorber contain only 0.1% HCl gas at a pressure of 730 mm Hg and 30 °C.

Calculate:

a. The volume of gas leaving the absorber per cubic foot of gas entering.
b. Mass of HCl gas entering per cubic foot of gas entering.
c. Mass of HCl gas leaving per cubic foot of gas entering.

▶ SOLUTION: Two methods of solution will be given. The first is chosen simply to indicate the application of Amagat's law; the second method is based on the molar approach, is preferred, and should regularly be used for two reasons: (1) economy of calculation and (2) avoidance of problems associated with changing gas composition. In both cases the logical choice for a basis of calculation is one cubic foot of entering gas, since all information required is referred to that amount.

METHOD 1 According to the data, the entering gas contains,

0.4 cu ft $HCl_{(g)}$	at	740 mm Hg and 50 °C,
0.6 cu ft air	at	740 mm Hg and 50 °C.

All of the air goes through the absorber unchanged except for a decrease in pressure to 730 mm and a decrease in temperature to 30 °C. Therefore, at discharge it will occupy

$$0.6\,\frac{(740)(303)}{(730)(323)} = 0.571 \text{ cu ft.}$$

Since the air constitutes 99.9% of the final volume, the total volume of gas leaving per cubic foot of gas entering is

$$\frac{0.571}{0.999} = 0.572 \text{ cu ft} \qquad \text{(answer to part a).}$$

The mass of HCl entering may be determined from its partial volume, 0.4 cu ft, reduced to standard pressure and temperature, converted to moles (all ideal gases occupy 359 cu ft at STP), and this number changed to pounds by the molecular weight of HCl (36.5 lb per lb-mole). This procedure can be accomplished in one computation as follows:

$$0.4\,\frac{(740)(273)}{(760)(323)(359)}\,36.5 = 3.34 \times 10^{-2} \text{ lb.} \qquad \text{(answer to part b).}$$

Finally, if the total volume at discharge (0.572) is converted to the partial volume of HCl by the discharge analysis (0.001), reduced to STP, changed to moles, and multiplied by the molecular weight to get mass, the answer to part (c) will be obtained.

$$(0.001)(0.572)\,\frac{(730)(273)}{(760)(303)(359)}\,36.5 = 5.01 \times 10^{-5} \text{ lb.}$$

METHOD 2 The moles of gas entering per cubic foot are:

$$1\Big/359\,\frac{(760)(323)}{(740)(273)} = 2.29 \times 10^{-3}.$$

Since volume analysis equals molar analysis there are:

$$(2.29 \times 10^{-3})(0.4) = 0.918 \times 10^{-3} \text{ moles HCl entering,}$$
$$(2.29 \times 10^{-3})(0.6) = 1.372 \times 10^{-3} \text{ moles air entering.}$$

The air represents 99.9% of the gas leaving; therefore

$$\frac{1.372}{0.999} = 1.373 \times 10^{-3} \text{ moles of mixture leaving.}$$

This can readily be converted to volume by

$$(1.373 \times 10^{-3})\,\frac{(760)(303)}{(730)(273)}\,359 = 0.572 \text{ cu ft.}$$

The masses entering and leaving can be easily calculated from the analyses and molecular weight of HCl.

$$\text{Mass HCl in} = (0.918 \times 10^{-3})(36.5) = 3.34 \times 10^{-2} \text{ lb,}$$
$$\text{Mass HCl out} = (1.373 \times 10^{-3})(0.001)(36.5) = 5.01 \times 10^{-5} \text{ lb.}$$

Example 3.4 (Dalton's law)

▶ STATEMENT: Hot air is being used to dry some wet wall board. The hot air enters the equipment at a total pressure of 768 mm Hg and 332 °F. The partial pressure of water vapor in the air is 25 mm Hg. At the exit of the drier, the temperature is 232 °F and the pressure is 760 mm Hg. In this air the partial pressure of the water vapor is 100 mm Hg. Calculate the volume of the exit gas per cubic foot of inlet gas.

▶ SOLUTION: If the partial pressure of water vapor is 25 mm Hg, that of the air is 768 − 25 = 743 mm Hg. The mass of air leaving will be the same as that entering. Since Dalton's law refers to an unchanging *total* volume, the volume of air at its partial pressure is 1 cu ft. We thus need only convert 1 cu ft at 743 mm Hg and 332 °F to V cu ft of air at (760 − 100 mm) = 660 mm Hg and 232 °F to determine the volume of inerts leaving and thus the *total* volume of all gases leaving.

$$V = \frac{(743)(692)}{(792)(660)} = 0.984 \text{ cu ft of total gas leaving.}$$

Example 3.5

▶ STATEMENT: The specifications of the process are exactly the same as in Example 3.4. Using these data calculate the weight of water removed from the wall board (and consequently picked up by the air) per 1000 cu ft of entering air.

▶ SOLUTION: The total number of moles of inlet air including its water content, may be calculated by the methods of Example 3.3, as follows:

$$1000 \times \frac{768}{760} \times \frac{492}{792} \times \frac{1}{359} = 1.748 \text{ moles.}$$

Since pressure ratios and mole ratios are identical as long as the ideal gas law applies, the mole fraction of water, N_w, in the inlet air is:

$$N_w = \frac{25}{768}.$$

Multiplying the total number of moles by the mole fraction of water

$$1.748\left(\frac{25}{768}\right) = 0.0569 \text{ moles of water entering the system.}$$

The moles of air which enter are then

$$1.748 - 0.0569 = 1.691 \text{ moles.}$$

Now this same number of moles of air leave the system and, at that point the partial pressure of air is the total pressure minus the partial pressure of water vapor or

$$760 - 100 = 660 \text{ mm Hg.}$$

Then the pressure ratio

$$\frac{100}{660}$$

is equal to the mole ratio of water to air. Multiplying the moles of air by this ratio

$$(1.691)\left(\frac{100}{660}\right) = 0.256 \text{ moles of water in exit gas.}$$

Obviously,

$$0.256 - 0.057 = 0.199 \text{ moles of water picked up by 1000 cu ft of entering hot air.}$$

Since the molecular weight of water is 18,

$$(0.199)(18) = 3.59 \text{ lb water.}$$

3.5 Mixtures of Gases and Vapors

If a glass of water is spilled on a table cloth, the latter will become dry after a period of time. If the spilled liquid had been acetone, the time required for drying under the same external conditions would be considerably less. If diethyl ether had been the fluid spilled, the requisite drying time would be even less. In the liquid state, the individual molecules are much closer together, and while mobility still exists, the molecules are much more restricted in their activity than they are in the gaseous state. However, from time to time some of the molecules at the surface of the liquid do escape into the gas phase above, and so the quantity of liquid gradually diminishes. This process by which "liquid" molecules become "gas" molecules is called vaporization or evaporation. The reverse process in which liquid is formed is called condensation. The more readily a liquid evaporates, the more volatile it is. Thus, we say that ether is more volatile than water.

It is common practice to distinguish between a substance that readily vaporizes (or condenses) at or near room conditions and those which would vaporize (or condense) only at low temperatures, high pressures, or both. Thus, oxygen is called a gas, but molecules of water in the gaseous state are spoken of as water vapor, not water gas. Thus, in a mixture there might be both gases and vapors. Ordinary air, for example, normally contains both oxygen and nitrogen (gases) and water (vapor). Although it is possible for such mixtures to behave nonideally, many important practical applications (humidification and air conditioning, for example) occur under relatively low pressure (at or near atmospheric) and at such temperatures that the ideal gas law is adequate. Thus, this special example of gaseous mixtures is worthy of attention at this point.

If a beaker of a liquid, such as water, is placed under a bell jar and maintained at constant temperature while the space is evacuated, the liquid will be observed to boil when a specific pressure is reached. This pressure will vary with the constant temperature at which the liquid is maintained, but for each temperature there is but one pressure at which free boiling occurs. This process may be shown graphically as in Figure 3.1. If the liquid is maintained at T_4 and the pressure lowered, a point D on the AC curve is reached. This is the pressure at which boiling takes place. It will be noted that the specific volume of the liquid has changed very little as the pressure was changed until boiling occurs. Then the formation of vapor results in an appreciable volume change. However, as long as any liquid remains, the pressure does not change. Stated another way, if the pressure remains constant, so will the temperature, since any additional input of energy will result in further vaporization of the liquid and not in its temperature increase.

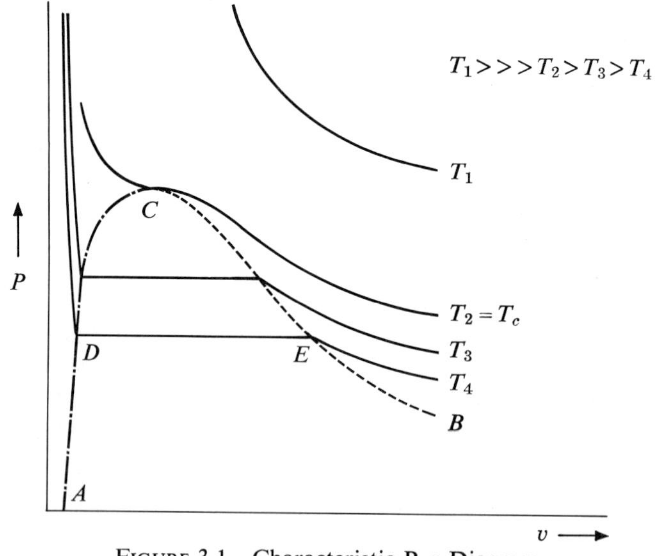

FIGURE 3.1 Characteristic P–v Diagram.
——————— Isotherms.
- - - - - - - Dew point curve.
– · – · – · – · Bubble point curve.

The pressure corresponding to the horizontal line DE is called the vapor pressure of the liquid at T_4. There is a vapor pressure corresponding to every temperature until point C (the critical point) is reached. At temperatures above T_C only the gas phase can exist, so the term vapor pressure has no meaning.

All liquids have a characteristic vapor pressure-temperature relationship. An important example is the one for water. Table 3.1 lists data for this substance. As will be shown later in Chapter 7, vapor pressure may often be computed from an equation such as

$$\ln p = A - \frac{B}{T - 43}, \tag{3.8}$$

where T is in degrees Kelvin and A and B are empirical constants. Other methods of representing vapor pressure data exist, but their discussion will also be reserved for Chapter 7.

Let us return to the example of water spilled on a cloth. The reason the cloth ultimately becomes dry is that the air above it has the capacity for retaining all the moisture as water vapor. However, suppose we consider once again the beaker of water under a bell jar. In this case the volume of air is limited. The water will still exert a vapor pressure corresponding to its temperature. For a time, water will continue to vaporize, but finally there will be sufficient water in the gaseous phase to exert a partial pressure equivalent to the vapor pressure of the liquid. When this condition is reached, the air above the liquid is said to be saturated with water vapor. Obviously, the higher the temperature, the higher the vapor pressure and thus, the greater the amount of water that may exist in the vapor form, but if the limiting situation results before all of the water has evaporated, no further vaporization will occur. (It is recognized that vaporization and condensation are dynamic processes. What actually occurs at saturation is that the rate of condensation from the vapor to the liquid state occurs at the same rate as vaporization from the liquid to the vapor state. Thus, the net result is the same as if the vaporization process had stopped and there was no condensation.)

Suppose the gas mixture contained a vapor but was not saturated with it. If this mixture is now slowly cooled, the vapor pressure is approached, since the lower the temperature the lower the vapor pressure. Finally, the partial pressure of the vapor will just equal the vapor pressure. Any further lowering of temperature will bring the vapor pressure below the saturated partial pressure, and to restore equilibrium, some vapor must condense. Experimentally, this point is usually determined by observing the formation of a dew on a polished surface, hence the name dew point. The line CB in Figure 3.1 is the locus of all dew points within the range of that diagram. Just as the vapor pressure is determined by the temperature, so the dew point is determined by the pressure for a fixed temperature, or by the temperature for a fixed pressure. Although this discussion has been concerned with water evaporating into air and the behavior of the resultant air-water vapor mixture, the principles are general and apply to any vapor-gas mixture. For example, one might be interested in the vaporization of benzene into an inert mixture

TABLE 3.1 *Vapor Pressure of Liquid Water*

t °C	Vapor Pressure—mm Hg	t °F	Vapor Pressure—in. Hg
0	4.579	32	0.180
5	6.543	40	0.248
10	9.209	50	0.363
15	12.788	60	0.522
20	17.535	70	0.739
25	23.756	80	1.032
30	31.824	90	1.422
35	42.175	100	1.932
40	55.324	110	2.596
45	71.88	120	3.446
50	92.51	130	4.525
55	118.04	140	5.881
60	149.38	150	7.569
65	187.54	160	9.652
70	233.7	170	12.199
75	289.1	180	15.291
80	355.1	190	19.014
85	433.6	200	23.467
90	525.76	212	29.922
95	633.90	220	34.992
100	760.00	230	42.308
105	906.07	240	50.837
110	1074.56	250	60.725
115	1267.98	260	72.134
120	1489.14	270	85.225
125	1740.93	280	100.18
130	2026.16	290	117.19
135	2347.26	300	136.44
140	2710.92		
145	3116.76		
150	3570.48		

Sources: *American Institute of Physics Handbook*, 2nd ed. (New York: McGraw-Hill Book Company), 1963; and J. W. Keenan and F. G. Keyes, *Thermodynamic Properties of Steam* (New York: John Wiley & Sons, Inc.), 1936. The precision indicated is that given in the sources listed.

of CO_2 and N_2, or the condensation of benzene from a mixture of benzene vapor, CO_2 and N_2. All of the preceding remarks on vaporization (or condensation) of water apply equally well to these mixtures.

In chemical engineering there are many processes in which air is used as a drying agent. Chemical engineers are also concerned with humidification (the addition of water vapor to the air) or air conditioning in order to provide any water-air mixture that may be needed for specified operating conditions. Thus, special terminology has developed for this case, and in the discussion which follows it will be noted, with its counterparts for any vapor-gas system.

The ratio $(p/p_s)100$ is normally known as the percentage relative saturation, but for the air-water vapor system, it is called relative humidity. The partial pressure of the vapor in the mixture is denoted by p, whereas p_s refers to the partial pressure of the vapor in a saturated mixture (p_s is, of course, the vapor pressure of the liquid at the temperature of the mixture).

All other terms used to define vapor-gas compositions involve the relationship of the quantity of vapor to that of the gas in which it is dispersed. Terms with which the reader is already acquainted are the mass ratio and the molal ratio. Using a as the subscript to designate the vapor and b as that to designate the gas, with p, n, and M_w to represent partial pressures, numbers of moles and molecular weights, respectively, we can write

$$\text{molal ratio} = \frac{p_a}{p_b} = \frac{n_a}{n_b}$$

and

$$\text{mass ratio} = \frac{p_a M_{w,a}}{p_b M_{w,b}}.$$

When the mixture is limited to water-air, special names have been assigned to these ratios, the molal ratio being known as molal humidity and the mass ratio being known simply as the humidity. Since the molecular weights of water and air are 18 and 29, respectively,

$$\text{humidity} = \left(\frac{18}{29}\right)(\text{molar humidity}).$$

One other term is in common use, percentage saturation, with its counterpart for water-air mixtures, percentage humidity. This quantity expresses the ratio of the actual mole ratio to that which would exist if the mixture were saturated with the vapor at the temperature of the actual mixture. Using the subscript s to designate the saturated condition percentage saturation can be written,

$$\text{percentage saturation} = \frac{p_a/p_b}{(p_a/p_b)_s}.$$

It is quite apparent that, if we try to distinguish between percentage saturation in molal and in mass units, the molecular weight ratios will appear in identical manner in both the numerator and the denominator of the percentage saturation ratio. The percentage saturation is then identical whether one uses the molal system or the mass system.

The various quantities just described are summarized in Table 3.2. Symbols are as defined above, with P being introduced to designate total pressure on the system. Subscripts a and b have been dropped, with p_b always being replaced by $P - p$, to which it is equivalent when $p = p_a$.

TABLE 3.2 *Quantities Representing Vapor-Gas Compositions*

General Case	Water-Air Mixtures	Symbolic Representation
Percentage relative saturation	Relative humidity	$100p/p_s$
Molal ratio	Molal humidity	$p/(P - p)$
Mass ratio	Humidity	$pM_{w,a}/(P - p)M_{w,b}$
Percentage saturation	Percentage humidity	$100 \dfrac{p/(P - p)}{p_s/(P - p_s)}$

Example 3.6

▶ STATEMENT: Water is present in air at 50 °C and a total pressure of 760 mm Hg in an amount such that the relative humidity is 50%. Water is to be removed by (1) cooling, (2) compression, or (3) a combination of both. In each case, calculate:

a. Fraction of the original water condensed.
b. Molal humidity and percentage humidity of the resulting gaseous mixture.

CASE 1 Cool to 20 °C at constant pressure.

CASE 2 Compress to 5 atm at constant temperature.

CASE 3 Compress to 5 atm and subsequently cool to 20 °C.

▶ SOLUTION: General information:

t °C	Vapor Pressure of H_2O, mm Hg
20	17.5
50	92.5

Partial pressure of water, initial air = (0.50)(92.5) = 46.25 mm. Moles of water per mole of original mixture = $\frac{46.25}{760}$ = 0.0608. Moles of air per mole of original mixture = 1 − 0.0608 = 0.9392, which is also equal to (760 − 46.25)/760.

CASE 1 After cooling, partial pressure of water = 17.5 mm. After cooling, partial pressure of air = 760 − 17.5 = 742.5 mm. Water in 0.939 moles of air = $(\frac{17.5}{742.5})(0.939) = 0.0221$ moles.

$$\begin{aligned}\text{Water removed by process} &= 0.0608 - 0.0221,\\ &= 0.0387 \text{ moles},\\ &= 0.697 \text{ lb}.\end{aligned}$$

Fraction of original water removed = $\frac{0.0387}{0.0608} = 0.637$. Molal humidity of gas after cooling:

$$\frac{17.5}{742.5} = \frac{0.0221}{0.9392} = 0.0236.$$

The first of these two ratios expresses moles of water per mole of B.D. ("bone dry") gas in terms of the pressures of each. The second uses directly the calculated quantities of each.

The humidity is the same type of ratio expressed in mass units. For a water-air mixture

$$\text{humidity} = \frac{18}{29}\,(\text{molal humidity}).$$

Therefore,

$$\text{humidity} = \left(\frac{18}{29}\right)0.0236 = 0.0146.$$

Since at the end of the cooling process, the air must be saturated with water vapor, the percentage humidity is 100.

CASE 2 In the final gas mixture, the partial pressure of water vapor is 92.5 mm since the final temperature is 50 °C, and the partial pressure of air is

$$5(760) - 92.5 = 3707.5 \text{ mm}.$$

The moles of water in the compressed gas must then be

$$\frac{92.5}{3707.5}\,0.939 = 0.0234.$$

The fraction removed during compression is

$$\frac{0.0608 - 0.0234}{0.0608} = 0.615.$$

Calculation of humidities follows as in Case 1 and will not be repeated here or in Case 3.

CASE 3 In the final gas mixture the partial pressure of water vapor is 17.5 mm and the partial pressure of air is

$$5(760) - 17.5 = 3782.5 \text{ mm}.$$

The moles of water in the final compressed and cooled gas are calculated as before.

$$\frac{17.5}{3782.5}\, 0.939 = 0.00434 \text{ moles.}$$

The fraction of the original water removed is

$$\frac{0.0608 - 0.0043}{0.0608} = \frac{0.0565}{0.0608} = 0.93.$$

The advantage of both compression and cooling for removal of condensible vapors from "permanent gases" is very apparent.

Example 3.7

▶ STATEMENT: A mixture of benzene in carbon dioxide has a percentage saturation of 75 under the conditions of $\frac{1}{2}$ atm and 71.2 °F. Calculate:

a. percentage relative saturation,
b. molal ratio,
c. mass ratio,
d. the same quantities if the mixture were benzene in air.

▶ SOLUTION: The vapor pressure of benzene at 71.2 °F is 100 mm Hg. At $\frac{1}{2}$ atm total pressure the partial pressure of CO_2 in a mixture of CO_2 and benzene, saturated with the latter, must then be

$$\tfrac{1}{2}(760) - 100 = 280 \text{ mm Hg.}$$

The saturated mole ratio is then

$$\frac{100}{280} = 0.357.$$

Let the actual partial pressure of benzene in the mixture be x mm of Hg. Then the molal ratio is:

$$\frac{x}{(380 - x)}$$

and

$$\frac{x}{(380 - x)(0.357)}$$

is the percentage saturation. Therefore,

$$\frac{x}{(380 - x)(0.357)} = 0.75,$$

from which

$$x = 80.2 \text{ mm Hg.}$$

All quantities can now be calculated:

a. Percentage saturation $= 80.2/100 = 80.2\%$.

b. Mole ratio $= \dfrac{80.2}{380 - 80.2} = 0.268$.

c. Mass ratio $= \dfrac{(80.2)(78)}{(380 - 80.2)(44)} = 0.474$,

where 78 and 44 are molecular weights of benzene and CO_2, respectively.

d. If the mixture had been benzene in air, rather than CO_2, the only quantity to change is the mass ratio, which becomes

$$\frac{(80.2)(78)}{(380 - 80.2)(29)} = 0.719.$$

3.6 Equations of State

Any equation relating the three primary thermodynamic state variables, P, v, and T is known as an equation of state. The simplest one is the so-called ideal gas law, $Pv = RT$. However, as has been discussed in Section 3.3, this equation does not adequately predict the behavior of real gases. To understand this problem better, let us look at a graphical representation of typical experimental data for real gases.

In Figure 3.1 pressure is plotted against specific volume. Apart from the envelope ACB, the curves represent lines of constant temperature, i.e., isotherms. Thus, any point on this diagram represents one piece of P-v-T data. Given the pressure and the specific volume of the gas, one can determine the temperature by interpolation between the plotted isotherms. Similarly, any other pair of these three variables will determine the third.

For any isotherm, both R and T are constant. Thus, if the ideal gas law is to hold, the appearance of the isotherm must conform to a curve generated by the equation $Pv =$ constant. Such a curve would be T_1. For a temperature such as T_4, the departure from the ideal is evident. Let us follow the isotherm T_4 from the right of the figure toward the left. As the specific volume decreases (caused by an increase of pressure), the molecules of the gas are forced closer together. Ultimately the force of attraction between them becomes so great that the gaseous state can no longer be maintained, and a droplet of liquid forms. This phenomenon occurs at the point where the isotherm touches the curve CB, and as has previously been noted, this is the dew point. Everyday experience (confirmed in theoretical thermodynamics by the phase rule) tells us that when two phases of a single substance exist together, both the pressure and the temperature cannot vary independently. Since the temperature has been fixed, the pressure is also fixed; therefore, the continuation of the isotherm is a horizontal straight line. Along this line the only way that the specific volume can decrease is by the continued condensation of the vapor to a liquid. This condensation becomes complete when the curve AC is reached. At this point only one phase exists, so that further decrease in specific volume requires an increase in pressure. Since liquids are almost incompressible, a very large increase in pressure is required for a very small decrease in specific volume; thus the isotherm rises sharply.

From the above discussion and that in Section 3.5, the meaning of the envelope ACB is clear. For every temperature there will be a specific pressure and specific volume at which condensation begins and another at which condensation is complete. We have already noted that the curve CB is the locus of all the dew points. If we had followed the isotherm from left to right, points on the curve AC would be observed as the conditions where the first bubble of vapor formed (and where the last bubble of vapor condenses in the opposite direction). Therefore, curve AC is known as the bubble point curve.

These two curves come together at C, the critical point. An isotherm passing through this point is called the critical temperature. Within the boundaries of the envelope ACB both the liquid and the gas phase exist To the right of CB there is only gas, and to the left of AC there can be only liquid. Far to the right in the gaseous region and at relatively low pressures, the isotherms conform fairly closely to the equation Pv = constant which is that of an equilateral hyperbola. As the isotherms approach the curve CB, the conformity lessens. Distortion is especially evident near the critical condition.

If one tries to devise an equation that will truly represent the behavior of actual gases, provision must be made for all of these complications. It is clear from the shape of the isotherms that such an equation will be at least a cubic in v. One of the earliest attempts to devise such an equation was made by van der Waals. It has been previously mentioned that the effective pressure is actually greater than the external or applied pressure due to the attractive forces that exist between molecules. Kinetic theory predicts that this added effect is inversely proportional to the square of the specific volume. Therefore, instead of P, van der Waals proposed that $(P + a/v^2)$ be used. The constant a varies with each gas, since the mass of each gas molecule is dependent upon its molecular weight.

It was also noted that gas molecules actually occupy a part of the total volume of the container; thus the free volume is not v but $(v - b)$, where b is another constant varying with each gas. Making these substitutions in the ideal gas law, one obtains van der Waals' equation, which may be written in either of the two following ways:

$$\left(P + \frac{a}{v^2}\right)(v - b) = RT \qquad \text{or} \qquad P = \frac{RT}{v - b} - \frac{a}{v^2}. \tag{3.10}$$

The units on the constants a and b must be observed carefully. Since b is a correction for specific volume, it must have the units of volume per mole, e.g., liters per gram-mole. Since the ratio a/v^2 must have the units of pressure, a must be expressed in pressure units multiplied by the square of specific volume, e.g., (atm)(ft)^6/(lb-mole)^2 or L^7/Mt^2 in fundamental units.

Reference to Figure 3.1 shows that the critical isotherm, T_2, is horizontally tangent to point C. At this point both the first and second derivatives of pressure with respect to specific volume are zero when the temperature is constant. That is, both the slope of the isotherm (the rate of change of P with v) and its rate of change are zero. If we take both the first and second derivatives of van der Waals' equation as indicated, set them equal to zero at the critical point, and combine them with the original van der Waals equation, also written for the critical point, the following relations result:

$$P_c = \frac{a}{27b^2}, \qquad T_c = \frac{8a}{27Rb}, \qquad \text{and} \qquad v_c = 3b. \tag{3.11}$$

These equalities should be verified by performing the mathematics outlined.

From these relations it is easy to show that the constants a and b in van der Waals' equation are related to the critical constants as follows:

$$a = \frac{27R^2T_c^2}{64P_c} \qquad \text{and} \qquad b = \frac{RT_c}{8P_c}. \tag{3.12}$$

It would, of course, be possible to express these constants in terms of v_c and P_c or T_c, but the critical pressure and critical temperature are usually known with greater accuracy than the critical specific volume; hence, the above relations are the most useful.

While van der Waals' equation represents experimental P-v-T data better than the ideal gas law, it is still of limited accuracy and fails particularly near the critical point. Over the years men have sought to obtain an equation that would express analytically the data obtained experimentally. The resulting equations have fallen into three categories. First are the general equations, designed to apply to any gas (given the appropriate constants); van der Waals' equation is of this class. The second type might be called restricted equations, because they apply only to certain kinds of compounds, for example, the lighter hydrocarbons. Then there are the specific equations that represent only one gas such as oxygen. The more limited the application, the greater is the expected accuracy for equations of similar complexity.

In the final analysis, theory has so far been unable to predict with sufficient accuracy the information we seek. Thus, most equations of state are essentially empirical, i.e., they are developed to fit the data we do have in the hope that they will adequately predict information we do not have for new regions of temperature and/or pressure and for other gases. Sometimes as many as eight or ten empirical constants are involved in the final equation.

There are so many equations of state that further discussion of them here would take us too far afield. Those interested in pursuing this topic further may wish to consult an article by Wallace R. Gambill in the October 19, 1959 issue of *Chemical Engineering* (pp. 195ff.), which gives an excellent summary of recommended equations and their use, together with an extensive bibliography of the subject. In this article will be found some 2-, 3-, and 4-constant equations, relatively simple to use and reasonably accurate, as well as some 5-, 8-, and 10-constant equations for more precise work or where broader ranges of pressure and temperature are encountered. Unfortunately, simplicity and accuracy do not go hand in hand.

An examination of any of these equations will show that P (and frequently T) may be solved for explicitly. However, even in the simple van der Waals equation, v is present to the third power, and in the more complex equations v is likely to occur with exponents of 6 or higher. This makes solving for v as the unknown quite tedious, which is unfortunate since that is the unknown we are most likely to be interested in for practical applications. Most frequently, the pressure and temperature of a reaction are specified by thermodynamic and kinetic considerations, and the chemical engineer wishes to know what size vessel is required to accommodate the reaction. With the advent of electronic computers, the problem is no longer as severe as it once was, but the difficulty is still there for those who are making only an occasional computation.

Therefore, it would be very convenient if we had an equation of state no more complex than the ideal gas law and as accurate as one of the more complicated equations. A compromise is found in the use of the compressibility factor, z, defined in the following equation:

$$Pv = zRT. \tag{3.13}$$

It will be seen that z is essentially a correction factor which makes the right side of the ideal gas law equal to the left side. If z can be predicted accurately, the resulting equation would be as easy to use as the ideal gas law. It is beyond the scope of this text to discuss how this problem has been solved, but such correlations are available (the article by Gambill, previously cited, is one such place where z-charts may be found). Equation 3.13 is thus very useful in engineering design work, and it is explored in considerable detail in textbooks on thermodynamics.

3.7 Mixtures of Real Gases

It has already been shown that when the ideal gas law applies, mixtures of gases can be handled as readily as individual gases. Real gases pose problems however, since the presence of molecules of different size and mass produces results unlike those obtained when all of the molecules are alike. There are various ways of attacking this problem. These different ways produce different results, and unfortunately at the present time there is no infallible way to predict which method is likely to produce the best answer. In this book only three of these approaches will be discussed, but it should be realized that other methods exist.

The first attack will be that of combining Amagat's law with an equation of state. Van der Waals' equation will be used because of its relative simplicity, but the complexity of the equation does not limit the principle. It will be recalled that Amagat's law states that the sum of the partial volumes (each obtained at the same total pressure and temperature as that of the mixture) will equal the total volume of the mixture. For any component, *i*, present in the mixture to the extent of n_i moles, van der Waals' equation, when combined with Amagat's law, becomes:

$$P = \frac{n_i RT}{V_i - n_i b_i} - \frac{n_i^2 a_i}{V_i^2}. \tag{3.14}$$

The constants a_i and b_i are the constants a and b for the *i*th component. The partial volume of the *i*th component is V_i expressed in units of L^3. The equation will be seen to be dimensionally consistent. This equation must be repeated for each component of the mixture, the constants a_i and b_i as well as n_i naturally being the appropriate values for the component for which the equation has been written. P, R, and T will be the same in all equations in accordance with Amagat's principle.

Assuming that P (the total pressure) and T (the temperature), as well as the composition of the mixture are known, each individual V_i may be solved for and summed to give V, the total volume of the mixture. When this is done, it will be noted that $V_i \neq N_i V$ as it would if the ideal gas law held. This fact was emphasized in Section 3.4.

While (with van der Waals' equation) several cubic equations must be solved (one for each component), the attack is straightforward.* However, suppose that the total volume of the mixture is given, along with the temperature and composition of the mixture, and that P is desired. Each of the equations now contains two unknowns, V_i and P. Since $V_i \neq N_i V$, it must be estimated. There exists, however, a limitation on the values chosen for the partial volumes; they must sum to the known V. If the estimated values of V_i are now substituted in Equation 3.14, P can be determined for each equation. Obviously there can be only one value of P for the mixture, so if the various equations yield differing values of the pressure, it means that the values of V_i were incorrectly chosen. Thus the estimated values must be revised until a single value of P is obtained for all equations. This procedure is a trial and

• • • • •

* See Supplement to Chapter 9 for a convenient method of solving cubic equations.

error solution, and it can become tedious if more than two gases are present in the mixture. Probably the best approach to the problem is to make as the first estimate the partial volume predicted by the perfect gas law, i.e., $V_i = N_i V$. When the various values of P are obtained, they can be averaged to get an approximation of the correct answer. Then by inspection of the departure of each P from the average, it can be decided whether to increase or decrease the assumed value of V_i for a particular component and by how much. When these changes in the estimates of V_i are made, care must be taken so that the sum of the partial volumes still equals the total volume.

Although this type of solution may be seen to be time consuming, particularly when the number of components is large, it must be remembered that it will seldom be required. First of all, pressure is seldom the desired unknown in a practical problem. Secondly, other approaches to problems of mixtures do exist, which are simpler, and probably just as good, if it does become necessary to determine P. When T is the unknown, a similar situation to the one just discussed exists, for again we do not know the individual partial volumes. So if the Amagat approach is to be used, the solution once again becomes a trial and error attack like the one for pressure. However, for the reasons just given, this approach is seldom needed for practical problems.

A second general approach to mixture problems for nonideal gases consists of combining Dalton's law with an equation of state. Any equation of state may be used, but van der Waals' equation will again be chosen for illustration. Since Dalton's law concerns partial pressures determined at the total volume and temperature of the mixture, the van der Waals equation for this case may be written as follows:

$$p_i = \frac{n_i RT}{V - n_i b_i} - \frac{n_i^2 a_i}{V^2} \tag{3.15}$$

and

$$P = \sum_{i=1}^{C} \frac{n_i RT}{V - n_i b_i} - \sum_{i=1}^{C} \frac{n_i^2 a_i}{V^2}, \tag{3.16}$$

where C is the total number of components in the mixture.

Again, either P, V, or T may be the unknown that is desired, although V is the one most frequently met in practice. If P is known, together with T and the composition, and V is desired, it is only necessary to put the known values in Equation 3.16 and solve for it. Should the individual p_i values be desired, they may be determined from Equation 3.15. Again, it will be observed that $p_i \neq N_i P$.

If P is the unknown, Equation 3.16 may again be used. In this case it will be observed that no trial and error solution is required. Therefore, Dalton's law is easier to apply for this case than Amagat's law. It must be noted, however, that Dalton's law often does not apply to nonideal gas mixtures.

When temperature is the unknown, Equation 3.16 is again used, with no trial and error problems arising.

Since neither Amagat's law nor Dalton's law may give satisfactory answers for some mixtures, other methods of solving such problems have been proposed. One of these will be chosen for the third and final method to be discussed in this section. It consists of using an equation of state with the constants appropriately blended. The difficulty in this approach is to know how to blend the constants. The simplest approach is to multiply each constant by the mole fraction of that substance and sum the results. For example, in van der Waals' equation, the constants a_{mix} and b_{mix} would be obtained as follows:

$$a_{\text{mix}} = N_A a_A + N_B a_B + N_C a_C + \cdots$$

and

$$b_{\text{mix}} = N_A b_A + N_B b_B + N_C b_C + \cdots.$$

In these equations A, B, and C represent the components whose mole fractions are given by N and for which the van der Waals constants are a and b. Once the blended constants are obtained, they are put into the equation and the latter solved for the mixture as if it were a single substance.

This blending procedure is quite suitable for van der Waals' equation. For the more precise equations of state, different blending procedures have been recommended. In a given equation of state the blending procedure may differ among the various constants. Therefore, information concerning the equation must be sought before one attempts to use this method of solution. With a good equation of state and a proper blending procedure, this approach is one of the best methods of handling problems of mixtures of real gases.

Some uncertainty always surrounds the solution of problems of the type just discussed. Until more experimental data for actual mixtures of varying types and complexity have been recorded, it will not be possible to make a firm recommendation regarding the best method of attack on these problems.

LIST OF SYMBOLS

Latin letters

A, B	Constants in the Antoine vapor pressure equation
a, b	Empirical constants in van der Waals' equation of state
B.D.	"Bone dry," i.e., free of moisture
M	Molecular weight
N	Mole fraction (n/n_{total})
n	Number of moles
P	Absolute pressure (any desired units)
p	Partial pressure; when a subscript, refers to a specified component

R	Universal gas constant (units depend on P, v, T)
STP	Standard temperature and pressure (freezing point of water and 1 atm)
T	Absolute temperature (°K or °R)
V	Total volume; when a subscript, refers to partial total volume of the specified component
v	Molar specific volume (L^3/n); when a subscript, refers to partial molar volume of the specified component
z	Compressibility factor

Subscripts

A, B, C	Indicate individual components of a system
a, b, c	Indicate individual components of a system
c	As subscript on P, v, T, indicates critical point
i	Indicates any component
s	Indicates saturated condition
w	Indicates water

PROBLEMS

NOTE Unless otherwise specified, the ideal gas law may be assumed to be an acceptable approximation.

3.1 The value of the gas law constant is 1544 in units of foot-pounds force per (pound-mole) (°R). You wish a value in liter atmospheres per (gram-mole) (°K). State two ways of obtaining the desired value. Calculate the value by the easier method.

3.2 One thousand cubic feet of flue gas, measured at 500 °F and 29.2 in. of mercury absolute, contains 15% CO_2 (by volume). Calculate the pounds of CO_2 in this quantity of flue gas.

3.3 Freon-12 is CCl_2F_2. Assuming the validity of the simple gas law, calculate the density, in pounds per cubic foot, of Freon-12 gas at 75 °F and 2 atm gauge pressure.

3.4 A cylinder 6 in. in diameter and 4 ft high contains O_2 under 1000 psi when the temperature is 25 °C. Calculate:
   a. pounds of O_2.
   b. pressure in the cylinder when stored in a room at 102 °F.
   c. reduction in weight of contents if gas is discharged until the residual pressure in the cylinder is 1 atm.

3.5 An automobile tire is filled at a temperature of 79 °F to a pressure of 24 psig. After driving, the tire temperature has risen to 150 °F. Calculate, assuming no volume change:
   a. pressure in the tire in pounds force per square inch, gauge.
   b. the absolute pressure in atmospheres.

3.6 Two gases are in separate 1-gal containers. O_2 is at 15 psig and 15 °C. N_2 is at 30 psig and 25 °C. Calculate the composition of the mixture if the two gases are placed in the same 1-gal container.

3.7 A gaseous mixture, measured at 40 psig and 30 °C, contains 0.274 lb-moles of HCl, 0.337 lb-moles of nitrogen and 0.089 lb-moles of oxygen. Calculate the density of the mixture, in pounds per cubic foot.

3.8 A gaseous mixture is reported to contain 8% CO_2, 0.5% CO, 4% O_2, 7% water vapor, and 80.5% N_2. There are 1000 cu ft of this mixture, measured at a gauge pressure of −5 in. of water (with a barometer at 29.78 in. of Hg), and a temperature of 400 °F. Calculate:
   a. the analysis on a dry basis (i.e., as if all the water were removed with the other constituents unaffected).
   b. moles of carbon, oxygen, nitrogen, and hydrogen present.
   c. pounds of carbon, oxygen, nitrogen, and hydrogen present.

3.9 A mixture of hydrocarbon vapors contains 50 lb of benzene (C_6H_6), 40 lb of toluene (C_7H_8), and 20 lb of xylenes (C_8H_{10}). The mixture is at 443 mm of mercury absolute and 300 °F. Calculate:
   a. cubic feet occupied.
   b. mole fraction of each component.

3.10 A gas analyzing 15% NH_3, with the remainder nitrogen and hydrogen in stoichiometric proportions of 1:3 is produced in an ammonia converter. Twelve thousand cubic feet of this gas, measured at 62 °F and 28 in. of mercury absolute, is produced per minute. Calculate the daily capacity of the converter, in pounds of NH_3.

3.11 Gases under a pressure of 100 atm (abs) and at a temperature of 400 °C are passed to a reactor containing 10.8 cu ft of catalyst. Only 20% of the gases react, but, in the reaction, only 0.6 moles of product are formed per mole of feed. Calculate the rate of feed, in cubic feet per minute, to produce 25 moles of product per hour.

3.12 A process yields 10,000 cu ft per day, measured at 550 °F and 29.2 in. of mercury absolute, of a mixture of dry hydrogen chloride gas in air. The HCl constitutes 6.2% of the mixture. Calculate:
   a. pounds of limestone, 92% $CaCO_3$, to neutralize the HCl.
   b. cubic feet of CO_2 liberated, when measured at 1.8 in. of mercury and 70 °F.

3.13 The reaction between HCl and $CaCO_3$, described in Problem 3.12 is conducted so that all the CO_2, together with all the initial air, issues from the apparatus, saturated with water at 70 °F. Calculate the volume of effluent gas.

3.14 A mixture of ammonia and air at a pressure of 745 mm of mercury absolute and a temperature of 40 °C, and containing 4.9% NH_3 is passed to an absorption tower at a rate of 100 cu ft per min. In the tower only NH_3 is removed from the gas stream, which emerges from the tower at a temperature of 20 °C, a pressure of 740 mm of mercury absolute and a concentration of 1.3% NH_3. Calculate:
   a. volume of gas leaving the tower in cubic feet per minute.
   b. pounds of ammonia absorbed in the tower per minute.

3.15 A nitrogen-hydrogen mixture, 75% of which is hydrogen, is compressed and sent to an ammonia converter. The gas, before compression, is at 150 °C and 25 atm. In one pass through the converter 10.2% of the gas is converted to ammonia, the unconverted gas being recycled. The plant produces 300 tons per day. Calculate:

a. specific gravity of the original mixture referred to air at dry, standard conditions.
b. average molecular weight of original mixture.
c. average molecular weight of gases from the converter.
d. cubic feet of the original mixture used per day.
e. flow rate in pounds per hour through the converter.

3.16 A flue gas has the following composition, by volume: 10.5% CO_2, 1.1% CO, 7.7% O_2, with nitrogen constituting the remainder. The gas is measured at 67 °F and 29.1 in. of mercury absolute. Calculate:

a. the composition by weight.
b. volume occupied by 1 lb of the mixture.

3.17 A gas mixture containing 10% CO_2, 22% H_2, 5% CO, 28% CH_4, 12% C_2H_4, 6% C_2H_6, 7% C_3H_6 and 10% C_3H_8 is produced in a cracking plant at a rate of 100,000 cu ft per hr (measured at standard conditions). In a subsequent operation the gas is compressed to 6 atm, its temperature rising to 90 °F. Calculate:

a. pounds of each gaseous component produced per hour.
b. partial pressure of each gaseous component after compression, expressed in atmospheres.

3.18 Wet air issues from a drier at the rate of 15,000 cu ft per hr, recorded at dry, standard conditions, and 50% saturated at 150 °F. Calculate the volume of wet air leaving the drier per hour. Total pressure = 1 atm.

3.19 On an uncomfortable day in summer the temperature is 80 °F with a partial pressure of water vapor equal to 20.4 mm of mercury. The barometer stands at 748 mm of mercury. Calculate:

a. the humidity.
b. the relative humidity.
c. the percentage saturation.
d. the relative saturation.

3.20 It is desired to adjust the humidity of the air in a room so that the maximum amount of water vapor present will not cause condensation on the walls, if the temperature of the latter should fall to 50 °F. The temperature of the air charged to the room is 70 °F, at a pressure of 14.50 psia. Calculate the maximum weight of water that may be present in the air charged to the room, reported in grains per cubic foot of dry air.

3.21 A drier is a device in which water is evaporated from a wet solid. In some driers, the water vapor is removed from the system in a stream of air. In a unit of this type, 25 lb of water per hr is removed from wet sand. The air enters dry, 75 °F, 29.62 in. of mercury absolute, at a rate of 7000 cu ft per hr. It leaves at 200 °F, 29.4 in. of mercury absolute. Calculate:

a. percentage relative humidity of the effluent air.
b. molal humidity of the effluent air.

3.22 A mixture of air saturated with water vapor is at 199 °F and a total pressure of 1 atm. The mixture is cooled and compressed, the final conditions being 80 °F and 5 atm. Calculate the ratio of the initial to final volumes.

Note that the cooling and compression may result in condensation of some of the water vapor. Do not count condensed water in the final volume.

3.23 Air at 25 °C and 760 mm of mercury absolute with a relative humidity of 50% is to be dried. Compression is contemplated as the means of effecting the drying. The gas is to be used at 80 °C and it will be cooled to this temperature after compression. The water content of the dried gas must not exceed 0.01%. Calculate the pressure that would be required to produce the desired gas.

3.24 A hot air drier discharges hot humid air to the atmosphere, at a temperature of 194 °F, a pressure of 29 in. of mercury, and a molal humidity of 0.50, at a rate of 10,000 cu ft per hr at actual conditions. It mixes with atmospheric air at a temperature of 68 °F, a pressure of 29 in., and a relative humidity of 40%. Assume that the temperature after mixing is also 68 °F.
Calculate the volume ratio of atmospheric air to discharged air in order barely to prevent the formation of clouds of "steam."

3.25 Five hundred cubic meters of a gas, containing water vapor, measured at 700 °K and 14.93 psig is cooled to 95 °F at which point in the process the pressure is 32 in. of mercury absolute. The dew point of the final mixture is 31 °C. Calculate:
a. final volume of the gas mixture in cubic feet.
b. kilograms of water in the final mixture.

3.26 Illuminating gas, at a temperature of 90 °F and a pressure of zero gauge enters a gas holder containing 14 grains of water per cubic foot. In the holder the gas is cooled to 35 °F without change in pressure. The barometer stands at 29.92 in. of mercury. Calculate the weight of water condensed from the gas in the holder, per 1000 cu ft of gas entering.

3.27 Atmospheric air at a relative humidity of 70% and a temperature of 35 °C is drawn into a dehydrating column from which it emerges at 25 °C and a dew point of 32 °F at a rate of 48,500 cu ft (dry, standard conditions) per min. The pressure is atmospheric throughout. Calculate:
a. pounds of H_2O removed per hour.
b. volume of air entering per minute.

3.28 Plot the vapor pressure of benzene in the range of 1 to 1000 mm of mercury, using the following coordinate systems:
a. $1/T$ as abscissa, $\log p$ as ordinate.
b. log of the vapor pressure of benzene as ordinate *vs.* log of the vapor pressure of water as abscissa, where each plotted point represents the vapor pressure of the two substances at a single temperature. The plot on these coordinates is known as an Othmer plot. (See Chapter 7 for details of this method of representing vapor pressure data.)

3.29 Construct $\log p$ *vs.* $1/T$ plots (from 1 to 1000 mm Hg) for:
a. bromine; b. sulfur dioxide; c. tungsten; d. helium.
Write the equations for the lines drawn.

3.30 The data listed below are obtained from the *Chemical Engineers' Handbook*. For each pair of points, calculate:
a. the constants A and B in Equation 3.8.
b. an estimate of the vapor pressure at the average temperature.

DATA

	Temperature °C	Vapor Pressure
		mm Hg
A. Ethyl benzene	74.1	100
	113.8	400
B. Cyclohexane	25.2	100
	60.8	400
C. Naphthalene	145.5	100
	193.2	400
D. 1-Propanol	52.8	100
	82.0	400
		atm
E. Acetylene	−84.0	1.0
	−50.2	5.0
F. *n*-Butane	−11.7	1.0
	7.5	2.0
G. Acetone	78.6	2.0
	113.0	5.0

3.31 Ethyl bromide exerts a vapor pressure of 200 mm of Hg at 4.5 °C and has a normal boiling point of 38.4 °C. Estimate its vapor pressure at 50 °C.

3.32 Ethylene glycol has a normal boiling point of 197.3 °C. At a temperature of 120 °C it exerts a vapor pressure of 40 mm of mercury. Calculate:
  a. The vapor pressure at 170 °C.
  b. The boiling point under a pressure of 80 mm of mercury absolute.

3.33 An alarm system in a laboratory is set to sound a warning when the concentration of mercury vapor reaches 40 milligrams per liter in air. Liquid mercury is heated in a closed container in which the air pressure is 1.032 atm, and the detector of the alarm system is inserted in the container. Calculate the temperature at which it will sound.

3.34 A mixture of benzene and air at a temperature of 24 °C and a pressure of 745 mm of mercury absolute is found to have a dew point of 11 °C. Calculate:
  a. percentage by volume of benzene in the gas.
  b. moles of benzene per mole of air in the gas.
  c. weight of benzene per unit weight of air in the gas.

3.35 The vapor pressure-temperature relationship for benzene may be closely approximated as follows:

$$\log_{10} p = -\frac{1782}{T} + 7.9622, \qquad T \text{ in °K}, \qquad p \text{ in mm Hg}$$

A mixture has the following analysis, by weight:

20.02%	C_6H_6,
74.40%	CO_2,
1.31%	O_2,
4.27%	N_2.

Calculate the dew point of the mixture:
a. at a total pressure of 1 atm, abs.
b. at a total pressure of 2 atm, abs.

3.36 One thousand cubic feet of gas, measured at 100 °C and 120 mm of mercury, gauge, is passed through a cooler. The partial pressure of toluene in the entering gas mixture is 445 mm of mercury, the remainder being air. The saturated gas mixture leaving the cooler measures 720 cu ft per hr, measured at 50 °C and the prevailing barometric pressure of 740 mm of mercury. The vapor pressure of toluene at 50 °C is 265 mm of mercury. Calculate pounds per hour of toluene condensed in the cooler.

3.37 Benzene is nitrated during passage of a mixture of benzene vapor and inert gas through mixed acid. The gas mixture (89% N_2, 10% CO_2, 1% CO) is bubbled through a tank of heated benzene, which evaporates into it until the benzene content of the vapor-gas mixture is 43% and its relative saturation 81.9%. The total pressure throughout this bubbler system is 2 atm abs. Calculate:
a. temperature at which the vapor-gas mixture leaves the bubbler system.
b. weight fraction of benzene in the mixture.

3.38 One gallon of gasoline (average molecular weight of 114) of specific gravity 0.71 is vaporized in a completely empty room 12 ft × 15 ft × 19 ft. The initial pressure in the room is 780 mm, and the temperature is 20 °C. The room is kept tight during vaporization and temperature is constant. Calculate the final pressure.

3.39 Air is saturated with benzene (C_6H_6) vapor at a pressure of 750 mm of mercury absolute and a temperature of 70 °F. Calculate:
a. composition of the mixture, by volume *and* weight.
b. composition of the mixture, by volume and by weight if CO_2 is substituted for the air.

3.40 A gas leaves a solvent recovery system saturated with benzene at 70 °F and −10 mm of mercury gauge. The barometer stands at 29.92 in. of mercury. The gas analyzes, on a benzene-free basis: 15% CO_2, 4% O_2, and the remainder nitrogen. This gas is compressed to 5 atm and subsequently cooled to 100 °F. Calculate the percentage of benzene condensed in the process.

3.41 Toluene and methanol are completely immiscible. If a mixture of the two is stirred vigorously and heated, the mixture will boil when the sum of the vapor pressures of the pure materials is equal to the total pressure above the mixture.

In a particular instance where the total pressure is 1000 mm Hg, calculate:
a. temperature at which boiling occurs.
b. composition of the vapor mixture evolved.

Also calculate:

c. the pressure of the system when boiling occurs at 25 °C.

3.42 In a distillation process concentrating 90% ethyl alcohol (by weight), benzene is fed with the alcohol. All of the benzene and all of the water, the latter entering in the alcohol mixture, leave the top of the column and 100% alcohol leaves the bottom. The benzene-water vapor mixture is condensed and the two components are separated due to their immiscibility.

In a particular operation it is noted that the water layer constitutes 4.29% (by weight) of the condensed liquids. Calculate:

a. vapor composition (mole fraction) leaving the top of the column.
b. vapor pressure of each component in the vapor leaving the top of the column.
c. total pressure at the top of the column.

NOTE A temperature must be determined such that the vapor pressures of alcohol and water are in the ratio indicated by the composition of the mixture. Then the sum of the vapor pressures equals the total pressure.

3.43 Prove the relationships noted in Equations 3.11 and 3.12.

3.44 Repeat Problem 3.4 using van der Waals' equation. The van der Waals constants for oxygen are:

$$a = 1.36 \times 10^6, \qquad \text{in units of atm}\left(\frac{\text{cm}^6}{\text{gm-mole}^2}\right);$$

$$b = 31.9, \qquad \text{in units of } \frac{\text{cm}^3}{\text{gm-mole}}.$$

3.45 Convert the van der Waals constants for oxygen to English units expressed:

a. P in atmospheres; v in cubic feet per pound-mole.
b. P in pounds force per square foot; v in cubic feet per pound-mole.

3.46 The critical constants for oxygen are:

−118.8 °C,
49.7 atm.

Calculate values of van der Waals' constants and compare those listed in Problem 3.44.

3.47 For a mixture of 25% N_2, 75% H_2, use the blending rule suggested in Section 3.7 to obtain estimates of the van der Waals constants for the mixture. Then calculate the density using only the ideal gas law, if the mixture exists at 150 atm and 200 °C. Finally, using this density, calculate:

a. pressure corresponding to this density and at 200 °C, using van der Waals' equation.
b. percentage deviation of the pressure calculated by the ideal gas law.

3.48 A mixture of 25% N_2,75% H_2 is measured at 150 °C and 25 atm. Using the blending rule suggested in Section 3.7 to obtain values of the van der Waals constants, calculate:

a. the density, using the ideal gas law.
b. the density, using van der Waals' equation.
c. the percentage deviation of the density by the ideal gas law from that calculated by van der Waals.

DATA

Van der Waals' Constants

Constant	N_2	H_2	Units
a	1.347	0.245	(atm)(liters/gm-mole)^2
b	0.0386	0.0266	liters/gm-mole

3.49 Repeat Problems 3.47 and 3.48 using Amagat's law in combination with van der Waals' equation instead of the method of mixed constants. Compare the answers obtained with those of the previous problems.

3.50 Repeat Problems 3.47 and 3.48 using Dalton's law in combination with van der Waals' equation instead of the method of mixed constants and compare the answers with those obtained by both preceding methods.

4 Material Balances (Steady State)

The earliest chemists appreciated the fact that while matter could be transformed from one form to another, it could be neither created nor destroyed (Lavoisier, 1743–1794). At the end of an experiment, there was as much total material available as in the beginning, despite different chemical composition. This experimental result gave rise to the so-called law of conservation of mass, abbreviated for convenience in the chemical industry to "material balances" or "mass balances." As with many empirical laws, there is an exception, in this case atomic fission, where the simple material balance does not hold due to the transformation of mass into energy; it is necessary to balance on mass and energy jointly. This complication can, without serious loss, be excluded from a basic study of material balances where the purpose, usefulness, and techniques are of primary concern. The exclusion of fission processes is applicable to all material in this text.

The purpose of the discussions in this chapter will be: the elucidation of the relationships between the law of conservation of mass and the equivalent statements contained in the mathematical expressions of material balances; the distinctions which may be drawn between steady and unsteady state processes; the uses to which material balances are put by the chemical engineer; and finally, methods of dealing with a variety of common material balance problems. The number of items of data required to provide a unique solution of a material balance problem is an important one. It is treated in the Supplement to this chapter. Its relegation to the end of the chapter is not intended to imply lack of importance; rather it indicates, properly, that we may study the principles of material balances and their solutions granted a supply of the right amount of information. Then, knowing how to solve the problems, we turn to the Supplement to determine how the correct specification is defined.

4.1 Material Balance Principle

The law of conservation of mass states that mass is neither created nor destroyed in any process. In mathematical form, the statement would read

$$\text{mass at input} = \text{mass at output} + \text{mass accumulated within the process.} \tag{4.1}$$

Several aspects of such a statement deserve special note. First, no restriction is made as to type of process. The statement applies to both physical and chemical processes. Second, it applies both to the total masses handled in the process and to the individual components. For example, if the physical process is a mixer, to which nitric and sulfuric acids are being charged, each of which contains some diluting water, the total amount of the two streams being charged will eventually be capable of being withdrawn as mixture. If any portion is not withdrawn, it constitutes accumulation in the process. Similarly, if attention is centered on the pure sulfuric acid alone, the amount withdrawn is exactly equal to the amount charged, corrected for the amount retained in the mixer. The same remarks obviously apply to the water and to the nitric acid. Thus, the statement applies to the total material and to each of the three components. The situation is not changed basically when chemical reaction is involved, but in this instance, attention must be centered on some component which is unchanged by the reaction. If we are concerned with the combustion of carbon in oxygen to give carbon dioxide, the total balance will state that the mass of carbon plus the mass of oxygen charged will exactly equal the amount of carbon dioxide withdrawn, granted that none is retained in the system as an accumulation. Also, the amount of carbon leaving in the carbon dioxide is exactly equal to the amount of carbon charged in its pure state, again granting no accumulation. Similar remarks apply to the oxygen. In summary, we can state that the balance may be applied to total material or to any of the unchanged constituent component substances or elements.

Third, it is quite apparent that the balance is applied to mass, not to rates. Since many chemical processes operate continuously, this restriction would appear to be disadvantageous. The use of a fixed time interval gets us out of this potential difficulty, for if a rate which is constant over a time interval is multiplied by that time interval, the resulting quantity is mass. The only problem here is the duration of the interval in which the rate is constant. If the rates are unchanged over long time periods, as in the case of the continuous process, the choice of time interval is relatively unimportant. However, if rates are changing rapidly, the time interval may have to be restricted to a differential, dt. Then the balance will produce a differential equation that offers no disadvantages except for additional complexity in solution.

Finally, some more detailed attention should be given to the accumulation term. This term is necessary for situations where there is no output, all material being held in the system, or where the output does not match the input due to the build-up of material in some reservoir within the system, or process. Without build-up in the process, the third term in Equation 4.1 becomes zero. These two different situations suggest that two different classifications of material balances might be appropriate. They will be called the steady and transient states.

4.2 Steady and Transient States

In the simplest possible situation, consider an empty tank to which water is pumped at a constant rate. If the tank constitutes the process (or system), it is obvious that there is input and no output. Therefore, during the period of filling, the accumulation term must account for all the water added. At the end of the filling period, if pumping is continued, the tank will overflow and the overflow is the output. It is apparent that the input and output masses are now equal and no accumulation term is required. Thus two distinct types of operation are encountered; in the first, operation is transient, limited in duration by a condition of the process, in this case the filling of the tank; in the second, operation is capable of being continued indefinitely, and the accumulation term in the balance becomes zero. This latter type is termed *steady state operation.* Where an accumulation term is required, the term *unsteady state* is common in chemical engineering. A more general term throughout all scientific and engineering processes would be the *transient state.*

The filling of a tank is so simple that the balance involves an input and an accumulation term, but no output. It is important to realize that transient problems generally include all three terms, but the steady state never involves any terms except the input and output. If the tank problem is slightly complicated, all three terms become involved in the transient situation. Bore a small hole in the bottom of the tank. Now pump water to the tank as before, but at a rate in excess of that at which it can escape through the hole. Now there is an input term for the water pumped to the tank, an output term for the water discharged through the hole, and an accumulation term for the increase in water contained in the tank.

In this chapter the subject of material balances will be limited to the steady state. After the principles and techniques have been treated in this limited situation, some aspects of the transient state will be presented in Chapter 5.

The Steady State • The primary requirement for the steady state is that accumulation shall equal zero. This condition is equivalent to a piece of equipment with no internal volume; in such a hypothetical unit there is no residence time; all chemical and physical changes occur instantaneously and material emerges at the exit (even though in a changed chemical and/or physical state) as fast as it is charged. Despite this hypothetical condition of zero equipment volume, diagrams should be drawn to give some idea of

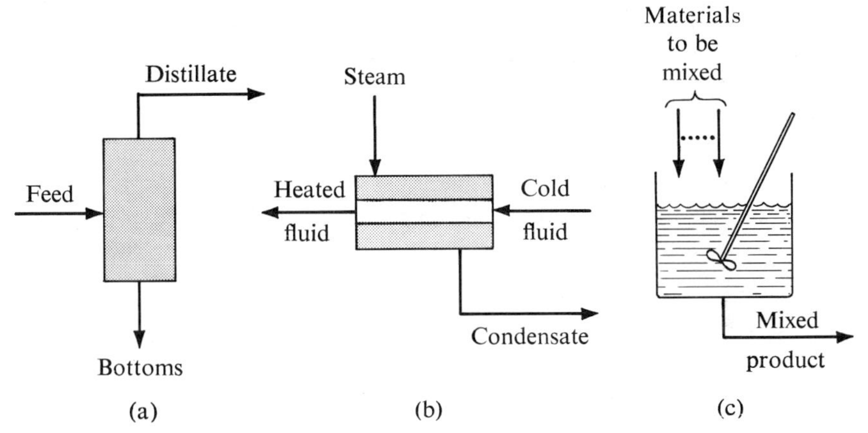

FIGURE 4.1 Characteristic representation of process elements: a. distillation column; b. steam heater; c. mixer.

the pieces of equipment which they represent. Thus, a distillation column or an absorption column would be drawn as a vertical rectangle, Figure 4.1a, a steam heater as a horizontal rectangle, Figure 4.1b, a mixing tank as a box with some indication of an agitator, Figure 4.1c, and so forth. The input and output streams should be in reasonably realistic locations. These objectives will be adhered to in all illustrative figures.

4.3 Use of Material Balances

Material balance calculations serve useful purposes in two general categories: design of equipment and evaluation of equipment performance. All chemical engineering calculations can be placed in these general categories and it is therefore worthwhile at this point to expand on the various uses of each.

Very few engineers will be employed in pure design work. An emphasis on problems encountered in design might then appear to be out of place. The knowledge required in effecting a design will, however, be exactly the same as would be required to evaluate and improve the performance of a unit. It is the same sort of knowledge required to develop a process from laboratory scale to full plant scale where the engineer must plan for installation of the most suitable units and for obtaining the maximum output from each at minimum cost. It is the same sort of knowledge required by the technical salesman in selling a product to be used in a specific process or in correcting the customer's use of the product in a specific piece of equipment. It is the same sort of knowledge required to predict the conversion (or separation) that may be expected if an existing piece of equipment is to be adapted to use in a new process, thereby saving the cost of purchasing a new unit.

A primary use of material balances occurs in *design calculations* where the balances are used to predict quantities of products when the specifications include complete information on the feed, information on any chemical conversion, and information on any physical separations. The feed will be specified normally for one does not often face a problem without knowing the materials available for use and the quantities to be processed. There may be a variety of materials available for a given process and the company may be interested in designs for the production of different quantities; each choice then constitutes an individual problem in which the charge is known, in material and in quantity. The material balance is used to predict the output, given this input information, along with knowledge, or assumptions, about the operation.

For example, take the very simple separation process in which 1000 lb of a solid containing 20% (by mass) of water is to be dried to produce a solid containing only 5% water. The process is represented schematically in Figure 4.2. The quantity of product and the amount of water to be removed (and condensed, the condenser not being shown) is to be predicted. The solution is very simple. Of the 1000 lb charged, 800 lb is dry solids. This mass must all be in the product stream where it represents 95% of the stream; the product is $\frac{800}{0.95} = 842.1$* lb of solids containing 5% of water. The

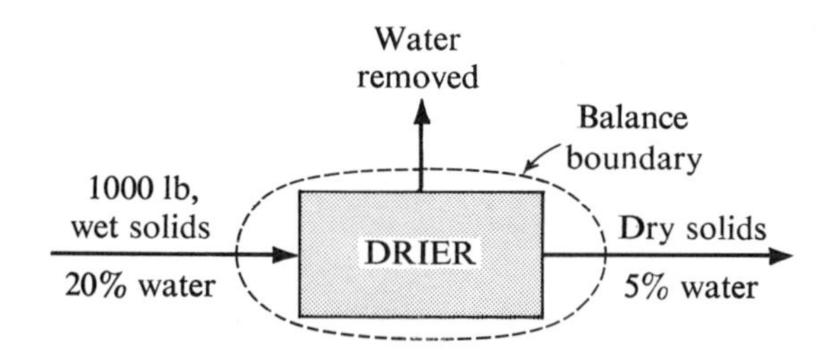

FIGURE 4.2 Flow sheet of drier.

• • • • •

* This calculation is one which is useful in many material balance problems. Given the quantity and analysis of one component of any stream, divide the mass by the mass fraction to obtain the total mass of the stream. The calculation may be rationalized as follows:

Let

$$X = \text{total mass of stream;}$$

if

$$A = \text{total mass of component 1—a known mass, and}$$
$$x_1 = \text{mass fraction of component 1—a known fraction,}$$

then

$$x_1 X = A,$$

$$X = \frac{A}{x_1}.$$

Obviously if A is in moles and x_1 in mole fraction, the same calculation will give X in moles.

difference 1000 − 842.1 = 157.9 lb of water, is removed as vapor and sent to the condenser, where all of it is condensed. The designer now knows the magnitude and composition of all streams, the first essential in sizing the drying equipment.

As a second illustration let 1 lb of pure carbon be charged to a furnace. The conditions are such that 95% of this carbon should burn to carbon dioxide and the remainder to carbon monoxide. By material balance you are to predict the quantities of these gases appearing in the flue gases emerging from the furnace. The solution is again simple. Here one must take into account the chemical reaction and the additional charge of oxygen to burn the carbon. From the combustion equations,

$$C + O_2 = CO_2$$

and

$$C + \tfrac{1}{2}O_2 = CO,$$

it is easy to calculate that 44 lb of CO_2 are formed from the combustion of 12 lb of carbon and 28 lb of CO are formed from the combustion of 12 lb of carbon. The actual carbon burned to CO_2 is 0.95 lb and that burned to CO is 0.05 lb. The CO_2 formed is then

$$\frac{44 \text{ lb } CO_2}{12 \text{ lb C burned}} \times 0.95 \text{ lb C burned} = 3.49 \text{ lb } CO_2,$$

and the CO formed is $(\tfrac{28}{12})0.05 = 0.12$ lb. The total mass of 3.61 lb indicates that 2.61 lb of oxygen were required to combine with the carbon to provide the mixture specified. If the fractions of CO and CO_2 were to be varied, the amount of oxygen required would also vary. In this problem the source of the oxygen is not specified, but it is clear that the 2.61 lb must be supplied in some form, either in air, which will involve a charge of nitrogen as well, or in the form of pure oxygen.

Material balances may also be used by the design engineer to locate the measuring instruments that will be required to indicate the performance of the equipment. For instance, in the case of the drier, some indication of the water concentration in the product should be available. Alternately, measuring the water from the condenser in a given time period, as well as the feed rate and concentration of water in the feed, would allow the *calculation* of the water concentration in the product. And measurements of all rates (feed product, and water), plus both water concentrations (feed and product), would provide a performance test by which losses might be estimated. (See subsequent remarks on performance tests.) The design engineer then decides how much control and analytical equipment he shall install, dependent upon his objective for evaluating the final plant equipment. Obviously he can save money by eliminating some measuring equipment and using material balances to supply the missing information.

In design, then, material balances may be used to predict quantities of products, given the concentrations desired. Alternately, for performance tests, they may be used to locate a minimum number of the positions for measuring instruments.

In addition to the use of material balances in design, there are uses in *evaluation* of *performance*, already alluded to above. For example, suppose that the design of the drier has included instruments to measure water concentration in feed and product and also to measure quantities of wet solids charged. If, for 1000 lb charged with water content at 20%, the outlet water content is found to be 5.6%, it is clear that the equipment is not operating up to specifications. The output rate is not 842.1 lb as specified but 847.5 lb and the water removed may be calculated by material balance to be 152.5 lb rather than the 157.9 lb intended.

Suppose, however, that the designer of the process decided to incorporate not only these instruments, but also others to measure the quantities of product and of condensed water as well. There are no remaining quantities or concentrations to be calculated by material balance. The component quantities in and out (such as the amounts of solids and of water in the drier problem) can be calculated from the data. They may not check exactly. In fact it would be surprising if they did since all of the measurements are subject to error. In addition to error, there is another source of discrepancy in the possibility of losses of water vapor through leaks, solids through spillage, and so forth. It is not possible to calculate these losses exactly since errors exist in the instruments. However, with a proper statistical evaluation of the precision of the instruments, estimates may be made of the losses. There will be no attempt to treat such problems here since applications of statistics constitutes a major study in itself.

Despite the exclusion from further discussion in this text of the situation where complete data are available on every stream, there are aspects of the complete performance test which are interesting and which are worth additional remarks. It should be fairly apparent that, under these conditions, one can pick out certain items of data and calculate others, comparing the calculated result with the reported value. Many combinations can be treated in this fashion. As illustration, use again the drier problem. Suppose that in a performance test 1000 lb of solids with 20% water content were charged to the drier, that 152 lb of water were recovered from the condenser, and that the product amounted to 830 lb, water content reported to be 5.5%. Now if the feed quantity and water content, and the product water content, are used, the water removed will be calculated to be 153.5 lb as opposed to the 152 lb reported. If, on the other hand, the outlet quantity and water content have been used along with the inlet water content, the water removed will be calculated to be 150.4 lb. Similar situations can be established to show that the dry solids contents can have a variety of values depending upon which items of data are chosen. Actually, there are more items of information than required; consequently there is no unique solution. The process is "over-specified."

In contrast to this situation, it is easy to imagine a test run in which an insufficient amount of data is taken and the process is underspecified. Then there is no way to calculate all the unknowns. (This situation is somewhat difficult to imagine in this simple case, but not hard to conceive of in a more complex unit.)

Effectively these remarks indicate that where too many unknowns remain, no solution can be effected, but when there are too few unknowns a myriad of solutions is possible. The establishment of the perfectly specified process is then a matter of concern. The problem can become very complex. It has been studied by a number of authorities and summarized by Kwauk [*AIChE J.*, *2*, 240 (1956)] who has presented a systematic approach to the establishment of the exact number of items of data which can and must be specified if a unique solution is to result. This method, and some of its applications, is presented as a supplement to this chapter. An understanding of the material in this supplement is not prerequisite to subsequent chapters of this book.

4.4 Material Balance Fundamentals

The concept of "material in = material out" (in the steady state process) is so simple that detailed discussion appears to be unnecessary. There are, nevertheless, a few vital points that should be emphasized, despite their relatively obvious nature, prior to a discussion of types of problems.

Independent Balances. Reference has already been made to the drying of solids and the combustion of carbon. These two simple processes can serve as illustration of two important points. First, in the physical separation of water from solids, use has been made of the fact that balances exist for total material and for either of the physical components. Thus we write

800 lb of solids in the wet feed = 800 lb of solids in the dried product;

200 lb of water in the wet feed = 42.1 lb of water in the dried product + 157.9 lb of water removed;

1000 lb of total material in = 842.1 lb of material out in the wet solids + 157.9 lb of water out.

These three equations represent two component balances, one on solids and one on water plus one total material balance. Obviously the three are not independent. Any one can be obtained from linear combinations of the other two (i.e., the sum of the first two equals the third). Thus for the two-component system, two independent material balances may be written. In general, for a C-component system, C independent material balances may be written, to be chosen from among the total of $C + 1$ possible material balances. In each of these balances, *mass in* is balanced against *mass out*.

In the combustion process, chemical change complicates the picture, but only slightly. Carbon, oxygen, carbon monoxide and carbon dioxide are all present, but there are not four components since the last two (the products) contain the same elements as found in the (pure) reactants. Treating the problem as in the case of the physical drying process, there are three balances which may be written:

1 lb of carbon charged = 0.95 lb of carbon in CO_2 + 0.05 lb of carbon in CO;

2.61 lb of oxygen charged = 2.54 lb of oxygen in CO_2 + 0.07 lb of oxygen in CO;

1 lb of carbon + 2.61 lb of oxygen, both charged = 3.49 lb of CO_2 + 0.12 lb of CO.

Again there are three material balance equations, one based on the carbon, one on the oxygen, and one representing the total material. Again they are not independent, any one being obtainable from linear combinations of the other two. This problem is then also a two-component situation. We shall now explore the general problem of components in both physical and chemical processes, so that the number of available independent material balances may always be stated unequivocally.

Counting the Components

Physical systems and chemically reactive systems will be considered separately.

Physical • The number of components is the number of distinct chemical species which exist in all parts of the element or unit. Since no chemical change occurs these species are all those which enter. As a simple example, a three-component mixture enters a complex of distillation columns; the components are three. As a further example, air and water vapor enter in one stream to an air drier, solids and liquid water in another; the components are three: air, water, and solids, regardless of the existence of water in two separate phases. In yet another example, nitric acid and sulfuric acid are mixed, each being somewhat less than 100% pure; the components are again three, water and the two pure acids. If the sulfuric stream had been a mixture of 100% acid and SO_3 (a mixture commercially known as "oleum") the process would not be purely physical, for the SO_3 would react with the free water charged with the nitric acid, and chemical combination would occur, negating the existence of a purely physical process.

The reader is now in a position to confirm the fact that there are two components in the drier discussed as the first illustration in Section 4.4.

Chemical • The number of *components* is not solely dependent upon the number of chemical *species* since all are not independent. Some, or all, of the species will be related through chemical equations. The number of components will then be the number of species *minus* the number of independent equations. Several examples will be considered in detail.

Example 4.1

▶ STATEMENT: Dissociation of $CaCO_3$ into CaO and CO_2. There are three chemical species and they are all related through the chemical equation

$$CaCO_3 = CaO + CO_2.$$

The components are then

$$3 - 1 = 2.$$

Note that an inherent restriction is involved in this case, for the molal ratio of CaO/CO_2 must be equal to unity. However, if CO_2 and CaO are to be mixed and $CaCO_3$ is formed (under favorable t and P conditions) by recombination, the components are still *two* with the inherent restriction removed, since no specification of equal molal quantities of CaO and CO_2 was stated.

Example 4.2

▶ STATEMENT: Combustion of carbon to CO and CO_2, the oxidizing agent being air and an excess of air being an allowable situation. The chemical species are now 5 (C, O_2, N_2, CO, and CO_2). The independent chemical equations are:

$$C + \tfrac{1}{2}O_2 = CO,$$

$$C + O_2 = CO_2.$$

The components are:

$$5 - 2 = 3 \text{ (C, O, N)}.$$

The situation here is not exactly analogous to that of the combustion of carbon in pure oxygen, as discussed in Section 4.4. The reader is now in a position to confirm the fact that only two components were involved in that case.

Example 4.3

▶ STATEMENT: Oleum is mixed with dilute nitric acid. The species are four (two acids, water, and SO_3). One chemical equation is involved,

$$H_2O + SO_3 = H_2SO_4.$$

The components are

$$4 - 1 = 3 \text{ } (HNO_3, SO_3, H_2O).$$

From these three examples two facts are apparent. First, inerts are counted among the species but will never be involved in the chemical equations. Second, without some systematic approach one might encounter some difficulty in guaranteeing that the equations are all independent. The independence of equations requires further study.

Number of Independent Equations • There is a variety of ways to determine the number of independent equations. Each has advantages and each will, at times, become quite tedious in use. Two will be briefly summarized here.

The first method is the one traditionally reported in standard texts. It is discussed in a book by Smith and Van Ness, *Introduction to Chemical Engineering Thermodynamics* (New York: McGraw-Hill Book Co., 1959), pp. 42–46. The second is suggested by Amundson, *Mathematical Methods in Chemical Engineering* (Englewood Cliffs, N.J.: Prentice-Hall, Inc., 1965), pp. 50–54. Yet a third, which will not be summarized here, but which also has considerable merit, is presented by Aris, *Introduction to the Analysis of Chemical Reactors* (Prentice-Hall, Inc., 1965), pp. 10–14.

In the *traditional analysis,* we proceed as follows. First, write an equation for the formation of each chemical species from its elements (in their normal states of aggregation; i.e., O_2, not O). Second, combine these equations in such fashion that species not represented in the process will be eliminated. Third, count the resulting equations. This number of equations is independent. In the case of Example 4.3 of this section

$$H_2 + \tfrac{1}{2}O_2 = H_2O, \tag{a}$$

$$S + \tfrac{3}{2}O_2 = SO_3, \tag{b}$$

$$S + 2O_2 + H_2 = H_2SO_4. \tag{c}$$

Sulfur, oxygen, and hydrogen are not involved in the process. They may be eliminated only by subtracting equations (a) and (b) from (c) to give the *one* independent equation

$$SO_3 + H_2O = H_2SO_4.$$

More complex examples are usually encountered in the case of competing reactions. Example 4.2 on p. 107 contained competing reactions, but no elimination was necessary since C and O_2 were reactants. A more definitive case could be made with the hydrogenation of CO to methanol (CH_3OH). Suppose that methane (CH_4) and water are also formed in the process. The species are five (CO, H_2, H_2O, CH_4, CH_3OH). The equations representing the formation of these species from their elements are:

$$C + \tfrac{1}{2}O_2 = CO, \tag{a}$$

$$H_2 + \tfrac{1}{2}O_2 = H_2O, \tag{b}$$

$$C + 2H_2 = CH_4, \tag{c}$$

$$C + 2H_2 + \tfrac{1}{2}O_2 = CH_3OH. \tag{d}$$

Not involved in the process are C and O_2. They can be eliminated between (d) and (a) to give

$$CO + 2H_2 = CH_3OH.$$

Combination of (b) and (c) will eliminate neither C nor O_2, but (b) + (c) − (a) will yield

$$CO + 3H_2 = CH_4 + H_2O.$$

Two independent equations remain and the components for this process are

$$5 - 2 = 3.$$

The logical designation of components is C, H, O; although other choices may be made, there is no other *good* choice for this problem.

The method is reasonably straightforward, but the labor of writing the synthesis equation for each species, as well as the subsequent effort of eliminating elements which do not appear in the reactions for the process, is time-consuming and tedious. When stated in simplest terms, *Amundson's method* is much quicker. In simplest terms (which are not entirely correct and will therefore require some precautionary remarks), this method states that the number of independent equations is the difference between the number of reacting species and the number of chemical elements in these species. Take the situation just discussed, the hydrogenation of CO. The number of species is five (H_2, CO, H_2O, CH_3OH, CH_4). The number of chemical elements is three (C, H, O). Therefore, the number of independent equations is the difference between these two numbers, or two as obtained before. The components, three in number, are the same as the number of chemical elements.

This procedure is so simple that one suspects that there may be circumstances, or processes, where its use would lead to trouble. Otherwise, why has it not been suggested before? The answer to this question lies in the fact that the simplest statement is not entirely true. The method depends on establishing the number of components from the rank of a matrix, and this rank is sometimes less than the number of chemical elements.

The elements of each row of the matrix are the subscripts on each chemical element as it appears in each of the species, the latter comprising the column designations. For example, in the same hydrogenation process, the matrix will be composed as follows.

	H_2	CO	H_2O	CH_3OH	CH_4
C	0	1	0	1	1
H	2	0	2	4	4
O	0	1	1	1	0

Now this 3 × 5 matrix is of rank three (all 3 × 3 determinants contained therein—ten in number—have nonzero values). Therefore, the number of chemical elements and the number of components in the reaction are identical.

On the other hand, take the other example, absorption of sulfur trioxide in water. This matrix will take the following form.

	H_2O	SO_3	H_2SO_4
H	2	0	2
S	0	1	1
O	1	3	4

This matrix is not of rank three. Its determinant has a value of zero. The rank is, then, less than three and the number of components must also be less than three. The rank is, in fact, two, for the third row is a linear combination of the first two (one-half the first plus three times the second). The same situation exists when we analyze the reaction

$$CaCO_3 = CaO + CO_2.$$

Further, in the combination of decompositions of magnesium and calcium carbonates, the matrix is as follows.

	$CaCO_3$	$MgCO_3$	CO_2	CaO	MgO
Ca	1	0	0	1	0
Mg	0	1	0	0	1
C	1	1	1	0	0
O	3	3	2	1	1

The last row is a linear combination of the first three (the sum of the first two plus twice the third), so that the rank is not four, but rather three. There are, then, three components (usually chosen as MgO, CaO, CO_2), and two independent reactions, the two decompositions.

In summary, the use of this method might include the following steps:

a. Count the number of reacting species, N_{sp}.
b. Count the number of elements in these species, N_e.
c. Make a first estimate of the number of independent equations, N_R, by the relationship,

$$N_R = N_{sp} - N_e.$$

d. Check the rank of the matrix in cases of doubt, or determine the number of independent equations by another method.

e. Determine the number of components by the rank of the matrix (i.e., ordinarily the number of reacting elements) and add one for each inert in the system.

Mass vs. Moles

It is obvious that the establishment of the balances is on the basis of the *masses* of the components. It is not possible to write the balances in terms of the moles of the various materials since stoichiometric proportions of reactants to products are not always 1:1. In fact, although the proportions of reactants *are* 1:1 in the combustion to CO_2, they are 2:1 in the combustion to CO and in *neither case* are the proportions of reactants to products 1:1. Thus an attempt to balance the total number of moles in against the total number of moles out would result in the following inequality:

1 mole of C + 0.95 moles of oxygen for the CO_2 + 0.025 moles of oxygen for the CO $\neq$ 0.95 moles of CO_2 + 0.05 moles of CO.

Balance of Atoms

In chemical reaction the atoms charged must equal the atoms in the product stream. Therefore, a balance on the number of atoms is equivalent to a mass balance; in this type of balance the component balances are the important ones; the total material balance is not particularly helpful and will rarely be written. For the combustion problem under discussion it is obvious that:

0.95 atoms of carbon for CO_2 + 0.05 atoms of carbon for CO = 0.95 atoms of carbon in CO_2 + 0.05 atoms of carbon in CO;

and

1.90 atoms of oxygen for CO_2 + 0.05 atoms of oxygen for CO = 1.90 atoms of oxygen in CO_2 + 0.05 atoms of oxygen in CO.

It is also obvious that:

1 atom of carbon + 1.95 atoms of oxygen = 2.95 total atoms in the products.

The last equation is written only as a result of writing the previous two, accounting for the earlier statements that the overall total material balance would not normally be written and that the $(C + 1)$ st material balance is not independent.

The component atomic balances are so obvious that their usefulness appears to be doubtful. Any such doubt would be entirely wrong. The use of the balance by atoms with subsequent conversion to mass is extremely useful in the presence of chemical reaction, particularly when one of the components forms multiple products or is incompletely consumed. These remarks will be illustrated after extension of the principle of balance by atoms to balance by arbitrary components.

Balance by Arbitrary Components

Consider the case of SO_3 being absorbed in water to form sulfuric acid. The acid formed would normally be written symbolically as H_2SO_4. However, for the purposes of examining the reaction and the utilization of SO_3 and H_2O in the product, it is reasonable to write H_2SO_4 as $SO_3 \cdot H_2O$. Then we can write the single independent reaction:

$$SO_3 + H_2O = H_2SO_4 = SO_3 \cdot H_2O.$$

There are three chemical species (H_2O, SO_3, H_2SO_4). Since there is one independent reaction, the number of components is

$$3 - 1 = 2.$$

A convenient choice of components is H_2O and SO_3. (Another less convenient choice might be H and S.)

Now we can say that 1 mole of SO_3 and 1 mole of H_2O will be consumed in the formation of 1 mole of sulfuric acid, no matter how the latter is written. Consider the question: how much SO_3 is required to produce 100 lb of sulfuric acid, 98% pure. There are 98 lb of pure acid to be produced. Since the molecular weight of H_2SO_4 is 98, 1.0 lb-mole of acid is to be formed. Balancing the number of moles of SO_3, 1.0 lb-mole of acid contains 1.0 lb-mole of SO_3 and the molecular weight of SO_3 is 80. Therefore, 80 lb of SO_3 are required. Also, 1.0 lb-mole of water reacts with the 1.0 lb-mole of SO_3 or 18 lb of water. Two additional pounds of water are used to dilute the mixture.

This problem might be complicated slightly by dissolving SO_3 in sulfuric acid rather than water. Industrially, this method is necessary since the heat of solution is so great that direct solution of SO_3 in water creates enough heat to vaporize some of the acid in the form of a mist. The mist is very difficult to precipitate. It has a tendency to leave the absorber and deposit on objects in the surrounding country, an obviously objectionable situation. Solution in an acid provides enough inert material (the acid) to take care of the same heat evolution without excessive rise in temperature. Suppose then that 100 lb of 90% acid are available and that this material is to be concentrated to 98% by addition of SO_3. Now how much SO_3 is required?

▶ SOLUTION: With the information given, a water balance may be written containing only one unknown. However, since water reacts with SO_3, the water balance must account for both the combined (i.e., reacted) water and the free water of dilution. In the 100 lb of 90% acid there are 10 lb of free water. The combined water may be obtained by considering the amount which must be contained in the 90 lb of pure acid. There are $\frac{90}{98}$ lb-moles of acid and, consequently the same number of moles of water, or $(\frac{90}{98})18$ lb. The sum of these two quantities is the water input. The output quantity is unknown. Designate it as X lb. Then the free water is $0.02X$ lb and the combined water is $(0.98X/98)18$ lb. The balance on the water component can now be written as:

$$\begin{array}{ccc} \text{Input} & & \text{Output} \\ 10 + \left(\frac{90}{98}\right)18 & = & 0.02X + \left(\frac{0.98X}{98}\right)18, \end{array}$$

and

$$X = 132.5 \text{ lb of product.}$$

The SO_3 added must then be 32.5 lb, by a total material balance.

An alternate approach is to let X = total pounds of 98% acid formed as before and Y = pounds of SO_3 added. Then write a total material balance and a component balance on the SO_3, as follows:

$$\text{Total:} \quad 100 + Y = X,$$

and

$$SO_3\!: \quad \left(\frac{90}{98}\right)80 + Y = \left(\frac{0.98}{98}\right)80X.$$

These two equations may be solved simultaneously to give the same result obtained above.

The main point to be noted in this example is the decision to use SO_3 and H_2O as the components whether they are real compounds or combined in the form of H_2SO_4. An important secondary point is the balancing of these components in both their pure and combined states, as with water in a diluted acid stream.

The components selected here are stable chemical compounds. This type of selection is not necessary. For example, nitric acid is formed by a process in which nitric oxide (NO) is oxidized and absorbed in water. Although there is no true N_2O_5 present, it is entirely reasonable to treat this process as

$$N_2O_5 + H_2O = 2HNO_3.$$

Then every mole of N_2O_5 requires two of NO and three atoms (or $\frac{3}{2}$ moles) of oxygen to produce it. The fact that the mechanism is not this simple or direct is unimportant when one is concerned with the *overall changes* that have taken place. And these overall changes are exactly what the material balances are concerned with. Therefore, mechanisms are not required and the three components selected for treating this system comprised of four species could be N_2O_5, O_2, H_2O (or more conveniently NO, O_2, H_2O, or even N, O, H).

Multiple Products from One Reactant

A basic mistake in the case of one reactant going to two (or more) products is an attempt to write an overall equation representing the entire operation. The correct procedure is to write the simplest possible chemical equations and then apply the pertinent molal (or mass) quantities. This procedure has already been illustrated for carbon burning partly to carbon monoxide and partly to carbon dioxide. As another example consider the hydrogenation of carbon monoxide. Presume that the following reactions are possible:

$$CO + 2H_2 = CH_3OH,$$
$$CO + 3H_2 = CH_4 + H_2O.$$

The first reaction is the one desired, but the second cannot be entirely excluded. Equilibrium and rate considerations lead the designer to assume that 80% of the CO will be hydrogenated to methanol and that the remainder will go to methane. On the basis of 10 moles of CO, 8 will go to CH_3OH and the other 2 moles will go to CH_4. The summation of the two reactions, the first multiplied by 8 and the second by 2 would read:

$$10CO + 22H_2 = 8CH_3OH + 2CH_4 + 2H_2O.$$

Nothing is gained by writing this composite. If the question is the number of moles of hydrogen required, it is obviously 16 for the first reaction plus 6 for the second and the two individuals are better than the composite since the individuals represent single reactions unsullied by specific requirements of an individual problem. Quite apparently, the other common mistake of adding the two basic equations to give

$$2CO + 5H_2 = CH_3OH + CH_4 + H_2O$$

has nothing to do with the case except in the unlikely event that equal amounts of CO are consumed by each reaction. In summary, the simplest individual equations should be written and then used in whatever ratio of reactants is specified by the problem statement (or the designer's knowledge).

4.5 Diagrams

One of the most important requirements in writing material balances is a proper and complete visualization of the process. It cannot be effectively accomplished without a diagram (except in the most trivial cases). The diagram has two functions. The first is to lay out for the engineer all the units around which the calculations are to be made and to incorporate on the diagram as much of the available information as feasible in a diagram of reasonable size. Such diagrams are called schematics or flow sheets. The second function is to provide a further visualization, this time of the boundaries of the balance position. The dashed line in Figure 4.2 represents such a boundary. Any flow stream which is cut by this boundary line must be considered in the material balances.

As problems become more difficult and the processes become more complicated, the establishment and the use of the flow sheet becomes more important. Two instances will be noted here. The first is the process that contains more than one element. In such a case, there is more than one position at which to take balances. Balances may be drawn around each individual element of the entire process, around combinations of any subset of the entire group of elements, or around the process as a whole. The positions must be clearly marked to provide an accurate analysis and to provide a means of rechecking where necessary. The second is the process that has complex stream flows such as recycle, by-pass, and purge streams. Both types of processes are discussed and illustrated in some detail at a later point in this chapter and in some of the illustrations in the supplement.

4.6 Types of Material Balance Calculations

There are *two fundamental types* of problems for which solution by material balances are useful. In the *first*, analyses of all streams are known along with the mass of one. This problem is best attacked by assigning a letter to each unknown mass, writing C independent material balances, and solving the C equations simultaneously for the C unknown masses. Quite clearly, the number of unknown stream masses must be C in number. An example of this technique is given in Example 4.19. Another example is given below.

Example 4.4

▶ STATEMENT: One thousand pounds of 90% sulfuric acid is to be charged to a batch reactor. A waste acid containing 40% sulfuric is available as well as an unlimited supply of 98% acid. If the waste acid is used to dilute the 98% acid, how much of each is required to provide the required charge for the reactor?

▶ SOLUTION: A schematic diagram is shown in Figure 4.3.
(The number of components is two, H_2O and H_2SO_4.)

BASIS There is no reason to choose any basis other than 1000 lb of the 90% acid.

Material Balances:

Total: $$1000 = X + Y. \tag{a}$$

Partial, H_2O: $$100 = 0.60X + 0.02Y. \tag{b}$$

Partial, acid: $$900 = 0.40X + 0.98Y. \tag{c}$$

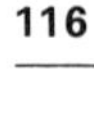

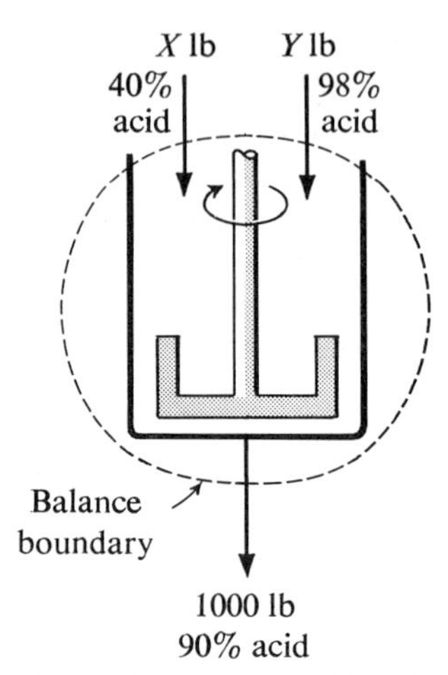

FIGURE 4.3 Flow sheet of a specific mixing process.

Any two of these equations may be chosen, but the arithmetic will be simplified if (a) is one of the two. Choose (a) and (b), the latter arbitrarily; then substituting (a) in (b):

$$\begin{aligned} 100 &= (0.60)(1000 - Y) + 0.02Y, \\ 0.58Y &= 500, \\ Y &= 862 \text{ lb of } 98\% \text{ acid}, \\ X &= 138 \text{ lb of } 40\% \text{ acid}. \end{aligned}$$

One reason for adding this example, simpler than that in the supplement to this chapter, is the failure here to specify completely the feed stream. In general, we proceed on the assumption that the feed stream is completely specified. The reason that this generality is violated in this instance is that the product stream is the one in which we are interested—as a feed stream to another process. There is then no real violation of the principle of always knowing mass and analysis of the feed, but in this case the feed stream to another unit is being prepared.

In the *second type* of problem, all product analyses are *not* given. The feed stream will ordinarily be completely specified, both as to mass (or rate) and composition. In the product streams a quantity (or rate) may be substituted for a composition item. Alternately, the action of the unit may be specified in such a manner that a product mass and/or composition may be calculated directly. The substitution of a mass for an item of composition is shown in Example 4.20. Another example is given below. In general, simultaneous equations will not be required in this type of specification. The unknown masses and compositions can be calculated progressively, granted of course that the correct sequence of calculations is used.

Example 4.5

▶ STATEMENT: Fifty pounds of an acetone-water solution analyzing 10% acetone (by mass) is to be produced in an absorption column. (Again the product, rather than feed stream, is the one completely specified.) An acetone-air mixture analyzing 20% acetone (by mass) is fed to the column and the acetone is scrubbed from the air by water fed to the top of the column. The acetone is not completely removed and the exit gas stream analyzes 3% acetone (by mass). Calculate the amount of acetone-air mixture fed to the column.

▶ SOLUTION: Figure 4.4 is a schematic diagram of this process. The number of components is three (air, acetone, water).

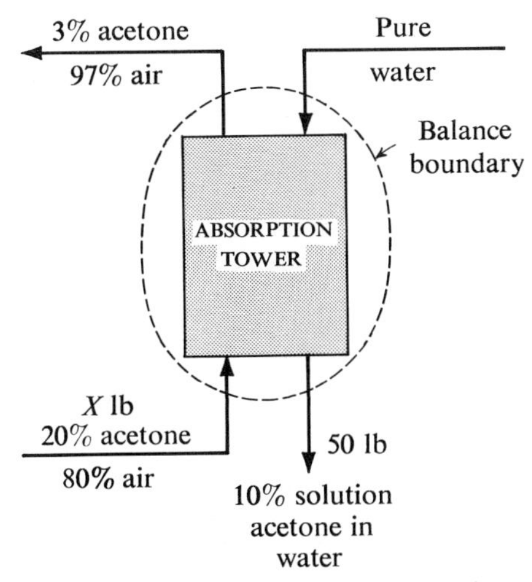

FIGURE 4.4 Flow sheet of acetone absorption process.

BASIS The desired 50 lb of product will be chosen as a basis.
It must contain: 5 lb acetone
45 lb water—the required water feed.

Let

$$X = \text{acetone-air mixture charged, lb.}$$

Then,

$$0.20X = \text{acetone charged, lb,}$$

$$0.80X = \text{air charged, lb,}$$

$$\frac{0.80X}{0.97} = \text{exit air stream, lb,}$$

and

$$\left(\frac{0.80X}{0.97}\right)(0.03) = (\text{lb air chg})\left(\frac{\text{lb mixture}}{\text{lb air}}\text{ at exit}\right)\left(\frac{\text{lb acetone}}{\text{lb mixture}}\text{ at exit}\right),$$

$$= \text{acetone in exit air stream, lb.}$$

Therefore, an acetone balance will state:

$$5 = 0.20X - \left(\frac{0.80X}{0.97}\right)0.03,$$

$$= \text{lb acetone charged in air minus lb acetone lost in exit gas,}$$

$$= 0.1752X,$$

and

$$X = 28.53 \text{ lb of acetone-air mixture charged.}$$

Note especially that the 20% and 3% acetone compositions of the two air streams are not on the same basis. The 20% is a percentage of the entire gas stream fed. The 3% is a percentage of a smaller stream, some of the original acetone in the feed having been removed in the scrubbing process. A common mistake is to subtract 3 from 20 saying that 17% of the original gas is removed by the scrubbing action. In light of the above remarks, this calculation is obviously incorrect; it has just been shown to be 17.52%.

Example 4.6

▶ STATEMENT: Pure methane is used as fuel in a furnace. It is burned completely to CO_2 and water vapor. Air is the oxidizing agent (21% oxygen, 79% nitrogen by volume*) and the air supplied is 50% more than theoretically

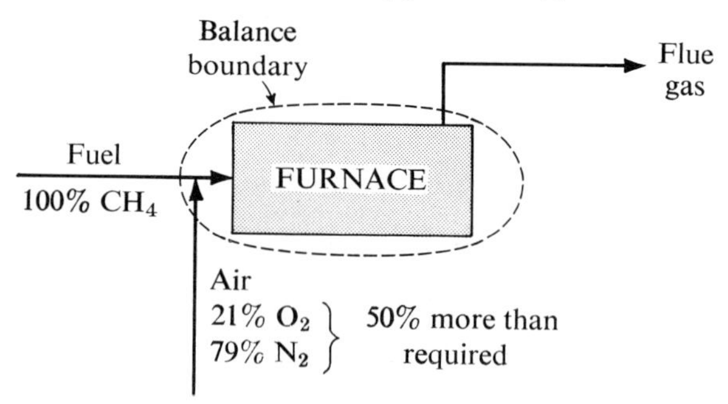

FIGURE 4.5 Flow sheet of methane-fired furnace.

• • • • •

* According to the *International Critical Tables*, the analysis of air is:

O_2	20.99%
N_2	78.03%
A	0.94%
CO_2	0.03%
H_2, He, Ne, Kr, Xe	0.01%
	100.00%

For the purposes of combustion calculations, all inerts can be lumped with nitrogen to give

O_2	20.99%
N_2	79.01%
	100.00%

required. Calculate the amount of flue gases produced and the composition (by volume) of these gases.

▶ SOLUTION: Figure 4.5 is a schematic representation of this process. The feed streams (methane and air) are completely specified. Although nothing about the product (flue gases) is specified, either mass (or moles) or composition, the action in the furnace is specified (complete combustion with 50% excess air). A direct calculation should be possible.

BASIS Since no quantity of methane is specified, 1 lb-mole will be chosen; the quantity of flue gases calculated will be per mole of methane; the required composition of the flue gases is independent of the quantity chosen. The mole is chosen, rather than the pound, inasmuch as all compositions are specified by volume, a standard procedure for gases (see Chapter 2), and the mole is more convenient in handling the chemical reaction.

Chemical Species

5 (CH_4, O_2, CO_2, H_2O, N_2)

Reaction

$$CH_4 + 2O_2 = CO_2 + 2H_2O \text{ (gaseous)}$$

Components

4 (C, O, N, H)

Requirement: 2 moles of oxygen
Supply: 3 moles of oxygen
Unused: 1 mole of oxygen (appears in flue gases)
Flue Gases: 1 mole CO_2 (C-balance)
2 moles water vapor (H-balance)
1 mole oxygen (unused) (O-balance)
X moles of nitrogen (N-balance)

The number of moles of nitrogen can be calculated from the number of moles of oxygen. Since the oxygen is 21% of the air stream, the total air supplied is $\frac{3}{0.21}$ moles. The nitrogen is 79% of this stream, or $(\frac{3}{0.21})0.79 =$ 11.28 moles. The total flue gases per mole of methane consist of 15.28 moles of mixture. The analysis of these gases is as follows:

$$
\begin{aligned}
CO_2 \quad & \frac{1}{15.28} = 6.54\% \\
O_2 \quad & \frac{1}{15.28} = 6.54\% \\
H_2O \quad & \frac{2}{15.28} = 13.08\% \\
N_2 \quad & \frac{11.28}{15.28} = 73.82\% \\
& \text{Total} = 99.98\%
\end{aligned}
$$

Frequently water vapor is condensed before flue gases are analyzed. The composition would then be reported on a "dry basis." In such a case, the total moles are 13.28 and the analysis of the flue gases would be reported as:

Flue Gas Analysis (Dry Basis)

CO_2	$\frac{1}{13.28}$	= 7.53%
O_2	$\frac{1}{13.28}$	= 7.53%
N_2	$\frac{11.28}{13.28}$	= 84.90%
	Total	= 99.96%

4.7 Linking Stream Quantities Through Inert (or Unchanged) Components

A frequently valuable aid in the solution of material balance problems is the "inert," or "unchanged," component.* The fact that it does not undergo change in the process makes it a useful link between input and output quantities. For example, in the simple drier of Figure 4.2 the solids undergo no change in the process and appear in only two streams, the input wet solids and the product stream. The solids are 80% of the feed, or 800 lb. Since the same 800 lb appear in the product stream and the analysis of that stream is known to be 95% solids, the total stream mass is immediately known to be $\frac{800}{0.95} = 842.1$ lb, with the water in the stream being 5% of this quantity or 42.1 lb. (The latter figure could be obtained by subtraction of 800 from 842.1, but the subtraction of large numbers can lead to appreciable errors in the difference so that the direct calculation of the quantity of water affords a check free of this difficulty.) Note that this unchanged solid material is *not an element.*

As a further example, in the simple combustion process (95% of the carbon burning to CO_2 and the other 5% to CO), the carbon formed a link between the amount of fuel and the amount of combustion products formed. In this instance, the unchanged material *is an element.*

• • • • •

* Another name for this component is the "tie element." It is a rather unfortunate choice since the so-called tie element does not necessarily involve a chemical element. It may be a pure material, such as water or oxygen. It may be a mixture, such as air. In some instances, generally in those processes which involve chemical as opposed to physical changes, it *may* be an element. The true element is, therefore, the exception rather than the generality.

Two generalizations might be drawn from these two simple examples. First, the unchanged material is most valuable when it appears in two streams only. Second, the process in question may be a purely physical separation or it may involve chemical change. The character of the unchanged material will be dependent upon this classification.

If the process is a physical separation, the unchanged material may be a pure solid (as in the case of the drier) or a mixture of solids (as in the case where filtration is promoted by use of a filter aid); on the other hand, if the process involves chemical change, the "unchanged component" may be a part of some substance in the input and output streams. In the combustion process, the component was an *element*, carbon; in the thermal decomposition of $CaCO_3$ to CO_2 and CaO, the link might be either of the latter two. In chemical processes, there may also be components of the feed streams which do not react and these inerts might well form suitable unchanged component links. An example of the true inert, despite chemical reaction, occurs in combustion where the air stream supplying the oxygen must bring in nitrogen; this nitrogen is unchanged and appears exclusively in the flue gases, providing a link between the quantities of air and of flue gases.

In the physical processes all components are always unchanged. It is not true, however, that such components shall appear in two streams only and if not, they are not suitable links. For instance, in a simple distillation process, as represented in Figure 4.1a, every component of the feed may appear in both the distillate and the bottoms. A 40% alcohol-in-water solution fed to a column would not normally be expected to produce pure alcohol as distillate and pure water as bottoms. The extent to which each is contaminated with the other is dependent upon the number of stages in the column. Theoretically, it is possible to obtain perfect separation only with equipment infinite in size and even then only when the vapor-liquid relationships have certain characteristics.* Since such equipment is impossible, perfect separation may not be expected to occur. Thus the distillation of a binary mixture will not produce a component which appears in two streams only, when dealing with practical equipment.

While on the subject of distillation processes, these remarks attributable to binary mixtures do not *necessarily* apply to multicomponent mixtures. Thus, if benzene is added to the ethanol-water mixture a constant boiling mixture is formed and substantially all the benzene fed will appear in the distillate. Substantially all the water fed will also appear in the distillate and the bulk of the alcohol then appears only in the two streams, feed and bottoms.

• • • • •

* It is not even theoretically possible to produce pure ethanol in distillation of a binary ethanol-water mixture. A constant boiling mixture is formed at atmospheric pressure, so that 95% (by volume) is the maximum concentration of distillate obtainable. Pure water as bottoms *is*, however, *theoretically* possible in this case.

In other multicomponent mixtures, some of the components may be so volatile that they do not appear in measurable amounts in the bottoms, while others may be so relatively nonvolatile that they do appear essentially only in the bottoms. In these instances, these components appear only in two streams and provide unchanged components between the two streams in question. This situation will be illustrated.

In some physical separations the unchanged material is sure to occur in two streams only. The drier has already been noted. If the unit is an air drier (see Figure 4.10, p. 138), two unchanged material streams exist, the solids connecting the wet and dry solids and the air connecting the wet and dry air streams. The absorption tower, Figure 4.4, is another instance of two unchanged material streams. Water is the unchanged material on the liquid side and air the unchanged material on the gas side. Other examples will occur to the reader in such processes as mixing, concentrating by mixing, crystallization, adsorption of a gaseous component on a solid adsorbent, and many more. Some representative examples follow.

Example 4.7

▶ STATEMENT: Pure ethyl alcohol is to be produced by distillation of a 40% alcohol feed (by mass). Benzene is added with the feed in order to form an azeotrope with the water. It is estimated that the distillate will contain 75% benzene, 24% water, with the remaining 1% being alcohol. The bottoms will be the product and consists of pure alcohol. You wish to calculate the feed rates of benzene and alcohol solution and the fraction of alcohol lost in the distillate. The unit is to produce 1000 lb of product per hr.

▶ SOLUTION: The process is represented schematically in Figure 4.6. There may be some doubt as to the sufficiency of data for the solution of this problem. However, the principles described in the supplement will show that six composition and rate items are needed for a unique solution to this three-component distillation. Two are used in specifying the complete composition of the distillate (the third mass fraction is not independent since the sum of the three must equal 1.00). Two are also used to specify that the product is pure, based on similar reasoning. One is used to specify the ratio of water to alcohol in the two-component feed. One is used to specify the rate of product desired. Therefore six items *are* specified and the problem is capable of a unique solution.

There are two possible methods of attack in this three-component process. The rates of the two feeds and of distillate may be treated as unknowns as indicated in Figure 4.6, with simultaneous equations obtained from material balances being solved for these rates. This method is presented first. Alternatively, since benzene appears only in two of the three major streams, and

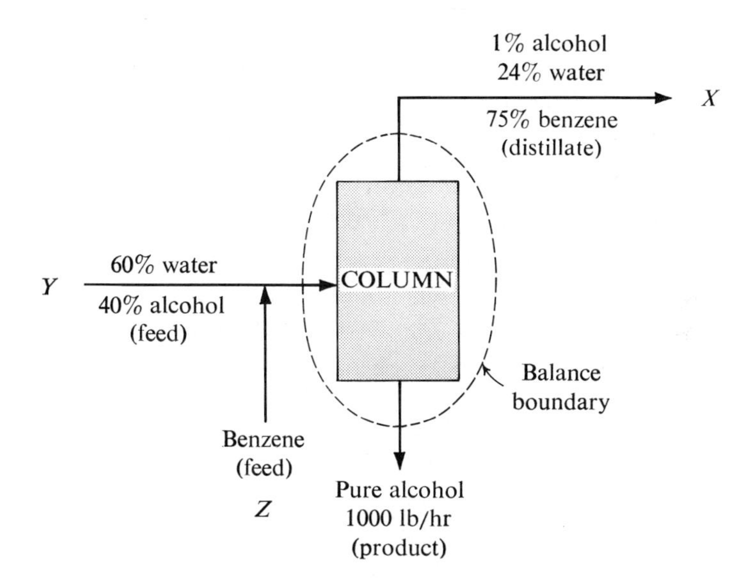

FIGURE 4.6 Flow sheet of azeotropic dehydration of ethyl alcohol.

since water also appears only in two streams, a solution employing an "unchanged component" approach can be employed. The latter approach presented second, will be seen to be the simpler, by far, in the present instance. It is possible only due to the appearance of two of the three components in two streams only. If one of the components appeared in two streams only, the problem would devolve into a two-component, two-unknown case, as illustrated in Example 4.8.

▶ SOLUTION A: By simultaneous equations.

BASIS 1000 lb product = hr.*

Designate mass rates of the various streams: alcohol feed-Y; benzene feed-Z; distillate-X.

• • • • •

* When a continuous process is encountered, information on quantities of input and output are expressed as rates, rather than masses. By the device of naming, or implying, time in the basis, the problem of deciding whether to balance on rate or on mass is avoided. The two equivalent forms of the basis chosen here indicate clearly that the choice is immaterial. If time is chosen, all rates over this time interval are automatically converted to quantities of mass, so that all balances are performed on a pure mass basis as fundamentally required (see Section 4.1, Equation 4.1). Alternately, if a specific mass of any stream or component is chosen, a time period is automatically implied even though not a neat interval such as 1 day, 1 hr, 1 min, 100 min, etc. Consequently, on the implied time interval, all calculated masses can be immediately reconverted to rates.

Material Balances

	Input Output	
Total material:	$Y + Z = X + 1000,$	(a)
C_6H_6 component:	$Z = 0.75X,$	(b)
H_2O component:	$0.60Y = 0.24X,$	(c)
C_2H_5OH component:	$0.40Y = 0.01X + 1000.$	(d)

Using (c) in (d), $X = 6670$ lb/hr.
Using X in (c), $Y = 2670$ lb/hr.
Using X in (b), $Z = 5000$ lb/hr.

The values of X, Y, Z satisfy (a), as a check on the arithmetic.

▶ SOLUTION B: Through unchanged linking materials.

BASIS

1000 lb	distillate
750 lb	benzene
240 lb	water
10 lb	alcohol

Benzene fed = 750 lb,

$$\text{Alcohol solution fed} = \frac{240 \text{ lb } H_2O}{(0.6 \text{ lb } H_2O/\text{lb soln})},$$

$$= 400 \text{ lb alcohol-water mixture fed,}$$

Alcohol fed = (0.40)(400) = 160 lb,
Alcohol to product stream = 160 − 10 = alcohol in feed minus alcohol in distillate,
= 150.

Since 150 lb product per 1000 lb of distillate

$$\frac{1000}{150} = 6.67 \text{ units of this size are required.}$$

All quantities multiplied by 6.67:

Alcohol feed: (6.67)(400) = 2670 lb/hr.
Benzene feed: (6.67)(750) = 5000 lb/hr.
Distillate: (6.67)(1000) = 6670 lb/hr.
(As by simultaneous equations.)

Fraction of alcohol fed which is lost in the distillate = 10/160 = 6.25%.

Example 4.8

▶ STATEMENT: Change the preceding example only to the extent of having a measurable amount of water appear in the product alcohol stream. Although, from an industrial point of view, a 1% water content is high, it will serve the present purposes of providing an example with reasonable numbers. It will also offer an interesting comparison of the rates with those obtained in Example 4.7.

▶ SOLUTION: The specification of the number of items has not changed. Two concentrations specify the product in this case as in Example 4.7. A unique solution is still possible.

Immediately apparent is the fact that the amount of benzene fed is the amount to appear in the distillate and that once the benzene is disregarded the problem becomes one in two components. Then the analysis of the distillate should be placed on a benzene-free basis.

Converting distillate composition to benzene-free basis:

Per 100 lb of distillate
1 lb of alcohol
24 lb of water

25 lb total—benzene-free

Benzene-free Compositions

$$\text{Alcohol:} \quad \frac{1}{25} = 4\%.$$

$$\text{Water:} \quad \frac{24}{25} = 96\%.$$

Now all streams are analyzed for alcohol and water only and the following material balances may be written:

Total material: $Y = X' + 1000$, where X' is the benzene-free rate.

Alcohol component: $0.4Y = 0.04X' + 990.$

Water component: $0.6Y = 0.96X' + 10.$

Any two of these equations may be used to obtain:

$$X' = 1640 \text{ lb/hr},$$
$$Y = 2640 \text{ lb/hr}.$$

Since

$$4X' = X, \qquad X = 6560 \text{ lb/hr},$$

and the benzene feed rate is $6560 - 1640 = 4920$ lb per hr.

Example 4.9

▶ STATEMENT: A four-component petroleum mixture, consisting of 20% propane (C_3), 20% isobutane (i-C_4), 20% isopentane (i-C_5), and 40% normal pentane (C_5) is fed to a column. All of the C_3 and none of the C_5 is expected to appear in the distillate. The distillate is expected to analyze 50% C_3, 45% i-C_4. Calculate the composition of the bottoms.

▶ SOLUTION: The process is represented schematically in Figure 4.7.

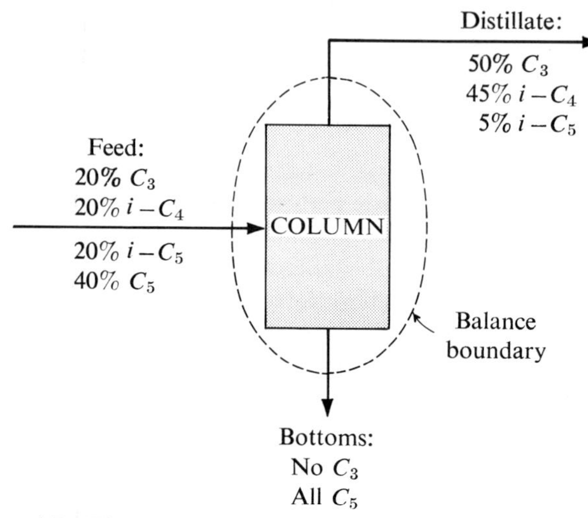

FIGURE 4.7 Flow sheet of distillation of four-component mixture.

BASIS One hundred pounds of feed is arbitrarily chosen since compositions, but no quantities, are required. The 20 lb of C_3 all appears in the distillate and constitutes 50% of this stream. It is therefore an unchanged component, existing in only two streams, and constitutes a link allowing the calculation of the mass of the distillate stream. Distillate $= \frac{20}{0.50} = 40$ lb.

All distillate compositions are known. Therefore,

$$\left.\begin{array}{l}(40)(0.45) = 18 \text{ lb } i\text{-}C_4 \\ (40)(0.05) = 2 \text{ lb } i\text{-}C_5\end{array}\right\} \text{other components of distillate.}$$

The components of the bottoms may be obtained by subtraction. The composition of this stream then follows directly.

Bottoms:			
	C_3:	zero	
	i-C_4:	20 − 18 = 2 lb	3.33%
	i-C_5:	20 − 2 = 18 lb	30.00%
	C_5:	40 − 0 = 40 lb	66.67%
		Total = 60 lb	100.00%

Example 4.10

▶ STATEMENT: In the previous example suppose that the specification is somewhat more realistic. The feed is the same and all of the C_3 still appears in the distillate and all of the C_5 in the bottoms. Now the C_3 will still be expected to be 50% of the distillate and under temperature conditions which will achieve these objectives, 90% of the i-C_4 is expected to appear in the distillate. Complete the analyses of distillate and bottoms.

▶ SOLUTION: The same arguments as in Example 4.9 establish the distillate at 40 lb. Since 90% of the 20 lb of i-C_4 is 18 lb, it constitutes 45% of the distillate. Of the 40 lb of distillate, all but 2 are accounted for, leaving this amount for the i-C_5. Calculations of the bottoms are as before.

Example 4.11

▶ STATEMENT: A problem in absorption of acetone from an air-acetone mixture has already been worked in Example 4.5. Since there are two unchanged materials in this situation, recalculation on this basis is interesting.

▶ SOLUTION: On the gas side, the air goes through unchanged. On the liquid side, the water goes through unchanged.

BASIS 100 lb of air-acetone mix charged to the unit

80 lb of air charged to the unit

$$\frac{80}{0.97} = 82.5 \text{ lb of gas leaving the unit}$$

$$100 - 82.5 = 17.5 \text{ lb of gas (acetone) absorbed.*}$$

Now it is necessary to transform to a temporary basis for similar calculations on the liquid side. On the natural basis of 50 lb of product, 5 lb of acetone leaves and none enters. Therefore the correct charge of air-acetone mix to provide 5 lb absorbed is all that is required. Then

$$\frac{5 \text{ lb acetone}}{(17.52 \text{ lb acetone/100 lb charge})} = 0.285 \text{ units of 100 lb each,}$$

which amounts to 28.5 lb air-acetone mix to be charged. The answer is, of course, the same as obtained before.

It is apparent that if the liquid fed to the top of the column is a waste liquor containing some acetone (e.g., 1%), the second unchanged material, water, is more useful. In this case the water leaving is 45 lb, which is 99% of the entering liquor. The entering liquor is then

$$\frac{45}{0.99} \text{ lb.}$$

• • • • •

* As noted earlier this subtraction of large numbers leaves something to be desired. It can be avoided in this case by the following calculation:

$$(0.03)(82.5) = 2.48 \text{ lb acetone leaving,}$$
$$20 - 2.48 = 17.52 \text{ lb acetone absorbed.}$$

Both figures, 17.5 and 17.52, were obtained by slide rule. A small gain in precision was obtained by avoiding subtraction of numbers of comparable magnitude. Had a calculator or computer been used, either of which is capable of carrying many significant figures, no difference of any consequence would be detected between the two methods of calculation.

The acetone in that stream is

$$(0.01)\left(\frac{45}{0.99}\right) = 0.45 \text{ lb},$$

and the acetone absorbed in the tower is $5 - 0.45 = 4.55$ lb. The air-acetone charge is then $(\frac{4.55}{17.52})100 = 25.95$ lb. This figure is lower than for pure water feed, since some of the required 5 lb of acetone is fed with the water and need not be obtained from the gas. The fraction of acetone removed from the gas fed is, of course, unchanged, since it depends only on the inlet and outlet gas compositions. Note that there is no claim that the size of the unit is the same for both situations.

Examples of an unchanged material occurring in chemical processes are plentiful. An example is found in the combustion of ammonia to form oxides of nitrogen for the manufacture of nitric acid. The nitrogen of the air does not burn to NO as does the nitrogen in ammonia; therefore the nitrogen in the air is a link between the air supply and the product gases. The combined nitrogen (in ammonia fed and in oxides of nitrogen in the product) links the ammonia fed to the product stream. Note that if an appreciable fraction of the ammonia decomposes, these simple two-stream links are not valid.

Any process that uses an excess of one reactant to force complete consumption of the other has a link through an "unchanged material," in this case the completely reacted material which appears only in the feed and product streams, although in different chemical forms. In the catalytic water gas reaction:

$$CO + H_2O \text{ (gaseous)} = CO_2 + H_2.$$

The object of this reaction is the formation of hydrogen. Since carbon dioxide is more easily removed (by scrubbing with water, with ethanolamines, or other solvents) than is carbon monoxide, complete conversion of CO is desired. Equilibrium and rate considerations indicate that completion of reaction can only be obtained by use of large excesses of steam. Steam being relatively cheap, such large excesses (as much as 1000%) are practical. Then the CO in the feed and the CO_2 in the product are linked through the carbon and these two streams are the only ones containing this material.

A classic example of the use of unchanged material to link two streams in the presence of chemical reaction is the combustion process. Nitrogen is used to link flue gases and air supply, provided that the fuel does not itself contain an appreciable amount of nitrogen (as for some gaseous fuels). Carbon is used to link the flue gases with the fuel. An oxygen balance is used to establish a not-immediately-evident hydrogen link between the flue gases and the fuel. Noncombustible inorganic ash is used to link the fuel charged to the ash pit residues. Not only do such calculations involve several links in one problem, but they also employ various types of links. The nitrogen is a truly inert material. The ash is a mixture of inert materials. The carbon is an element involved in the reactions. The hydrogen is a portion of a compound (H_2 in H_2O). Combustion processes will be illustrated in Examples 4.12, 4.13, and 4.14.

Example 4.12

▶ STATEMENT: A furnace is fired with a coke containing nothing but carbon and inorganic ash. The carbon constitutes 90% of the coke. The ash pit residue, after being wet down with water, analyzes 10% carbon, 40% ash, and the rest water. The flue gases analyze 14% CO_2, 1% CO, 6.4% O_2, and the remainder nitrogen. In analyzing the performance of this furnace, calculate: (a) moles of flue gas per ton of coke charged to the furnace; (b) the percentage of excess air used; (c) percentage of the combustible charged which is lost in the ash.

▶ SOLUTION: Combustion problems have many similarities. A general organization of the solution will then be applicable to many combustion problems. The form used below is recommended, primarily for its ability to maintain a number of necessary component balances in simple presentation. The process is indicated schematically in Figure 4.8.

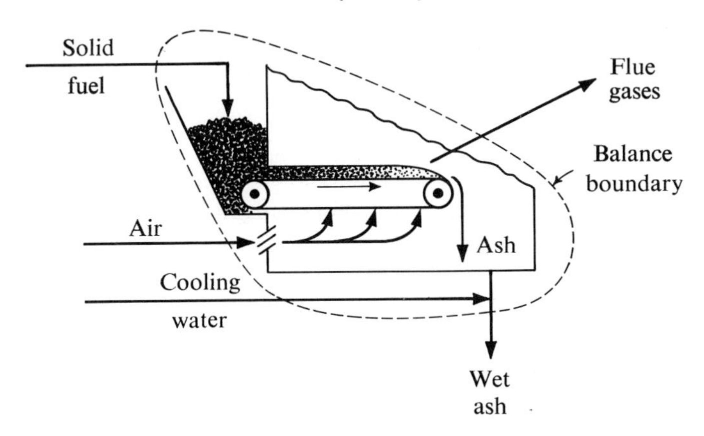

FIGURE 4.8 Flow sheet of coal-fired furnace.

The number of chemical species is 6 (C, ash, O_2, N_2, CO_2, CO). The number of independent reactions is 2:

$$C + O_2 = CO_2,$$
$$C + \tfrac{1}{2}O_2 = CO.$$

The number of components is 4, which may logically be chosen to include C, O, N, ash. We have deliberately neglected the water used to wet down the hot ash.

BASIS The greatest amount of information in one stream appears to be located in the flue gas analysis. It contains contributions from both air and flue gas and does not include any contributions from ash pit losses. Therefore *100 moles of flue gas* will be chosen.

Flue Gas

	Moles in Flue Gas	Atoms of C	Moles of O_2	Moles of N_2
CO_2	14.0	14.0	14.0	—
CO	1.0	1.0	0.5	—
O_2	6.4	—	6.4	—
N_2	78.6	—	—	78.6
Total	100.0	15.0	20.9	78.6

Nitrogen Link

$$\frac{78.6}{0.79} = 99.5 \text{ moles of air supplied;}$$

$(0.21)(99.5) = 20.9$ moles of oxygen supplied.
This quantity checks the amount found in the flue gas.

Carbon Link

The 15.0 moles of carbon in flue gas = 180 lb.

But 180 lb is not the amount of carbon charged due to some losses in the ash pit. Calculate the loss per 100 lb *charged* using the *ash link*.

Ash Link

Per 90 lb C charged,
10 lb ash charged,

$$\frac{10}{0.40} = 25 \text{ lb ash pit residue containing}$$

10 lb ash,
12.5 lb water,
2.5 lb C.

$90 - 2.5 = 87.5$ lb C burned/100 lb coke *charged.*

Therefore, for 180 lb carbon as noted in the flue gas, $(\frac{180}{87.5})100 = 206$ lb of coke had to be charged, thus producing the 100 moles of flue gases used as a basis for the calculations.

For 2000 lb charged, $(\frac{2000}{206})100 = 971$ moles of flue gases.

Excess Air

Excess air is *defined* as the supply over and above the theoretical requirement divided by the theoretical requirement. Note that the divisor is not the amount supplied, for this would not relate the excess to the fuel but rather to the operation. Thus, the only way to obtain a fixed basis independent of operation is to relate the excess to the requirement, the latter always being fixed by the amount of fuel *charged*, and independent of the operation (i.e., different amounts of air supplied). Retaining the basis of 100 moles of flue gases, the coke charged is 206 lb. The carbon in this mass of coke is:

$$(0.9)(206) = 185.4 \text{ lb [also} = \text{C in flue gas plus carbon in ash pit}$$
$$= 180 + (2.06)(2.5) \text{ except for rounding errors],}$$
$$= 15.45 \text{ moles.}$$

The 15.45 moles of carbon require 15.45 moles of oxygen for complete combustion and 20.9 moles of oxygen were supplied. The excess is

$$20.9 - 15.45 = 5.45 \text{ moles.}$$

The percentage excess is $5.45/15.45 = 35.2\%$.

Combustible Lost

2.5 lb of C were lost per 100 lb of coke charged.

The 100 lb of coke contained 90 lb of carbon.

The percentage loss is $\dfrac{2.5}{90} = 2.78\%$.

Example 4.13

▶ STATEMENT: Complicate the situation in Example 4.12 by charging a coal containing 90% carbon and 6% ash with the remaining 4% being divided between hydrogen and moisture in some unknown fashion. The flue gases analyze, *on a dry basis*, 12% CO_2, 1% CO, 7.4% O_2, and the rest nitrogen. The ash pit residue will be as before. Complete the coal analysis and calculate the percentage of excess air.

The number of chemical species in this case is 8 (C, ash, O_2, N_2, CO_2, CO, H_2, water). The number of independent reactions is three, one each to burn C to CO and CO_2 and one to burn hydrogen to water. The number of components is five, which may logically be chosen to include C, O, N, H, ash.

▶ SOLUTION: The greatest amount of information is still definitely centered in the flue gas analysis. Therefore,

BASIS 100 moles of *dry* flue gas.

Flue Gas

	Moles in Flue Gas	Atoms of C	Moles of O_2	Moles of N_2
CO_2	12.0	12.0	12.0	—
CO	1.0	1.0	0.5	—
O_2	7.4	—	7.4	—
N_2	79.6	—	—	79.6
Total	100.0	13.0	19.9	79.6

Nitrogen Link

$$\frac{79.6}{0.79} = 100.7 \text{ moles of air supplied.}$$

$$(0.21)(100.7) = 21.15 \text{ moles of oxygen supplied.}$$

Hydrogen Link

Oxygen present in flue gas = 19.9 moles
Oxygen unaccounted: 1.25 moles
Hydrogen burned: 2.50 moles (5.0 lb)

$$(2H_2 + O_2 = 2H_2O).$$

Carbon Link

The 13.0 moles of carbon in the flue gas = 156.0 lb. But 156 lb is not the amount of carbon charged due to ash pit loss. Therefore, per 100 lb of coal charged,

Ash Link

90 lb of C

6 lb of ash which constitutes 40% of the ash pit residue.

$$\frac{6}{0.4} = 15 \text{ lb ash pit residue containing 1.5 lb carbon.}$$

$90 - 1.5 = 88.5$ lb carbon burned per 100 lb coal *charged.*

$$\left(\frac{156}{88.5}\right)100 = 176.3 \text{ lb coal charged per 100 moles of flue gas.}$$

Summary of components in coal:

$(176.3)(0.90) = 158.6$ lb carbon in the coal
[also $\doteq 156 + (1.76)(1.5)$],
5.0 lb hydrogen—as calculated above
$(176.3)(0.06) = 10.6$ lb ash.

Subtotal = 174.2 lb—moisture-free coal,
2.1 lb moisture.

Therefore,

$$\begin{aligned} \text{percentage of hydrogen} &= 2.84 \\ \text{percentage of moisture} &= \underline{1.19} \\ & 4.03 \end{aligned}$$

The figure should be 4.00 since the carbon and ash comprise 96.00% according to information supplied. Rounding, accumulated errors, and analytical errors can easily account for the 1% error incurred.

Excess Air

The oxygen requirement is an amount sufficient to burn all the carbon charged (158.6 lb = 13.22 moles), plus that for all the hydrogen charged (2.5 moles). The total requirement is therefore,

$$13.22 + 1.25 = 14.47 \text{ moles.}$$

The supply is 21.15 moles. The excess is then,

$$21.15 - 14.45 = 6.70 \text{ moles.}$$

The percentage of excess is 6.70/14.45 = 46.4%.

A coal may be much more complicated than indicated in Example 4.13. The main complications will lie in the content of nitrogen and/or sulfur. Since the nitrogen is very difficult to analyze, it is frequently estimated at 1.7% of the total combustible. The combustible consists of two parts, carbon and hydrogen, but these two are frequently combined into two other approximate groups: a) volatile combustible matter (VCM) which can be distilled from the coal in the absence of air, and by fixed carbon (FC) remaining after the distillation. An analysis which includes these two items, along with moisture and ashy residue can easily be performed by heating at successively higher temperatures, the last heat being conducted in the presence of air to oxidize all the carbon. This analysis is known as the *proximate analysis.*

Sulfur is present in the originally inorganic matter. We can, with little error, assume that it is all present as pyrites, FeS_2. It is then easy to correct for the fact that this material oxidized to Fe_2O_3 and SO_2. If we assume that these are the products, the ash-as-weighed weighs less than the original inorganic matter, for sulfur with weight of 128 in 2 moles of FeS_2 is replaced by 3 atoms of oxygen, of weight 48. If we report the sulfur separately, we must exclude the weight of sulfur in the ash reported (known as ash-corrected). Therefore ash-corrected is equal to ash-as-weighed minus three-eighths of the percent sulfur.

There is some disagreement as to the method of treating SO_2 in gas analysis. Some authorities state that it is not reported since it is dissolved in the water which acts as the leveling fluid in the gas analysis apparatus. Others state that it is all dissolved with CO_2 in the caustic pipette and therefore reported as CO_2. When the gas sample is contained over mercury, this latter view must be correct. When the sample is contained over water, there is doubt that time is allowed for complete solution of SO_2 in water, particularly in view of the relatively low water solubility of this gas and its low concentration in the gas stream. We have then, in this text, chosen to consider that SO_2 is reported with the CO_2. The molal ratio of SO_2 to CO_2 is ordinarily so low that the choice is not material.

A second way of reporting coal analysis is the *ultimate analysis*, in which elemental quantities are reported, with the exception of water. The elements are C, H, S, net H, combined water and free moisture. The last is the same as moisture in the proximate analysis. The combined water can be broken down into H and O. The former is added to net H and listed as total H, with the oxygen reported individually. It will be apparent that both of these analyses contain approximations but, due to the complex nature of coal, the two give a very good picture of the material for combustion calculations.

For ease of analytical procedures, only carbon and sulfur are analyzed in elemental form for the ultimate analysis. Then, with the ash pit analysis, on a proximate basis, and with the proximate analysis of the coal and the flue gas analysis, it is quite easy to complete the ultimate analysis by material balances.

Example 4.14

▶ STATEMENT: *Coal Analysis.* The following percentages are reported for a coal burned under a boiler. We can easily deal with the material balances if the flue gas analysis is given, and if we can handle the losses through the grate. We therefore illustrate the latter only.

Proximate Analyses			Ultimate Analysis	
Coal		Ash Pit	Coal	
5.2	Moisture	zero	75.0	C
30.0	VCM	5.0	2.0	S
60.0	FC	15.0		
4.8	Ash	80.0		

▶ SOLUTION: Assume that any VCM which appears in the ash pit is due to original coal falling through the grate. Then any FC came from this source or from coal that had been distilled on the grate, producing carbon in the process. This carbon did not have time to burn before falling through the grate.

BASIS 100 lb of coal charged to the boiler
4.8 lb ash,

$$\frac{4.8}{0.80} = 6.0 \text{ lb residue,}$$

$$\left.\begin{array}{l}0.3 \text{ lb VCM,}\\ 0.9 \text{ lb FC.}\end{array}\right\} 1.2 \text{ lb total combustible.}$$

Correcting for the sulfur,

$$4.8 + \left(\frac{5}{8}\right)2.0 = 6.05 \text{ lb original ash.}$$

$$4.8 - \left(\frac{3}{8}\right)2.0 = 4.05 \text{ lb ash-as-weighed corrected for sulfur being independently reported, ultimate analysis.}$$

Calculating the carbon in the ash pit residue:

$$0.3\,\frac{30.0 + 60.0}{30.0} = 0.9 \text{ lb of original coal through the grate.}$$

But,

1.2 lb of total combustible matter through the grate.

Therefore,

$$1.2 - 0.9 = 0.3 \text{ lb of carbon present due to coking of coal.}$$

Further,

$$\frac{75.0}{30.0 + 60.0} = 0.833 = \text{fraction of carbon in original combustible matter (VCM + FC).}$$

Therefore,

$(0.833)(0.9) = 0.75$ lb of carbon in combustible which fell through the grate.
0.3 lb of carbon in ash due to coking.
1.05 lb of carbon in the ash pit = total loss of carbon.
$1.20 - 1.05 = 0.15$ lb hydrogen lost in the ash pit.

Example 4.15

▶ STATEMENT: One means of producing hydrogen in large quantities is to control the combustion of carbon to promote the formation of CO, rather than the spontaneously formed CO_2, and subsequently to react the CO with steam in the catalytic water gas reaction:

$$CO + H_2O = CO_2 + H_2.$$

Several large scale means of forming CO are in common use. One is the gas producer in which carbon is first burned to CO_2 and later reduced to CO as follows:

$$C + O_2 = CO_2,$$
$$CO_2 + C = 2CO.$$

Another is the water gas process in which the carbon bed is first heated by burning part of the charge to CO_2 and exhausting the gaseous products; the heat in the fuel bed is then utilized to sustain the following reaction:

$$C + H_2O \text{ (steam)} = CO + H_2.$$

Operating on pure carbon and without appreciable side reactions, the gas producer generates a gas of approximately 35% CO and 65% N_2. (The reader might well check this statement from knowledge of combustion reactions.) The water gas reaction generates a gas of approximately 50% CO and 50% H_2. You are asked to mix these gases in correct proportion so that, after the mixture has been subjected to the catalytic water gas reaction, the N_2 and H_2 will be in correct stoichiometric proportion for production of ammonia.

▶ SOLUTION: Essentially, this is a problem of a mixer in which a mole of input CO is equivalent to a mole of H_2. Choose producer gas as a basis since both nitrogen and equivalent hydrogen are represented.

BASIS 100 moles of P.G.,
35 moles CO,
65 moles N_2.

Nitrogen Link

This source of N_2 is the only one and these 65 moles should be accompanied by 3(65) = 195 moles of hydrogen.

H_2 available from CO in the P.G. charged = 35 moles

H_2 required from W.G. = 160 moles

Moles of H_2/mole of W.G. = 1.
(one-half existing and one-half by converting CO)

Moles of W.G. required = 160.

The required feed ratio is 5:8::P.G.:W.G.

4.8 Recycle and By-Pass Streams

A process involving a recycle stream is shown in Figure 4.9a. One involving a by-pass stream is shown in Figure 4.9b. Occasionally recycle streams will contain purge streams as shown in Figures 4.9c and 4.10. These types of streams can occur in either physical separations or in processes involving chemical reactions.

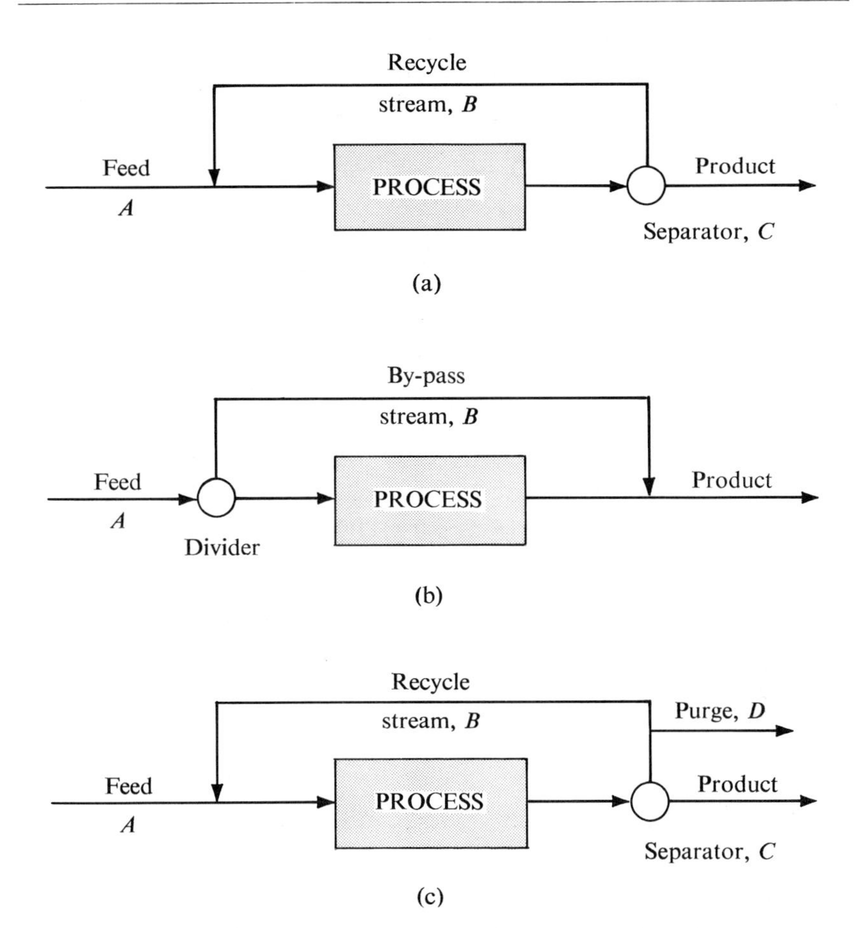

FIGURE 4.9 Characteristic representations of recycle and by-pass streams.

Consider first a physical process such as an air drier. Occasionally in an air drier (Figure 4.10) the designer wishes to have a high air velocity, providing good mass transfer rate for removing water from the solid to the air stream but also wishes to avoid heating the very large quantities of air which would be required for high rates of flow. He may recycle some of the hot air at the outlet (stream B in Figures 4.9a and 4.9c), purge a part of the effluent humid air (stream D), and add some fresh, dry air (stream A). He thus maintains a high flow rate, heats only the fresh air stream, and keeps the water content of the air stream (i.e., the humidity) to a reasonable level.

In petroleum refineries most of the streams are quite complex, containing numerous components. Cracking, reforming, and distillation steps do not yield pure compounds, but rather complex mixtures. The desired conversions are rarely accomplished in a single pass through a reactor. The products are

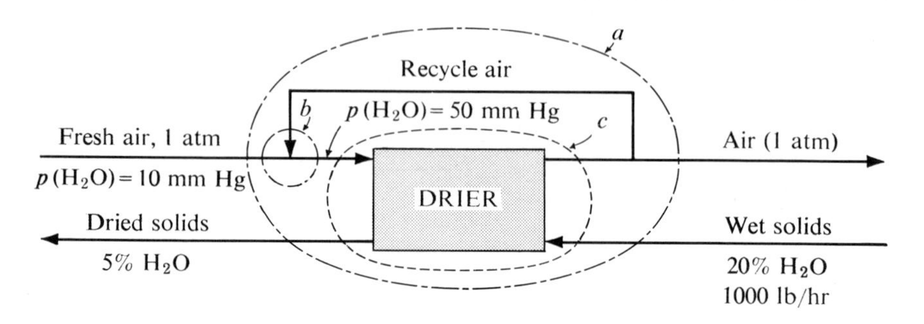

FIGURE 4.10 Flow sheet of air drier with partial recycle of used air stream.

separated from the unreacted material in distillation columns and the unreacted portions returned to the various reactor units. Multiple recycle streams can result in this type of operation. [See Nagiev, *Chem. Eng. Progress*, *53*, No. 6, 297 (1957).]

In catalytic reactors of the type producing ammonia from N_2 and H_2, or methanol from CO and H_2, only a fraction of the gases react. The products are separated and the unreacted portion returned and added to fresh feed.

Each of these operations has resulted in at least one recycle stream. If inerts are present in the feed, such as argon in the nitrogen-hydrogen mixture fed to an ammonia converter, a bleed stream is inserted in the recycle stream to prevent unlimited build-up of such a component (Figure 4.9c).

Recycle calculations should not give difficulty although they frequently do. The source of trouble, when encountered, is generally a failure to analyze the balance positions correctly. There are four possible balance positions as illustrated in Figure 4.11: (a) a normal balance (the dot-dash line) which

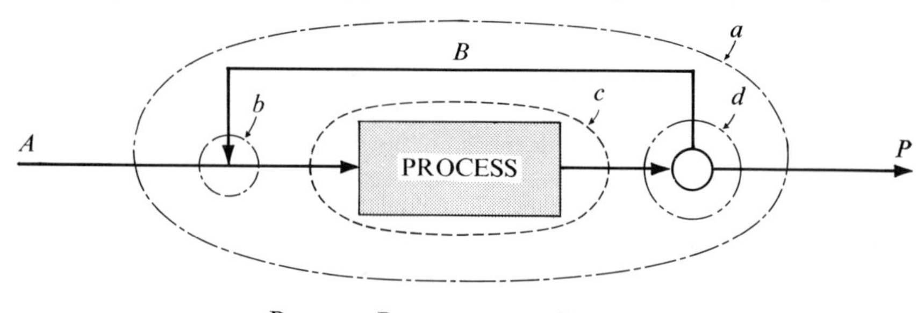

BALANCE POSITIONS WITH RECYCLE

a —·—·— Outside recycle
b ——·—— Mix point
c -------- Inside recycle
d —·-—--- Divider point

FIGURE 4.11 Balance boundaries in recycle processes.

completely ignores the presence of a recycle stream; (b) the mix point (the dot-double-dash line) where feed and recycle join to form total feed to the process; (c) a balance immediately around the process (dashed line) that includes the recycle material into and out of the process; (d) the stream divider point (dash-double-dot line) where a portion of the outlet stream is bled off to provide a product stream while the remainder is returned to the mix point as recycle. All four are not independent. Any three will give the fourth.

Of the four positions, only (a) completely ignores the recycle stream. In drawing this balance boundary line, it is impossible to cut the recycle stream. If the boundary is incorrectly drawn, as in Figure 4.12, the recycle leaves the

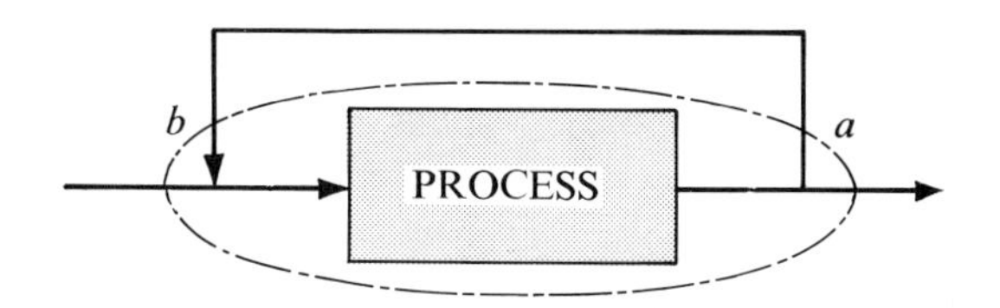

FIGURE 4.12 Poorly located overall balance boundary in recycle process.

boundary at point *a* and returns at point *b*. These streams would neutralize each other in the balance. All other balance positions, (b), (c), and (d), are so located that the recycle stream enters or leaves at least once. At the mix and divider points recycle enters once and leaves once, respectively. In balance type (c) the recycle material both enters and leaves so that the quantities of both of these streams are greater than those at *A* and *P*.

From the point of view of production, balance (a) is most important since it deals with net input and output. From the point of view of equipment size, the other balances are important. They include the recycle stream and the equipment must be designed large enough to handle these streams. Complete analysis of a process involving recycle will usually require material balances at three of the four possible positions.

In general the remarks on recycle apply to by-pass situations. It is apparent that by-passing the process reduces the equipment size rather than increasing it as is the case with a recycle stream.

Example 4.16

► STATEMENT: In an air drier, part of the effluent air stream is to be recycled in an effort to control the inlet water content (i.e., humidity). Figure 4.10 is a schematic representation of the process. The solids contain 20% water and 1000 lb per hour are to be handled. The dried solids are to contain a maximum of 5% water. The partial pressure of water vapor in the fresh air is 10 mm of mercury and in the outlet must be limited to 200 mm of mercury.

The recycle stream is to be regulated so that the partial pressure of water input in air to the drier shall be 50 mm of mercury.

Calculate the rates of fresh air, recycle air, and total air to the drier. Assume that the process operates at a constant total pressure of one standard atmosphere.

▶ SOLUTION: The number of components (three) is unchanged by the introduction of a recycle stream. Calculate first the rate of water removal from the solids. Use the *normal balance* position, *a*, since the recycle stream is not involved in the solids streams.

BASIS 1 hr or 1000 lb of solids fed,

800 lb of solids—*link* to dried solid,

$$\frac{800}{0.95} = 842.1 \text{ lb dried solids containing } 42.1 \text{ lb water.}$$

$$1000 - 842.1 = 157.9 \text{ lb water removed,}$$
$$= 8.78 \text{ moles water removed.}$$

Use the same balance position to calculate the rate of *fresh* air feed—dry air as *link* between inlet and outlet.

Let

$$X = \text{moles of fresh air with mole fraction of water,}$$
$$= \frac{10}{760} = 0.0132.$$

$$\text{Moles of water input-air stream} = 0.0132X.$$
$$\text{Moles of air input-air stream} = 0.9868X.$$

Ratio, water-to-air, in effluent air stream, calculated

$$\text{from specification of } p_w = 200 \text{ mm,}$$
$$= \frac{200}{760 - 200} = 0.3575.$$

$$\text{Moles of water leaving in air stream} = (0.3575)(0.9868X),$$
$$= 0.352X.$$

Net moles of water picked up by air stream is $0.352X - 0.013X$; this value is known to be 8.78 from the initial balance.

$$\therefore\ 0.352X - 0.013X = 8.78 \text{ moles water removed from solids.}$$
$$X = 25.9 \text{ moles of air required/hr.}$$

Now calculate, by balances at the *mix point* (b, dot-double-dash line), moles of recycle to be mixed with the 25.9 moles of fresh air in order to obtain the proper inlet partial pressure of water. Retain the original basis.

Let

$$Y = \text{moles of recycle mixture.}$$

The mole fraction of water in the air stream entering the drier is known to be $\frac{50}{760}$. X = moles of fresh air as before, this quantity now being known to be 25.9. The total air stream entering the drier: $X + Y$ moles. The moles of water in this air stream is $(10X + 200Y)/760$.
Therefore,

$$\frac{(10X + 200Y)/760}{X + Y} = \frac{50}{760};$$

whence,

$$10X + 200Y = 50X + 50Y,$$

and

$$150Y = 40X.$$

Since X is already known to be 25.9 moles per hr,

$$\therefore\ Y = 6.9 \text{ moles/hr (recycle).}$$

The input air rate to the drier is $25.9 + 6.9 = 32.8$ moles per hr.

A third balance position, that *around the drier* and within the recycle stream (c, dotted line) could be used to calculate the total effluent air rate.

Example 4.17

▶ STATEMENT: An ammonia converter is fed a stoichiometric mixture of nitrogen and hydrogen (i.e., 1:3). In the converter 10% of the reactants are converted to ammonia. The ammonia formed is removed in a condenser. The unconverted gas is recycled to the converter. In order to avoid unlimited accumulation of inert rare gases (largely argon), a bleed stream is incorporated in the recycle stream. Calculate the fraction of recycle gas that must be bled off if the argon entering the converter is to be limited to 0.5%.

▶ SOLUTION: The number of chemical species is 4 (N, H, A, NH_3). There is one independent equation,

$$N_2 + 3H_2 \rightarrow 2NH_3.$$

The components are, therefore, three and may be logically listed as N, H, A.

Although rare gases can be lumped with nitrogen in combustion problems where nitrogen is also an inert, they must be taken into account when nitrogen is a reactant, as in the present case. Air truly analyzes 20.99% oxygen, 78.03% nitrogen, and 0.95% argon, lumping all the inert gases into the last category. CO_2, in amount equal to 0.03% can be disregarded here, having been removed in the purification of the reactant gases. Thus, 100 moles of nitrogen and hydrogen in stoichiometric proportion will contain 25 moles of nitrogen, 75 moles of hydrogen, and will also contain argon in the following amount:

$$25\left(\frac{0.95}{78.03}\right) = 0.304 \text{ moles.}$$

The process is represented schematically in Figure 4.13. The first balance will be around the converter inside the recycle (dashed line). The *basis* will be

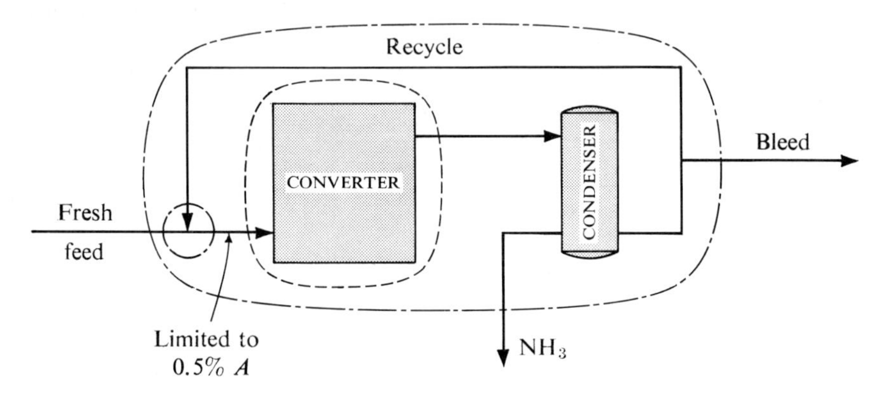

FIGURE 4.13 Flow sheet of an ammonia synthesis process.

100 moles of $N_2 - H_2$ mixture (argon-free) and therefore containing 0.503 moles of argon, as fed to the reactor.

Gas	Input Moles	Converted Moles	Output Moles
N_2	25	2.5	22.5
H_2	75	7.5	67.5
A	0.503	—	0.503
NH_3	0	—	5.0

The ammonia is assumed to be completely removed in the condenser,and the recycle gas (prior to purge) is as follows:

Gas	Moles	%
N_2	22.5	
H_2	67.6	
A	0.503	0.556
	90.5	

The concentration of argon will not, of course, be affected by the bleed stream. It can then be used at the mix point (dot-double-dash line) to calculate the ratio of fresh and recycle streams.

Let

X = feed rate, prior to addition of recycle;
Y = recycle rate.

Argon Balance

$$\frac{0.304X}{100.304} + 0.00556Y = 0.503.$$

Overall Balance

$$X + \qquad Y = 100.5.$$

Solving simultaneously:

X = 22.3 moles,
Y = 78.2 moles.

The bleed stream may now be determined by an argon balance over the whole process (dot-dash line).

Argon in 22.3 moles fresh feed = (0.003)(22.3) = 0.067 moles.

This amount of argon must be removed in the bleed, analyzing 0.556% argon. The total bleed stream is then

$$\frac{0.067}{0.00556} = 12.0 \text{ moles.}$$

The total recycle gas was previously calculated to be 90.5 moles. The required fraction bled is $\frac{12.0}{90.5}$, which is 13.3%.

Note that the bleed could have been calculated by subtracting the moles of recycle gas (78.2) from the amount of gas leaving the condenser (90.5), a balance around the stream divider at the bleed point. The 12.3 moles so obtained does not quite check by 3% due to accumulated errors. The magnitude of the error is due to the wide range of concentrations, and consequent roundings incurred in additions and subtrắctions.

It might appear that balances have been made on a molal basis rather than on a mass basis in this example. While partly true, no violation of the law of conservation of mass has occurred. The simultaneous equations for fresh feed and recycle came from a straight molal balance, *but* since the mix point is a physical process, no moles are made or lost. The molal and mass balances are then identical. The analysis around the converter violates no constant mass principle. The moles in are 100.5 while those out are 95.5. The lack of molal balance results from proper use of the chemical reaction. The reduction in moles reflects the maintenance of a proper balance based on the numbers of atoms of nitrogen and hydrogen. Fifty atoms of nitrogen entered; 45 left as nitrogen, and 5 as ammonia. The same treatment shows a proper accounting for hydrogen. This general treatment of numbers of moles is often helpful in balances involving chemical reactions.

4.9 Units Composed of Several Elements

In discussing the principles of material balance calculations, we have largely concentrated on balances around a single element of equipment. Practical problems will usually involve units consisting of a number of these elements. Complete specification of all streams will be the normal objective for material balance calculations.

If there are n elements of equipment in a unit, there are, in simplest terms, $n + 1$ potential balances which may be drawn, one around each element and one around the unit as a whole. Of these $n + 1$ balances, only n are independent and the $(n + 1)^{st}$ will provide no new information.

In establishing the independent balances, it will be obvious that the number of ways in which one can select the $n + 1$ possibilities is greater than indicated in the preceding paragraph. Pairs of elements might be treated as one small unit, for example. Then the balance around this unit and the balance around *one* of its constituent elements provide exactly the same amount of information on all input and output streams as would be obtained if one had balanced individually around the two elements. This statement is obvious from the following argument. The two-element unit has three potential balance points, one around each element and one around the unit which the two comprise. Of these three balances, only two are independent and any two may be selected.

The best sequence to be used in solving for the unknowns in the various balances is dependent upon the available data. There are, however, two general types of data, and two general categories of methods. In one, known specifications are such that the order in which the balances are selected is immaterial. In this category, the available data are largely concentrated at one end of the unit. An illustration of this situation appears in Example 4.18, which immediately follows this discussion. In the second category, some of the data pertain to inlet and some to outlet streams in such fashion that the unknowns in certain of the balances cannot be solved until other specific balances have first been completed. Illustrations of this situation appear in Examples 4.19, 4.20, and 4.21. A method for establishing an unambiguous choice of the first balance boundary position is given in the Supplement to this chapter, Section S4.8.

Example 4.18

▶ STATEMENT: A diagrammatic representation of a unit in which ammonia is oxidized to nitric oxide and the nitric oxide is further oxidized and absorbed in water is shown in Figure 4.14.

The following information is given:

a. The ammonia is burned in air, 75% excess air being used, based on the combustion to nitric oxide;

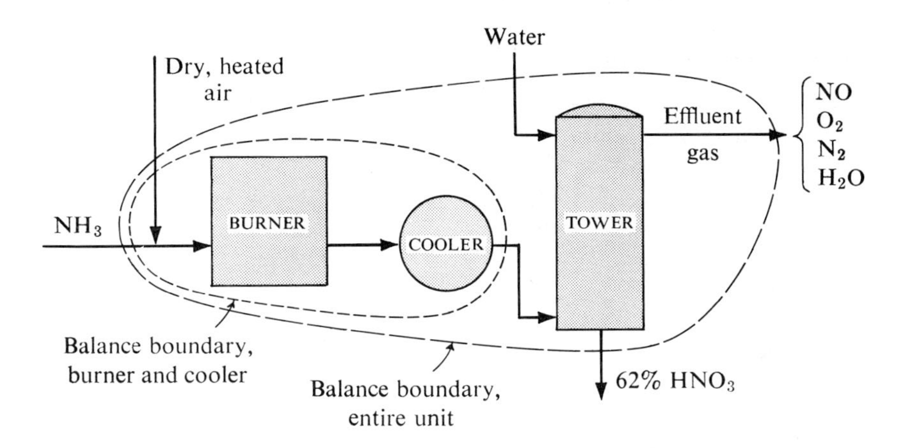

FIGURE 4.14 Flow sheet of an ammonia oxidation process for manufacture of nitric acid.

b. Two per cent of the ammonia may be expected to decompose to nitrogen and hydrogen, the latter being burned to water;
c. Gases leave the absorption tower at 35 °C, saturated with water vapor at the pressure of the system, 4 atm;
d. Ten per cent of the NO oxidizes to N_2O_4 before the burner gases enter the absorber, but N_2O_4 in the effluent gases may be considered to be negligible;
e. One per cent of the nitrogen oxides are lost in the effluent gas.

The following information is required, all on a basis of a ton of ammonia charged to the unit:

a. Production of 62% nitric acid;
b. Total percentage loss of available nitrogen;
c. Moles of gas to be handled in the absorber;
d. Water required to be fed to the absorber.

▶ SOLUTION: Of the required information, only item *c* involves a stream inside the unit. This item will result in two balances being required, a balance around the entire unit to provide items *a*, *b*, and *d*, and a balance around the burner and cooler to provide item *c*. Obviously the latter may also be obtained by a balance around the absorber. There is no reason to perform one balance prior to the other. The two balances selected each involve the inlet stream about which the maximum amount of information is available. Additional information given is concentrated at the end of the process or after the cooler.

Balance 1, Around Burner Plus Cooler

BASIS 1.000 moles of ammonia burned to NO.

Using the information on the amount of ammonia decomposed, we find,

$$\frac{1.000}{0.98} = 1.020 \text{ moles of ammonia charged.}$$

The 0.020 moles of decomposed ammonia yield,

0.020 *atoms* of nitrogen,
0.060 *atoms* of hydrogen, burned to 0.030 moles of water.

Oxygen required can be determined on the basis of 1.020 moles of ammonia charged and from the reaction,

$$NH_3 + 1.25O_2 = NO + 1.5H_2O.$$

The oxygen required is (1.020)(1.25) and the oxygen supplied is (1.020)(1.25) $\times$ (1.75) = 2.231 moles. The nitrogen accompanying this oxygen in the air stream is (2.231)($\frac{79}{21}$) = 8.389 moles.

The oxygen used is a combination of that consumed in the combustion of the mole of NH_3 plus that consumed in the combustion of the hydrogen from the decomposed NH_3.

For NH_3:	(1.000)(1.25) =	1.25 moles
For H_2:	(0.030)(0.5) =	0.015 moles
Total		1.265 moles

The gases from the burner may now be determined:

NO:	1.000 moles	
H_2O:	1.500 + 0.030 =	1.530 moles
O_2:	2.231 − 1.265 =	0.966 moles
N_2:	8.389 + 0.010 =	8.399 moles
	Total =	11.895 moles

The oxidation reaction,

$$2NO + O_2 = N_2O_4$$

will change the amounts of O_2 and NO, and will add N_2O_4 as an additional component. The O_2 used will be exactly the same as the N_2O_4 formed so that the change in total number of moles will be equal to the loss in NO. Since 10% of the NO is oxidized, this loss is 0.100 moles. The total gases to the absorber are, then,

$$11.895 - 0.100 = 11.795 \text{ moles.}$$

These calculations are on the basis of 1.020 moles of NH_3 charged, or (1.020)(17) = 17.34 lb. In a ton there are $\frac{2000}{17.34}$ units of this size, or 115.3 units. The absorption tower will be required to handle a charge of

$$(115.3)(11.795) = 1360 \text{ moles } (1.36 \times 10^3).$$

Balance 2, Around the Entire Unit

The basis will be the same. The nitric acid production will be calculated first.

Only 0.010 of the available mole of NO are lost. Therefore, 0.990 moles of NO are converted to HNO_3, obviously yielding 0.990 moles of HNO_3. The molecular weight of this material is 63. The weight of HNO_3 is (63)(0.990) = 62.4 lb. These 62.4 lb constitute 62% of the dilute acid stream, which will total $\frac{62.4}{0.62}$ = 100.6 lb. (The water in this stream is 38.2 lb, or 2.122 moles, a figure which will be required later in determining the amount of water to be fed to the tower.) We know, from the first balance, that 115.3 units of the basis chosen will constitute a ton of NH_3 charged. Therefore, the nitric acid produced will be

$$(115.3)(100.6) = 11{,}600 \text{ lb.}$$

Next, we shall calculate the amount of water to be charged to the tower. This calculation necessarily involves a water balance around the unit. Water enters due to combustion of NH_3 and H_2 (calculated in Balance 1), as well as in the charge to the tower. Water leaves as diluent in the dilute acid (just calculated), as combined water, and in the effluent gas stream. Calculation of the last two quantities will then allow completion of the balance. We shall attack the water in the gas stream by first obtaining the effluent *dry* gas, subsequently obtaining the amount of water contained therein.

Let X = moles of oxygen consumed in oxidizing NO to HNO_3. Then:

Dry Gas Leaving—Moles

NO	0.010
O_2	$0.966 - X$
N_2	8.399
Total	$9.375 - X$

The net reaction, not necessarily mechanistic, is

$$H_2O + 2NO + 1.5O_2 = 2HNO_3.$$

For the 0.990 moles of NO oxidized, there will be $\frac{3}{4}(0.990)$ moles of oxygen used, or 0.742 moles. The total dry gas is

$$9.375 - 0.742 = 8.633 \text{ moles.}$$

The vapor pressure of water in this gas, saturated at 35 °C, is

$$42.2 \text{ mm Hg} = 0.0555 \text{ atm.}$$

The moles of water in the gas stream will be

$$8.633 \times \frac{0.0555}{4.000 - 0.0555} = 0.121 \text{ moles water vapor,}$$

since

$$(\text{dry moles})\left(\frac{\text{partial pressure water}}{\text{partial pressure dry gas}}\right) = \text{moles water.}$$

The water leaving the system may be summarized as follows:
Combined in the formation of

HNO_3	0.495 moles = $\frac{1}{2}$ moles NO oxidized
Diluting HNO_3	2.122 moles
In gas stream	0.121 moles
Total	2.738 moles

The water from combustions has been previously calculated to be 1.530 moles. The net requirement as water to the tower is 2.738 − 1.530 = 1.208 moles. For the 115.3 units constituting the ton of NH_3 charged, there are

$$(1.208)(115.3)(18) = 2507 \text{ lb of water charged.}$$

Finally, we deal with the loss of available nitrogen. The losses are of two sources, unburned (or decomposed) ammonia and unabsorbed NO. In moles, these amount to 0.010 and 0.005, respectively, as indicated by previous calculations, or a total of 0.015 moles. The total moles charged is $\frac{1.020}{2}$ = 0.510 moles. The percentage loss is

$$\frac{0.015}{0.510} = 2.94\%.$$

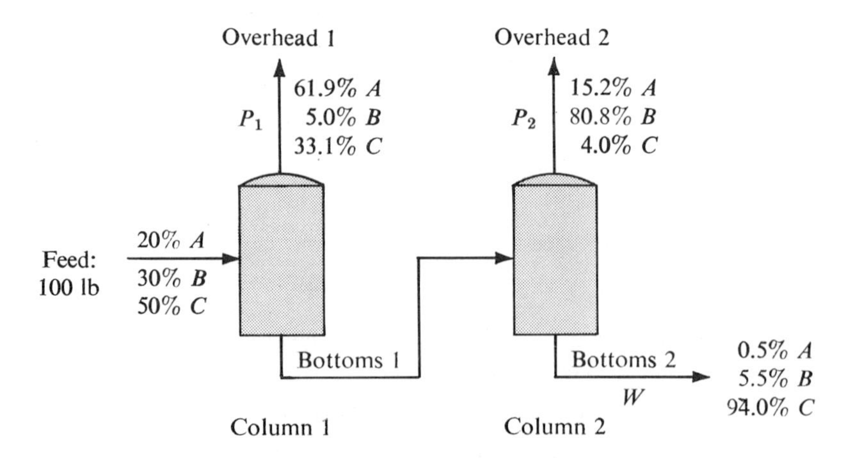

FIGURE 4.15 Flow sheet of a two-column distillation unit operating under specific conditions.

Example 4.19

▶ STATEMENT: A two-column still is separating a three-component mixture. In this physical process, take the case where analyses of all streams entering and leaving are given. The problem may be stated as follows:

Feed: 20% A, 30% B, 50% C.
Overhead from first column: 61.9% A, 5% B.
Overhead from second column: 15.2% A, 80.8% B.
Bottoms from second column: 0.5% A, 5.5% B.
Bottoms from first column will be feed to the second column.
Calculate the percentage recovery of the various components in each output stream.

▶ SOLUTION: Since percentage recoveries are required, a specific quantity of feed is not required. It is convenient to take 100 lb as a basis for the calculations. The diagram of the equipment is then as shown in Figure 4.15.

BASIS 100 lb feed
Material Balances:

1. Total $100 = P_1 + P_2 + W.$ (a)
2. Partial on A: $20 = 0.619P_1 + 0.152P_2 + 0.005W.$ (b)
3. Partial on B: $30 = 0.050P_1 + 0.808P_2 + 0.055W.$ (c)

Solving simultaneously, by a standard elimination procedure:

	P_1	P_2	W	Feed	Check*	
	1	1	1	100	103	(a)
	0.619	0.152	0.005	20	20.786	(b)
	0.050	0.808	0.055	30	30.913	(c)
(a) ÷ 1	1	1	1	100	103	(d)
(b) ÷ 0.619	1	0.2456	0.0081	32.3102	33.5719	(e)
(c) ÷ 0.050	1	16.16	1.1	600	618.26	(f)
(d) − (e)	0	0.7545	0.9919	67.6898	69.4281	(g)
(d) − (f)	0	−15.16	−0.1	−500	−515.26	(h)
(g) ÷ 0.7545		1	1.3148	89.713	92.000	(i)
(h) ÷ −15.16		1	0.00660	32.982	33.984	(j)
(i) − (j)			1.3082	56.731	58.016	(k)
(k) ÷ 1.3082			1	43.43		(l)

• • • • •

* Note how rounding errors accumulate in this type of solution. The check column is the sum of all elements in the same row. When a row is operated upon [e.g., (a) ÷ 1] operate on the check column in the same fashion and then confirm that the new value in this column is truly the sum of all other elements in the same row. This equivalence is the check on the arithmetic performed.

From (l) $W = 43.43,$
From (j) $P_2 = 32.98 - (0.0066)(43.43) = 32.69,$
From (d) $P_1 = 100 - 43.43 - 32.69 = 23.88.$

Components in Each Stream

	P_1		P_2		W	
	Lb	Fraction of Charge	Lb	Fraction of Charge	Lb	Fraction of Charge
A	14.8	0.740	5.0	0.250	0.2	0.010
B	1.2	0.040	26.4	0.880	2.4	0.080
C	7.9	0.158	1.3	0.026	40.8	0.816

Example 4.20

▶ STATEMENT: A two-column still for a three-component mixture with analysis of one component in each output stream and fraction of another component to be recovered in each stream also specified. The problem can be stated as follows:

Feed: 20% A, 30% B, 50% C.
Overhead from first column: 75% of A charged to unit with B ≤ 5%.
Overhead from second column: 80% of B charged to unit with C ≤ 4%.
Bottoms from second column: 80% of C charged to unit with A ≤ 1%.

Complete the analyses of each output stream.

▶ SOLUTION: As in Example 4.19, these specifications use only eight of the available nine degrees of freedom with one quantity yet to be chosen. Since percentage recoveries are part of the data, a specific quantity of feed is not required. It is again convenient to take 100 lb as a basis for the calculations. The diagram of the equipment is then shown in Figure 4.16.

BASIS 100 lb feed

Partial Material Balances

$$A: 20 = 15 + x_{A2} P_2 + 0.01 W, \tag{1}$$

$$B: 30 = 0.05P_1 + 24 + x_{B3} W, \tag{2}$$

$$C: 50 = x_{C1}P_1 + 0.04P_2 + 40. \tag{3}$$

Composition Restrictions

$$\sum_{i=1}^{3} x_{ij} = 1,$$

where $j = P_1, P_2, W$ and $i = A, B, C.$

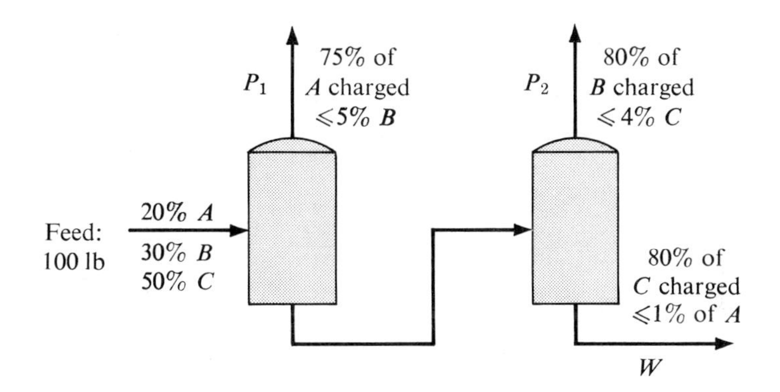

FIGURE 4.16 Flow sheet of a two-column distillation unit operating under specific conditions.

Applying the restriction to each stream:

$$P_1: \frac{15}{P_1} + 0.05 + x_{C1} = 1, \tag{4}$$

$$P_2: x_{A2} + \frac{24}{P_2} + 0.04 = 1, \tag{5}$$

$$W: 0.01 + x_{B3} + \frac{40}{W} = 1. \tag{6}$$

Solving (4), (5), (6) for unknown compositions, substituting into (3), (1), (2), respectively, and solving simultaneously, gives

	P_1	P_2	P_3
Total Lbs	25.04	29.74	45.20
x_A	0.599	0.153	0.010
x_B	0.050	0.807	0.105
x_C	0.351	0.040	0.885
Lbs A	15.00	4.55	0.45
Lbs B	1.25	24.00	4.75
Lbs C	8.79	1.19	40.00
	25.04	29.74	45.20

It is interesting to note (see Supplement to this chapter) that the analysis of the degrees of freedom will, in either of the two-column still problems, indicate that the calculation *must* be made by balancing around the unit, with the individual elements being insoluble due to insufficient data. This conclusion is drawn as follows. The individual elements have six degrees of freedom each. For the first column, three degrees of freedom are used to specify the feed (rate and two independent compositions) and two degrees of freedom are used to specify the composition of stream P_1. These five degrees of freedom are an insufficient specification. For the second column there are two degrees of freedom used for each of the output streams, or a total of four, again insufficient. This use of the degrees of freedom to locate useful material balances is a great advantage to the performance analyst.

If the problem had been specified somewhat differently, it would have been desirable to start with the individual elements rather than the unit as a whole. For example, if one of the degrees of freedom used in specifying compositions in the second column had been used to specify the quantity of overhead product desired from the first column, all information on the innerstream could have been calculated directly and subsequently used to complete the total and component quantities for the second column. Example 4.21 treats this case.

Example 4.21

▶ STATEMENT: The problem:
Feed: 20% A, 30% B, 50% C.
Overhead, first column: 62% A, 5% B, 25% of feed.
Overhead, second column: 15% A, 80% B.
Bottoms, second column: 0.5% A.
Complete all analyses.

▶ SOLUTION: The degrees of freedom used in the first column specification are now six and completion of the data on this unit appears to be the easiest attack. The diagram of the equipment is now as shown in Figure 4.17.

BASIS 100 lb feed

$$P_1 = 25 \text{ lb containing: } 15.50 \text{ lb } A, \; 1.25 \text{ lb } B, \; 8.25 \text{ lb } C.$$

Since innerstream, W_1, is 75 lb (i.e., 75% of feed) and contains the remaining feed components not found in P_1, the components of W_1 are

$$4.50 \text{ lb } A, \; 28.75 \text{ lb } B, \; 41.75 \text{ lb } C.$$

Around the second column, the total material balance is

$$75 = P_2 + W_2.$$

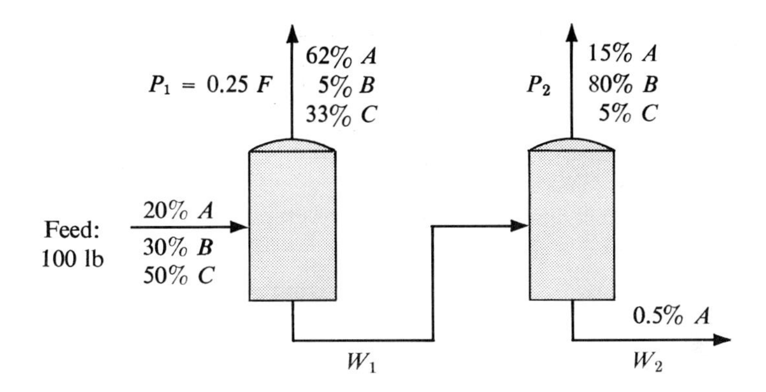

FIGURE 4.17 Flow sheet of a two-column distillation unit operating under specific conditions.

The partial balance on A, using the limiting value of 0.5% A in stream W_2 is

$$4.50 = 0.15\,P_2 + 0.005\,W_2.$$

Solving simultaneously,

$$W_2 = 46.6 \text{ lb},$$
$$P_2 = 28.4 \text{ lb}.$$

The analysis of stream P_2 is complete. Components are, therefore

$$\left.\begin{aligned} A &= (0.15)(28.4) = 4.260 \text{ lb},\\ B &= (0.80)(28.4) = 22.720 \text{ lb},\\ C &= (0.05)(28.4) = 1.420 \text{ lb}.\end{aligned}\right\} 28.4 \text{ lb}.$$

The components of stream W_2 are obtained by subtraction

$$\left.\begin{aligned} A &= 4.40 - 4.260 = 0.140 \text{ lb},\\ B &= 28.75 - 22.720 = 6.030 \text{ lb},\\ C &= 41.75 - 1.420 = 40.330 \text{ lb}.\end{aligned}\right\} 46.5 \text{ lb}.$$

LIST OF SYMBOLS

(Exclusive of chemical symbols.)

Latin letters

C	Number of physical or chemical components
N_e	Number of chemical elements
N_R	Number of independent chemical equations
N_{sp}	Number of chemical species
x_i	Mass, or mole fraction, component i
X, Y, Z	Unknowns, usually quantity or rate

(Alternate symbols employed in examples are appropriate to the specific problem; e.g., P_i may stand for mass or rate of the ith product stream.)

PROBLEMS

*Section A—Problems Containing No Unnecessary Data**

4.1 Carbon is burned with air with all the carbon burning to CO_2. Calculate:
  a. flue gas analysis when percentage of excess air is zero.
  b. flue gas analysis when percentage of excess air is 50.
  c. flue gas analysis when percentage of excess air is 100.

4.2 A fuel composed entirely of methane and nitrogen is burned with excess air. The flue gases analyze: 7.5% CO_2, 7.0% O_2, and the remainder nitrogen. Calculate:
  a. fuel gas analysis.
  b. percentage of excess air.

4.3 A coke has the following composition: carbon 87.9%, ash 12.1%. The ash, wet down in the ash pit, analyzes: ash 45.0%, carbon 13.2%, water 41.8%. The flue gases produced by firing this coke contain 12% CO_2 and 3% CO. Calculate:
  a. percentage of excess air.
  b. analysis of flue gas.

4.4 A carbureted water gas, having the composition: C_2H_4 17%, CH_4 21%, H_2 32.9%, CO 26.1%, and CO_2 3.0% is burned in a furnace. Both gas and air for combustion enter at 70 °F and 28.5 in. Hg and are substantially dry. The flue gases analyze CO_2 12.2%, CO 0.4%, and leave the stack at 680 °F and 27.5 in. Hg. Calculate:

  a. cubic feet of air supplied per cubic feet of gas fed.
  b. percentage of excess air.
  c. cubic feet of flue gas per cubic feet of gas fired.

4.5 A coke, containing 89.5% carbon and 10.5% ash is burned in a domestic heater, using 35% excess air. The refuse removed from the ash pit (after being wet down to minimize dust) is found to contain 45% moisture, 12% carbon, and 43% ash. Calculate:

  a. moles of dry flue gas per pound of fuel *charged.*
  b. moles of dry flue gas per pound of fuel *burned.*
  c. percentage of total heating value of fuel lost as unburned combustible in the refuse.

4.6 The nitrogen-hydrogen mixture for ammonia synthesis may be prepared by mixing producer and water gases, converting the CO to CO_2 by the following reaction:

$$CO + H_2O\,(\text{steam}) = CO_2 + H_2.$$

• • • • •

* This statement does not carry the complementary guarantee that no data are missing. Additional required items of data will be readily found in handbooks.

The carbon dioxide is then easily scrubbed from the gaseous mixture. Calculate:

a. proportions in which the original mix of producer and water gases should be made.
b. moles of carbon dioxide recovered per mole of final synthesis gas mixture.

DATA

Composition	Producer Gas %	Water Gas %
CO	20	50
CO_2	2	—
N_2	78	—
H_2	—	50

4.7 The feed to a distillation column contains 36% benzene by weight, the remainder being toluene. The overhead distillate is to contain 52% benzene by weight, while the bottoms are to contain 5% benzene by weight. Calculate:

a. the percentage of the benzene fed which is contained in the distillate.
b. the percentage of the total feed which leaves as distillate.

4.8 The feed to a continuous two-column still consists of a mixture of 30% benzene (B), 55% toluene (T), and 15% xylene (X). The overhead stream from the first column is analyzed and contains: 94.4% B, 4.54% T, and 1.06% X. The bottoms from the first column is fed to the second column. In this second column it is planned that 92% of the original T charged to the unit shall be recovered in the overhead stream and that the T shall constitute 94.6% of the stream. It is further planned that 92.6% of the X charged shall be recovered in the bottoms from this column and that the X shall constitute 77.6% of that stream. If these conditions are met, calculate:

a. analysis of each stream leaving the unit.
b. percentage recovery of benzene in the overhead stream from the first column.

4.9 The continuous feed to a two-column distillation unit consists of 50% A, 30% B, and 20% C. The streams leaving the unit in a performance test are found to analyze as follows:

	Overhead First Column %	Overhead Second Column %	Bottoms Second Column %
A	4.5	6.9	95.5
B	9.1	90.1	4.1
C	86.4	3.0	0.4

Calculate, per 1000 lb of feed:

a. weight of each overhead distillate and weight of bottoms from the second column.
b. weight of bottoms from the first column.

4.10 An old process for the manufacture of hydrochloric acid consisted of heating nitre cake ($NaHSO_4$) and salt ($NaCl$) in a Mannheim furnace. The HCl was removed as a gas. The nitre cake was converted to salt cake (Na_2SO_4). In such a process 5800 lb of salt and a stoichiometric quantity of nitre cake were charged per 24 hr. Both reactants may be considered to be pure. The reaction was 98% complete. Calculate:

a. tons of salt cake produced per day.
b. analysis of the salt cake.
c. pounds of nitre cake charged per day.

In the HCl recovery process, 89% of the gas was absorbed and the acid strength was 31%. Calculate the number of tons of acid made per day.

4.11 A 13% solution of caustic soda is required by a paper mill. The solution is made by reacting soda ash (Na_2CO_3) and slaked lime ($Ca(OH)_2$) in a wet process. The charge is 1000 lb of soda ash, dissolved to form a 20% aqueous solution, plus 699 lb of slaked lime, which has been slurried into a "milk of lime" containing 30% by weight of suspended solids. The solubility of calcium hydroxide may be taken to be 0.1 gm per 100 ml of water under the conditions of the process. Calcium carbonate is formed in the reaction. It is filtered from the caustic solution on an Oliver filter. Calculate:

a. the reactant which is in excess.
b. pounds of caustic solution produced per batch of 1000 lb of soda ash.

4.12 A sludge, having the composition 32% H_2SO_4, 18% H_3PO_4, 35% $CaHPO_4$, 15% H_2O, is neutralized with a gas mixture analyzing 14% O_2, 20% N_2, 55% NH_3, and 11% CO_2. The neutralization takes place in a rotating inclined cylinder, and the gas emerging contains 3% NH_3 on a dry basis. The neutralization reactions are as follows:

$$2NH_3 + H_2SO_4 = (NH_4)_2SO_4,$$
$$2NH_3 + H_3PO_4 = (NH_4)_2HPO_4,$$
$$2NH_3 + 3CaHPO_4 = Ca_3(PO_4)_2 + (NH_4)_2HPO_4.$$

The heat evolved in the reactions evaporates all the water from the neutralization product, yielding a 5000-lb cake of mixed salts. Calculate:

a. volume of gas required (STP).
b. analysis of the dry cake.

4.13 There are 58 gm of 25° Bé sulfuric acid required. The only sulfuric acid available is 66° Bé. Calculate:

a. the milliliters of 66° acid and of water to be mixed.
b. the milliliters of the 25° acid formed.

For density-concentration data, see table of the Manufacturing Chemists' Association in the *Handbook of Chemistry and Physics.*

4.14 In the sulfuric acid industry, oleum is a term used for 100% acid containing free, unreacted SO_3, dissolved in the acid. A 20% oleum, for example, is 20 lb of SO_3 in 80 lb of 100% acid, per 100 lb of mix.

Oleum can also be designated as a percentage of sulfuric acid in excess of 100%. It is figured as the pounds of 100% acid which would be produced by adding enough water to 100 lb of oleum to dissolve all the free SO_3.

Using these definitions, calculate:

a. pounds of 25% oleum that can potentially be produced per 100 lb of sulfur.
b. percentage of sulfuric acid corresponding to 25% oleum.
c. price the consumer can afford to pay for 25% oleum in competition with 98% acid at $23.50 per ton, based upon the following conditions. Freight rate: 2.5¢ per ton-mile. Distance of purchaser from shipping center: 200 miles.

4.15 A contact sulfuric acid plant produced 10 tons per day of 20% oleum* and 40 tons per day of 98% acid. The 98% tower is fed 97% acid obtained by diluting part of the output of the tower. The oleum tower is fed with 18% oleum, obtained by mixing 98% acid and 20% oleum.
Calculate:

a. tons of 98% acid fed to each tower per day.
b. tons of 20% oleum recirculated to the oleum tower per day.

4.16 The spent acid from a nitrating process contains 43% H_2SO_4 and 36% HNO_3. This dilute acid is to be strengthened by the addition of concentrated sulfuric acid containing 91% H_2SO_4 and concentrated nitric acid containing 88% HNO_3. The product is to contain 41.8% H_2SO_4 and 40.0% HNO_3. Calculate the quantities of spent and concentrated acids that should be mixed to yield 1000 lb of the fortified mixed acid.

4.17 A soap factory tries to maintain a moisture content of 24.6% in its white floating cake soap. One batch of 10 tons is analyzed and found to contain only 12.7% moisture. Since it is undesirable to sell a product that is off-standard, this lot of soap is rehumidified to the desired average moisture content by standing for a couple of days in a room containing 1,000,000 cu ft of air at 80 °C and 80% relative humidity and 29.5 in Hg pressure. Radiators keep the temperature constant but no steam jets are open, so that the relative humidity drops. The humidifying room has vacuum-tight walls. Calculate the final relative humidity in the room.

4.18 In the modern process for manufacture of HNO_3, ammonia is oxidized to NO, which is in turn further oxidized, and the oxidation products absorbed in water to form nitric acid. The process operates at a pressure of 100 psig. Ammonia, at a rate of 10 CFM is mixed with air, at a rate of 150 CFM

• • • • •

* See Problem 4.14.

(both measured at 70 °F and 100 psig) and passed through a platinum oxidation catalyst. The bulk of the ammonia (95%) is burned, the reaction being:

$$4NH_3 + 5O_2 = 4NO + 6H_2O.$$

The remaining 5% of the ammonia decomposes, the hydrogen subsequently burning to water, the combined reactions being:

$$2NH_3 + \tfrac{3}{2}O_2 = N_2 + 3H_2O.$$

The gas containing the NO is cooled to 80 °F at which temperature all the NO is oxidized to NO_2. Assume that the water condensed is removed from the system without absorbing any oxides of nitrogen. These gases are fed to the scrubber, where the NO_2 is further oxidized and absorbed in water. The gases leaving the scrubber contain 0.5% NO_2 and no NO and are saturated with water vapor at 25 °C. The acid made in the scrubber is 60% HNO_3. Calculate:

a. the total volume of gases entering the scrubber per hour.
b. composition of the gases entering the scrubber.
c. output of 60% HNO_3 per 24-hr day.

4.19 A glycerol plant is treating a 10% glycerol (trade name for glycerine) solution with a 3% NaCl content with butyl alcohol in a solvent extraction tower. The alcohol fed to the tower contains 2% water. The raffinate

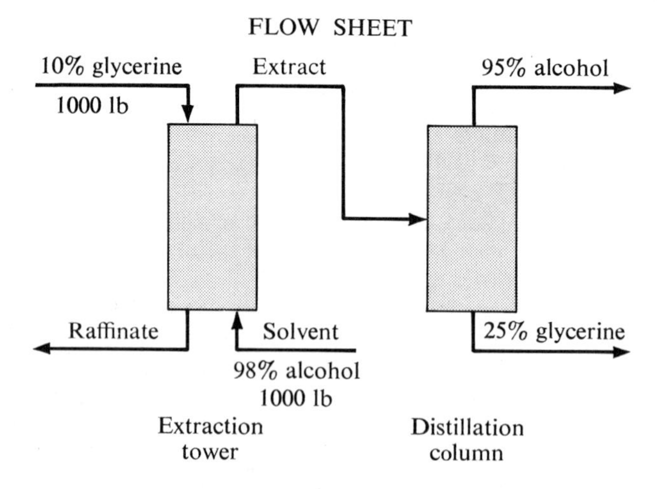

leaving the tower contains all the original salt, 1.0% glycerine and 1.0% alcohol. The extract from the tower is sent to a distillation column. The distillate from this column is the alcohol, containing 5% water. The bottoms from the column is 25% glycerine and the rest water.

The two feed streams to the extraction column amount to 1000 lb per hr each. Calculate the output of glycerine per hour from the distillation column.

NOTE The process is hypothetical only.

4.20 Fifty grams of a slightly soluble salt is added to 100 gm of water and mixed thoroughly until the solution is saturated at 75 °F. The mixture is then filtered and the wet filter cake obtained is dried and weighed. The dry weight is 49.0485 gm. Calculate the solubility of the salt in grams per 100 gm of water, noting that the liquid adhering to the wet cake has the same concentration of salt as does the clear filtrate. Weight of wet cake = 53.9 gm.

4.21 1. One thousand pounds of KCl is dissolved in sufficient water to provide a solution saturated at 90 °C. At this temperature the solubility of KCl is 53.8 parts per 100 parts of water. This solution is cooled to 20 °C, at which temperature the solubility of KCl is 34.7 parts per 100 parts of water. Calculate:

a. weight of water required for solution.
b. weight of the crop of crystals obtained, assuming that there was no evaporation during cooling.
c. weight of the crop of crystals obtained, assuming that $7\frac{1}{2}\%$ of the original charge of water was evaporated during cooling.

2. Repeat these calculations for the case where the original 1000 lb of KCl was dissolved to form a 30% solution instead of a saturated one.

4.22 Ten thousand pounds of a solution of soda ash in water, 30% Na_2CO_3, is cooled slowly to 20 °C. The crystals formed are the decahydrate, $Na_2CO_3 \cdot 10\,H_2O$. The solubility of the decahydrate at 20 °C is 21.5 parts of Na_2CO_3 (calculated as the anhydrous salt) per 100 parts of water. During the cooling 3% of the solution is lost by evaporation. Calculate:

a. weight of the crop of crystals.
b. percentage of Na_2CO_3 in the crystals.

4.23 $NaHCO_3$ is fed to a unit drying and calcining in one step. The calcining follows the reaction:

$$2NaHCO_3 \longrightarrow Na_2CO_3 + H_2O + CO_2.$$

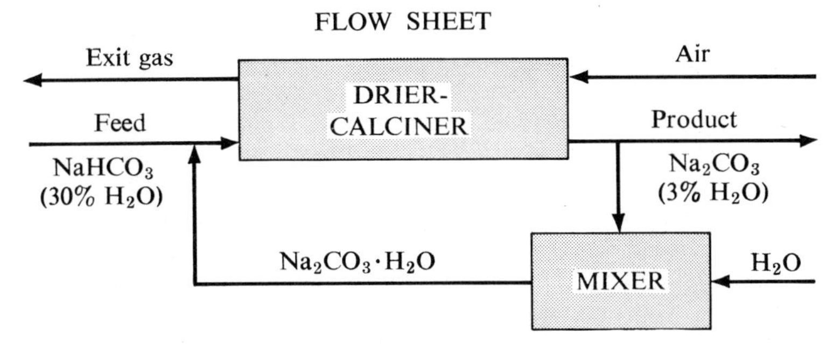

The drying air enters with a dew point of 15 °C. The exit gases (air + moisture picked up + CO_2) leave at 300 °F with a relative saturation of 5%. The $NaHCO_3$ is 70% solids, 30% H_2O when fed.

The Na_2CO_3 leaves with a moisture content of 3%, all $NaHCO_3$ being decomposed. To improve the character of the solids, 50% of the dry material is moistened to produce $Na_2CO_3 \cdot H_2O$ and recycled. Calculate, per ton of Na_2CO_3 product:

a. total cubic feet of air fed (DSC).
b. total volume of gases (at exit conditions) leaving the unit (assume $p = 760$ mm).
c. water fed to moisten recycle material.
d. weight of wet $NaHCO_3$ fed.

4.24 Acetylene is manufactured by treating calcium carbide with water:

$$CaC_2 + H_2O = CaO + C_2H_2.$$

The calcium oxide is treated with carbon in an electric arc to provide the carbide:

$$CaO + 3C = CaC_2 + CO.$$

Losses of calcium oxide are supplied from a lime kiln in which calcium carbonate is decomposed. The reaction goes to completion. The raw limestone ($CaCO_3$) is not pure but averages 92% carbonate with the remaining 8% inerts.

In the electric furnace, a 10% excess of CaO is used and the reaction is 94% complete. The acetylene reaction is substantially complete if a 25% excess of water is used.

Carbon monoxide formed in the furnace is to be used for the production of hydrogen by the water-gas reaction, or can be burnt as a fuel. The quantity produced is, therefore, of considerable importance.

The acetylene evolves saturated with water, because of the excess used, at 110 °F and 750 mm of mercury total pressure. Calculate:

a. raw limestone required per 10 tons of pure acetylene.
b. tons of CO per ton of acetylene.
c. tons of water vapor in the acetylene stream per ton of acetylene.
d. tons of water that must be removed from the recycle CaO per ton of pure CaO.

Section B—Problems Containing Some Unnecessary Data

4.101 A solution of caustic soda, 10% by weight, is concentrated in an evaporator, where 1000 lb of water is vaporized and removed. The boiling temperature is 140 °F. The final solution of caustic is 26.8%, and the density must be $1.226^{20/4}$. Calculate the weight of the feed.

4.102 The pigment, lithopone, is an equimolal mixture of zinc sulfide and barium sulfate. It is made by reacting solutions of zinc sulfate and barium sulfide. The products are so insoluble that the reaction is essentially complete. A 10% $ZnSO_4$ solution, specific gravity $= 1.107^{60/60}$, is mixed with a 15% BaS solution, specific gravity $= 1.170^{60/60}$. The resulting slurry of lithopone has a specific gravity of $1.24^{25/4}$. This slurry is filtered and washed, the water content of the cake being 25%. If 100 lb of dry lithopone is desired, calculate the volumes, in cubic feet, of the two reactant solutions and of the final slurry.

4.103 Pure CO_2 can be made by dissolving CO_2 from lime kiln, or flue, gases. The process can be economical only if the CO_2 content is appreciable. The solvent can be an alkali carbonate from which the CO_2 can be recovered by heating. The solvent may then be reused. A plant of this sort is fed a flue gas containing 16.2% CO_2, 4.1% O_2, and the remainder N_2. The CO_2 content of the effluent gases is 9.5%. The fraction of the CO_2 fed, which is recovered in the solvent, is calculated to be 0.47. Fifteen per cent of the absorbed CO_2 is lost in the stripping process (boiling from the solvent). Calculate:

a. fraction of CO_2 recovered in the absorber.
b. fraction of CO_2 recovered overall.

4.104 Limestone is essentially a mixture of the carbonates of magnesium and calcium, usually with some small amount of impurities. Upon calcining (heating to the decomposition point of the carbonates), both carbonates dissociate into the oxides and CO_2. When a certain limestone containing 100 lb of the carbonates is calcined, 44.8 lb of CO_2 is formed. If it is mixed with the combustion gases from the fuel that has provided the heat for the calcining action, the CO_2 is 39% in this gaseous mixture. Calculate the ratio of the carbonates in the limestone.

4.105 A high grade clay, containing 21% moisture must be dried to a moisture content of 2.5%. A countercurrent air drier is used (see Figure S4.1). Ten thousand cubic feet per hour of air, measured at 75 °F, 755 mm of mercury, and having a relative humidity of 10%, are heated to 400 °F and are then blown through the drier, emerging at 180 °F, 740 mm of mercury and 10% relative humidity. Calculate:

a. capacity of drier in tons of bone dry clay per day.
b. capacity of drier in tons of clay as produced per day.

4.106 A gas producer manufactures a low calorific industrial gas, which should theoretically consist of CO and N_2. The reactions, which take place consecutively, are:

$$C + O_2 = CO_2,$$
$$CO_2 + C = 2CO.$$

In units of this type the carbon source is coal or coke and air is the source of the oxygen.

In a special application, a lime kiln gas is to be mixed with the air fed to the producer so that the composition of the producer gas may be raised to 50:50 in CO and N_2. The air is saturated with water vapor at 15 °C (i.e., the dew point). The kiln gas analyzes 70% CO_2, 29% N_2, and 1% O_2. The fuel is a coke consisting of 90% C and 10% ash. The "ash," removed from the producer, contains unburned carbon amounting to 10% of the total ashy residue. Calculate the ratio of lime kiln gas to dry air.

4.107 A metallurgical plant separates a mixture of galena and quartz into two fractions by a wet flotation process. The original mixture contains 45% by volume of galena (dry basis), and is made up into a slurry containing only 5% solids. Ten tons of the rich fraction (95% by weight galena, dry basis) are obtained per day. The lean fraction contains 80% quartz by weight on a dry basis. Calculate the daily weight of feed.

4.108 A catalyst is prepared from a mixture of 40% solution of sodium silicate (Na_2SiO_3), a 26% solution of sulfuric acid, and a 20% suspension of aluminum oxide (Al_2O_3). The final product is intended to contain 89% SiO_2, 11% Al_2O_3. The gel is precipitated and washed to remove all soluble material, after which its water content is 85%. It is then dried. After drying, its analysis is 83.5% SiO_2, 10% Al_2O_3, 6.5% H_2O. Production is 50 tons per day of this material. The reactants are mixed in such proportions that there is no loss of either aluminum or silicate ion. Calculate, per ton of product:

a. weight of sodium silicate solution.
b. weight of water removed in the drier.

4.109 A solution of hydrogen chloride in benzene (5.24% by weight, 3.89% by volume) is extracted with an aqueous solution of sodium hydroxide (3.00% by weight, 1.418% by volume). Some of the hydrogen chloride dissolves in the water solution and reacts with all of the sodium hydroxide to form sodium chloride. No unreacted hydrogen chloride is found in the water solution. Benzene is not soluble in the water solution. The benzene solution contains, after extraction, some hydrogen chloride (4.45% by weight, 3.30% by volume). Feed is 5000 lb per hr of the benzene solution.
Calculate:

a. proportion of the hydrogen chloride extracted.
b. mole fraction of sodium chloride in the water solution after extraction.
c. average molecular weight of the benzene solution before extraction.

4.110 A benzene-air gas mixture passes through a 500-lb batch of mixed acid (51.8% H_2SO_4, 40.0% HNO_3, 8.20% H_2O), and its benzene content is completely consumed in the reaction

$$C_6H_6 + HNO_3 = C_6H_5NO_2 + H_2O.$$

The nitrobenzene layer is insoluble in the mixed acid, but the water is absorbed by the acid. The sulfuric acid content, after the reaction, is 58.1%. Calculate the fraction of the nitric acid consumed.

4.111 Hydrogen, measured at 150 °F, 37 in. of Hg, is burned with dry air. Calculate, on both wet and dry bases:

a. flue gas analysis when percentage of excess air is zero.
b. flue gas analysis when percentage of excess air is 50.

4.112 Methane (CH_4) is burned with dry air. The flue gas generates steam at 125 psig and leaves the flue at 400 °F. Calculate, on both wet and dry bases, assuming complete combustion of both C and H_2:

a. flue gas analysis when percentage of excess air is zero.
b. flue gas analysis when percentage of excess air is 50.

4.113 A gaseous fuel is being fired under a boiler. It is known to consist of hydrocarbons and nitrogen. The nitrogen is 36% of the fuel, the hydrocarbons making up the remaining 64%. The dry flue gases analyze: 11.20% CO_2, 4.00% O_2, the rest N_2. The volume of dry flue gas is 24 times the volume of the fuel gas fired, each corrected to standard conditions. Combustion is complete. Both fuel and air enter saturated with water vapor at 78 °F. Barometer is 756 mm Hg. Calculate:

a. the empirical formula of the hydrocarbons in the fuel.
b. percentage of excess air used.

4.114 From an oil-fired burner the Orsat flue gas analysis is: 11% CO_2, 0.3% CO, and 5% O_2. The air enters at 82 °F, 30 in. Hg, and 40% relative humidity. The oil contains nothing but carbon and hydrogen. The manufacturer specifies this oil as $C_{10}H_{22}$. Do you agree? Calculate percentage of excess air.

4.115 An oil-fired drier is using a fuel containing 84.7% C, 14.8% H, and 0.5% S, which is burned using 10% excess air. Secondary air is mixed with the flue gas and the resulting mixture is used in a drier which reduces the moisture content of clay from 15% to 2%. The flue gas-air mix enters the drier at 475 °F. A gas analysis made on the gas leaving the drier shows 9% CO_2 (dry basis). Calculate:

a. moles dry secondary air per mole of flue gas.
b. cubic feet of secondary air (DSC) per cubic feet of flue gas (SC).

4.116 To a dry water gas containing 47% H_2, 45% CO, and 8% N_2, hydrocarbons are added, the latter obtained by cracking an oil containing 92% C and 8% H (by weight) at a temperature averaging 1500 °F. The oil breaks down to form CH_4, C_2H_6, H_2, and a coke deposit considered to be 99.5% C and 0.5% H_2. The gas, after being mixed with the cracked hydrocarbons, is cooled to 75 °F, 30 in. Hg, and stored. It analyzes 10% CH_4, 17% C_2H_6, and 32.85% H_2. CO and N_2 are not analyzed after mixing. If the oil has a specific gravity of 0.94, calculate, per million cubic feet of original water gas, measured at 105 °F and 30 in. Hg:

a. gallons of oil used.
b. pounds of coke deposited.

NOTE Do not try to write equations for hydrocarbon decomposition. Make balances on carbon and hydrogen alone.

4.117 A gas producer is fired with coal of the following ultimate analysis: C, 66.3%; H, 3.5%; N_2, 1.5%; combined water, 13.5%; moisture, 8%; S, negligible; ash, 7.2%.

The producer gas contains: CO_2, 7.5%; CO, 24.0%; H_2, 16.5%; N_2, 50.2%; O_2, 0.6%; CH_4, 1.2%. Neglect tar and soot. The gas contains 3.43 grains of water per cubic foot of hot gas. Air supply is at 75 °F, and 29.75 in. Hg, and 40% relative saturation. Steam is mixed with the air fed to the producer. The ash contains: moisture, 1.1%; C, 3.1%; ash, 95.8%. The heating value of the coal is 12,200 Btu per lb. Calculate:

a. cubic feet of producer gas at 1070 °F and 29.65 in. Hg per pound of coal fired.
b. pounds of steam per cubic foot air used.

4.118 A coal of the following proximate analysis: moisture, 4.0%; volatile matter, 32.6%; fixed carbon, 53.1%; ash, 10.2% is burned in a furnace. The ash analyzes: volatile matter, 3.1%; fixed carbon, 16.0%; ash, 80.9%. The ash, after being wet down with water, contains 5.1% fixed carbon. The ultimate analysis of the coal is known only in part as follows: sulfur, 1.3%; carbon, 74.8%. Consider all the S as originally in the ash as pyrites, FeS_2. The Orsat analysis of the flue gas is: CO_2, 12.1%; CO, 0.2%; O_2, 7.2%; the remainder N_2. Assume no N in the coal. Air enters saturated at 60 °F and 29.7 in. Hg. Calculate:

a. volume of air and flue gas per 100 lb of coal charged.
b. percentage of excess air used.
c. complete ultimate analysis of the coal.

4.119 A plant is producing the decahydrate of Na_2SO_4. The process consists of dissolving the crude salt cake (Na_2SO_4), purifying the solution, and concentrating the purified solution in an evaporator. The concentrated liquor is sent to crystallizers, from which the crystallized product is obtained.

The crude salt cake is 94% pure. The original solution is 8.26% Na_2SO_4. Per 5 tons of crude salt cake, 1200 lb of water is added in the purifier and 0.3% of the original Na_2SO_4 is lost in this part of the process. The evaporator is operated under vacuum so that the boiling point of the solution is 60 °C and the concentrated solution contains 40 parts of Na_2SO_4/100 parts of water. The crystallizer is operated so that the final temperature is 52 °F. The decahydrate is separated from the mother liquor in a centrifuge. The mother liquor can be discarded or recycled. Solubility data are as follows:

60 °C: 45.3 parts Na_2SO_4 per 100 parts of water.
11.1 °C: 10 parts $Na_2SO_4 \cdot 10\,H_2O$ per 100 parts of water.

1. Calculate, on the basis of discarding the mother liquor:
   a. pounds of water evaporated per 5-ton charge.
   b. tons of decahydrate produced per 5-ton charge.
2. Repeat the calculations, on the basis of recycling the mother liquor.

4.120 A peat has the following composition: moisture, 88.0%; volatile combustible matter, 8.05%; fixed carbon, 3.18%; ash, 0.77%. If the material is dried until it contains only 10% water, it can be marketed locally as a domestic fuel at a price of $6.00 per ton. A process for drying is worked out at a temperature such that moisture only is removed, none of the volatile combustible matter being affected. The cost of handling the peat and of removing the water is estimated to be 2.5¢ per 100 lb of water removed. Calculate:

a. price per ton which can be paid for the wet peat as delivered at the drying plant in order to break even.
b. reduction in selling price which could be accomplished if the handling and water removal costs are reduced to 2.5¢ per 1000 lb of water removed.

4.121 The gas from a butanol fermenter contains CO_2 and H_2 in the ratio of 4:7. This gas leaves the fermenter at 30 °C, at atmospheric pressure and saturated with water vapor. This gas is to be fed to a methanol converter so that an adjustment in the ratio of CO_2 to H_2 from 4:7 to 1:3 is obtained. The adjustment may be accomplished by scrubbing out some of the CO_2. Water is the solvent, the tower operating at 25 atm. Calculate, on the basis of data and assumptions listed below:

a. pounds of water condensed in the cooler which follows the gas compressor and precedes the scrubber.
b. gallons of water fed to the scrubbers per hour.
c. size of the water main to the tower.

DATA AND ASSUMPTIONS

a. pressure of the system: 25 atm.
b. temperature of gas after compression: 400 °C.
c. temperature of gas after cooling: 50 °C.
d. scrubbing water enters at 20 °C and gas leaving the tower is at the same temperature.
e. concentration of CO_2 in effluent liquid from the tower: 1.0 lb per 40 gal.
f. velocity of water desired in feed main: 6.4 ft per sec.
g. solubility of H_2 in water: zero.
h. feed rate: 1,000,000 cu ft per day of gas, measured after the compressor and before the cooler.

Section C—Problems Unspecified as to Amount of Necessary Data Available

4.201 In a double effect evaporator, steam is used as the heating medium in the first effect. The water vaporized in that effect becomes the steam heating medium to the second effect. The water vaporized in the second effect goes to a condenser. The liquor to be concentrated may be routed through the two effects forward or backward. For the former, the feed goes to the first effect; after partial concentration there, the solution is then passed to the second effect, where the concentration is completed. Backward feed goes first to the second, and then to the first, effect.

In a double effect evaporator, with forward feed, the solution to be concentrated is 55,000 lb per hr of a solution containing 10% $CaCl_2$. The concentration desired is 50%. Steam pressure to the first effect is 25 psig and the vacuum on the condenser is 4 in. of mercury absolute.

Two-thirds of the evaporation is to be done in the first effect. Calculate:

a. pounds of water evaporated in each effect.
b. concentration of $CaCl_2$ to the second effect.

4.202 Chlorine gas containing 2.4% oxygen is flowing through an earthenware pipe in an electrolytic caustic plant. In order to measure the rate of flow, air is introduced into the pipe line at a known rate of 115 CFM, measured at the temperature and pressure of the gas in the pipe at the point of mixing. After mixing is complete, further down the line, a sample of gas is removed and analyzed. It contains 10.85% oxygen. Pressure at mixing: 780 mm Hg; at sampling: 772 mm Hg. Calculate the CFM of the original chlorine gas.

4.203 Distilled water is to be added to a sodium chloride solution to dilute it from 25% to 7%. All materials are at and will end at 20 °C. Calculate:

a. milliliters of each material to give a final volume of 125 ml.
b. same quantities by assuming volume additivity. Why is this method so nearly correct in this case?

4.204 Ethyl alcohol is available in a plant as an aqueous solution containing 58% by weight of alcohol. The density of this solution is 25.56° Bé (lighter than water scale). How many gallons of this solution must be added to a reaction batch in order to provide a charge containing 1000 lb of pure alcohol?

4.205 A batch of leather, upon emergence from an air drier, weighs 900 lb and contains 7% of moisture. During drying, the leather lost 59.1% of its original wet weight. The air to the drier enters at 786 mm Hg, 200 °F, and 5% relative humidity. Calculate:

a. pounds of water removed per pound of bone-dry leather.
b. percentage of bone-dry (moisture-free) leather in the original stock.
c. percentage of the original water removed in the drying process.

4.206 In a wet spinning process, cellulose acetate in acetone is forced through spinnerets to form fibers from which the solvent is evaporated into a stream of warm air. The acetone is recovered by scrubbing the gaseous mixture with water.

Ten thousand cubic feet per hour of mixture, 6% acetone, is fed to the scrubber. The acetone concentration is reduced to 0.5%. Twelve hundred pounds per hour of water act as the scrubbing agent. The tower temperature and pressure are essentially constant at 22 °C and 30 in. of mercury. Neglect loss of water as vapor in the outlet gas. Calculate the concentration of acetone solution produced.

4.207 Sea water contains 65 parts per million (by weight) of bromine in the form of bromides. In the Ethyl-Dow recovery process, 0.27 lb of 98% sulfuric acid is added per ton of water, together with the theoretical Cl_2 for oxidation, e.g.,

$$MgBr_2 + Cl_2 \longrightarrow MgCl_2 + Br_2.$$

Finally, ethylene, C_2H_4, is united with the bromine to form $C_2H_4Br_2$. Assuming complete recovery, calculate, per pound of bromine fed:

a. weights of acid, chlorine, and dibromide.
b. the volume of ethylene (cu ft at 75 °F and 765 mm).

4.208 A spent chromium-plating solution contains 62% H_2SO_4, 15% CrO_3, 23% H_2O. To fortify 100 lb, 20 lb of strong sulfuric acid and an unspecified amount of chromic acid containing 52% CrO_3 will be added. The fortified acid must contain 48% H_2SO_4, 24% CrO_3, and 28% H_2O. Calculate:

a. cubic feet of chromic acid solution required.
b. concentration of sulfuric acid.

4.209 A petroleum refinery produces a hot gas stream containing 13% CO_2, 18% CO, 20% H_2, 18% N_2, 16% H_2O, 11% CH_4, and 4% HF. As the gas passes at standard atmospheric pressure through a cooler it is cooled to 25 °C. Some water condenses and dissolves all the HF. The gas stream, saturated with water vapor, then passes through a spray tower in which all the CO_2 is scrubbed out by 5% Na_2CO_3 solution according to the equation:

$$Na_2CO_3 + H_2O + CO_2 = 2NaHCO_3.$$

No component of the gas stream other than CO_2 is absorbed in the solution. The volume of gas stream so treated is 25,000 cu ft (STP) per day and the Na_2CO_3 solution fed is 32,000 lb per day. Calculate:

a. composition of the gas stream after leaving the tower.
b. composition of the scrubbing solution leaving the tower.

4.210 Sulfur dioxide is scrubbed from a gas mixture having the analysis: 15% SO_2, 5% O_2, 1% CO, 79% N_2. The gas leaving the scrubber contains 0.5% SO_2. The scrubbing solution enters as 10.0% Na_2CO_3. The only reaction is:

$$Na_2CO_3 + H_2O + SO_2 = NaHSO_3 + NaHCO_3.$$

The effluent solution contains 10.0% $NaHCO_3$. Calculate the gallons per hour of scrubbing solution per 5000 cu ft per hour of gas mixture fed.

4.211 The grainer was, at one time, a common form of concentrator and crystallizer for manufacturing common salt from brine. Water was evaporated from brine although the latter did not boil. The following data were obtained in a performance test on such a unit:

Heating surface of the steam coils: 1700 sq ft.
Average temperature, entering brine: 120 °F.
Average temperature, brine and crystals in grainer: 180 °F.
Average steam temperature in coils: 225 °F.
Average temperature of condensate from coils: 220 °F.
Quality of entering steam: 100%.
Weight of condensate, 24 hr: 80,000 lb.

Pounds salt produced, plus adhering brine, 24 hr: 30,000 lb.
Percentage of water in wet salt: 21.0% (based on total yield of salt and brine).
Percentage of NaCl in entering brine: 26.0%.

Calculate:

a. brine fed per day.
b. water fed per day in the brine.
c. water evaporated per day.
d. dissolved salt in product, per cent.
e. dissolved salt in product, pounds per day.
f. crystallized salt in product, pounds per day.

4.212 Salt can be made from sea water by a process of solar evaporation. In the first stage of the process, all of the $CaCO_3$ and of the $CaSO_4$ is precipitated. The partially concentrated brine is then pumped to a crystallizing pond where crystalline NaCl is obtained on further evaporation. This process is allowed to proceed until 91% of the NaCl has been crystallized and the mother liquor has a density of 1.262 and a concentration of 523 gm of solid per liter of solution.

The original sea water has a solids content of 3.7% and a specific gravity of 1.025. On the basis of total solids, the various dissolved salts can be listed: NaCl 77.76, $MgCl_2$ 10.88, $MgSO_4$ 4.74, $CaSO_4$ 3.60, K_2SO_4 2.46, $CaCO_3$ 0.34, $MgBr_2$ 0.22.

Assuming that the losses of $MgSO_4$, $MgCl_2$, and $MgBr_2$ in the final crystallized NaCl are negligible, calculate:

a. total water evaporated per ton of brine.
b. concentration of the final mother liquor.

4.213 In a fermentation process the reactions are:

$$C_{12}H_{22}O_{11} + H_2O = 2C_6H_{12}O_6$$

and

$$C_6H_{12}O_6 = 2C_2H_5OH + 2CO_2.$$

These reactions occur successively and quantitatively. The CO_2 is evolved under such conditions that it carries appreciable quantities of water and alcohol vapors. These quantities may be figured to be 80% of saturation. The alcohol in these vapors is largely removed in a scrubber for return to the beer still of the distillation system handling the fermenter liquor.

The scrubber produces a 1.5% alcohol solution by weight, and the exit gases contain only 0.03% alcohol. Both the scrubber and the fermenter operate at essentially 1 atm.

The fermenter operates at 60 °F. The reactions may be assumed to be complete. The liquor produced is 12% alcohol by weight. The charge to the fermenters averages 4000 lb per hr. Calculate:

a. strength of liquor to the beer still after addition of the solution from the scrubber.
b. fraction of total output saved in the scrubber.
c. charge to beer still per 1000 lb of sugar.

DATA The average vapor pressures of water and alcohol may be assumed to be 15 and 22 mm of mercury, respectively.

4.214 Tannin can be made by extracting this naturally occurring substance from certain tree barks. A plant has been using mangrove bark with 30% tannin content. It pays 8¢ per lb of tannin for the bark. The tannin content of the extracted waste bark is 1.5%. The cost of the extraction process is 0.2¢ per lb of raw, unextracted bark handled.

This plant is now offered hemlock bark with 10% tannin content at a considerably reduced price. The waste bark content of tannin will not be affected, nor will the handling charges. The products from the two sources will be of the same grade and will command the same market price. Calculate the price the plant can afford to pay for the hemlock bark in order to break even.

4.215 The gas from an ammonia converter analyzes: 22.7% N_2, 68.1% H_2, 9.2% NH_3. The pressure in the system is 5000 psia. The temperature is reduced in a cooler, operating at constant pressure, until the vapor pressure of ammonia is 130 psia. Condensed ammonia is removed. The effluent gas from the condenser is recycled. Calculate:

a. fraction of the ammonia that will be removed in the cooler.
b. analysis of the gas leaving the cooler.

4.216 In an ammonia oxidation plant, the converter is fed a mixture of ammonia and air at 600 °C and 100 psig, in which the ammonia constitutes 10% of the mixture. Ninety-six per cent of the ammonia is converted to NO and the remainder decomposes, with the hydrogen being burned to water vapor. Calculate the analysis of the gases leaving the converter on both wet and dry bases.

4.217 The gas in Problem 4.216 passes to a cooler, where nitric acid, 40% concentration, is recovered. The gases leaving the cooler at 40 °C contain the same number of moles that were contained in the air supplied for combustion of the ammonia. Calculate the number of pounds of 40% acid made per 100 moles of ammonia charged to the converter.

4.218 The total recovery of nitric acid from the plant described in Problems 4.216 and 4.217 is 95% of that theoretically attainable, based on the ammonia charged. The 40% acid from the cooler is fed to the main absorption tower. The acid made in this tower is 63%. Calculate:

a. water fed to the absorption tower.
b. total yield of acid per 100 moles of ammonia charged to the converter.
c. percentage recovery of the NO fed to the tower.

4.219 Benzene vapor, from an air-benzene mixture, is being adsorbed on porous carbon. The gas entering contains 18% benzene *by weight*, and the gas leaving contains 1% benzene *by volume*. The total pressure of the system is substantially constant and the temperature is also constant, the values being 30.1 in. of mercury and 77 °F, respectively. Calculate the percentage of benzene *not* adsorbed.

4.220 Carbon is burned with air, 95% burning to CO_2 and the remainder to CO. Calculate the flue gas analysis when the percentage of excess air is 25.

4.221 The flue gas from a gas-fired furnace consists of 8.0% CO_2, 0.3% CO, and 6.5% O_2, by Orsat* analysis. The fuel consists solely of CH_4 and C_2H_6, and is fed at a rate of 1000 CFM (DSC). Calculate:

a. percentage of excess air used.
b. analysis of fuel.

A new operator is accustomed to coke-fired furnaces. When he gets the 8% CO_2 reading on the Orsat he wants to reduce the air to the burners in order to raise the CO_2. From your calculations and from knowledge of flue gas analyses resulting from combustion of pure carbon, explain why his view is incorrect.

4.222 Your company is burning a fuel known to consist entirely of N_2 and CH_4. The flue gas analysis shows 8% CO_2, 0.2% CO, 6% O_2, and the rest N_2. The flue gas temperature is 400 °F at the sample point. Calculate:

a. the analysis of the fuel.
b. percentage of excess air.
c. the ratio of fuel to air, both entering saturated at 20 °C and 760 mm.

4.223 A natural gas from Oregon containing 87% methane and 13% N_2 is used for heating a kiln, excess air being 25%. The gas fed to burners at 20 °C has a relative saturation of 80% and is under a pressure of 3 in. of water above atmospheric pressure. The air for combustion enters at a temperature of 25 °C and with a partial pressure of water of 20 mm. The gases entering the stack are at a temperature of 620 °F, and the stack draft is 2 in. of water measured at the base of the stack. Barometer stands at 30 in. of mercury. Calculate:

a. cubic feet of air per 100 cu ft of fuel gas.
b. cubic feet of flue gas per 100 cu ft fuel gas.

4.224 A furnace is fired with a fuel oil containing nothing but carbon, hydrogen, and 3% ash. There is no unburnt combustible. The flue gas analysis is reported as: 10.7% CO_2; 7.5% O_2; nitrogen constitutes the remainder. Calculate:

a. empirical formula of the original oil, ash-free.
b. moles of dry flue gas at 1200 °F, 740 mm of mercury pressure, per pound of oil fired.

4.225 A fuel which is known to consist of methane (CH_4), ethane (C_2H_6), and nitrogen is burned with 20% excess air. The flue gas analyzes: 8.40% CO_2, 1.20% CO, 4.20% O_2, and the remainder nitrogen. Calculate the analysis of the fuel gas.

• • • • •

* An Orsat apparatus provides analyses on a dry basis.

(The reader will note that, since the composition of the fuel gas is unknown, the percentage of excess cannot be known a priori. The problem is, therefore, an exercise in the use of material balances, but is not a practical industrial problem.)

4.226 A flue gas analyzes (by Orsat):* 7.0% CO_2, 1.0% CO, 7.0% O_2. Two fuel gases are mixed in the furnace producing this flue gas:

A	*B*
80% CH_4	60% CH_4
20% N_2	20% C_2H_6
	20% N_2

Calculate:

a. proportion in which the two gases are mixed.
b. percentage of excess air used.

4.227 A boiler furnace is fired with a fuel oil whose empirical formula is $(CH_2)_n$. An Orsat analysis (dry basis) of the stack gas shows 7% O_2 and no CO. Calculate the percentage of excess air used.

4.228 An oil-fired furnace is fired with a fuel oil containing 87.3% C, and 12.7% H. The heating value of this oil is 18,000 Btu per lb. Combustion is complete, 25% excess air being used. Dry, saturated steam at 100 lb gauge pressure is used for atomization of the oil. This steam amounts to $\frac{1}{4}$% of the weight of the oil. The enthalpy of steam above 32 °F at 100 lb gauge is 1189 Btu per lb. Air for combustion enters at 20 °C, 60% relative humidity, and a total pressure of 30.2 in. Hg. The flue gases leave the stack at a temperature of 520 °F and the barometer is 753 mm. Calculate:

a. moles of air per pound of oil fired.
b. cubic feet air per pound of oil fired.
c. moles of flue gas per pound of oil fired.
d. cubic feet flue gas per pound of oil.

4.229 The ultimate analysis of a coal indicates 74.46% carbon, 11.0% oxygen, and negligible sulfur and nitrogen. The "proximate analysis" indicates 2.98% moisture, 33.62% volatile combustible matter (VCM), 56.16% fixed carbon (FC), and 7.24% ash. (These fractions are obtained by heating at progressively higher temperatures, in air. Therefore, they do not represent anything but an empirical characterization. They are useful in relating the fuel and the refuse.) The refuse removed from the ash pit has an average analysis of 10.0% FC, 6.0% VCM, and 84.0% ash. The

• • • • •

* See footnote to Problem 4.221.

fuel has a heating value of 13,840 Btu per lb. Thirty per cent excess air is used and the flue gas contains 2.0% CO. Calculate:

a. Orsat* analysis of the flue gas.
b. percentage of heat lost due to unburned combustible.

4.230 A #6 fuel oil is burned in a fire-tube boiler. The oil, whose organic components are solely carbon and hydrogen, must be heated to 160 °F to exceed its flash point and allow combustion. The dry flue gases from this boiler analyze: 12% CO_2, 1% CO, and 6% O_2.

The CO may be formed in the combustion but is more likely to be obtained by reduction of CO_2 in contact with carbon deposited on the tubes of the boiler. Calculate the composition of the oil on the basis of each of the following assumptions:

a. no interaction exists between combustion products and deposited carbon.
b. the CO is formed only from contact with carbon on the tubes and no further decomposition occurs.
c. the CO is formed by reaction with carbon and 10% of the oil is decomposed.

Also calculate the percentage of excess air in each case.

4.231 A boiler is fired with a coal containing 79.90% carbon and 6.50% ash. Sulfur, nitrogen, and moisture are negligible. The part of the coal which is not analyzed is hydrogen and oxygen. The portion of the hydrogen that is just sufficient to react with all the oxygen will be so assigned and these two materials will be reported as "combined water." The remaining hydrogen, which requires oxygen from the air during combustion, can be calculated from the flue gas analysis. This hydrogen will be reported as "net hydrogen."

Air enters the combustion zone at 68 °F, dry. Combustible left in the refuse is negligible. Flue gases go to the stack at 700 °F and 750 mm pressure.

The average flue gas analysis reports: 13.0% CO_2, 6.4% O_2, and no CO. Calculate:

a. percentage of net hydrogen in the coal.
b. complete analysis (known as ultimate analysis) of the coal.
c. percentage of excess air used.
d. volume of flue gases going to the stack per pound of coal charged.

4.232 A boiler is fired with a coal containing 78.0% C and 9.0% ash. Elimination of combustible matter from the ash is complete. Sulfur and nitrogen in the coal are negligible. The average Orsat flue gas analysis reports:

• • • • •

* See footnote to Problem 4.221.

12.9% CO_2, 6.0% O_2, and 1.5% CO. Air enters at 88 °F and 29.6 in. Hg, with a relative humidity of 29%. Calculate:

*a. weight ratio of net hydrogen to carbon and percentage of net hydrogen.
b. percentage of excess air.
*c. ultimate analysis.

4.233 A bituminous coal is burned in a furnace with air at 85 °F, 90% saturated, 29.2 in. Hg. The furnace gases leave at 572 °F. The ash leaves the furnace moisture free at 520 °F, and when analyzed contains 22.3% moisture, 12.3% volatile matter, and 41.1% fixed carbon. The proximate analysis of the coal is: fixed carbon 56.34%, volatile matter 37.75%, moisture 2.97%, ash 2.94%.

A partial ultimate analysis on the ash-free and moisture-free basis is: carbon 84.43%, nitrogen 2.00%, sulfur 0.82%. The Orsat analysis of the flue gas is: CO_2, 12.0%; CO, 1.2%; O_2, 6.2%; and the rest nitrogen. Calculate:

a. the net hydrogen content of the coal, neglecting the hydrogen remaining in the ash and neglecting combustion of S.
b. the complete ultimate analysis of the coal.
c. the percentage of excess air used, based on the total combustible charged.

4.234 The process of dry distillation of coal to produce coal gas and coke is known as coking. In a specific instance a plant is coking a coal containing 32% volatile matter and obtaining a coke containing 4.2% volatile matter. We assume that the composition of the volatile matter in the coke is the same as in the original coal. The coal gas analyzes: CO_2, 2.2%; unsaturates, 4.0%; O_2, 0.8%; H_2, 46.5%; CO, 6.3%; CH_4, 32.1%; and nitrogen, 8.1%. This gas has a heating value of 570 Btu per cu ft at DSC, and an average molecular weight of 21.20. One thousand tons of coal is charged to the ovens per day.

The coal costs $3.50 per ton delivered to the plant. The gas sells for an average of 60¢ per 1000 cu ft (DSC). The total charges against the process are $9000 per day. Value of by-products is negligible. Assume that there is no air leakage into the coke ovens (i.e., that all the gas comes from volatile matter distilled from the coal). Calculate:

a. the lowest price for which the coke may be sold per ton in order for the plant to break even.
b. the same if the price of the coal decreases by 20%.

4.235 A producer gas has the following analysis: CO_2, 4.1%; CH_4, 3.2%; C_2H_4, 2.9%; O_2, 0.2%; CO, 27.4%; H_2, 12.6%; and N_2, 49.6%. There is

• • • • •

* Defined in Problem 4.231.

no loss of combustible due to tar or unburned combustible in the refuse. The coal used in firing the producer contained 73.8% C and no N. Calculate:

a. moles of dry producer gas per pound of coal.
b. moles of dry air per pound of coal.
c. weight ratio of carbon to net hydrogen in the coal.
d. moles of water (both combined water and steam) decomposed per 100 moles dry producer gas.
e. moles of water (both combined water and steam) decomposed per 100 lb coal charged.

4.236 A natural gas is cracked at high temperature to produce a degraded gas and some coke (pure carbon) residue. The compositions of the two gases involved are as follows:

Composition	Natural Gas %	Cracked Gas %
CH_4	82.3	23.8
C_2H_6	8.3	2.2
C_2H_2	—	2.3
C_2H_4	—	6.5
C_3H_6	—	1.4
C_3H_8	3.7	—
H_2	—	60.2
N_2	5.7	3.6

Calculate:

a. increase in gaseous volume (DSC) per 100 standard cu ft of natural gas.
b. quantity of coke produced per 100 cu ft (STP) of natural gas.

4.237 In an experimental plasma jet project, a coal in a stream of argon is fed to the unit. The coal contains 1.1% moisture and 7.5% ash. On an ash-and-moisture-free basis, the coal contains:

C—81.9%
H— 4.9%
O—10.2%
S— 1.5%
N— 1.5%

This stream is mixed with the plasma stream, also pure argon, at a temperature of 8000 °C.

The products are analyzed and reported by weight on a moisture-free basis:

H_2 — 3.9%
CH_4 — 0.6%
C_2H_2—15.4%
Diacetylene—Trace
CO —24.3%
Solids—45.3%

Calculate:

a. analysis of the solids in the product.
b. comparison of these solids to the original coal to see whether these solids represent unchanged coal.

4.238 Pyrites (FeS_2) is burned in air under such conditions that the iron oxide formed is Fe_3O_4 and all the sulfur burns to SO_2. Calculate the flue gas analysis when percentage of excess air is 50.

4.239 In a contact sulfuric acid plant, sulfur is burned in air to SO_2, then oxidized catalytically to SO_3, and the SO_3 absorbed in dilute sulfuric acid.

The sulfur is burned with 25% excess air (based on conversion to SO_2 only), no greater excess being desired in order to maintain a high temperature in the burner. Then secondary air is added to provide an 80% excess in the converters (based on SO_2 to SO_3). It is reasonable to assume that no SO_3 is formed in the burners. Calculate:

a. composition of gas leaving the burners.
b. composition of gas entering the converters.

4.240 Zinc blende (mineral name for ZnS) is roasted, in air, to provide ZnO and SO_2. A particular blende contains: Zn, 55.3%; As, 1.82%; Fe, 5.54%; Pb, 0.64%; S, 34.7%; and the remainder unreactive gangue. All metallics are in the forms of their normal sulfides, iron being in the form of the natural pyrites. In the roasting, all metallics form their normal oxides, iron going predominantly to Fe_2O_3. The arsenic oxide volatilizes.

Essentially no sulfur is left in the cinder in this roasting. The gases analyze (As-free and SO_3-free): SO_2, 12.00%; O_2, 4.34%; N_2, 83.66%. Secondary air will be added to provide 130% excess for the oxidation of SO_2 to SO_3. Calculate:

a. whether any SO_3 was formed in the roaster.
b. amount of secondary air supplied, per 100 lb of blende.
c. analysis of gas to converters.

4.241 In the combustion of pyrites (FeS_2), either ferrous or ferric oxide may be formed. Since ferric oxide is a mild catalyst for the oxidation of SO_2 to SO_3, a small quantity of the trioxide may be expected in the gas from the roaster. However, it will not be reported in the usual analysis, which is then on an SO_3-free basis.

Flue gases from one such roaster contain 8.8 mole % SO_2, 9.5 mole % oxygen, and the remainder nitrogen. Four per cent of the sulfur will be converted to SO_3.
Calculate:

a. ratio in which the two combustion reactions will take place.
b. excess air used, assuming complete combustion of the pyrites by the combination of the two reactions.
c. tons of SO_2 formed per ton of pyrites charged.
d. tons of cinder formed per ton of pyrites charged.

4.242 Sulfur and atmospheric air are fed to a Glens-Falls sulfur burner. The air is at a pressure of 753 mm of Hg and the water vapor pressure is 13 mm of Hg. The combustion gases contain no SO_3 and 2% oxygen. Sulfur vapors, in the form of S_4, are also present, with a partial pressure of 70 mm of Hg. There is no appreciable change in total pressure over the unit.

After leaving the burner, the combustion gases go to a secondary combustion chamber to which secondary air is added in amount necessary to burn all the sulfur vapor and to provide sufficient oxygen for the subsequent oxidation of SO_2 to SO_3. Calculate:

a. moles of primary air per mole of sulfur fed to the Glens-Falls burner.
b. moles of secondary air, if the exact requirement is to be met.
c. moles of secondary air if the gas from the secondary combustion chamber is to contain 8% SO_2, on a dry basis, and if no SO_3 has been formed at this point.

4.243 FLOW SHEET

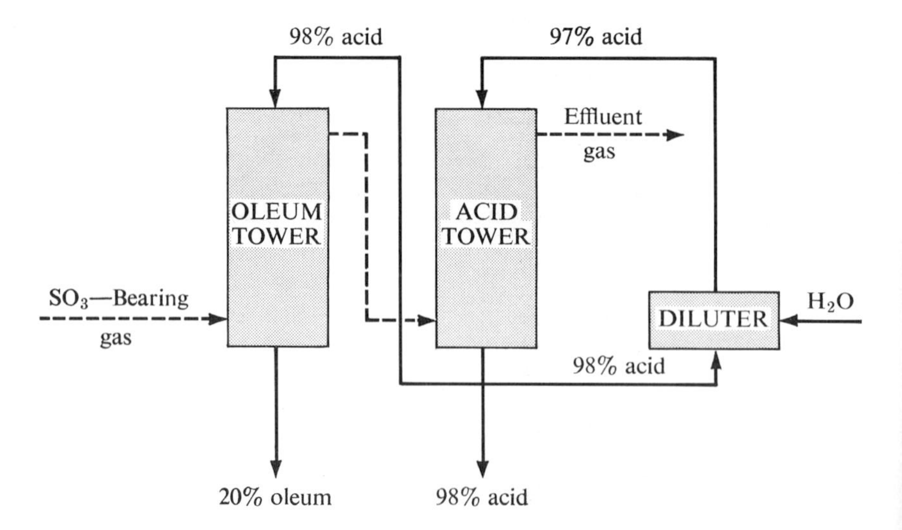

4.243 A contact sulfuric acid plant produces 10 tons of 20% oleum* and 40 tons of 98% acid per day (see flow sheet, opposite page). The oleum tower is fed with a portion of the acid from the 98% tower. The 98% tower is fed with 97% acid, obtained by diluting part of the 98% output. The gas fed to the oleum tower analyzes 10.8% SO_3. Calculate the amount of acid to be fed to each tower per day.

4.244 A performance test is run on the Glover tower in a lead chamber sulfuric acid plant. The quantities of gas entering the bottom of the tower and of the Gay-Lussac acid fed at the top are measured during the test period. These quantities are reported below, along with an analysis of the gas entering, analyses of the chamber acid, the nitric acid, and the Gay-Lussac acid fed to the top of the Glover tower, and an analysis of the gas leaving the top of the Glover tower. The concentration of sulfuric acid leaving the bottom of the tower is 78% by weight.

	Input				Output
	Gas	HNO_3	H_2SO_4	Gay-Lussac Acid	Gas
SO_2	8.36	—	—	—	6.44
O_2	9.86	—	—	—	8.75
N_2	80.10	—	—	—	74.59
N_2O_3	—	—	—	0.885	—
NO	—	—	—	—	0.77
H_2O	1.68	36	37	22.1	9.45
H_2SO_4	—	—	63	77.0	—
HNO_3	—	64	—	—	—
Pounds	540	—	—	580	—

Calculate:

a. percentage of SO_2 fed to the tower that is converted to sulfuric acid.
b. the loss of combined nitrogen reported as pounds of HNO_3 (100%).
c. weight of nitric acid fed to the tower.
d. weight of 78% acid leaving the bottom of the tower.
e. weight of chamber acid fed to the top of the tower.

4.245 A chamber sulfuric acid plant (see flow sheet) is burning sulfur averaging 96.4% purity, the impurity being inert. The gases leaving the burner contain 10.1% SO_2 and, from averages taken over a long period of time, it is known that 4% of all the sulfur burned is converted to SO_3 in the burner.

The gases leaving the Glover tower contain 7.1% SO_2, 11.5% O_2, and 81.4% N_2.

• • • • •

* Defined in Problem 4.14.

FLOW SHEET

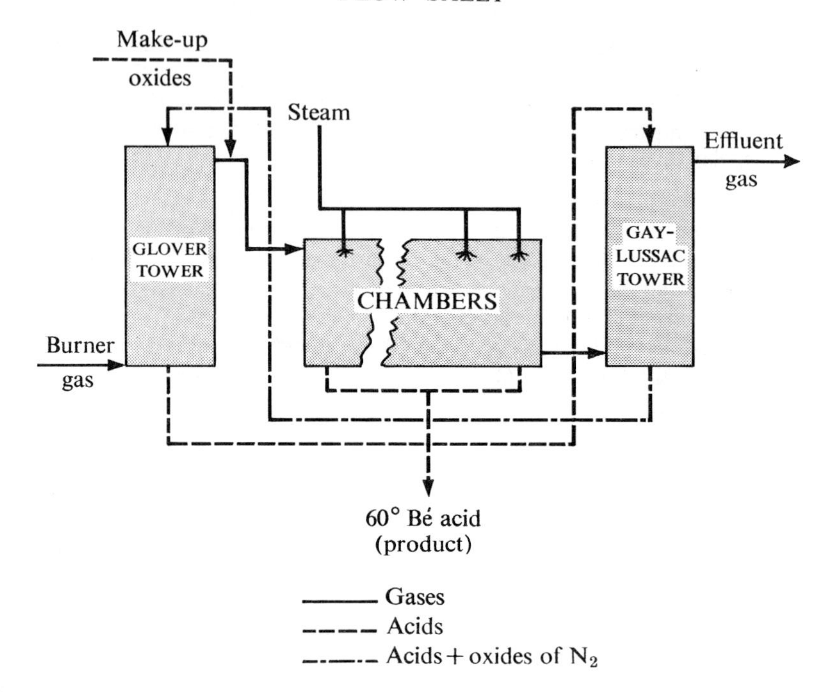

The gases leaving the last chamber contain 0.3% SO_2, 10.1% O_2, and 89.6% N_2. All analyses are on an oxides-of-nitrogen-free basis.

Since the material balances do not check over the Glover tower and over the chambers, and since the process is being run under slightly less than atmospheric pressure, it is concluded that air is leaking into each of the units. Per 100 lb of sulfur fired, calculate:

a. moles of air supplied to the burner.
b. moles of air leaking into the Glover tower and into the chambers.
c. percentage of total SO_3 which is made in the burner, in the Glover tower, and in the chambers.

4.246 You purchase a product from a manufacturer at a price of 15.2¢ per pound provided that the water content does not exceed 5%. Above 5% there is an adjustment in the price so that your net cost per bone-dry pound is reduced 1% for each per cent increase of water content (or fraction thereof).

You receive a shipment of 15,300 lb total weight and analyze the water content to average 8.1%.

The freight rate is $2.25 per 100 lb, paid by the customer. Calculate the adjusted price for which you should be billed.

4.247 A two-column unit is used to purify a hydrogen stream of the small amount of an impurity consisting entirely of H_2S. As shown in the flow

FLOW SHEET

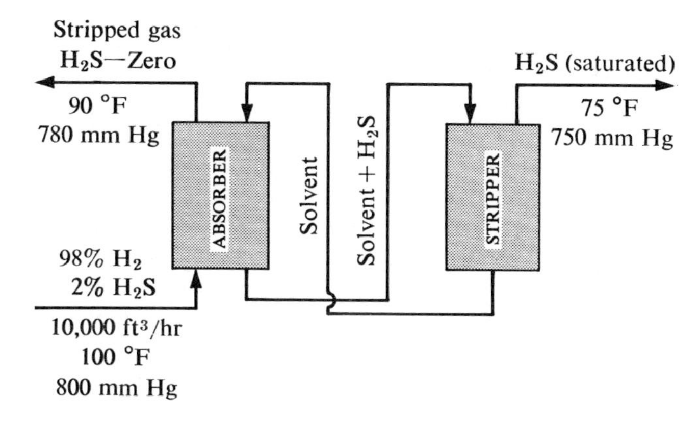

sheet, the first column acts as an absorber for the H_2S, while the second strips this dissolved gas from the solvent stream. The solvent is then recycled to the absorber. Calculate:

a. the flow rate of H_2S leaving the stripper.
b. the flow rate of absorbing solvent to the absorber.

0.1% H_2S in solvent to absorber; 0.5% H_2S in solvent to stripper

4.248 In a purification process crude sulfur is treated with liquid carbon disulfide which dissolves the sulfur and leaves the impurities as an insoluble residue. After separation, the carbon disulfide is evaporated into a stream of air and the sulfur crystallizes.

Since carbon disulfide is an expensive solvent, the success of the process will depend upon the effectiveness of its recovery. In the first stage of the recovery, the stream of air and carbon disulfide vapor will be cooled, and then passed to a condenser cooler. The temperature of the cooler can easily be maintained at 50 °F. The allowable loss of carbon disulfide, in the air stream leaving the condenser, is 0.005%. If the gas stream from the evaporator contains 20% carbon disulfide, calculate:

a. the pressure required in the compressor.
b. percentage recovery of the disulfide.

4.249 Methanol may be catalytically oxidized to formaldehyde, the reaction producing a mole of water per mole of formaldehyde (CH_2O). Since formaldehyde will also oxidize to formic acid ($HCOOH$), a certain amount of the formaldehyde is lost in the secondary reaction. In such a process, the effluent gases from catalytic oxidation contain:

Substance	Per Cent
CH_2O	6.6
$HCOOH$	0.4
CH_3OH	13.9
O_2	13.8
N_2	65.3

Calculate:

a. percentage loss of methanol as formic acid.
b. percentage conversion of methanol.
c. percentage of excess air supplied.

4.250 A fuel oil containing 0.5% S, 87.2% carbon, 12.3% hydrogen is burned completely with 15% excess air. This gas is then cooled to 225 °F and introduced into a dryer in which the moisture content of a cotton fabric is being reduced from 40% to 5.5%. Of the gas entering the dryer, 40% is recirculated and the remainder is exhausted to the atmosphere. The vapor pressure of water in the exhaust gas is 140 mm and the whole system operates under a pressure of 760 mm. Calculate the number of pounds of water evaporated per pound of oil burned.

4.251 In an atmospheric process for the catalytic hydration of ethylene to ethyl alcohol, only a fraction of the ethylene is converted. As in other processes of this type, the product is condensed and removed after each pass through the converter and the unconverted gases are recycled. The condenser may be assumed to remove all of the alcohol and the recycle gases will be saturated with water vapor at 100 °F. The conversion of ethylene per pass through the converter is 4.5%. The molar ratio of water to ethylene in the feed to the converter, after mixing the recycle gas with fresh feed, is 0.55. Calculate:

a. moles of recycle per mole of fresh feed.
b. analysis of fresh feed.
c. analysis of ethyl alcohol product.

4.252 In an ammonia synthesis plant the high pressure converter operates at 1000 atm and 1000 °F. A stoichiometric mixture of nitrogen and hydrogen is carefully maintained since only a fraction of the feed is converted to ammonia and the remainder is recycled. Failure to attain the true stoichiometric mixture would result in a build-up of one of the reactants in the recycle stream.

The process for preparing the mixture frequently involves partial oxidation of fuel oil to obtain a stream of CO and H_2. The CO is converted to CO_2 and an accompanying mole of hydrogen. The nitrogen is obtained from the air by exhausting the oxygen in the reaction with fuel oil. The main impurities, CO_2 and H_2O, can easily be removed by absorption and condensation, respectively. Small amounts of methane exist in the mixture. Being more difficult to remove, they end up in the reactor so that a typical feed analysis might be: 74.0% H_2; 24.5% N_2; 1.2% CH_4; and 0.3% argon. The argon, which in combustion processes acts exactly the same as nitrogen, now is an important impurity in this process.

In the manufacture of the ammonia, 65% of the entering mixture is converted to ammonia, and 75% of the ammonia formed is recovered from the cooler as liquid ammonia. A fraction of the recycle gas is purged to prevent build-up of methane and argon.

The stream to the converter, after mixing fresh feed with recycle, is to be controlled so as to contain a maximum of 18% methane. Calculate:

a. fraction of the unconverted gas which is purged.
b. percentage loss of converted ammonia in the purge gas.
c. fraction of the reactive feed gases recovered as liquid ammonia.

4.253 A plant manufacturing chemical caustic uses a continuous, three-stage Dorr leaching system, operating as shown in the accompanying figure. The reactants are slaked lime, $Ca(OH)_2$, 95% pure, and soda ash, Na_2CO_3, 98% pure. Impurities may be considered to be inert to the process and totally insoluble. The caustic solution is to be 8% NaOH. Production amounts to 100 tons per day of the solution.

FLOW SHEET

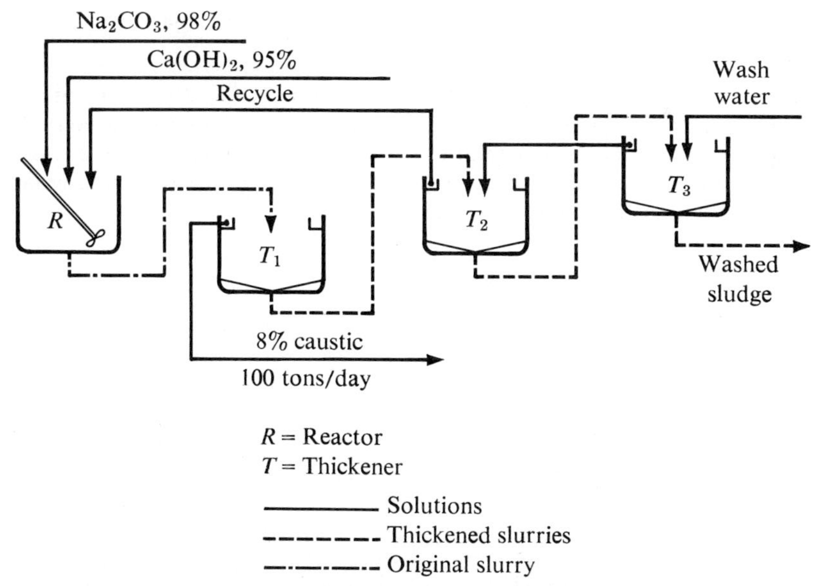

R = Reactor
T = Thickener

——— Solutions
- - - - Thickened slurries
—·—·— Original slurry

The sludge leaves each thickener as a mixture of insoluble solids and caustic solution, the latter at the concentration established in the individual thickener. Although the ratio of solids to liquid undoubtedly varies with concentration, the pounds of water per pound of solids may be taken as 2:1 as a first approximation. Calculate:

a. NaOH lost from the third thickener.
b. saving resulting from use of the third thickener.

DISCUSSION If an exact solution is attempted, the simultaneous equations will not all be first order. To avoid the difficulties inherent in this situation, try an iterative approach. First, assume that the amount of reactants is based upon the production of NaOH, disregarding the amounts required to provide the loss. Then correct the amounts of reactant by the first estimates of the loss. Repeat this procedure until the error in the last iteration is less than 0.1%, in terms of the estimated loss.

4.254 A plant produces phosphoric acid by treating phosphate rock with sulfuric acid. The rock charged amounts to 120 tons per day and the acid produced analyzes 22% P_2O_5. The rock contains 75% $Ca_3(PO_4)_2$, with the remainder inert to the reaction. Ninety-five per cent of the phosphate is converted and 97% of that converted is recovered in the washing and leaching of the inert solids. The sulfuric acid used is 60° Bé and the amount is 120% of that theoretically required. Calculate:

a. tons of rock treated per day.
b. tons of sulfuric charged per day.
c. tons of wash water used per day, if leached solids contain a pound of water per pound of waste solids.
d. concentration of phosphoric and sulfuric acids in product.
e. tons of water to be evaporated to produce a final acid containing 45% H_3PO_4.

4.255 Phosphate rock is comprised of 75% $Ca_3(PO_4)_2$, 20% SiO_2, and 5% inerts. The phosphate reacts with sand (SiO_2) and coke (carbon) in an electric furnace. Phosphorus vapor, CO_2, and CO are the gaseous products. A slag is formed, comprised of $CaSiO_3$, the inerts, unburnt carbon, and unconsumed sand.

DATA

	ΔH_{f298} kcal/mole	B. Pt. °C	Fusion °C	Approx. C_p, T in °K Pcu/(mole)(°C)*
CO_2	−94.052			$18.04 + 0.0000447T$
CO	−26.416			$6.89 + 0.001436T$
P	0 (solid)	280	44.1	Solid: 5.5(0–44 °C) Liquid: 2.8 (44–280 °C) Gas: 2.0
SiO_2	−203.4		2230	$10.90 + 0.0055T$
$Ca_3(PO_4)_2$	−988.9		975	
Inerts			2000	0.25 Btu/(lb)(°F)
C	0		Decom-position	$2.673 + 0.00262T$
$CaSiO_3$	−378.6		1540	$27.95 + 0.00206T$

Latent heat of vaporization of phosphorus—11.88 kcal/gm-mole
Latent heat of fusion of phosphorus — 6.615 kcal/gm-mole
Molecular Weights:
P—31; Ca—40; Si—28; C—12; O—16.

* All C_p expressions have been truncated to simplify the calculations.

Quantities charged are figured as follows:

a. sand should be added in sufficient quantity to provide a 5% excess over the stoichiometric supply for the phosphate.
b. coke is added in 30% excess over the quantity required if all were to burn to CO_2; CO_2:CO in the gaseous products is in ratio of 9:1.
c. twenty per cent of the rock is unreacted and recycles to the furnace.

All feeds are at one atmosphere and 25 °C. All products leave at one atmosphere, the gas at 300 °C and the slag at 1300 °C. Other data are provided above. Calculate quantities and analyses of all streams, per ton of rock charged.

SUPPLEMENT TO CHAPTER 4

Degrees of Freedom in Process Specification

The number of items of data that the "designer" may specify are termed degrees of freedom. They amount in number to the difference between the number of variables and the number of restrictions. This concept is entirely rational. The variables represent the total possible number of items of data; i.e., the total potential unknowns. The number of restrictions is the total number of formal relationships. There is now some number of "unknowns," say N_v, and some number of equations (relationships), say N_r. If $N_v - N_r > 0$, the problem is insoluble without $N_v - N_r$ additional items of information. If the "designer" specifies these $N_v - N_r$ conditions, the number of unknowns and the number of equations is identical and a unique solution is available. Thus, the $N_v - N_r$ specifications available to the designer are termed degrees of freedom.

S4.1 Definition of Terms

Streams

Any line entering or leaving a process will be designated a "stream," the number of such streams being represented symbolically as N_s.

Variables

All items of information required to designate a stream completely will be called variables. In general, the temperature, pressure, flow rate, and complete analysis will be required. For complete analysis of a C-component mixture, $C - 1$ percentages are required, the Cth not being independent. Thus, every stream can be specified with $C + 2$ items of data; that is, the variables per stream are $C + 2$. It should be noted that while flow rate applies to a continuous process only, the quantity of material charged or removed would be an equivalent requirement in a batch process. For complete specification of a stream, its phase must be known, but this requirement does not add a variable since the phase is not independent of temperature, pressure, and concentration.

Restrictions

Any mathematical relationship imposed on the process constitutes a restriction. For example, the existence of a process indicates the applicability of material balances, C in number for a C-component system. Other types of restrictions are discussed in Section S4.2.

Elements and Units

Many chemical processes are quite complicated, containing many elements such as condensers, heaters, stream splitters, mixers, reactors, distillation plates, and many others. These elements can be combined into units, such as the distillation column, consisting of heater, plates, condenser(s), and splitter. The units can then be combined into processes. The degrees of freedom can be calculated for any part of the hierarchy. It is frequently easier to establish them for the smaller units, or elements, and then to combine them for the larger combinations of equipment. This procedure will be discussed and illustrated in Sections S4.3, S4.4, S4.5, and S4.7.

S4.2 Method

The essentials of the method are very simple. They can be summarized in the following simple terms. Count the variables and the restrictions, subtract the latter from the former, and establish thereby the number of degrees of freedom. Details of this procedure will be discussed after a few words regarding the conclusions, which are drawn from the calculated number of degrees of freedom.

The use of the degrees of freedom will vary, depending upon the orientation of one's interests, either toward design or toward evaluation of performance. If it is design, the designer knows that *he* must specify a number of items of data equal to the degrees of freedom if a unique solution is to be found. He may distribute the degrees of freedom among the variables as he sees fit, presumably to give him the easiest solution. Ordinarily he will start by specifying completely all feed streams, since it is difficult to conceive of a large practical problem where one does not know the composition, conditions, and quantity of material to be reacted or separated. He will then probably choose to designate the fraction converted, for chemical reaction, or the purity of a product (i.e., the separation desired) in a physical process. He will distribute the other degrees of freedom as he sees fit to effect a solution. Any of his specifications may be arbitrary (e.g., reflux ratio in distillation). The design will vary if such an arbitrary specification is changed. The cost of the unit will also change. Therefore, when such arbitrary choices are involved, a series of designs is mandatory to obtain the most economical unit for the job assigned. This iterative procedure is common in practice but rarely encountered in all its ramifications by the student since the time consumed in repetitive calculations is difficult to justify in competition with that needed for the accumulation of new knowledge.

When interest is centered on the evaluation of performance, the engineer will proceed differently. He will compare the degrees of freedom against the numbers of items of data that are known. If the numbers are identical, he has a unique solution. If the number of degrees of freedom exceeds the numbers of items of data, he cannot effect a complete solution although he may be able to calculate a portion of the stream quantities and compositions. He would then investigate subdivisions of the process to find out whether certain small units or elements may be completely specified. In this fashion, completion of the performance data through material balance calculations on the elements might provide the missing items of information necessary to evaluate the process as a whole. Lastly, if the number of degrees of freedom is less than the number of items of data, there is no unique solution for any missing items. Any conclusions from calculations performed with the over-specified process should involve statistical analysis; interpretations of such calculations rely heavily on the judgment of the engineer making the report.

We now turn to the detailed technique for applying this method.

Components

The number of components present in any system, chemical or physical, has been discussed in Section 4.4. The reader is referred to that section for a clarification of the number of components in any situation, chemical or physical.

Counting the Variables

The potential variables are in three categories: those pertaining to input and output masses; those pertaining to input and output energies; and those pertaining to repetitious elements. We shall consider these in the order named.

In any stream supplying or removing mass, complete specification of the stream requires $C + 2$ items of information, as indicated in Section S4.1. Now, in the physical system, the number of components and the number of species are the same in each stream, but each stream does not necessarily contain all of the species present in all other streams. We shall, however, make the decision to count all the species in the *system*, designate this number as N_{sp} and state that the potential number of variables depending on mass, in a system that contains a number of species, N_{sp}, is

$$N_v = N_s(N_{sp} + 2). \tag{S4.1}$$

In this expression N_v is the number of potential variables and N_s is the number of streams. Where all species are not present in all streams, we shall recognize this fact by specifying that these species have zero concentration and including this information among other types of restrictions upon the system.

If physical systems only were to be considered, we could write C in place of N_{sp}, where C is the number of components. However, in chemical systems we have seen that these two quantities are not the same. Definition of N_v in terms of N_{sp} rather than in terms of C will allow general use of Equation S4.1 in both types of system.

Input and output energy streams have no characteristics related to composition, temperature, or pressure. They are completely specified by a designation of the quantity of energy, in appropriate units. Each energy stream adds, then, exactly one variable. Prior to a consideration of the application of energy balances in Chapter 9, energy streams will not be involved in the systems.

The repetitious use of physically identical elements of equipment suggests an interest in, and concern with, the design of the internal aspects of a unit (e.g., the plates in a distillation column). The subject of staged operations is so concerned. Material and energy balances are not, their main concern being with the overall changes effected in such units. For the present, therefore, we have no use for the variables denoting such repetitious use of elements. The variables are frequently termed "redundancy variables." They exist solely to provide the designer with an opportunity to choose the number of such repetitious elements and they therefore number one for all the elements of a single type. We shall not deal with this type of redundancy variable.

Counting the Restrictions

This procedure is more complicated than the counting of variables, partly due to a larger number of categories of restrictions and partly due to somewhat more complex situations within categories. We shall first consider only restrictions due to elements and subsequently note an additional category as elements are combined into units. The categories of restrictions are as follows: material balances, energy balances, physical equilibrium, chemical equilibrium, and inherent.

Material Balances

For a C-component system there are $C + 1$ balances which can be written, one for each of the C-components and one total material balance. Only C of these $C + 1$ will be independent and C is then the number of restrictions due to material balances. The number of material balances is related to the number of components, *not* to the number of species.

Energy Balance

When information sufficient for the establishment of an energy balance is available, there is *one restriction* provided by the one energy balance which can be written around any process. This restriction will not be required prior to Chapter 9. When more than one form of energy exists the number of restrictions will require modification to account for the various forms. (See Section S9.1B.)

Chemical and Physical Equilibrium

Concentrations of components in a mixture may be interrelated if either chemical or physical equilibrium exists. For example, if one distills a mixture of methanol and water, the vapor composition will be related to the liquid composition at any given temperature and pressure provided that the two phases are held in contact long enough for the equilibrium relationship between phases to be established. Similarly, in a reversible chemical reaction, such as the dissociation of calcium carbonate (limestone) into the oxide (lime) and carbon dioxide, the fraction of the carbonate dissociating will be governed by the temperature of the process, provided that the three phases are retained in contact for a time sufficient to establish the equilibrium concentrations.

The establishment of equilibrium will usually be a long process. It is reasonable to ask whether the limiting concentrations of physical or chemical equilibrium can then ever be important in industry, where the rate of through-put governs the amount of material produced for sale. The answer is technically "no—an equilibrium process is fundamentally uneconomical." There are, however, some circumstances where it is desirable to assume that a process would effect some conversion, or some separation, if equilibrium *were* attained, and then rate the process as to its ability to attain it. Such a rating is termed an efficiency and indicates the fraction of that separation (or conversion) attainable at equilibrium actually attained in the real process. An important example of this treatment is the *staged operation*, two instances of which are the plate distillation column or the mixing stage of an extraction process. Useful design procedures for staged processes in distillation, extraction, absorption, leaching, stripping, washing, extractive distillation, etc., result from this treatment. They are primarily concerned with the *internal* workings of the equipment. The present discussion is primarily concerned with the *external* streams to and from any type of unit, including the staged unit. Therefore, there is seldom any need for consideration of physical equilibrium restrictions in a study of the applications of the basic laws of conservation. The same arguments hold for chemical equilibrium.

For present purposes both types of restriction can be ignored. The student is warned of the importance of physical equilibrium in staged operations and is referred to Buford Smith's *Design of Equilibrium Stage Processes* (New York: McGraw-Hill Book Co., 1963), the original article by Kwauk, and Examples 10.1 and 10.2 in Section 10.1 of this text.

If compositional restrictions due to chemical equilibrium are encountered, the student is referred to Smith and Van Ness, *Introduction to Chemical Engineering Thermodynamics* for a discussion of the phase rule in the presence of chemical reactions.

Inherent Restrictions

Among inherent restrictions, one of the most important is the missing species. If the system contains a total of three species and if one stream contains only two of the three, there is one restriction imposed by the necessity of noting this zero concentration. For example, consider an element in which ammonia is absorbed into water from an ammonia-air mixture. The water enters pure but all of the ammonia is not removed from the mixture during the contact with the water phase. There are three species (ammonia, air, and water). But the inlet water does not contain two of them, the outlet liquid does not contain one (air), the inlet and outlet gas streams do not contain one (water). The total number of restrictions due to zero concentrations is then five.

As an example of the same problem in a chemical system, consider a very simple combustion where pure carbon is burned in air to form a flue gas consisting of CO_2, O_2, and N_2. Now the number of species in this system is four (C, O_2, N_2, CO_2). The fuel contains only one so that three zero concentrations are implied. The air stream has two more zero concentrations (C, CO_2) and the flue gas has one (C). The total number of zero concentrations is six.

It is important to realize that this specification of restrictions due to zero concentrations results from the definition of variables according to Equation S4.1, where each species is potentially present in each stream.

A *second type* of inherent restriction is found in the combustion example, this time in the form of a fixed ratio. The ratio of oxygen to nitrogen in air is established. As soon as one specifies that air constitutes the stream the composition is implied. The composition can be established by any single item of information concerning this two-component mixture, the percentage of either component, or the ratio of one to the other.

For present purposes, there is a *further inherent restriction* on any process since solely a material (and no energy) account is being made. Thus, information on temperature is unnecessary, and the same is true of pressure. To all intents and purposes, all temperatures and pressures are considered to be identical and independent of any specific chosen values. Thus the two variables counted in each stream for these conditions are unnecessary and $2N_s$ inherent restrictions are imposed. This introduction and subsequent removal as restrictions is thought to be better than simply ignoring their existence. A complete process analysis, involving an energy balance, will include such information and will require the use of these two variables. The present case is then an exception and is treated as such, with the generality of the method maintained by introducing these $2N_s$ items as variables and removing them as inherent restrictions when the calculations are limited to material balances.

Another inherent restriction is found in composition identities of elements, which effect no composition changes, such as total condensers and pumps, Figure S4.1. In each of these cases one would say that there are two streams,

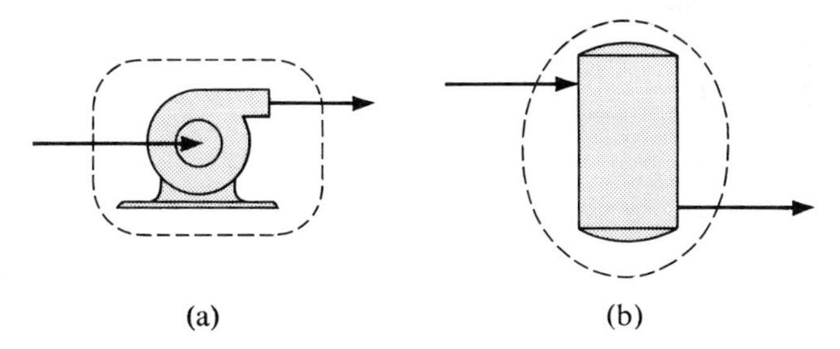

FIGURE S4.1 Two stream elements: a. pump; b. total condenser.

one in and one out and that the number of variables was $2(C + 2)$. But the compositions in and out are identical so that there are $C - 1$ composition restrictions. In such a unit, the rates in and out are also equal, providing an additional inherent restriction, and a total of C. Note that these C-inherent restrictions are, in this instance, the same ones which would be imposed by drawing C material balances. Care must therefore be taken not to count inherent restrictions twice, once in their inherent category and once as material balances.

When two streams leave a single element they must be at identical temperatures and pressures (i.e., those of the element from which they emerge) even though their compositions are not the same. Thus for the two streams, one can specify the temperature and pressure of one, and the temperature and pressure of the other must be the same. Since the number of variables attributed to the temperature and pressure conditions of these two streams was originally two each, or four, this number is reduced to the realistic two by the use of the inherent restriction. This type of restriction is important only when temperature and pressure are important parts of the analysis (as when the energy balance is to be employed). It is not pertinent when the material balance only is involved, since all pressures and temperatures have been ignored in this instance.

Counting the Degrees of Freedom

Since we are dealing with numbers, of variables, of restrictions, and of degrees of freedom, each will be designated by the letter N, with appropriate subscript, v, r, and df, respectively. Then we can write

$$N_{df} = N_v - N_r, \tag{S4.2}$$

and the resulting N_{df} is to be treated by the designer or the performance analyst as indicated earlier.

S4.3 Combining Elements into Units

A unit may be composed of several elements. The degrees of freedom of the latter may be quite easy to analyze while those for the unit may be quite difficult. In such instances one can combine the degrees of freedom for the elements according to the following rule:

$$N_{df}^u = \sum_{i=1}^{n} N_{df,i} - N_r^u, \tag{S4.3}$$

$$i = 1, 2, \ldots, n.$$

Here n is the number of elements combined into the unit, the superscript u indicates that the number refers to the unit rather than to the element. The number of new restrictions encountered in the unit, N_r^u, and not previously encountered in the elements is due solely to the fact that elements have been combined; all other restrictions applicable to the elements are embodied in the $N_{df,i}$. New restrictions, N_r^u, are inserted to correct for the duplication in counting the variables of streams which appear in each element and become innerstreams (i.e., connecting streams), not cut by the balance boundary line around the unit. Such a duplication (i.e., the same stream counted in each of two adjacent elements) produces $N_{sp} + 2$ too many variables in the unit. Consequently, if the number of innerstreams is represented as N_I, we can designate the number of new restrictions in the unit, N_r^u, as

$$N_r^u = N_I(N_{sp} + 2).$$

There will be occasions when Equation S4.3 will appear at first glance to give erroneous results. Discussion of situations where this occurs will be left until later (see Section S4.7) when they may be referred to specific examples.

S4.4 Examples and Applications of Degree of Freedom Calculations

A number of calculations of degrees of freedom will now be considered in detail. In each instance the number of species, the number of components the number of restrictions, and the number of degrees of freedom will be presented. Finally, possible reasonable assignments of the degrees of freedom will be suggested.

The examples deal first with purely physical processes; in this Section, S4.4, degrees of freedom are calculated for a variety of such systems. Section S4.5 treats a special problem in combining elements when zero compositional restrictions are encountered. Subsequent to these remarks relating exclusively to physical processes, Sections S4.6 and S4.7 deal with degrees of freedom in processes involving chemical reactions. Section S4.8 offers a very practical use of degree of freedom calculations. Section S4.9 deals with the very real problem of impossible specifications, the situation where the proper variables are specified, but the specification chosen results in a physical impossibility.

Pumps, Total Condensers

Figure S4.1 represents this type of two-stream element, heating and cooling being ignored.

$$N_v = 2(N_{sp} + 2) = 2N_{sp} + 4$$

Restrictions

Ignoring t and p: 2×2.
Composition and rate identity of streams A and B: N_{sp}.
Material balance, energy balance and equilibrium: zero, material balances being included in identity of A and B.
Total: $N_r = N_{sp} + 4$.

Degrees of Freedom

$$N_{df} = (2N_{sp} + 4) - (N_{sp} + 4) = N_{sp}.$$

Since the process is a purely physical one,

$$N_{sp} = C.$$

The designer has at his disposal the specification of the composition of the mixture ($C - 1$ degrees of freedom) and the rate of input. The performance analyst must have these items supplied. Note that these specifications provide complete information on the feed stream.

Stream Divider (Splitter)

Figure S4.2 represents this type of unit, heating or cooling again being ignored. There are three streams. Therefore,

$$N_v = 3(N_{sp} + 2) = 3N_{sp} + 6.$$

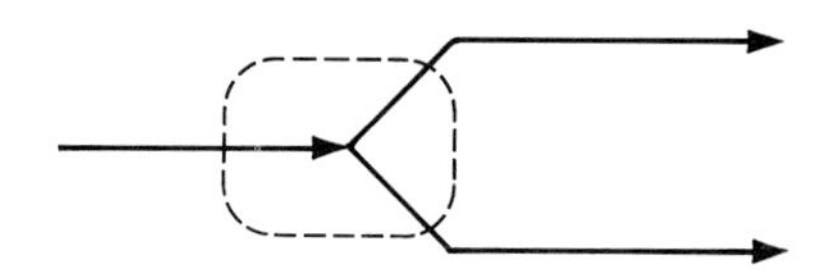

FIGURE S4.2 Stream divider.

Also, since the process is a purely physical one,

$$N_v = 3C + 6.$$

Restrictions

Ignoring t and p: 3×2.
Composition identity of streams B and C: $C - 1$.
Material balances: C.
Energy balance and equilibrium: zero.
Total: $N_r = 2C + 5$.

Degrees of Freedom

$$N_{df} = (3C + 6) - (2C + 5) = C + 1.$$

The designer has at his disposal: feed composition $(C - 1)$; feed rate (1); ratio of B to C (1) *or* rate of either of B or C, specification of either defining the other by the material balances. The performance analyst needs the same information.

Mixers

Figure S4.3 represents this type of element, heating and cooling being ignored, and the number of streams to be mixed being generalized as M. The process being a purely physical one, $N_{sp} = C$, and

$$N_v = (M + 1)(C + 2) = CM + 2M + C + 2.$$

Restrictions

Ignoring t and p: $2(M + 1) = 2M + 2$.
Material balances: C.
Energy balance and equilibrium: zero.
Total: $N_r = 2M + C + 2$.

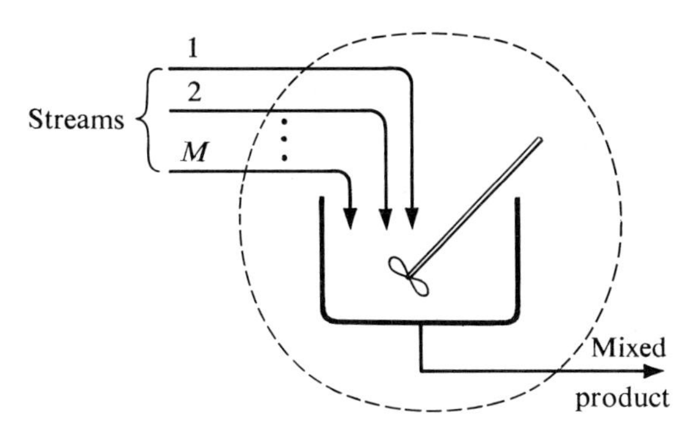

FIGURE S4.3 Mixer.

Degrees of Freedom

$$N_{df} = (CM + 2M + C + 2) - (2M + C + 2) = CM.$$

The designer can specify the composition and rate of each of the feed streams. The analyst must have the same information.

Distillation Column with Single Feed

Figure S4.4 represents this type of unit with the usual restriction of ignoring heat effects. In this case the number of stages is not in question, so that the

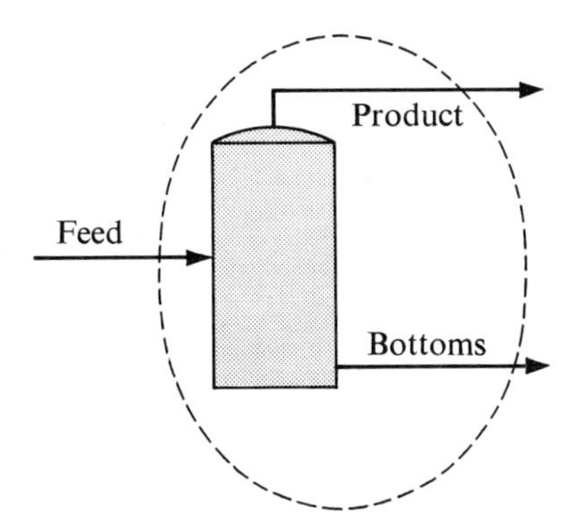

FIGURE S4.4 Generalized distillation column.

situation is one of performance only. The unit is not equivalent to a splitter, since the compositions of the two product streams are not identical. There are three streams, so that, noting from the physical character of the process that $N_{sp} = C$,

$$N_v = 3(C + 2) = 3C + 6.$$

Restrictions

Ignoring t and p: $3 \times 2 = 6$.
Material balances: C.
Energy balance and equilibrium: zero.
Total: $N_r = C + 6$.

Degrees of Freedom

$$N_{df} = (3C + 6) - (C + 6) = 2C.$$

The analyst must know the feed composition and rate, leaving C-items of information to be chosen. If the components number two, one mass per cent in each output stream could be specified, a relatively standard type of specification since the effectiveness of the separation is thereby designated. Both output rates should *not* be specified since they are related through the material balances. One of them and one composition of one of the output streams is a reasonable specification. If the components number three, one stream composition can be completely specified, with $C - 1$ degrees of freedom. The other degree can be assigned to a rate or to one composition of the other stream.

Drier

In the case of the drier of the type represented in Figure 4.2, there are three streams, with components and species identical and equal to two. Then,

$$N_v = 3(C + 2) = 12.$$

Restrictions

Ignoring t and p: $3 \times 2 = 6$.
Material balances: $C = 2$.
Inherent: The solid component cannot appear in the vapor stream, reducing the number of variables by one: 1.
Total: $N_r = 9$.

Degrees of Freedom

$$N_{df} = 12 - 9 = 3.$$

These three degrees of freedom would usually be used to specify the feed rate, the water content of the feed, and the water content of the dried product.

Air Drier

Figure S4.5a represents this type of unit. The air is introduced to carry away water vaporized from the solids. The unit can be considered in the form shown or can be treated as a combination of two elements as shown in

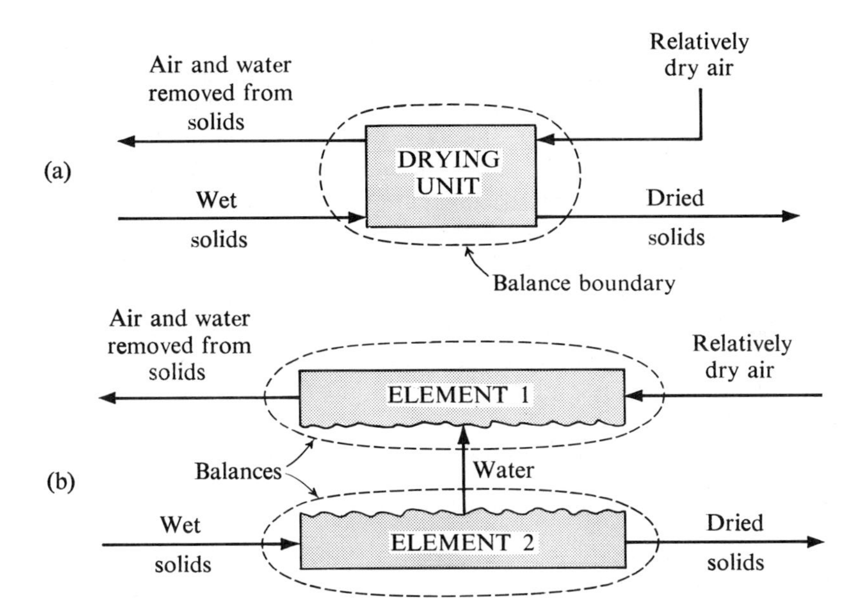

FIGURE S4.5 a. Air drier. b. Air drier unit treated as combination of two elements.

Figure S4.5b. First consider the unit as a whole. There are four streams and three species. Therefore, again recognizing that $N_{sp} = C$,

$$N_v = 4(C + 2) = 4C + 8 = 20.$$

Restrictions

Ignoring t and p: $4 \times 2 = 8$.
Material balances: $C = 3$.
Inherent: Restricting each of the four streams to two of the three components, equivalent to specifying four percentages as equal to zero: 4.
Total: $N_r = 15$.

Degrees of Freedom

$$N_{df} = 20 - 15 = 5.$$

The three specifications for the previous type of drier are still required. In addition the air rate and water content of either the inlet or output air stream would be normal.

Air Drier—Analyzed by Elements

Since the solid phase consists of two components and the air phase also consists of two components, with the water component common between them, the entire operation can be treated as the sum of two individual elements. In the first, water is removed from the solids as in the drier shown in Figure 4.2. In the second, water is added to the air stream, and the element can be considered to be a mixer. The drier element has three degrees of freedom as determined on p. 194. The mixer has been found to have MC degrees of freedom. In the present instance, $C = 2$ and $M = 2$ so that $N_{df} = 4$. There is the special restriction in this mixer that one of the streams shall be a pure component (i.e., water). Thus N_{df} is reduced by one and there are three degrees of freedom for this type of unit. Now if the two units are combined, Equation S4.3 states that the degrees of freedom of the unit so formed shall be the sum of the degrees of freedom of the elements, with correction for double counting of any common streams. The water is a common stream. Since it is a pure component, its C degrees of freedom are unity. Therefore,

$$N^u_{df} = (3 + 3) - 1 = 5.$$

This number agrees with that from the preceding analysis.

Distillation Columns in Series

Figure S4.6 represents a two-column system, a normal arrangement for the separation of a three-component mixture. According to previous analysis of

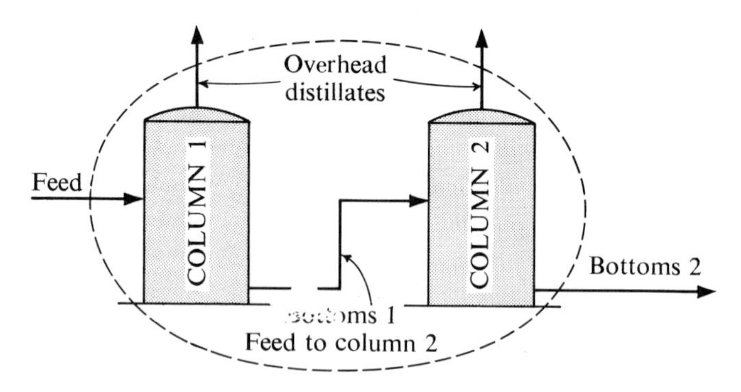

FIGURE S4.6 Two-column distillation unit.

a single column, Figure S4.4, the degrees of freedom are $2C$, or 6 with a three-component mixture. Each element of the two-column unit will have $N_{df} = 6$. The unit will have,

$$N^u_{df} = 12 - N^u_r.$$

The restriction added by combining elements into units constitutes the duplicate counting of one innerstream with C degrees of freedom. Therefore, for the unit.

$$N^u_{df} = 12 - 3 = 9.$$

These degrees of freedom might be distributed as follows: one for feed rate and two *each* for compositions of the four streams. Since all streams are three-component, the composition of every stream is completely specified and the material balances, three in number, will contain only the output rates as unknowns, producing a simple set of first-order simultaneous equations.

A second type of distribution of degrees of freedom could be the following: three to specify the feed completely as before; each output stream specified to contain a maximum composition of one component and a fraction of the charge for one other component.

Illustrations of material balances dealing with each of these specifications have been provided in Chapter 4.

S4.5 Combination of Elements with Compositional Restrictions

The case of multiple distillation units is easy. When, as with absorption or extraction units there are components which are consistently zero, much greater care must be exerted in applying the rules suggested for calculating the degrees of freedom. A simple case at point would be the absorption of ammonia from an air-ammonia mixture in water. It would ordinarily be permissible to consider that the air is insoluble and that the liquid streams would contain zero percentage of this component throughout the system. In many cases, it would also be permissible to consider that the gas stream contained a negligible amount of water vapor.

If we now consider multiple absorption or extraction elements combined into one unit, as in Figure S4.7, we should be able to analyze the degrees of freedom for a single element and obtain the degrees of freedom for the unit by combination according to Equation S4.3. We can indeed do so, but not without careful analysis of the problems posed by the zero compositions.

The multiple unit will occur if the total tower height required to effect a given separation is beyond physical, constructional reason. In such instance the total tower height will be broken into a number of units, probably of identical height, and we shall call this number of units n. Now two different sets of degrees of freedom are theoretically possible. The first relates to the overall changes in composition on a balance boundary which cuts only the two inlet and two outlet streams, boundary A in Figure S4.7. In this case, the summation of the degrees of freedom for the unit should quite obviously be the same as the number of degrees of freedom for each individual element. The second, a more reasonable and more revealing request of the analysis, is the number of degrees of freedom required to define completely the system of n elements including compositions, rates, etc., at the break points between the elements.

To attack these two problems we shall first analyze the single tower element

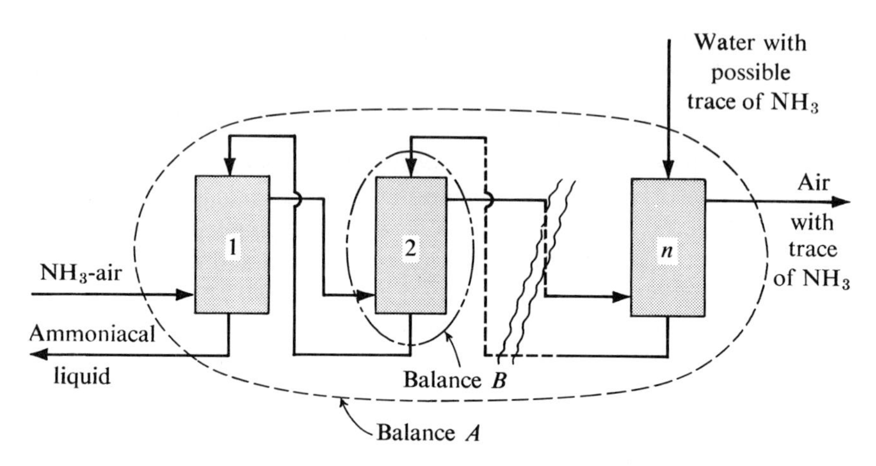

FIGURE S4.7 Multi-column absorption unit.

and then the general situation for the multiple unit, where the complete system is to be specified. We shall subsequently justify the equality of the degrees of freedom for the unit and the individual elements when only the overall changes (i.e., boundary A) are required.

For any one element, boundary B in Figure S4.7 is representative. The number of species, N_{sp}, is equal to the number of components, C. The number of total variables, N_v, is $4(C + 2)$, since there are always four streams, two input and two output. As usual we ignore at this point the 8 specifications for temperature and pressure in the four streams, treating them as restrictions. There are C available material balances. Finally, in the four streams, there are z zero compositions, stated in general terms. If the solvent is always zero in the gas streams and the inert gas is always zero in the liquid stream, these z restrictions would be four in number. If the solvent enters pure, one more zero concentration is added. The degrees of freedom for the element would be

$$N_{df} = N_v - N_r = 4(C + 2) - 8 - C - z,$$
$$= 3C - z.$$

For simplicity in analyzing the combined elements, let us assume that zero concentrations will be consistent throughout the tower. This assumption merely says that, granted there is no solvent in the gas stream at entrance, there will be none at exit, *and*, granted that the inlet solvent contains no inert gas, the exit liquid stream will also contain none, *and*, the solvent does not enter the last tower completely devoid of solute. Then we can change the definition of z to z_i, each z_i representing one consistent zero concentration which applied to each of two streams. The degrees of freedom for the element will be rewritten as

$$3C - 2z_i.$$

Now when we combine n of these elements to form a single unit,

$$\sum_{i=1}^{n} N_{df,i} = 3Cn - 2z_i n.$$

There are two innerstreams between each of the elements, or $2(n - 1)$ innerstreams. Disregarding temperatures and pressures as we have already done for the single element, there are $2C(n - 1)$ variables which have been double counted in these innerstreams. However, counting all $C - 1$ compositional variables in these streams repeats the zero compositional restrictions already taken out in dealing with the single element. Therefore, between each pair of elements, z_i restrictions have been double counted and this number should be removed from the $2C(n - 1)$ innerstream restrictions. The net restrictions due to combining elements is then

$$N_r^u = 2C(n - 1) - z_i(n - 1),$$

and the degrees of freedom for the unit, according to Equation S4.3 are

$$\begin{aligned} N_{df} &= 3Cn - 2z_i n - 2C(n - 1) + z_i(n - 1), \\ &= Cn + 2C - z_i(n + 1). \end{aligned} \tag{S4.4}$$

This expression is entirely general. It does not account for zero solute in the solvent stream fed to the tower since this restriction is not general to all types of processes, but it does account for all other consistent zero concentrations.

Close inspection of this result will indicate that it is equivalent to analyzing the element without taking account of zero concentrations and subsequently reducing the number of degrees of freedom for the unit by the number of zero concentrations in *all* $2(n + 1)$ streams.

In the instance where the interest is concentrated solely with the input and output streams (balance boundary A), with the break points between the n elements already defined, $C - 1$ compositions must be defined in one of the streams at each of the $n - 1$ break points between elements. Establishment of the compositions also establishes the flow rates in these streams. Of the previously calculated degrees of freedom, $C(n - 1)$ have thus already been used. But $z_i(n - 1)$ zero concentrations previously counted as restrictions have been re-included in this number. Therefore, with this type of specification, the degrees of freedom for the unit have devolved into

$$Cn + 2C - z_i(n + 1) - C(n - 1) + z_i(n - 1).$$

This expression simplifies to $3C - 2z_i$, which is identical with the expression for the single element. This identity is correct, since the only undefined points are at the end of the system as in the case of the single element. It represents a trivial case for multi-element analysis.

Example S4.1

Take the case represented by Figure S4.7. Each element has $3C - 2z_i$ degrees of freedom and $z_i = 2$ if water is assumed to be negligible in the gas streams; air is certainly insoluble in any measurable amounts in water or ammonia solutions. The components are three in number and

$$N_{df} = 9 - 4 = 5.$$

These degrees of freedom might be used by the designer as follows: one for inlet gas flow rate; one each for ammonia concentration in the two gas streams, thereby defining the percentage recovery of the ammonia; one each for ammonia concentrations in the two liquid streams. Alternate specifications will occur to the reader.

Now if two of these elements are used in series, according to Equation S4.4,

$$\begin{aligned} N_{df} &= 2C + 2C - 3z_i = 4C - 3z_i, \\ &= 12 - 6 = 6. \end{aligned}$$

This number is only one higher than that for the single element and it may be used to specify the composition of one of the streams at the break point between the two elements.

This increase of one degree of freedom per added element will be found to be perfectly general for any number of elements in a three-component system. For a four-component system, the increase is two per added element since this number is required to define the composition of one stream at the break points.

S4.6 Degrees of Freedom in the Presence of a Chemical Reaction

When a chemical reaction or reactions are involved in the process, proceed normally. Name the variables in each element as the product of (number of streams) times (number of species plus two). Count as restrictions the number of material balances plus the number of zero concentrations in each stream (i.e., the number of species missing from each stream) plus any fixed ratios in physical mixtures (e.g., the ratio of O/N which must exist in an air stream) plus any temperatures and pressures that may be ignored. The number of degrees of freedom will then be the number of variables minus the number of restrictions, as before. Note that the number of material balances is the number of true components, not the number of species. It is only in this aspect of the calculation that the correction for the number of independent equations is involved.

Example S4.2

▶ STATEMENT: Take the case of pure carbon burning in an excess of air to form carbon dioxide only.

$$N_{sp} = 4\ (C, O_2, N_2, CO_2).$$

▶ SOLUTION: The number of independent equations is one:

$$C + O_2 = CO_2.$$
$$C = 3\ (C, O, N).$$

Number of variables $= N_s(N_{sp} + 2) = 3 \times 6 = 18.$

Restrictions (*based on species*)

Ignoring t and p:	6
Material balances:	3
Zero concentrations:	
In carbon	3 (O_2, N_2, CO_2)
In air	2 (C, CO_2)
In flue gas	1 (C)
Fixed ratios:	
O_2/N_2 in air	1
Total restrictions	16

$$N_{df} = N_v - N_r = 18 - 16 = 2.$$

These two degrees of freedom might be assigned in at least two manners:

a. quantity of flue gas and percentage of CO_2.
b. quantities of carbon and of air.

Example S4.3

▶ STATEMENT: Take the case of decomposition of calcium carbonate containing a small amount of an inert (e.g., a silicate which will not decompose during the course of the reaction) with the carbonate being nearly, but not quite, completely decomposed.

▶ SOLUTION:

$$N_{sp} = 4\ (CaO, CO_2, CaCO_3, \text{inert}).$$

The number of independent reactions is one:

$$CaCO_3 = CaO + CO_2,$$
$$C = 3\ (CaO, CO_2, \text{inert}).$$

Number of variables $= N_s(N_{sp} + 2) = 3 \times 6 = 18.$

N_r (based on species):

Ignoring t and $p = 2N_s$:	6
Material balances:	3
Zero concentrations:	
In limestone	2 (CaO, CO_2)
In lime	1 (CO_2)
In gas	3 (CaO, $CaCO_3$, inert)
Fixed ratios:	None
	15

$$N_{df} = N_v - N_r = 18 - 15 = 3.$$

These degrees of freedom might be assigned in the following manner:

a. quantity of feed and per cent inert; per cent unburned $CaCO_3$.

b. quantity of gas; per cent inert in the feed; per cent $CaCO_3$ in lime.

Example S4.4

▶ STATEMENT: This problem is the same as Example S4.3 except that the feed consists of a mixture of magnesium and calcium carbonates. The decomposition reactions will be assumed to be complete.

▶ SOLUTION:

$$N_{sp} = 6\ (MgCO_3,\ CaCO_3,\ CaO,\ MgO,\ CO_2,\ \text{inerts}).$$

The number of independent reactions is two:

$$CaCO_3 = CaO + CO_2,$$
$$MgCO_3 = MgO + CO_2,$$
$$C = 4 \text{ in general } (CaO,\ MgO,\ CO_2,\ \text{inerts}).$$

Number of variables $= N_s(N_{sp} + 2) = 3 \times 8 = 24.$

N_r (based on species):

Ignoring t and p, $2N_s$:	6
Material balances:	4
Zero concentrations:	
In limestone	3 (CaO, MgO, CO_2)
In lime	3 ($CaCO_3$, $MgCO_3$, CO_2)
In gas	5 (all except CO_2)
Total restrictions	21

$$N_{df} = N_v - N_r = 24 - 21 = 3.$$

These degrees of freedom might be distributed in the following manner:

a. quantity of feed and percentages of two of the feed components

or

b. quantities of gas and lime plus the per cent inert in the latter.

Example S4.5

▶ STATEMENT: Consider an identical situation as in Example S4.4 except that decomposition of neither carbonate goes to completion, and these small amounts of unburned carbonates leave in the lime stream.

▶ SOLUTION:

$$N_{sp} = 6 \text{ (as before since no new species are introduced).}$$

The same two independent reactions are involved. The same four components are involved.

$N_v = 3 \times 8 = 24$, as before.

N_r (based on species):

Ignoring t and p:	6
Material balances:	4
Zero concentrations:	
In feed	3 (CaO, MgO, CO_2)
In lime	1 (CO_2)
In gas	5 (all except CO_2)
Total restrictions	19

$N_{df} = 24 - 19 = 5.$

These degrees of freedom might be assigned in the following manner:

a. three to the feed as before; one each to specify the percentage of each carbonate unburned, or

b. quantities of gas and lime plus per cent inert in the latter as before; percentage of each carbonate unburned, as in assignment a.

S4.7 Combining Elements, Chemical Reaction(s) Present

The rule stated in Equation S4.3 is applicable. The total number of degrees of freedom for the individual elements is corrected by the number of restrictions that have been counted double when the same streams appear in two of the elements. Two examples are presented.

Example S4.6

▶ STATEMENT: A fuel, consisting solely of methane and nitrogen is burned in air to give a flue gas containing no CO and no unburned methane. You are to calculate the degrees of freedom for this process where the gas analysis is by Orsat (i.e., the water is considered to be entirely condensed and removed).

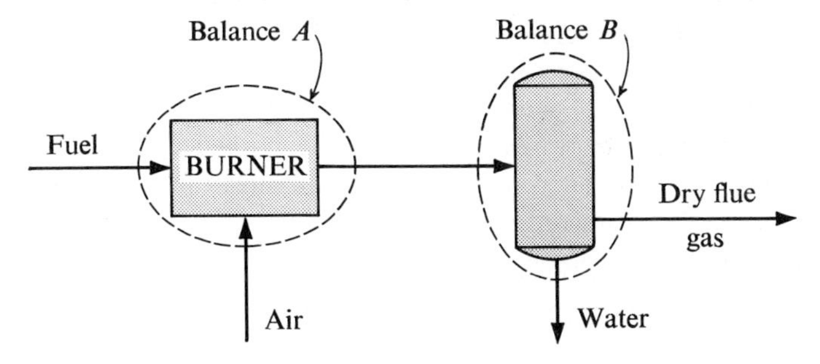

FIGURE S4.8 Combustion unit treated as combination of burner and condenser elements.

▶ SOLUTION: Application of previous principles will show that this process has three degrees of freedom. We propose to show here that the same answer would be obtained by considering the overall unit to be composed of a burner and a condenser, as indicated in Figure S4.8.

A. Burner: $N_{sp} = 5$ (CH_4, CO_2, O_2, N_2, H_2O)
Independent reactions, one:
$CH_4 + 2O_2 = CO_2 + 2H_2O$,
$C = 5 - 1 = 4$.
Streams $= 3$.
$N_v = 3(5 + 2) = 21$.

N_r:	
Ignoring t and p:	6
Material balances:	4
Zero concentrations:	
In fuel	3
In air	3
In flue gas	1
Fixed ratios:	
O_2/N_2 in air	1
Total restrictions	18

$N_{df} = 21 - 18 = 3$.

B. Condenser: $N_{sp} = 4 = C$ (physical system)
$N_s = 3$
$N_v = 3(4 + 2) = 18$

N_r:	
Ignoring t and p:	6
Material balances:	4
Zero concentrations:	4 (3 in water and 1 in dry gas)
Total restrictions	14

$N_{df} = 18 - 14 = 4$.

The sum of the degrees of freedom for the two elements is seven and there is one innerstream with $C = 4$. Therefore, the net degrees of freedom is three, which agrees with the overall analysis of the unit.

Example S4.7

▶ STATEMENT: An ammonia reactor, Figure S4.9, is fed a mixture of nitrogen and hydrogen, contaminated with argon, the ammonia formed is removed in a condenser and the unconverted gas is recycled. To prevent build-up of argon, a small part of the recycle stream is removed from the system.

▶ SOLUTION: Consider this unit to be composed of four elements: a mix point where fresh feed and recycle mix; the reactor; the condenser; and a stream divider at the point where the bleed stream is removed. Consider each of these elements separately. Since some of the elements deal with four species (N_2, H_2, NH_3, A) and some with only three (N_2, H_2, A), some decision which will be applied consistently needs to be made regarding the number of species. If all four are used, zero concentration of ammonia can, wherever ap-

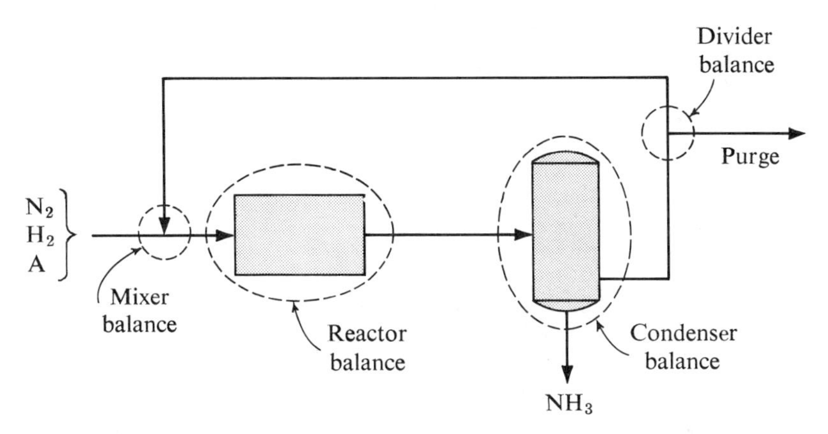

FIGURE S4.9 Ammonia synthesis unit treated as combination of four elements: reactor, condenser, divider, mixer.

plicable, be used as a specification rather than as a restriction. Alternately the actual number of species can be used in each element if care is taken in combining elements to see that these variable numbers of species are properly accounted in summing the elements. Errors are much less likely to be introduced in the first of these two methods and we shall then use four species consistently in this example. For the time being, completeness of removal of ammonia in the condenser will not be assumed.

Analyzing the individual elements:

A. Mix point
From Section S4.4, $N_{df} = CM$, where C is four and M is the number of streams mixed, two. Therefore, $N_{df} = 8$.

B. Stream divider
From Section S4.4, $N_{df} = C + 1 = 5$.

C. Condenser
In Section S4.4, a total condenser was analyzed. The partial condenser of interest here requires a different analysis.

$N_{sp} = 4 = C;\ N_s = 3.$

$N_v = N_s(C + 2) = 18.$

N_r:		
Ignoring t and p:		6
Material balances:		4
Zero concentrations:		
	NH_3 stream	3
Total restrictions:		13

$N_{df} = 18 - 13 = 5.$

D. Reactor

$N_{sp} = 4$; $N_s = 2$; independent reactions = 1.
($N_2 + 3\ H_2 = 2\ NH_3$).
$C = 4 - 1 = 3$ (N_2, H_2, A).
$N_v = 2(4 + 2) = 12.$
N_r: Ignoring t and p: 4
Material balances: 3
Total restrictions 7
$N_{df} = 12 - 7 = 5.$

Summation for elements: $8 + 5 + 5 + 5 = 23.$
Number of innerstreams counted twice = 4.
Number of redundancies in these innerstreams = $4\ N_{sp} = 16.$

$N_{df} = 23 - 16 = 7.$

Without making any assignment of these degrees of freedom, analyze the unit as a whole. There are only three streams: fresh feed, condensed ammonia, and bleed. Again assuming four species, with one independent reaction, we obtain:

$N_v = N_s(N_{sp} + 2) = 18.$
N_r: Ignoring t and p: 6
Material balances: 3
Zero concentrations:
Condensate 3
Total restrictions 12

$N_{df} = 18 - 12 = 6.$

Clearly the number of degrees of freedom do not agree between the two analyses, those around the unit being less than the sum of those around the elements, a situation which did not occur in Example S4.6. Some explanation is in order. In Example S4.6, there was no recycle stream, while in the present unit one exists. In the overall analysis of a unit in which a recycle stream exists no information is required about this stream in order to complete the material balances around the unit as a whole. But when balances around all elements must also be completed, additional information must be provided about this stream. Consequently, extra degrees of freedom will appear in the analysis, and subsequent summation, of degrees of freedom around the elements. In this instance the extra degree of freedom obtained by adding the elements is required to specify recycle rate. Temperature, pressure, and composition of this stream are each identical to the same property in the purge and need not be respecified in the recycle.

S4.8 Choice of Sequence When Many Balance Boundaries Exist

When a system, or unit, consists of more than one element, the potential balance boundaries number at least one more than the number of elements. A choice will have to be made as to the first balance to be completed (see Section S4.9). Frequently this choice may be made by inspection and intuition. In cases of doubt, analysis of the degrees of freedom will provide an unequivocal decision. Quite obviously, if the analysis is made and an element is not completely specified, some other element which is completely specified should be sought. This approach can easily be illustrated.

Consider a system consisting of two elements, an absorption tower and a distillation column. The feed to the first is 20% mixture of acetone in air, obtained from the dry spinning of acetate filaments. This mixture is to be absorbed in pure water, fed at the top, with the intent of reducing the acetone in the effluent gas to one quarter of a per cent. The acetone-water solution leaving the bottom is to be distilled in the column with the intention of producing 99% acetone in the distillate and 0.5% acetone in the bottoms. For the sake of this example suppose that past experience indicates that 1100 lb of distillate per 100 moles of gas fed to the tower is the objective.

There are three obvious balance boundaries, one about the tower, one about the column, and one about the whole system. The three are, of course, not independent, but any one may be chosen as the first point of attack. We now analyze the degrees of freedom associated with each and compare the specifications.

Start with the tower. For four streams and three components, the number of variables is $4(3 + 2) = 20$. The restrictions for this case include eight for temperature and pressure, three for material balances, and five for zero compositions (two for the pure water feed, and one for each other stream). The restrictions total sixteen, leaving four degrees of freedom. Two compositions and one rate (the 100 moles of feed) are specified, leaving one degree of freedom unsatisfied. The balances cannot be completed on this element.

In like manner, there are four degrees of freedom for the column (see Section S4.4). Two compositions and one rate are specified, again leaving one degree of freedom unspecified. The balances cannot be completed on this element.

One might be led to believe that, with each element underspecified, the same condition must prevail with the balance about both elements. Because of the common stream between them, this conclusion would be incorrect. In the present instance, the degrees of freedom for the combined elements will be

$$8 - N_r^u.$$

The new restrictions for the unit, N_r^u, are the number of variables for the common innerstream, this number having been counted twice. For the two-component mixture, the single innerstream, and with the recollection that temperature and pressure have already been disregarded,

$$N_r^u = 2.$$

The degrees of freedom for the unit will be

$$8 - 2 = 6.$$

The number of specifications is the sum of these counted for the two elements, or six, since none of the specifications involved the innerstream. The unit is, then, completely specified, and the material balances around this unit can be completed.

Completion of one set of balances will provide new specifications around the other units or elements (elements in the present case). The addition of these specifications will allow the completion of data on compositions and rates of all streams.

The solution follows for the very simple material balances in the case at point.

BASIS 100 moles of gas fed,
20 moles of acetone.

$$(0.0025)\left(\frac{80}{0.9975}\right) = 0.202 \text{ moles of acetone in effluent.}$$

$$19.80 \text{ moles of acetone absorbed} = 1148 \text{ lb.}$$

Let X = lb of water fed to the absorption tower, and B = lb of bottoms from the column.

Then, by an acetone balance, around the column,

$$1148 = (0.99)(1100) + (0.005)B;$$

$$B = \frac{59}{0.005} = 11{,}800 \text{ lb.}$$

By a water balance, also around the column,

$$X = (0.01)(1100) + (0.995)(11{,}800),$$
$$= 11{,}750 \text{ lb.}$$

Evaluation of the last item, X, supplies an additional specification needed to complete the material balances on the tower. The rate and concentration of feed to the column may then be determined. However, the information calculated for stream B also allows completion of the balances about the column. Either set of balances may be used to obtain the rate and concentration of column feed. We shall use the balance about the column as the simpler.

$$\text{Total feed} = 1100 + 11{,}800 = 12{,}900 \text{ lb.}$$

$$\text{Total acetone fed} = 1148 \text{ lb.}$$

$$\text{Concentration of feed} = \frac{1148}{12{,}900} = 8.91\%.$$

S4.9 Impossible Specifications

One very important point has not been apparent in any of the previous remarks. The analysis of degrees of freedom tells the designer the *number* of specifications required to provide a unique solution. There is, however, nothing in the number of degrees of freedom to indicate to the designer *how* to specify, once the decision is made of *what* to specify. It is entirely possible then that the designer will make a specification and subsequently find a unique but impossible solution for this specification. For instance, in Example 4.20 change the fraction of the feed required in the stream P_1 from 25% to 20%. The calculated values of P_2 and W_2 will then be 49.6 and 30.4 pounds, respectively, and these quantities are entirely satisfactory *until* one starts to calculate components. Then the quantity of component B in W_2 must be -10.68 lb since stream P_1 has 1.0 lb and stream P_2 contains 39.68 lb of this component. The designer has, in this instance, made a specification which is entirely capable of providing a unique, but not a realistic, solution. His only recourse is to change his specification. Since there are six items specified and any one may be changed, he may have a tedious task to make a practical assignment of the degrees of freedom.

For example, even while retaining the compositional specification there is only a narrow range of fractions of feed in P_1 which will provide a practical solution. At or below 20% too large a quantity of B appears in P_2. At and above 32.5% the quantity of A required in stream P_1 is greater than the amount fed.

The same problem will not face the performance analyst. He specifies the minimum number of items of data which must be provided. The performance cannot produce an impractical solution. Negative quantities in such a calculation represent losses of material or errors in the instruments.

LIST OF SYMBOLS

(Excluding chemical symbols.)

Latin letters

C	Number of physical or chemical components
M	Number of streams mixed (mixing operation only)
N_{df}	Number of degrees of freedom

N_r	Number of restrictions
N_s	Number of streams
N_{sp}	Number of chemical species
N_v	Number of variables
n	Number of elements in a unit
p	Pressure
t	Temperature
z	Number of zero concentrations

Subscript

i	Refers to number of elements in a unit ($i = 1, 2, \ldots, n$)

Superscript

u	Used to indicate a unit rather than an element

PROBLEMS

It will be apparent to the reader that any problem in Chapter 4 also constitutes a legitimate problem for analysis of degrees of freedom. Two types of problems result. First, when a single element is involved, the analysis will tell whether the data are insufficient, exactly sufficient, or overabundant with respect to specifying a unique solution. If insufficient, no further progress can be made. If overabundant, a performance test is indicated and the unspecified rates and compositions can take on a multitude of values depending on the portion of the data chosen to provide this additional information. Thus, if there are ten specifications and eight degrees of freedom, there will be $\binom{10}{8} = 45$ possible solutions. Discrepancies among the various calculated quantities and between reported data and calculated quantities will provide information on the precision of the reported data. This sort of tedious repetitive calculation is not of educational value, and overspecified situations have, therefore, been intentionally omitted in Chapter 4.

Second, when several elements are combined into a single unit or process, one can prove on which element it is most advantageous to perform the first balance.

Both situations exist in the problems in Chapter 4, providing a sufficiently varied opportunity to study calculations of degrees of freedom in material balance situations. Very few problems are, then, added at this point. Those listed below are intended largely to show the type of question that should be asked of the analysis.

S4.1 A distillation column unit is represented schematically in the accompanying diagram. Its elements are also indicated. Calculate the degrees of freedom for each of the three units and show that the summation of the elements gives a number consistent with that for the unit. The system is two-component.

S4.2 A furnace is fired with an oil containing nothing but carbon, hydrogen, and ash. There is no unburnt combustible. The dry flue gas analysis shows 10.7% CO_2, 7.5% O_2, and no CO. The empirical formula of the oil, on an ash-free basis, is desired.

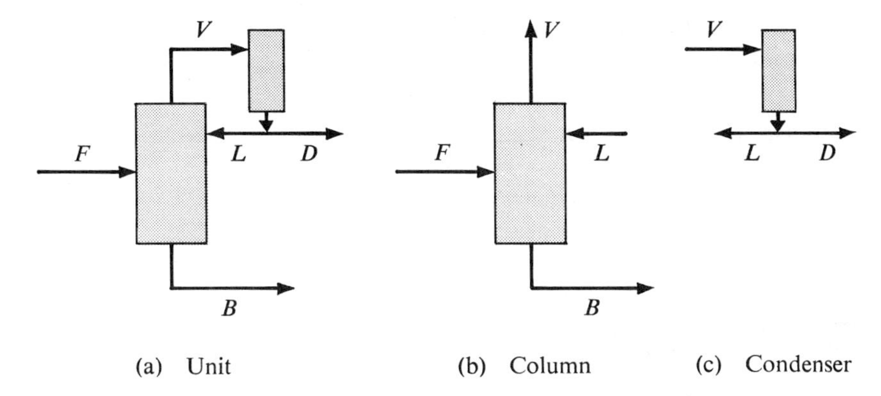

(a) Unit (b) Column (c) Condenser

Show, by analysis of degrees of freedom, that a unique situation is possible. Indicate the best manner for handling the information on ash content of the oil.

S4.3 Analyze Problem 4.2 two ways.

First, treat the unit as a single element, with the water removed from the flue gas in the same element in which fuel is burned (i.e., a four-stream element).

Second, treat the unit as composed of two elements. In the first, fuel and air are burned to provide a "wet" flue gas. The second acts purely as a splitter, wet flue gas constituting the only entering stream and dry flue gas and water constituting the effluents.

Prove that the degrees of freedom for the unit will be the same by either approach.

S4.4 Problems 4.8 and 4.9 represent two different types of specifications for identical units. Prove that both are fully specified. Show that the balance boundaries for the first material balances are different as a result of the change in utilization of degrees of freedom.

S4.5 In Problem 4.104 show that the specifications provide a unique solution. In a unit of this sort, list two other ways in which the degrees of freedom might have been employed, each to give a unique solution. Try to list cases where the data might be most easily obtained.

S4.6 In Problem 4.245 show that the unit is overspecified without the assumption of an air leak.

S4.7 In Problem 4.15, prove that the first balance must be taken around the diluter.

Why can this problem be worked without any information regarding the gas streams?

S4.8 In the modern process for manufacture of HNO_3, ammonia is oxidized to NO, which is in turn further oxidized and the oxidation products absorbed in water to form nitric acid. The process operates at a pressure of 100 psig.

Ammonia, at a rate of 10 CFM, is mixed with air, at a rate of 150 CFM (both measured at 70 °F and 100 psig) and passed through a platinum oxidation catalyst. The bulk of the ammonia (95%) is burned, the reaction being:

$$4NH_3 + 5O_2 = 4NO + 6H_2O.$$

The remaining 5% of the ammonia decomposes, the hydrogen subsequently burning to water, the combined reactions being:

$$2NH_3 + \tfrac{3}{2}O_2 = N_2 + 3H_2O.$$

The gas containing the NO is cooled to 80 °F at which temperature all the NO is oxidized to NO_2. These gases are fed to the scrubber, where the NO_2 is further oxidized and absorbed in water. The gases leaving the scrubber contain 0.5% NO_2 and no NO. The acid made in the scrubber is 60% HNO_3.

Note that water is condensed between the oxidation and absorption elements. It may be assumed here that no oxides-of-nitrogen are absorbed in this condensed water.

Analyze the degrees of freedom in the various elements, or combinations of elements, with the following objectives:

a. show that, with the specifications given, the first balance must be drawn around the oxidation element.
b. show whether all the material balances may then be completed with the specifications given.
c. show where the boundary for the second set of material balances should be located.

S4.9 A recycle stream is involved in Problem 4.23. By analysis of degrees of freedom, show the proper location for the first material balance.

S4.10 Analyze the degrees of freedom in Problem 4.255, with the intent of confirming the number of specifications given. If the specification of carbon charged is changed to state that the theoretical amount required is supplied, there is no change in the utilization of degrees of freedom. Is the problem still workable? Does the solution represent a realistic situation?

5 Material Balances Involving Transients

Chapter 4 dealt with material balances in the steady state only. That state was defined as one in which all rates and concentrations are invariant with time. It will be apparent to the reader that there are other interesting situations where this restriction will not apply, and where the accumulation term, which has zero value in the steady state, must be included in the balance.

5.1 Examples of Time-Dependent Processes

As a first example, consider delivering a fluid to a tank. If the tank is initially empty, the process starts with an input but no output. The balance is reconciled by having the accumulation over a chosen time period equal the input over the same time period. If the feed rate is $R(t)$, then from t_1 to t_2, the material balance is

$$\int_{t_1}^{t_2} R(t)\,dt = \text{amount accumulated in the tank from time } t_1 \text{ to time } t_2.$$

$$\text{Input} = \text{accumulation.}$$

Note that there is no need for the feed rate to be constant with time.

If this process is continued until the tank overflows, then the accumulation term becomes zero, since the tank can hold no more, and the overflow must exactly equal the input, whether the latter is constant or variable with time. At this point, the steady state balance is again attained. This sequence of operations, an unsteady (transient) state followed eventually by a steady state, is typical of the sequence encountered industrially in the start-up of any process or unit.

As a second example, let the situation just discussed be slightly complicated. Suppose that there is an outlet from the tank but, for some physical reason, the outflow at the beginning of the process is not equal to the input. This situation can readily exist if the outlet line is not large enough to carry off all of the material fed. Now there is an inlet rate, $R_1(t)$, and an outlet rate, $R_2(t)$, these two rates being unequal and neither being necessarily constant with respect to time. Over the time inverval t_1 to t_2 the material balance reads

$$\int_{t_1}^{t_2} R_1(t)\,dt = \text{change in mass within the tank} + \int_{t_1}^{t_2} R_2(t)\,dt.$$

$$\text{Input} = \text{accumulation} \qquad + \text{output.} \tag{5.1}$$

This process might also devolve into the steady state, and on this occasion, it is not even necessary for the tank to overflow. Suppose that the outlet rate is dependent upon the height of liquid in the tank, a supposition which has physical justification as shall be shown when dealing with energy balances in Chapter 9. Let the input rate be invariant with time. Then, at the start, the tank fills as the input exceeds the output through the open outlet pipe. But, as the tank fills, the rate of output increases. It is conceivable, if the tank height is sufficient, that the output rate may catch up to the input rate. Under such conditions, the unsteady state has devolved into the steady state, since the accumulation term has now become zero.

In both of these examples the balances have been written over a period of time, and rates have been converted to mass by multiplication by the time interval. All terms in the material balances have thereby been correctly written in units of mass. The accumulation terms have not however been defined in mathematical language. It is one of the objectives of this chapter to correct this last omission. In order to do so, the balances will have to be written as differential equations and subsequently integrated. They will, at least initially, be written with input and output terms expressed as rates multiplied by time intervals, as in the expressions already presented.

Before progressing to the formulation of the problems in terms of the differential equations, some representative physical and chemical situations involving the unsteady state will be discussed. Next, position-dependent conditions within a unit operating at steady state will be included. Finally, methods of formulation of the problem into mathematical language (the ordinary differential equation) will be discussed. This last is the true objective of this chapter.

It is certainly true that the formulation of the problem is of no assistance unless the resulting differential equations can be solved. The solution of the equations is not the province of this text but rather belongs in the realm of the reader's mathematical competence. Many of the equations developed can be solved with mathematical machinery no more complex than ordinary integral calculus. Some will, however, require a basic knowledge of ordinary differential equations. For those whose mathematical education has not progressed to this point, a supplement to this chapter suggests standard and straightforward methods of solution applicable to the situations encountered in this chapter and many others common in chemical engineering.

Representative unsteady state situations include the purging of a tank, batch distillation, and batch reaction. Each of these processes will be briefly described, and each will be involved in subsequent formulation of examples or in problems appended to this chapter as possible exercises.

The Tank Purging Problem

Consider a storage tank, not necessarily full, containing a solution—a mixture—of two or more components. Suppose that some feed stream is delivered to the tank. For simplification, consider that the solution in the tank originally is a salt dissolved in water and that the feed is pure water. Further suppose that there is an output stream, whose rate will, again for the purposes of simplicity only, be supposed to equal the rate of the input stream. In other words, the tank is full at all times, and rate of the output stream is equivalent to that of the input stream. Now the total volume in the tank is constant, but the concentration of the salt will be continuously diminishing. There is then no accumulation term in the *total* material balance (if we can neglect density changes) but there is a (negative) accumulation of the *salt component*. Analysis of this situation requires, first, the formulation of the component balance in the form of a differential equation and second, its solution (i.e., integration) to provide an expression for concentration as a function of time.

Batch Distillation

If two materials of different volatility are charged to a still, and heat is applied so that the mixture boils, the vapor withdrawn from the still will, due to the different volatilities of the components, be of different composition from that of the liquid in the still. The still composition therefore varies with time and the vapor composition does likewise. If the relationship between the liquid and vapor compositions is known, an equation may be written relating the composition of the liquid in the still (or of the vapor produced) to the time of the distillation process or to the amount of material in the still at any time. The still may, or may not, have more feed added during the process.

Consider a very specific process, the distillation of a mixture of alcohol and water. The alcohol is more volatile than water under most conditions of composition at atmospheric pressure. Therefore, the vapor contains a greater percentage of alcohol than did the liquid in the still. As the distillation progresses the liquid in the still decreases in total mass (the unsteady state total material balance) and the alcohol is distilled at a rate faster than that of the water (the unsteady state component balance). If fresh feed is added during the distillation period, the situation is somewhat complicated in that the direction of change of overall mass in the still cannot be simply predicted, but the problem is not basically changed. The objective is still to write and solve an equation relating composition of the liquid in the still to the time of distillation, *or* to the mass of material remaining in the still, *or* to the amount of vapor distilled.

The Batch Reactor

The problem is slightly changed in form, but the principles are the same. The total mass in the reactor is unchanged with time, but the reacting materials decrease with time and the products increase with time. The component balances must then be written as differential equations while the total balance is invariant with time, a situation analogous to the tank-purging problem.

Gaseous, liquid, or solid reactions, or combinations of any of the three, may be considered. For example, consider the decomposition of limestone, $CaCO_3$, into lime, CaO, and carbon dioxide, CO_2. Charge a batch of limestone into a heated reactor maintained at fixed temperature. During the course of the reaction, total material is constant. The input is the charge of limestone and the output consists of two streams, one solid and one gaseous. The solid is the mixture of $CaCO_3$ and CaO at any time. The gas is the CO_2 which has been liberated at that time. The sum of the two is equal to the charge. The component balances include the time-dependence. The rate of decrease of $CaCO_3$ is, of course, exactly the same as the rate of increase of either CaO or CO_2, in terms of moles. Solution of one equation for one component as a function of time will then provide, through stoichiometry, equations for both of the other chemical species present in the system.

5.2 Varying Concentrations (Position-Dependent)

Now consider operation at the steady state of a continuous unit which has some height, or length. Suppose that the feed stream(s) enter at some fixed point(s) in the apparatus and that product stream(s) leave at some other point(s). To be more specific, choose for initial study a physical unit such as an absorption tower, Figure 4.4. A gas stream enters at the bottom and leaves the top depleted in at least one component, while a solvent enters at the top, flows countercurrent to the gas stream through the tower, and leaves at the bottom containing the amount of the soluble gaseous components that were picked up (i.e., transferred to the solvent) in the tower. It is intuitively obvious that the pick-up of solute by the solvent is accomplished progressively in the passage of the liquid stream down through the tower. Consequently, the composition of the liquid stream is changing continuously from top to bottom. Coincidentally, the composition of the gas stream flowing up the column must be changing in a complementary manner. Thus, we have a unit operating at steady state, with input and output streams unchanging with respect to rate and composition, but with all internal streams changing continuously as a function of their position within the unit. It is not hard to make an analogy between this situation, with rates and compositions being height-dependent, to our previous unsteady state situation with rates and compositions being time-dependent.

The balances around the whole system involve only the external, steady state, time-invariant material balances, for total material and for each component. The balances at some general internal position are height- (or length-) dependent whether for total material or for a component.

Any continuous, steady state process in which the compositions of concurrent (or countercurrent) streams continuously change from one end of the unit to the other end may be characterized by the general description used with the absorption tower. There are numerous mass transfer situations to which these remarks apply. Examples will be encountered in processing units effecting extraction, drying, crystallization, distillation, leaching, and many others. In addition, some examples will subsequently be encountered in heat balances, the simplest cases involving steam heaters, coolers, and heat interchangers in general. Analysis of interchangers is actually simpler than analysis of mass transfer units since temperature scales are identical in both streams, while in mass transfer, the concentrations in the two streams are expressed in different units, due to the different components in the two streams. These remarks will be amplified when the equations that apply to the two cases are presented.

The basic objective in analyzing the *internal* changes in steady state equipment is the formulation of the compositional changes into mathematical expressions *at any general point* (height or length) in the apparatus. Since the compositions are continually changing, these expressions necessarily involve ordinary differential equations with the height or length as the independent variable. Solution of the equations then provides expressions for the concentrations as a function of height, or length, of the unit. With such an equation we can predict the height (or length) of unit necessary to effect a given desired separation, or to predict the separation which may be expected in an existing unit.

The reader may be disturbed by the change from an independent variable expressed in time units for the batch process to one expressed in terms of height or length for the continuous, co- or countercurrent, unit. He may rationalize the two as being nearly equivalent in the following terms. Concentrations throughout the tower can still be treated as time-dependent since the time for any stream to reach any position in the tower will be dependent upon the cross section of the tower and the velocity of the stream through the tower. Thus, length-dependent concentrations can be thought of as being also time-dependent granted that the internal velocities are known. In the steady state these internal velocities depend upon the mass feed rate, the changes in these rates effected by transfer to the other phase which has already occurred, and the portion of the cross section of the tower utilized by the stream in question. The analogy of length-to-time-dependence is mainly useful as a means of indicating that there is no real basic change between the methods required for the establishment of the differential equations for the batch and continuous cases.

5.3 General Treatment*

The immediate problem to be attacked is the establishment of the differential equations applicable to either case. Ability of the reader to solve the equations will be assumed. (Also see Supplement to Chapter 5.)

Take a simple time-dependent case to establish a general method of attack. Use as a typical example, the unit shown in Figure 5.1. This figure might

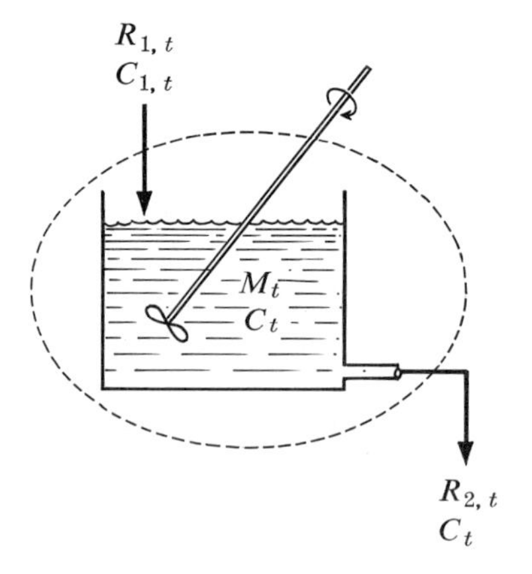

FIGURE 5.1 Perfectly stirred tank—purging or accumulating a component.

• • • • •

* To simplify mathematical expressions, symbolic representations in the remainder of the chapter will parallel those of Bird, Stewart, and Lightfoot, *Transport Phenomena* (John Wiley & Sons, Inc., 1960). The system will, in general, be as follows. A variable which is time dependent will be subscripted with a t. Thus $R(t)$ will be written R_t. If it is necessary to indicate that such a variable is measured at a specific time, e.g., at $t + \Delta t$, it will be written $R_{t|t+\Delta t}$. A product of two variables, R_t and C_t, measured at time t, will be written $(RC)_{t|t}$. If R is a constant, this same quantity will be written $RC_{t|t}$. If several rates, R_i, are each time-dependent they will be written $R_{i,t}$, where $i = 1, 2, \ldots$. The rate $R_{1,t}$ measured at time $t + \Delta t$ will be written $R_{1,t|t+\Delta t}$. Other variations, as they occur, should not cause any difficulty.

apply to a tank of perfectly stirred fluid with accumulation or depletion of the mass of fluid in the tank due to unequal rates of feed and effluent. These feed and effluent rates may also be time-dependent. They will be represented as $R_{i,t}$; in situations where rates are constant, the indication of the time dependence will be removed; the constant rate will then be represented as R_i.

In the general case, the several $R_{i,t}$ will not be equal and an accumulation will result in the tank. The material balance, which must include such a term, will be written as

$$\text{Mass in} = \text{mass out} + \text{accumulation.}$$

A generally suitable means of expressing each of these terms is necessary, taking into account the fact that input and output must be written in terms of mass rather than rates.

Expression of Total Input and Output

Inputs and outputs are expressed as rates, but material balances treat masses. Over any time interval, Δt, the total mass in (or out) may be written in terms of known rates as $\bar{R}_i \, \Delta t$, where $\bar{R}_i$ indicates the average rate over the interval Δt. Thus,

$$\bar{R}_i \, \Delta t = \int_t^{t+\Delta t} R_{i,t} \, dt \doteq \tfrac{1}{2}(R_{i,t|t+\Delta t} + R_{i,t|t}) \, \Delta t.$$

It can be readily seen from Figure 5.2 that

$$\lim_{\Delta t \to 0} \tfrac{1}{2}(R_{i,t|t+\Delta t} + R_{i,t|t}) = R_{i,t},$$

regardless of the form of the function $R_{i,t}$.

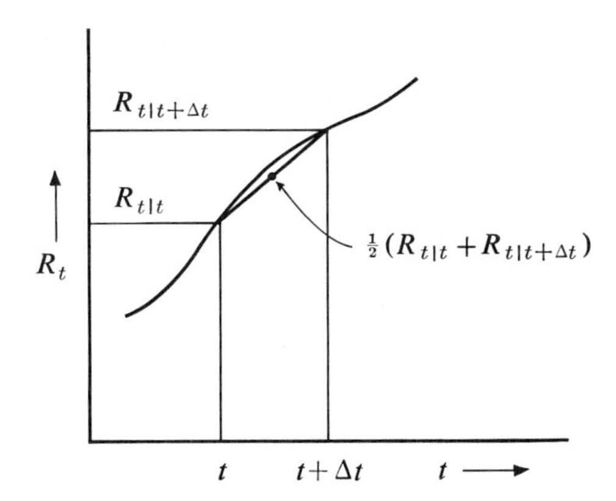

FIGURE 5.2 Averaging rates over a time interval, Δt.

Expression of Accumulation

The accumulation is the difference between the amount of the component in the tank at the beginning and at the end of the time interval. The definition implies that mass in the tank at the beginning shall be subtracted from that at the end. The accumulation, in these terms, is expressed as

$$(MC)_{t|t+\Delta t} - (MC)_{t|t}.$$

The Material Balance

Input, output, and accumulation terms form the material balance defined by Equation 5.1:

$$\underbrace{\tfrac{1}{2}[(R_1C_1)_{t|t} + (R_1C_1)_{t|t+\Delta t}]\,\Delta t}_{\text{Input}} = \underbrace{\tfrac{1}{2}[(R_2C)_{t|t} + (R_2C)_{t|t+\Delta t}]\,\Delta t}_{\text{Output}} + \underbrace{(MC)_{t|t+\Delta t} - (MC)_{t|t}}_{\text{Accumulation}}. \quad (5.2)$$

When we divide through by Δt and take the limit as $\Delta t \rightarrow 0$, the second term on the right becomes the definition of $d(MC)/dt$ and the balance, in the form of an ordinary differential equation, becomes

$$\underset{\substack{\text{Input}\\ \text{rate}}}{(R_1C_1)_t} = \underset{\substack{\text{Output}\\ \text{rate}}}{(R_2C)_t} + \underset{\substack{\text{Accumulation}\\ \text{rate}}}{\frac{d(MC)_t}{dt}} \quad (5.3)$$

While Equation 5.2 is the true mass balance and 5.3 is a balance based on rate of mass flow, both are obviously correct since one is directly derived from the other.

Equation 5.3 may be simplified by expanding the last term and eliminating one differential through the use of the total material balance. Expanding,

$$\frac{d(MC)_t}{dt} = C_t\frac{dM_t}{dt} + M_t\frac{dC_t}{dt}. \quad (5.4)$$

The total material balance will be of the same form as Equation 5.3, but without any C-terms. It therefore states,

$$\frac{dM_t}{dt} = R_{1,t} - R_{2,t},$$

which may be substituted in the first term on the right in Equation 5.4. Equation 5.3 then becomes

$$M_t\frac{dC_t}{dt} + R_{1,t}[C_t - C_{1,t}] = 0. \quad (5.5)$$

The Solution

The first two terms contain the dependent variable, C_t, and the third term is a function of the independent variable only. The equation is, then, a first order, ordinary differential equation with nonconstant coefficients, and containing a particular term. A general solution will depend upon knowledge of the forms of the functions $R_{1,t}$, M_t, and $C_{1,t}$ and on evaluation of the integration constant, K, from initial conditions. The general solution is*

$$C = K \exp\left[-\int\left(\frac{R_1}{M}\right)_t dt\right] + \exp\left[-\int\left(\frac{R_1}{M}\right)_t dt\right]\int\left(\frac{R_1C_1}{M}\right)_t\left[\exp\int\left(\frac{R_1}{M}\right)_t dt\right] dt. \tag{5.6}$$

Simplifications will result from limitations on the generality of the process.

CASE 1 Let $R_{1,t} = R_{2,t}$. Then, from the total material balance, $dM_t/dt = 0$, and M_t is a constant. The coefficients in Equation 5.6 are obviously simplified and the integrations are more readily performed.

CASE 2 Let $R_{1,t} = R_{2,t} = R$, a constant. Now the coefficients may be written as R/M, a ratio that has the dimension of $(\text{time})^{-1}$; this ratio is then usually treated as the reciprocal of the average residence time of any particle of mass in the tank. R/M is usually written as $1/\theta$, where θ is the residence time. All integrations except the last are readily performed. Difficulties with the last will depend on the form of the time-dependence specified by $C_{1,t}$.

CASE 3 In addition to the specification in Case 2, let $C_{1,t} = C_1$, also a constant. Now all the terms in Equation 5.6 are readily integrated, but the integration can equally well be performed from Equation 5.5. Rewrite that equation as

$$\frac{d[C_t - C_1]}{dt} + \frac{1}{\theta}[C_t - C_1] = 0. \tag{5.7}$$

This equation is readily integrated to give

$$C_t - C_1 = Ke^{-t/\theta}.$$

CASE 4 In addition to the specifications in Case 3, let $C_1 = 0$. The resulting simplification in the solution to Equation 5.7 is obvious. It is also apparent that Equation 5.5 has been made homogeneous by this limitation.

Other combinations of these specific cases will be apparent to the reader.

5.4 General Forms of Transient Problems

A general method of formulating ordinary differential equations in a transient situation has been presented. Despite the rather specific process

• • • • •

* See Supplement to Chapter 5.

analyzed, the situation was general in that it contained inputs, outputs, and an accumulation in the total and component balances. The solution of the final differential equations varied according to the characteristics of the process. We shall now generalize types of problems and then complete this chapter by writing the differential equations applicable to specific processes.

Inspection of the general material balance in differential form, as indicated in Equation 5.3, immediately suggests that at least three general types of process might be encountered, one containing all three terms and two others in which either the input or output term does not exist. There are, however, two other cases that should also be included. In one there is a differential term, without input or output so that an auxiliary statement is required to define the consumption (or formation) of the component of interest. The batch reactor to which nothing is added during the course of the reaction and from which nothing is withdrawn would represent such a process. In the second there is no accumulation term, due to the existence of the steady state, but a differential equation can be written in terms of the change in composition of a phase with respect to position in the unit (see previous remarks in Section 5.2). There are, then, five specific cases that should be recognized.

All Three Terms Present

Equation 5.3 contains three terms, already illustrated, those related to output and accumulation, both being functions of C_t, providing the complementary function, and the input term on the left, which is not a function of C but of t only, providing the particular integral. Standard methods of solution are available. Two of these, integrating factors and the method of undetermined coefficients are noted in the Supplement to this chapter. The input is frequently designated the "forcing function." The solution to the equation is obviously dependent upon this term, including the special case in which the term is zero.

Input Term Zero

In this instance there is no forcing function and the solution may be obtained by the simple procedure of separating variables. (See Section S5.3.)

Output Term Zero

This situation is relatively rare in practical cases, unless the inlet is also zero. When encountered, the solution may also be obtained by separating variables.

Both Input and Output Terms Zero

There is obviously a zero value of dM/dt, so that the total material balance may be applied only to the material charged at the beginning of the process and the material removed at the end. In itself, it is trivial. If, however, C is the concentration of one component, $d\,(MC)/dt = M\,dC/dt$ and dC/dt does not have zero value. This rate may be expressed as some function of the concentrations of the reacting components and, in the case of reversible reactions, of the products. The proportionality constants will be functions of temperature.

No Accumulation Term

There are input and output terms, with rates and compositions dependent upon the position chosen in the unit. These rates and compositions are fixed by the steady state conditions that exist throughout the equipment; any differences between them at the chosen positions must be occasioned by transfer of material from another phase. If a rate expression can be written in terms of the concentrations of the component in question in the two phases, a differential equation will result. This general procedure will be illustrated in Section 5.7.

5.5 The Simple Still

In a simple still operation, a mixture is first charged to the still. It is then heated and the vapor formed passes continuously to a condenser through which it is removed from the system. During the distillation there is the option of feeding more of the original charge continuously or of operating with no further additions to the still. We shall consider both situations.

Certain generalities should be noted before the equations for these transient cases are formulated. First we recall the purpose of a distillation is the separation, or partial separation, of the components in the charge. Separation can occur only when the vapor-liquid relationships indicate that one component is more volatile than the others, or, in the case of a simple binary mixture of two components, one component is more volatile than the other one. If this condition pertains, we expect that the vapor is richer in the more volatile component than is the liquid in the still at any time. Expressing mole concentrations in mole fractions, the mole fraction of the more volatile component in the distillate is greater than the mole fraction of the same component in the still. Although both of these mole fractions are expressed in identical units, moles of more volatile divided by total moles, we shall use different symbols designed to distinguish between compositions in the liquid and compositions in the vapor. The mole fractions will be designated x and y, respectively. Second, we note that, as the composition of the liquid in the still changes, the composition of the vapor being withdrawn must change in a similar manner; i.e., x and y will vary in the same direction but the change in one does not have to be directly proportional to the change in the other. Third, we note that equations involving both of these variables will not be capable of solution unless we have some means of relating the two. To do so, we usually resort to the equilibrium relationship between liquid and vapor. This assumption of equilibrium is made in full awareness of the fact, noted first in Chapter 4, that industrial processes can rarely be run at economical rates of production while still providing time of contact between phases sufficient for its establishment. The assumption is justified in two ways. First, there is no other means of relating x and y without empirical analysis of the equipment. Second, the establishment of equilibrium between liquid and vapor can be nearly obtained under proper conditions for good contacting of vapor and liquid; the ordinary simple still provides this type of contact.

The material balances may be written in terms of moles and mole fractions, or in terms of mass and mass fractions. Consistency between the concentration units and mass units is all that is required. Since there is no chemical change involved, the use of moles is justified if desired. The determining factor in the choice will be the form of the data relating x and y, which may be available in terms of either mole fractions or mass fractions.

The Simple Still, Fed Continuously

The process is represented in Figure 5.3. The mass in the still at any time will be designated M_t. The mass *rates* of feed into and of vapor removed from the still will be designated F_t and V_t, respectively. The concentrations will be

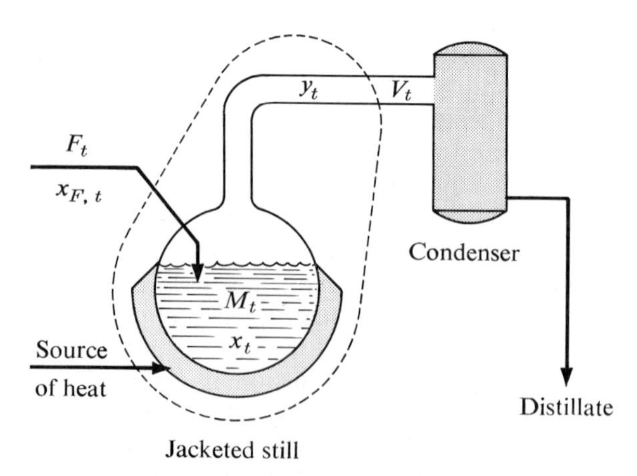

FIGURE 5.3 Simple batch still, with continuous feed but no removal of liquid residue.

in mass fractions. We shall assume that the boiling provides sufficient agitation so that the liquid in the still is of homogeneous composition at all times. This assumption is one of perfect mixing and is equivalent to that employed in the case of the tank purging but provided by different mechanical means.

Now write the terms required in a total material balance and in a component balance, using the most volatile component (more volatile of the two in the case of a binary mixture).

Inputs

Total: $\frac{1}{2}(F_{t|t} + F_{t|t+\Delta t})\,\Delta t.$
Component: $\frac{1}{2}[(Fx_F)_{t|t} + (Fx_F)_{t|t+\Delta t}]\,\Delta t.$

Outputs

Total: $\frac{1}{2}(V_{t|t} + V_{t|t+\Delta t})\,\Delta t.$
Component: $\frac{1}{2}[(Vy)_{t|t} + (Vy)_{t|t+\Delta t}]\,\Delta t.$

Accumulations

Total: $M_{t|t+\Delta t} - M_{t|t}.$
Component: $(Mx)_{t|t+\Delta t} - (Mx)_{t|t}.$

Adding outputs and accumulations and equating to inputs, dividing by Δt and taking the limit as Δt approaches zero, the two balances will state:

Total: $$\frac{dM_t}{dt} + V_t = F_t.$$

Component: $$\frac{d(Mx)_t}{dt} + (Vy)_t = (Fx_F)_t.$$

Using the total balance to simplify the component balance in the manner noted in Section 5.3, the component balance becomes

$$M_t\left[\frac{dx_t}{dt}\right] + V_t[y_t - x_t] = F_t[x_{F,t} - x_t]. \tag{5.7}$$

A general solution requires knowledge of the relationship between the two compositional variables, y and x. When mixtures are ideal, and Raoult's law holds over the entire range of compositions,

$$y = \frac{\alpha x}{1 + (\alpha - 1)x},$$

where α is the ratio of the vapor pressures of the pure components. In other limited circumstances and over restricted ranges, Henry's law may be applicable. The $x - y$ relationship can then be stated very simply,

$$y = kx,$$

where k is a constant.

When analytical relationships between y and x are not available, empirical relationships are frequently known for simple mixtures. This type of information will allow numerical or graphical integration of Equation 5.7.

The Simple Still, Batch Charge Only

If we limit the process in such a manner that no charge is to be added during the distillation, we are restricting Equation 5.7 to the condition that $F_t = 0$. The term on the right side of the equation is eliminated. Rearranging the other terms we may write

$$\frac{dx_t}{y_t - x_t} = -\left(\frac{V}{M}\right)_t dt. \tag{5.8}$$

In this expression, $V_t\, dt$ is a differential mass of distillate. Since there is no charge to the system during the time period in which this differential mass is removed from the system, the total material balance tells us that the mass in the still is decreased by the same amount. We can then write

$$-V_t\, dt = dM_t.$$

Equation 5.8 then may be written

$$\frac{dx}{y - x} = \frac{dM}{M}. \tag{5.9}$$

This form will be recognized by those familiar with the elementary principles of distillation theory as the well-known Rayleigh equation for batch distillation of binary mixtures. In Equation 5.9, we have omitted all indications of time functions since we have eliminated time considerations and refer now only to vapor and liquid compositions corresponding to any mass in the still. The general solution is

$$\int \frac{dx}{y - x} = \ln M + \text{const.}$$

The complete solution may be obtained by evaluation of the constant through knowledge of the initial conditions (i.e., values of M and x for the charge).

These two distillation processes illustrate the following cases: (a) all three terms in the balance having nonzero values, and (b) no forcing function or, more precisely, the forcing function having zero value. We now proceed to the cases of (c) no output stream and (d) no input or output streams.

5.6 The Batch Reactor

Consider a batch reactor. It is agitated sufficiently so that uniformity of concentration within the reactor may be assumed to be maintained at all times. The reactor may be fed during the course of the reaction. A system of this type is represented in Figure 5.4.

We intend to write the total and component material balances, in the usual manner. Since the procedure of writing the mass balances, dividing through by Δt to get rate expressions, and obtaining differential equations by taking the limit as Δt approaches zero is now well established, we shall take the short cut of writing directly in rates. These rates are, of course, mass rates. The symbols shall be F for feed rate, M for mass in the reactor, x for mass fraction of the component in which we primarily are interested, and the new symbol, r, for rate of output of this component. There is obviously no true output. The component which we consider disappears however by reaction and is thereby effectively removed from the system. It can then be treated as an output and is so represented by a dotted line in the figure.

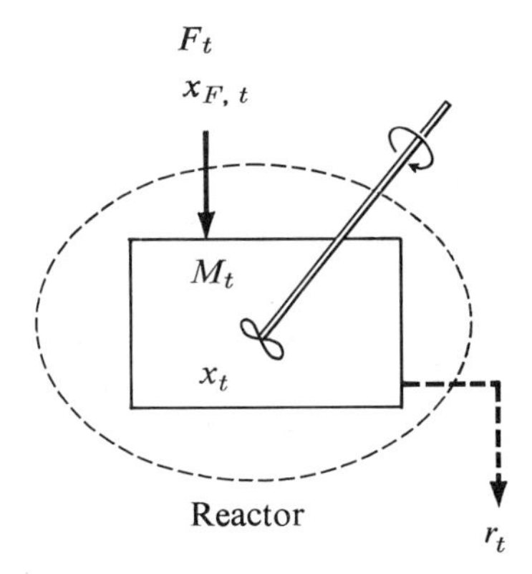

FIGURE 5.4 Batch reactor.

At this point there is no necessity to define the reaction. In kinetic terms it may be first order, second order, nth order and it may be reversible or nonreversible. Definition of the reaction *will* be required to effect a solution of the differential equations, but not for the writing of the balances.

The various terms in the material balances may be written as follows, in terms of rates.

Inputs

Total: F_t.
Component: $(Fx_F)_t$.

Outputs

Total: Zero.
Component: r_t.

Accumulations

Total: $\dfrac{dM_t}{dt}$.

Component: $\dfrac{d(Mx)_t}{dt}$.

The balances will now state:

Total: $\dfrac{dM_t}{dt} = F_t$.

$$\text{Component: } \frac{d(Mx)_t}{dt} + r_t = (Fx_F)_t. \tag{5.10}$$

Then the component balance can be written

$$M_t \frac{dx_t}{dt} + r_t = F_t[x_{F,t} - x_t]. \tag{5.11}$$

Equation 5.11 is a general statement for this type of equipment. If r is expressed in terms of x, it may be readily solved. As might be expected, the main difficulties with the solution lie in the proper expressions for r for various types of reactions. The solution of Equation 5.11 requires a statement of r in terms of mass in the reactor and the mass fraction of the component. The resolution of this problem as related to both cases will not be attempted until some elementary discussion of kinetic relationships has been presented.

Elements of Kinetics

There is, unfortunately, no guarantee that reaction rates can be deduced from the stoichiometry of the reaction. However, in the absence of information to the contrary, general rate equations are usually written according to the stoichiometry in the following manner. Given the chemical equation

$$aA + bB + \cdots = pP + qQ + \cdots,$$

the rate of disappearance of component A is often presumed to be

$$-\frac{dC_A}{dt} = kC_A^a C_B^b \cdots, \tag{5.12}$$

where C_i is a concentration in moles per unit volume, N_i/V, and k is a rate constant in appropriate units and the reaction is not reversible.

In general, substituting N_i/V for each C_i, and multiplying both sides by V, Equation 5.12 may be written

$$-\frac{dN_A}{dt} = kN_A\left(\frac{N_A}{V}\right)^{a-1}\left(\frac{N_B}{V}\right)^b \cdots; \tag{5.13}$$

then, for constant volume reactions, the V's may be combined in the constant. Equation 5.13 is general. Since most industrial processes are at constant pressure rather than at constant volume, the equation for the constant volume case need not be given special attention.

A reaction with one reactant only (dissociation, polymerization, etc.) is commonly termed *unimolecular*. It might be represented as

$$A = pP$$

or

$$A = pP + qQ + \cdots.$$

The dissociation of a polymer would be an example which is stoichiometrically correct.

A reaction with two reactants only is commonly termed *bimolecular*. It might be represented as

$$A + B = pP + qQ + \cdots$$

or as

$$2A = pP + qQ + \cdots.$$

Many common examples exist.

A reaction with more than three molecules required is generally considered to be unlikely since the chances of all three molecules colliding simultaneously is relatively small. Where the stoichiometry demands such combinations, the mechanism usually will be found to involve some intermediate, formed from no more than two of the molecules, which immediately reacts with another molecule to give the products. A series of reactions, all of bimolecular order or less would then be involved.

In terms of Equations 5.12 and 5.13, the rate of disappearance of component A in unimolecular and bimolecular reactions can be written respectively

$$\text{Unimolecular:} \quad -\frac{dN_A}{dt} = kN_A \tag{5.14}$$

and

$$\text{Bimolecular:} \quad -\frac{dN_A}{dt} = \frac{kN_AN_B}{V}. \tag{5.15}$$

Applications

If we examine Equations 5.10 and 5.11, we see that the term r is expressed in units of mass of the component per unit time. Equation 5.13 and any derived expressions such as 5.14 and 5.15 are not expressed in these units but rather as moles of the component per unit time. In order to substitute for r, Equations 5.13 through 5.15 must be multiplied by the molecular weight of the component, W_m. If it becomes necessary to distinguish between components, a letter (e.g., A, B, P, etc.) will be added to the subscript on W_m.

Now if the proper kinetic expression is substituted into Equation 5.10 we see that another complication has been introduced, a compositional variable different from the x_t already present. For example, with first-order kinetics, the substitution, $r_t = W_{m,A}kN_A$ results in the expression,

$$\frac{d\,(Mx)_t}{dt} + W_{m,A}kN_A = (Fx_F)_t.$$

However, since $d\,(Mx)_t/dt = W_{m,A}\,dN_A/dt$, we may rewrite Equation 5.10 in the form

$$\frac{dN_A}{dt} + kN_A = \frac{(Fx_F)_t}{W_{m,A}}. \tag{5.16}$$

When the reactor is not fed during the time interval considered

$$\frac{(Fx_F)_t}{W_{m,A}} = 0,$$

since

$$F = 0$$

and Equation 5.16 is identical to Equation 5.14.

Bimolecular kinetics could be handled in a fashion similar to that used in Equation 5.16 by substituting kN_AN_B/V for kN_A.

First-Order Kinetics • Assume a reaction of the type

$$A = pP + \cdots,$$

and assume that first-order kinetics apply. Then recalling that we defined r as the rate of *removal* of component A from the system,

$$r = kW_{m,A}N_{A,t}.$$

Equation 5.16 now applies. This first order, ordinary differential equation may be solved given values of k and the time-dependent functions F_t and $(x_{F,A})_t$. If the process is isothermal, k is a true constant, and the equation has constant coefficients.

Equation 5.16 applies to the case of any first-order process in which the rate of the forward reaction very greatly exceeds the rate of the reverse reaction. It also applies to the situation where reversibility is impossible due to the immediate removal of one of the products.

If the process is reversible, with forward reaction rate constant, k_1 and reverse rate constant k_2 and with both processes unimolecular, as represented by the typical equation

$$A \rightleftarrows P,$$

then component A disappears due to the forward reaction and is formed by the reverse reaction. The net rate of disappearance of A can be expressed as

$$\frac{-dN_{A,t}}{dt} = k_1 N_{A,t} - k_2 N_{P,t}. \tag{5.17}$$

The quantity N_P can be expressed in terms of the original quantities of component A, $N_{A|t=0}$, and of P, $N_{P|t=0}$ and the quantity of A remaining, N_A. The quantity of A consumed is $N_{A|t=0} - N_{A,t}$. It is also the quantity of N_P formed. If added to $N_{P|t=0}$ it represents the amount of N_P at any time. Therefore,

$$N_{P|t} = (N_P + N_A)_{t=0} - N_{A,t}.$$

The rate of disappearance of A, in terms of $N_{A,t}$ and constants only, is

$$\frac{-dN_{A,t}}{dt} = N_{A,t}(k_1 + k_2) - k_2(N_P + N_A)_{t=0}. \tag{5.18}$$

Second-Order Kinetics • Assume a reaction of the type

$$A + B = pP + qQ + \cdots,$$

and assume that second-order kinetics apply. Then the rate of disappearance of component A is

$$-\frac{dN_{A,t}}{dt} = \frac{kN_{A,t}N_{B,t}}{V}, \tag{5.18a}$$

as previously noted. In order to maintain only one variable, $N_{A,t}$, we can write $N_{B,t}$ in terms of $N_{A,t}$. The quantity of component A consumed is $N_{A|t=0} - N_{A,t}$. The quantity of $N_{B,t}$ consumed is the same. Therefore, if $N_{B|t=0}$ the original quantity of component B is

$$N_{B,t} = (N_B - N_A)_{t=0} + N_{A,t},$$

which will be noted to reduce, as it should, to $N_{A,t}$ for the reaction

$$2A = pP + qQ + \cdots.$$

The rate of disappearance of component A then may be written

$$-\frac{dN_{A,t}}{dt} = \frac{kN_{A,t}[(N_B - N_A)_{t=0} + N_{A,t}]}{V}, \tag{5.19}$$

which is an equation in one variable at constant volume.

In the instance where $(N_B \gg N_A)_{t=0}$,

$$(N_B - N_A)_{t=0} + N_{A,t} \doteq N_{B|t=0} \doteq N_B.$$

Then

$$-\frac{dN_{A,t}}{dt} = \frac{kN_{A,t}N_B}{V} = \frac{k'N_{A,t}}{V}, \tag{5.20}$$

and the process, although truly second order, will act like the first order.

With these limited remarks on kinetic processes, the reader should have little difficulty handling elementary batch kinetic processes, granted that isothermal conditions pertain so that k will remain a time-independent parameter of the system.

5.7 The Steady State Tower-Position Dependent Concentrations*

The steady state exists and the process is represented in Figure 5.5. In this figure the tower is schematically represented with G_1 moles per unit time of a gas containing y_1 mole fraction of a solute gas charged to the tower. G_2 moles per unit time leave at the top, with y_2 mole fraction of the solute gas unabsorbed. The rates of the absorbent stream are named L_i with mole fractions of solute, x_i. In each stream, inlet conditions are subscripted 1 with outlet conditions subscripted 2. Each of these variables is time-invariant due to the assumption of the steady state. An increment of height is designated Δz, and any general position in the tower is z linear units above the tower bottom, where $z = 0$.

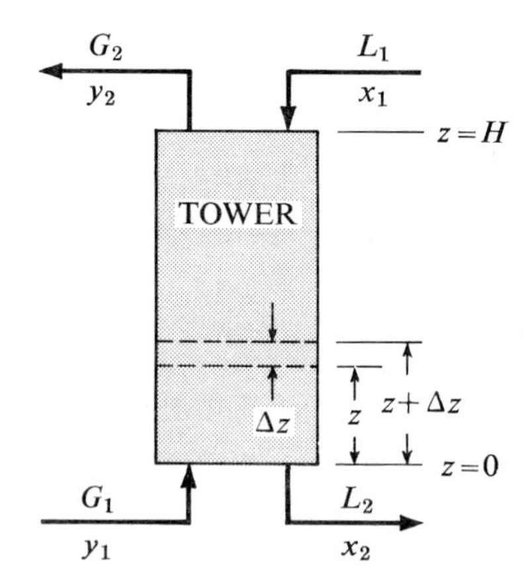

FIGURE 5.5 Continuous (absorption) column, operating in steady state.

In Figure 5.6, the mole fractions of each stream throughout the tower are represented diagrammatically. Mole fractions at general height, z, are y_z and x_z for gas and liquid streams, respectively; similarly $y_{z+\Delta z}$ and $x_{z+\Delta z}$ apply at height $z + \Delta z$.

• • • • •

* The reader may find this section easier to follow after a study of the analogous heat balance situation. This suggestion of a relatively disorganized approach is based upon the greater simplicity of driving force expressions in temperature units (heat balances) over those in concentration units (mass balances). The principles of position-dependent variables are identical in each.

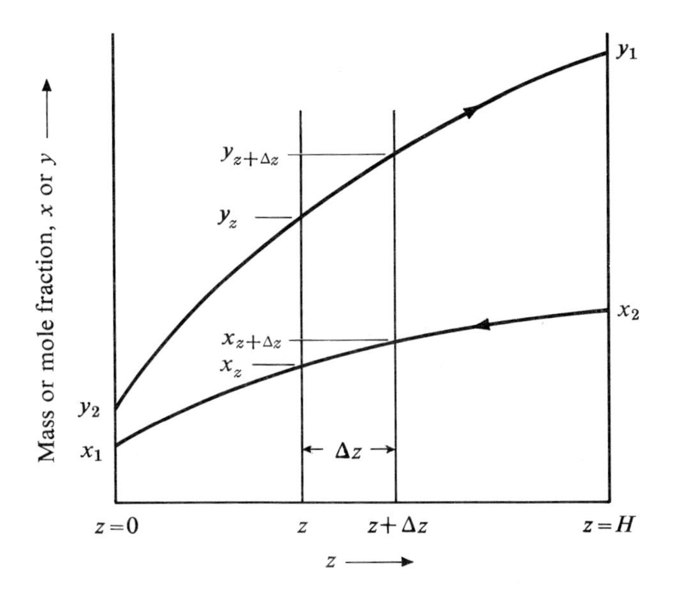

FIGURE 5.6 Compositions of two streams as functions of position in tower.

In order to design a column (i.e., specify the required height to remove a specific quantity of solute from G_1) one must set up an expression in which the relationship of y_z to z, or (x_z to z) is correctly stated; this expression must be integrated from $z = 0$ to $z = H$.

Make a material balance on a section of the tower of height Δz. This three-component system has only two components in each stream. Material balances can then be easily drawn on half of the column as described in Section S4.1 and as indicated in Figure 5.7. For the section there indicated, quantities of *solute* can be written as

Input: $(Gy)_z$.
Output: $(Gy)_{z+\Delta z}$ + solute in liquid.

As in the batch reactors where an action takes place within a section of the system, some independent expression is required for the second term of output. Assume that the *rate* of solute transfer is proportional to a driving force (i.e., difference in concentrations between the two phases) and the interfacial area of gas-liquid contact. Then using a characteristic rate constant, K, and an area per unit height, a, the average rate of solute removal can be expressed as

$$\tfrac{1}{2}Ka[(y_z + y_{z+\Delta z}) - (x_z + x_{z+\Delta z})]\,\Delta z^*.$$

The material balance solution becomes:

$$(Gy)_z = (Gy)_{z+\Delta z} + \tfrac{1}{2}Ka[(y_z + y_{z+\Delta z}) - (x_z + x_{z+\Delta z})]\,\Delta z.$$

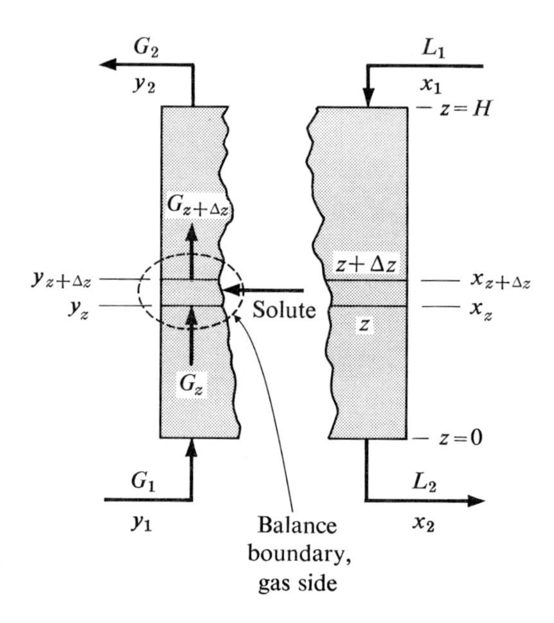

FIGURE 5.7 Gas and liquid streams segregated for convenience in analysis.

Performing the usual operations of dividing by Δz and taking the limit as $\Delta z \to$ zero:

$$\lim_{z \to 0} \frac{(Gy)_{z+\Delta z} - (Gy)_z}{\Delta z} = -Ka[y_z - y_z^*];$$

$$\therefore \quad \frac{d\,(Gy)_z}{dz} = -Ka[y_z - y_z^*]. \tag{5.21}$$

Equation 5.16 would be much easier to handle if G were not a variable. This condition would be met if G had been defined as moles of inert gas

• • • • •

* y_z is expressed in units of (moles of solute)/(moles of solute + moles of gas); x_z is expressed in units of (moles of solute)/(moles of solute + moles of solvent). These quantities cannot be subtracted without some provision for reconciling units (see p. 235).

Ka is expressed in units of

$$\frac{\text{moles of solute transferred}}{(\text{unit time})(\text{unit height})(\text{unit}\,\Delta y)}.$$

Therefore, y^* will replace x in Equation 5.21.

only, G_a, and mole fractions had been replaced by mole ratios of solute-to-inert, Y. Then Equation 5.21 would read

$$\frac{G_a\,dY_z}{dz} = -Ka(Y_z - Y_z^*)$$

and

$$\frac{dY_z}{dz} = -\left(\frac{Ka}{G_a}\right)(Y_z - Y_z^*). \tag{5.22}$$

In order to solve the problem of reconciling units of Y and X (also y and x) in the driving force expressions, some assumption will be required. The one ordinarily chosen states that Y^* (or y^*) is the gas phase concentration in *equilibrium* with the liquid phase concentration at any point. Note that this problem did not arise in distillation, where y and x are expressed in identical units, all components being present in each phase. (Nor will it arise in the heat balance where both terms in the driving force expression are in temperature units.) We need, therefore, two things:

a. an equilibrium relationship, which can be written in the general form $Y_z = \phi(X_z)$ or $y_z = \phi_1(x_z)$, and
b. a statement of X (or x) in the liquid phase corresponding to the Y (or y) at any general point in the tower.

Then the problem of obtaining an ordinary differential equation in one variable can be handled in three steps:

1. write Y_z as a function of X_z by a material balance on the solute.
2. solve this relationship for X_z in terms of Y_z.
3. use the equilibrium relationship to establish Y_z^* at the various X_z.

Obtaining Values of Y*

STEP 1 Cut the column at any arbitrary position, z, and write a component material balance from this position to *either* end. In Figure 5.8, the upper end of the column has been chosen arbitrarily.

Input: $L_sX_1 + G_aY_z$, where L_s = quantity of solvent.
Output: $L_sX_z + G_aY_2$.
Balance: $G_a(Y_z - Y_2) = L_s(X_z - X_1)$. (5.23)

STEP 2 Solve Equation 5.23 for X, the variable which must be eliminated from the driving force.

$$X_z = \frac{G_a}{L_s}(Y_z - Y_2) + X_1. \tag{5.24}$$

STEP 3 Suppose that the equilibrium relationship between Y and X is of the simple form

$$Y_z^* = QX_z, \qquad \text{where } Q \text{ is a constant;} \tag{5.25}$$

then from Equation 5.24,

$$Y_z^* = Q\frac{G_a}{L_s}(Y_z - Y_2) + QX_1. \tag{5.26}$$

Now Equation 5.22 may be integrated, since all variables are expressed in terms of Y and z only. Writing Equation 5.26 as

$$Y^* = \alpha Y + \beta, \qquad \text{where } \alpha \text{ and } \beta \text{ are constants,}$$

Equation 5.22 becomes, after separating variables,

$$\frac{dY_z}{(1-\alpha)Y_z - \beta} = -\frac{Ka}{G_a}\,dz. \tag{5.27}$$

The solution is

$$\frac{1}{1-\alpha}\ln\left[(1-\alpha)Y_z - \beta\right] = -\frac{Kaz}{G} + \text{const.}$$

If $Y_z^* = \phi(X_z)$ is a more complicated function than represented in 5.25, a solution may still be effected but numerical or graphical methods may be required.

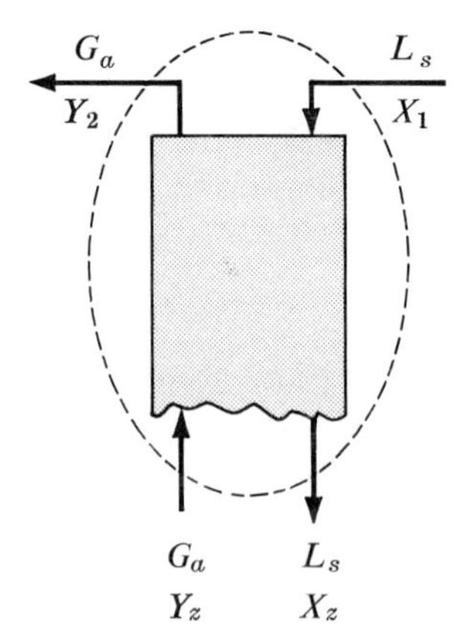

FIGURE 5.8 Upper half of steady state tower.

Equation 5.25 is an expression of Henry's law, and Q as the Henry's law constant. The use of α and β as constants assumes an isothermal process which is not far removed from the truth in many absorption processes. If heats of solution are so great as to invalidate this assumption, there is no change indicated in the above equations. However, $Y_z^* = \phi(X_z)$ can only be predicted at a known temperature. Simultaneous mass and heat balances are required for the solution of the problem.

We chose to write the driving forces in terms of gas phase concentrations. It is fairly obvious that we could equally well have chosen the liquid phase. In this case the driving forces would be in terms of X and the change in concentration in the differential term should be in the same units. Equation 5.22 would then become

$$\frac{dX_z}{dz} = \frac{-K_L a}{L_s}(X_z^* - X_z). \tag{5.28}$$

The primary differences between Equations 5.22 and 5.28, other than the obvious use of X in place of Y, are the substitution of L_s, the molal rate of the solvent, for G_a, the molal rate of the inert gas, *and* a change in the mass transfer coefficient. In Equation 5.28, $K_L a$ has the units of (moles of solute transferred)/(unit time)(unit height)(unit ΔX).

The *types of problems* associated with tower design are basically two in number. First, given the compositions and rates of the two input streams and the composition of solute in one of the output streams, how tall a tower must be designed. Second, given an existing tower (i.e., given height) and the compositions and rates of the two input streams, what are the outlet concentrations. The specifications in the second case might be varied to substitute the outlet gas concentration for the input solvent rate. Other variations of this sort will be obvious to the reader, in terms of the material in the Supplement to Chapter 4.

Example 5.1

▶ STATEMENT: A storage tank, perfectly mixed, contains M pounds of solution of concentration C_o of a salt-water solution. It acts as a source of supply to a process. The tank is full initially and is fed solution continuously from a salt and water mixer at a rate of R pounds per minute. You are to predict the concentration in the tank one hour after the mixer has failed to supply the concentration C_o. Either of two cases is important:

a. the concentration from the mixer remains constant during the hour in question but the concentration dropped suddenly to $C_o/2$ at the beginning of the hour.
b. at the beginning of the hour the concentration from the mixer is C_o but during the hour this concentration falls off linearly, following a time-dependence of the form

$$C_{1,t} = \alpha + \beta t,$$

where $\beta = -0.01$ concentration units per minute.

Determine the concentration in the tank after 60 min when

$$C_o = 0.25, \qquad \text{expressed as concentration units,}$$

and

$$R = \frac{M}{60}, \qquad \text{expressed as pounds per minute.}$$

▶ SOLUTION: The applicable differential equation, disregarding changes in density, and recognizing that the continuously full tank means that M_t in Equation 5.3 is constant, is readily obtained from Equation 5.3.

$$\underset{\text{Input}}{RC_{1,t}} = \underset{\text{Output}}{RC_t} + \underset{\text{Accumulation}}{\frac{M\,dC_t}{dt}}.$$

PART (a)

$$C_{1,t} = \frac{C_o}{2}.$$

$$\frac{dC_t}{dt} + \frac{R}{M}\left(C_t - \frac{C_o}{2}\right) = 0.$$

Since $dC_t = d(C_t - C_o/2)$, the equation is homogeneous. Variables may be separated and each side integrated to give

$$\ln\left(C_t - \frac{C_o}{2}\right) = -\left(\frac{1}{\theta}\right)t + \text{const},$$

where $\theta = M/R$ and represents the average time of contact, or residence of any particle fed. From this expression we obtain

$$C_t - \frac{C_o}{2} = Ke^{-t/\theta}.$$

K may be evaluated from the initial conditions; at $t = 0$, $C = C_o$. Therefore,

$$\frac{C_t}{C_o} = \frac{1}{2}(1 + e^{-t/\theta}).$$

When $\theta = 60$ and $t = 60$, $C_t/C_o = 0.684$, and $C_t = 0.171$ concentration units.

PART (b)

$$C_{1,t} = \alpha + \beta t.$$

Then,

$$\frac{dC_t}{dt} + \frac{R}{M}C_t = \frac{R}{M}(\alpha + \beta t). \tag{5.29}$$

Replacing M/R by θ, the homogeneous part of the solution is $Ke^{-t/\theta}$. We designate this as $C_{H,t}$, and now obtain the portion of the solution due to the particular term by the method of undetermined coefficients, designating this part as $C_{p,t}$. (See Supplement to Chapter 5.)

The form of the terms on the right-hand side of the differential equation tells us that the particular integral should contain two terms, one a constant, and one linear in t. We assume a solution:

$$C_{p,t} = P + Qt.$$

Substituting this trial solution into the left-hand side of Equation 5.29 we obtain:

$$\frac{dC_{p,t}}{dt} = Q,$$

$$\frac{C_t}{\theta} = \frac{P}{\theta} + \frac{Q}{\theta}t.$$

Adding these terms we find that the left-hand side of the equation, for this trial solution, equals

$$\left(Q + \frac{P}{\theta}\right) + \left(\frac{Q}{\theta}\right)t.$$

In order for this expression to equal the right-hand side of Equation 5.29, the following equalities of coefficients must exist:

$$Q + \frac{P}{\theta} = \frac{\alpha}{\theta},$$

$$\frac{Q}{\theta} = \frac{\beta}{\theta}.$$

Therefore,

$$P = \alpha - \beta\theta,$$

and

$$Q = \beta.$$

The general solution is the sum of the complementary and particular integrals:

$$C_t = C_{H,t} + C_{p,t} = Ke^{-t/\theta} + (\alpha - \beta\theta) + \beta t.$$

At the initial conditions where $C_t = C_o$ when $t = 0$,

$$K = C_o - \alpha + \beta\theta.$$

But, in the cases to be considered, $\alpha = C_o$, so that K is equal to $\beta\theta$ alone. Therefore, the complete solution is

$$C_t = \beta\theta(e^{-t/\theta} - 1) + \beta t + C_o. \tag{5.30}$$

After one hour of operation under these conditions,

$$C_t = \beta\theta\left(\frac{1}{e} - 1\right) + 60\beta + C_o.$$

For the condition that $\beta = -0.01$,

$$C_t = C_o - 0.221.$$

Then $C_t = 0.029$, granted that the solution is applicable over the entire hour. However, it is quite apparent that the concentration from the mixer to the tank will be zero after 25 min of operation. Therefore, Equation 5.30 holds only for 25 min after which we must apply a differential equation with zero concentration in the feed stream.

After 25 min, we find, from Equation 5.30,

$$C_t = 0.204.$$

For the last 35 min, the process operates with $C_1 = 0$. The right-hand side of Equation 5.29 then disappears and the differential equation is homogeneous. The solution is

$$\frac{C_t}{C_o} = e^{-t/\theta}.$$

Substituting $t = 35$ and $C_o = 0.204$, we find C_t after 60 *total* min of operation to be 0.114 concentration units.

Example 5.2

▶ STATEMENT: In the liquid phase, trimethyl amine (TMA) and normal propyl bromide (NPB) combine in equal molal proportions to give a quaternary ammonium salt. J. M. Smith, in *Chemical Engineering Kinetics* (McGraw-Hill Book Co., 1956), shows that this reaction proceeds according to second-order kinetics and reports a rate constant, at 140 °C, of 1.67×10^{-3} liters/(gm-mole)(sec).

The process appears to be promising to your company provided 99% consumption of the bromide is reasonable. You are asked to predict the time to attain this completion for a 5-liter laboratory kettle, starting with a stoichiometric mixture of reactants, each 0.2 molar in a benzene solvent.

▶ SOLUTION: Since the calculation is exploratory, assume as a first approximation that the volume is unchanged during mixing and during reaction. Then, from Equations 5.16 and 5.18a, we can write, using the first letter of each coded name, and omitting the subscripts, t, on the numbers of moles, N_i,

$$-\frac{dN_N}{dt} = \frac{kN_N N_T}{V}.$$

Due to the equal molal proportions

$$N_T = (N_T - N_N)_{t=0} + N_N = N_N, \qquad \text{since} \qquad (N_T = N_N)_{t=0}.$$

Therefore,

$$-\frac{dN_N}{dt} = \frac{kN_N^2}{V}.$$

Separating variables,

$$\frac{dN_N}{N_N^2} = -\frac{k\,dt}{5}.$$

Integrating,

$$\frac{1}{N_N} = \frac{kt}{5} + \text{const.}$$

The constant is obtained from the initial conditions, $t = 0, N_N = N_{N|t=0} = 0.5$. The complete solution is

$$\frac{1}{N_N} - \frac{1}{N_{N|t=0}} = \frac{kt}{5},$$

$$N_N = 1\bigg/\left(\frac{kt}{5} + \frac{1}{N_{N|t=0}}\right).$$

Substituting the known k, the fact that $N_{N|t=0} = 0.5$ (each reactant alone is 0.2 and they are mixed in equal proportions without volume effects), and N_N is (0.01)(0.5), we find that $t = 594{,}000$ sec $= 165$ hr $= 6.86$ days. The process operated in this fashion is impractical. We abandon it or try another approach, as in Example 5.3.

Example 5.3

▶ STATEMENT: Since consumption of the NPB is the prime objective, and since the product will be easily separated by dissolving the ammonium salt in water, we could consider a large excess of TMA as a possible means of forcing a faster conversion. Then, using as a first approximation that $N_T = N_{T|t=0}$, pseudo first-order kinetics may be assumed.

Let the charge be in the ratio of 19:1, favoring the TMA. Again calculate the time required to obtain 99% conversion of the NPB.

▶ SOLUTION: Again referring to Equations 5.16 and 5.18a we can write:

$$-\frac{dN_N}{dt} = \frac{kN_{T|t=0}\,N_N}{5}.$$

This differential equation is homogeneous with the complete solution:

$$\frac{N_N}{N_{N|t=0}} = \exp\left(\frac{-kN_{T|t=0}t}{5}\right).$$

By the specification of the problem $N_N/N_{N|t=0} = 0.01$, from which it is easy to calculate the time as 1.45×10^4 sec, or 4.02 hr. A vast improvement in time requirement is apparent, but the process does not, even under these conditions, appear to be commercially feasible.

Example 5.4

▶ STATEMENT: The height of an absorption tower is required. The tower is to be designed to handle:

a. 10,000 CFM of gas, measured at 20 °C and 1.00 atm, containing 15% of solute A.
b. effluent gas is to contain only 1% of A.
c. solvent is fed pure.
d. effluent solution from the tower shall be 20 mole % of A.

For this system the transfer coefficient, Ka, may be taken to be 10 moles of A transferred/(foot of height)(minute)(unit change in Y). The equilibrium between gas and liquid phase compositions may be expressed, for all conditions to be encountered in the tower, as

$$Y_z^* = 0.3X_z.$$

▶ SOLUTION: The situation is as represented in Figure 5.5. The first step in the solution will be the completion, by material balances, of all entering and leaving quantities. The second step, also by material balance, will be the development of a general relationship of Y_z and X_z at any general point in the column. Finally, this expression will be used to establish Y_z^* in Equation 5.22 and the equation will be integrated to give the desired height.

BASIS The basis of 1.0 min or 10,000 cu ft at the inlet conditions is as satisfactory as any. We convert immediately to moles.

$$\frac{10{,}000 \text{ ft}^3}{359(293/273) \text{ ft}^3/\text{mole}} = 26.0 \text{ moles of gas charged.}$$

(26.0)(0.15) = 3.90 moles of A charged in gas stream.
Moles of A charged in liquid stream = 0.
Total charge of A = 3.90 moles.
Moles of inert gas: 26.0 − 3.9 = 22.1.

The inert gas is constant throughout the column and represents G_a in Equation 5.22, for subsequent use. It also provides the means of calculating the moles of A in the effluent gas stream:

$$\text{Total effluent gas: } \frac{22.1}{0.99}.$$

$$\text{Total } A \text{ in effluent gas: } \left(\frac{22.1}{0.99}\right)(0.01) = 0.223 \text{ moles } A.$$

By material balance, moles of A charged (3.90 moles) are equal to the moles of A lost in effluent gas (0.223 moles) plus moles of A in the liquid product. The last figure is then 3.68 moles. Since the liquid is 20 mole % A, the total moles of solution is 3.68/0.20 = 18.90, and the total moles of solvent charged is 15.22. This latter quantity is L_s in Equation 5.23.

Now drawing a material balance for the solute around the top of the column, as indicated in Figure 5.8, we find:

$$X_z = \frac{G_a}{L_s}(Y_z - Y_2) + X_1.$$

$G_a/L_s = 22.1/15.22 = 1.45$, from previously calculated quantities. Also X_1 is known to be zero and $Y_2 = 0.01/0.99 = 0.0101$. Therefore,

$$X_z = 1.45 Y_z - 0.0146.$$

Since $Y^* = 0.3X$,

$$Y_z^* = 0.435 Y_z - 0.00438,$$

and

$$Y_z - Y_z^* = 0.565 Y_z + 0.00438.$$

Now applying Equation 5.22,

$$\int_{0.1765}^{0.0101} \frac{dY_z}{0.565 Y_z + 0.00438} = -\frac{10}{22.10} \int_{z=0}^{z=H} dz,$$

where $0.1765 = 0.15/0.85$. Integrating,

$$\frac{1}{0.565} \ln \frac{0.00571 + 0.00438}{0.0996 + 0.00438} = -\frac{1}{22.10} H,$$

$$H = 9.14 \text{ ft}.$$

LIST OF SYMBOLS

(Excluding chemical symbols.)

Latin letters

C	Concentration: mass fraction (tank purging); moles per unit volume (kinetics)
e	Exponential operator
F	Feed rate, distillation, mass or moles per unit time
G	Gas rate, moles per unit time
G_a	Inert gas rate, moles per unit time
K	Constant of integration *or* absorption capacity factor, moles absorbed per (unit area)(unit time)(unit concentration difference)
k	Constant
L	Liquid rate, moles per unit time
L_s	Solvent liquid rate, moles per unit time
M	Mass
N	Moles
P	Constant
Q	Constant

R_i Mass rate, component i
R_t Mass rate as a function of t
$R(t)$ Mass rate as function of time
$R_{i,t}$ Mass rate, component i, as a function of t
r Rate of disappearance of chemical species, kinetics
t Time
V Reactor volume
$W_{m,i}$ Molecular weight, component i (component designation omitted when no loss in clarity results)
X Mole ratio, solute to solvent
x Mole fraction in liquid
x_F Mole fraction in liquid feed
Y Mole ratio, solute to inert gas
Y^* Mole ratio, solute to inert gas, in physical equilibrium with mole ratio, X, in liquid
y Mole fraction in vapor or in gas
y^* Mole fraction in gas in equilibrium with mole fraction, x, in liquid
z Height

Greek letters

α Constant
β Constant
θ Residence time, M/R

PROBLEMS

5.1 A tank containing 200 gal of saturated salt solution (3 lb of salt per gal) is to be diluted by the addition of brine containing 1 lb of salt per gal. If this solution runs into the tank at the rate of 4 gal per min and the mixture runs out at the same rate, when will the concentration in the tank reach 1.01 lb per gal?

5.2 A salt solution in a perfectly stirred tank is washed out with fresh water fed at a rate such that the time of contact is 10 min. Calculate:

a. time in minutes to wash out 99% of the salt originally present.
b. percentage of the original salt removed by the addition of one tank full of fresh water.

5.3 In a homogeneous organic reaction, a miscible liquid catalyst is used in a concentration of 2% in one of the reactants. This reactant is mixed with the catalyst in a well-stirred 10,000-gal tank.

It is now decided to change the type of catalyst and the change will be made by mere substitution of the new catalyst for the old as the reactants and catalysts, in proper proportions, are continuously pumped into the holding tank at a total rate of 15 gal per min. This rate is the same at which material is being withdrawn from the tank and passed to the reactor. Calculate the time in minutes required for concentration of the new catalyst to reach 1.9%.

5.4 The air in a large meeting room is changed every two minutes. During a convention, the CO_2 concentration prior to the use of this room is 0.04%. After 1 hr of use, the concentration has quadrupled. Calculate:

a. CO_2 concentration after 6 hr continuous use with the same size audience.
b. ventilation rate required to maintain the concentration at 0.1% CO_2 after attainment of the steady state with the same audience.

5.5 A tank is filled with 100 gal of brine whose concentration is 0.5 lb per gal. The tank is perfectly stirred and fresh water is fed at a rate of 3 gpm. The effluent from the tank, also at 3 gpm, passes to a second perfectly stirred 100-gal tank initially filled with pure water. Calculate:

a. brine concentration in the second tank at the end of 1 hr.
b. brine concentration in the second tank at the end of 1 hr if the second tank were originally filled with brine whose concentration was 0.25 lb per gal.
c. brine concentration in the second tank at the end of 1 hr if the second tank were originally filled with brine whose concentration was 0.5 lb per gal.

5.6 A tank of 100-gal capacity is filled with brine whose concentration is 0.5 lb per gal. While the tank is being stirred so vigorously that perfect mixing may be assumed, fresh water is added at the rate of 2.5 gal per min.

The effluent from this first tank passes to a second tank of identical dimensions, initially filled with brine whose concentration is 0.25 lb per gal. Calculate the time when the concentration in the second tank is a maximum.

5.7 Three perfectly stirred tanks, each of 10,000-gal capacity, are arranged so that the effluent of each passes as feed to the next tank in series. Initially the concentration in each tank is C_0. Pure water is then fed to the first tank at a rate of 50 gpm. Calculate:

a. time required to reduce the concentration in the first tank to $C_0/10$.
b. concentrations in the other two tanks at this time.
c. a general equation for the nth tank in a cascade system of this type.

5.8 Two tanks, with capacities of 1000 gal and 900 gal, respectively, are each perfectly stirred and are connected so that effluent from the first passes to the second.

Both tanks are initially filled with a solution whose concentration is 1 lb per gal. Pure water is fed to the first tank at a rate of 100 gal per min. Calculate:

a. the complete solution to the differential equation representing concentration in the second tank as a function of time.
b. the time at which the concentration in the second tank is one-half its initial value.
c. proof of the statement that "the solution of this tank purging problem approaches that for two tanks of equal size when the size of the second tank approaches that of the first."

5.9 In an arrangement of tanks described in Problem 5.8, the second tank is initially filled with pure water. Calculate:

a. time at which concentration in the second tank is a maximum.
b. time at which concentrations in the two tanks are identical.

5.10 In an arrangement of tanks as described in Problem 5.8, the feed to the tanks has an initial concentration of 1.5 lb per gal.

At the end of 1 hr the metering pump that maintains this concentration breaks down. The breakdown is not discovered for another hour during which time pure water is fed to the tanks at the initial water rate. The density change is negligible. Calculate:

a. the concentration in each tank when the breakdown is discovered.
b. the concentration which should exist in each tank at the end of the second hour.

5.11 A tank, 3 ft in diameter and 4 ft high, is half full of water. Water is fed to the tank through a 2-in. pipe, at a velocity of 16 ft per sec. Water leaves the tank through a 1-in. line in the bottom of the tank, where a valve restricts the flow so that the velocity in the 1-in. line may be expressed as

$$u = kh,$$

where

u = velocity, ft per sec.
h = height of water in the tank, ft.
k = a dimensional constant, with units of reciprocal seconds, and a value of 8.

Using material balances, set up the ordinary differential equation for this situation. Solve the equation and evaluate the constant, or constants, for the boundary conditions stated. Prove that the solution is the correct solution by substitution back in the O.D.E. (ordinary differential equation). Answer the following questions:

a. will the tank fill or empty?
b. if the tank fills, how much time is required to reach 99% full, or to reach the steady state if the latter occurs before the tank is full.
c. if the tank empties, how much time will be required to empty the tank, or to reach the steady state if the latter condition occurs before the tank is empty?
d. if the steady state occurs, what is the level in the tank?

5.12 A cylindrical tank, holding water, is equipped with several drain lines, each going to another piece of equipment. At one time, two of them are open, one consisting of a 1-in. diameter line centered 1 in. from the bottom, and the other consisting of a 2-in. diameter line centered 1 ft from the bottom. The velocity in the two lines may be approximated by the relationship:

$$V = 0.9a\sqrt{2gh},$$

where

V = flow rate, cubic feet per sec.
a = cross-sectional area of the line.
g = gravitational constant.
h = height of liquid above center line of outlet.

Obviously these quantities must be expressed in consistent units.

If no water is added to the tank, calculate the time required to drain half the volume of water originally present.

If water is added at a rate of 50 cu ft per min, calculate:

a. the height of water after 1 hr.
b. the time to reach steady state, if any.

5.13 Two vertical tanks, each 4 ft deep and 4 ft in diameter, are connected at the bottom by a short horizontal pipe 2 in. in diameter.

Initially the pipe is plugged, one tank is full and the other tank is empty. When the plug is pulled, the velocity through the pipe is given by

$$u = 0.8(2gh)^{1/2},$$

where u is the velocity in feet per second and h is the difference between the levels in the two tanks. Calculate the time for the levels in the two tanks to become equal.

5.14 One pound mole of benzene is heated in an open dish from 10 to 70 °C. The temperature rise is uniform and the time required is 50 min. (Note that this specification eliminates the need for a heat balance.)

The rate of vapor loss may be represented as kp moles per min; p is the vapor pressure of benzene and k is a constant.

The measured loss of benzene during this heating period is 0.5 moles. Obtain an equation from which the value of k might be evaluated assuming that latent heat of benzene is constant and equal to 7350 cal per gm-mole.

5.15 In a first-order rate process, A reacts to form B. The amount of A present at the end of 1 hr is 480 lb and at the end of 3 hr is 120 lb. Calculate the amount of A initially present.

5.16 If in the simple reversible reaction, $A = B$, both forward and reverse reactions are first order, then the rate equation for the disappearance of species A is

$$-\frac{dc_A}{dt} = k_1 c_A - k_2 c_B,$$

where k_1 and k_2 are the rate constants for the forward and reverse reactions, respectively. Obtain:

a. the general form of the integrated equation.
b. complete solution for the initial condition $c_A = 1$, $t = 0$.

5.17 A simple batch still is charged with a mixture of A and B, analyzing 20% A. The mixture is distilled until the liquid remaining in the still analyzes 10% A. Calculate the fraction of the initial charge which was distilled when the equilibrium x-y relationships are:

A. $y = 3x$.

B.

y	0.60	0.52	0.40	0.23	0.
x	0.20	0.15	0.10	0.05	0.

5.18 A simple still, as in Problem 5.17, is fed continuously during the distillation. The feed rate is double the vaporization rate and the feed is of the same composition as the original charge. When $y = 3x$, calculate:

a. Composition in the still relative to that of the original charge after 1 hr if the feed rate is such as to provide a quantity equal the original charge in that time.
b. The minimum composition which can ever exist in the still regardless of the length of time of operation.
c. The conditions corresponding to part (a) if the feed rate is half that of the vaporization.

5.19 You are to calculate the height of a column needed to perform the following absorption:

The gaseous mixture is composed of a solute, A, and an inert gas, B. The feed stream amounts to 10,000 CFM at 30 °C and 1 atm and analyzes 10% A. The exit gas is to be reduced in component A so that the latter amounts to only 0.1% of the stream. The process can be assumed to operate at constant pressure of 1 atm.

The solvent is fed pure and the exit liquor concentration desired is 20 mole %.

The gas-liquid equilibrium relationship can be satisfactorily approximated as $Y^* = 0.32X$ over the entire range of operation; Y is expressed as moles of A per mole of B and X is expressed as moles of A per mole of solvent.

Under the operating conditions described a capacity coefficient can be estimated to be 10 moles of B transferred per [(min)(unit height)(unit difference in Y)].

5.20 Repeat Problem 5.19 when $Y = 0.1X + 0.2X^2$.

5.21 Repeat Problem 5.19 when the following data represent the Y-X relationship:

Y:	0.0	0.03	0.06	0.09	0.15	0.30.
X:	0.0	0.20	0.40	0.60	0.80	1.00.

SUPPLEMENT TO CHAPTER 5

Solution of Ordinary Linear Differential Equations

A number of types of differential equations have been encountered in this chapter. In general their solution is quite straightforward and elementary. For those who have studied only differential and integral calculus, the solution of some of the examples and some of the problems will not be readily apparent. A few elementary principles of ordinary linear differential equations are, therefore, presented here briefly, with no intention of comprehensive coverage. The methods summarized will provide solutions for most of the problems encountered in this text and for many types of chemical engineering problems. For more rigorous proofs and for more extensive discussion reference should be made to any standard text. Sources include T. Apostol, *Calculus, Vol. II* (Blaisdell Publishing Co., 1962); Mickley, Sherwood, and Reed, *Applied Mathematics in Chemical Engineering* (McGraw-Hill Book Co., 1957); and S. Ross, *Differential Equations* (Blaisdell Publishing Co., 1964).

S5.1 Definitions

Notations

$$\frac{dy}{dx} = y' = Dy,$$

where D is obviously d/dx;

$$\frac{d^2y}{dx^2} = y'' = D^2y, \text{ etc.,}$$

which results in the general statement,

$$\frac{d^ny}{dx^n} = y^{(n)} = D^ny.$$

Differential Equations • equations involving derivatives or differentials of one or more dependent variables with respect to one or more independent variables.

Ordinary Differential Equations • equations involving derivatives of one or more dependent variables with respect to a single independent variable, and consequently not involving any partial derivatives.

Order • defined by the order of the highest derivative in the equation. That is, if the equation contains terms in d^ny/dx^n, the order is determined by the highest value of n.

Degree • the power of the *highest derivative* after the equation has been rationalized and cleared of fractions. For example, Mickley, *et al.* report the differential equation representing the curvature of a liquid surface to be

$$\frac{d^2y}{dx^2} = f(x)\left[1 + \left(\frac{dy}{dx}\right)^2\right]^{3/2}.$$

The order is 2, because of d^2y/dx^2; the degree is 2, because d^2y/dx^2 is squared in rationalizing.

Linear Differential Equations • when (a) every dependent variable and every derivative involved occurs to the first degree only, and (b) neither products nor powers of dependent variables exist nor products of dependent variables with differentials. Both of the following equations are nonlinear, the first because of violation of condition (a) and the second because of violation of condition (b).

$$\text{(a)} \quad \frac{dy}{dx} + y^2 = f(x).$$

$$\text{(b)} \quad \frac{d^2y}{dx^2} + y\frac{dy}{dx} + ay = 0.$$

Homogeneous Equation • one in which all the terms contain y in some form; i.e., all terms contain $D^n y$, where $n = 0, 1, 2, \ldots$.

Constant Coefficients • no coefficients of the $D^n y$ shall contain either variable.

General Solution • a solution of the differential equation which contains unknown constants.

Complete Solution • all unknown constants have been evaluated from boundary conditions.

Boundary Conditions • conditions of x for which y, or some $D^n y$ are known from the specification of the problem. The commonest is a known y at $x = 0$; this boundary condition is usually referred to as initial conditions. The known boundary conditions must be of a number equal to the order of the equation.

Examples

$$K_1 y' + K_2 y = 0, \tag{A}$$

$$y'' + Ky' + K_1 y = x^2, \tag{B}$$

$$y''' + x^3 y = \sin x, \tag{C}$$

$$x(y')^{3/2} + y^{1/2} = f(x), \tag{D}$$

$$y'' + x^2 y' + y^2 = e^{ax}. \tag{E}$$

	A	B	C	D	E
Order	1	2	3	1	2
Degree	1	1	1	3	1
Linearity	yes	yes	yes	no	no
Homogeneity	yes	no	no	no	no
Ordinary	yes	yes	yes	yes	yes
Constant coefficients	yes	yes	no	no	no

S5.2 The General Ordinary Linear Differential Equation

Three notations have been suggested. The same equation, in all three notations, is:

$$a_n(x)\frac{d^n y}{dx^n} + a_{n-1}(x)\frac{d^{n-1}y}{dx^{n-1}} + \cdots + a_1(x)\frac{dy}{dx} + a_0(x)y = f(x),$$

$$a_n(x)y^{(n)} + a_{n-1}(x)y^{(n-1)} + \cdots + a_1(x)y' + a_0(x)y = f(x),$$

$$[a_n(x)D^n + a_{n-1}(x)D^{n-1} + \cdots + a_1(x)D + a_0(x)]y = f(x).$$

The last of the three is the neatest, since it can be compacted to

$$L_n(D)y = f(x). \tag{S5.1}$$

This last notation states that a linear combination of nth order (i.e., L_n) derivatives of y (i.e., $(D)y$) is equal to a function of x. This statement is exactly the same as that describing the first three notations.

In these general expressions, the coefficients are all listed as functions of x. This specification is consistent with the definition of an ordinary differential equation. Only in the instance that all $a_i(x)$ are constants (i.e., all may be written as a_i) does the equation have constant coefficients.

Of the many possible methods of attack on the O.D.E., only three will be considered here. They are chosen for simplicity and adaptability to problems encountered in this text and in chemical engineering in general.

In the order in which they will be presented, the first of the three methods is named *separation of variables.* It is limited to homogeneous, first-order equations but does not require that the coefficients be constant. The second is also limited to first order, but removes the restriction of homogeneity. It also has no restriction on the coefficients. The method is known for its attack through *integrating factors.* The third method *is* limited to constant coefficients, but has no restrictions to first order or to homogeneity. It is called the method of *undetermined coefficients.*

S5.3 Separation of Variables

When the first-order O.D.E. is homogeneous, it is frequently possible to separate variables and integrate each side. If the equation can be rearranged to appear in the form

$$f_1(x)\,dy + f_2(y)\,dx = 0,$$

it is readily apparent that it may also be written as

$$-\frac{dy}{f_2(y)} = \frac{dx}{f_1(x)}, \tag{S5.2}$$

and each side can be integrated. For example, the equation

$$y' + \frac{y}{a - x} = 0$$

may be written

$$\frac{dy}{y} = -\frac{dx}{(a - x)}.$$

Integrating each side,

$$\ln y = \ln (a - x) + K = \ln (a - x) + \ln K_1,$$

where K is a constant of integration; then

$$y = (a - x)e^K = K_1(a - x).$$

Even when this simple separation is not readily apparent, a *transformation of variables* may result in an equation in x and a new variable $v = \phi(y, x)$ which will result in

$$f_1(x)\, dv + f_2(v)\, dx = 0.$$

Then this equation meets the requirement of separable variables. After integration and statement of the solution for v in terms of x, $\phi(y, x)$ can be substituted for v, regaining the original variables. For example, if one has the equation

$$(xy^2 - y)\, dx - x\, dy = 0, \tag{S5.3}$$

direct separation of the variables is not possible since the coefficient of dx cannot be made into a function of y alone. Suppose that a new variable, v, be defined,

$$v = xy.$$

Then

$$dv = y\, dx + x\, dy,$$

and

$$dy = \frac{(dv - y\, dx)}{x}.$$

Substituting this expression in the original Equation S5.3, and also substituting v/x for y in the coefficient of dx, we obtain

$$\left(\frac{v^2}{x}\right) dx - dv = 0.$$

Separating variables and integrating each side,

$$\ln x = -\left(\frac{1}{v}\right) + K.$$

Replacing v by xy and rearranging, we obtain

$$y = \frac{1}{Kx - x \ln x}.$$

Choice of Transformation

Unfortunately there are no hard and fast rules for establishing suitable transformations. The two most frequently used are $v = xy$ and $y = xv$ (or $v = y/x$). One of these two is sure to work when only the terms xy or y/x appear as coefficients. Note that xy did appear in the coefficient of dx in the example, since it could have been written $y(xy - 1)$. A few examples follow, with transformations which will effect separation, written in terms of the general homogeneous equation:

$$f_1(x, y)\, dy = f_2(x, y)\, dx.$$

$f_1(x, y)$	$f_2(x, y)$	v
$y - xy - x^2$	x^2	y/x
$(x - y)^2$	a	$x - y$
$2y^2 - 3x$	$2xy$	x/y^2
$x^2 - 2y^3$	$3xy^2$	x^2/y^3
x^2	$xy + 2y^2$	y/x
x	$y + x^2 \sin (y/x)$	y/x
$1 + y^2e^{2x}$	y	ye^{-mx}, where m is an unknown constant.

Thus, although no rules may be established for selecting the transformation, some guidelines can be drawn for initial attempts. If we divide $f_1(x, y)$ by $f_2(x, y)$, or vice versa, we find the ratio y/x existing exclusively when the transformation $v = y/x$ applies. In the second case, there is no change effected in $f_1(x, y)$ by the division so that the function itself affords the transformation. In the third and fourth cases, y/x does not appear exclusively. In these two cases the division gives $(y/x) - 3/y$ and $(x/y^2) - 2y/3x$, respectively. Dividing the first term by the second in each case produces the transformation suggested. The last example is taken from Apostol.* The procedure suggested above would indicate that a reasonable transformation

• • • • •

* T. Apostol, *Calculus, Vol. I* (Waltham, Mass.: Blaisdell Publishing Co., 1962), p. 244.

might be $v = y^2e^{2x}$, a transformation which will be found to be successful. With v expressed as the square root of this function,

$$v = ye^x,$$

the variable will also be separable.

S5.4 Integrating Factors for Linear, First-Order O.D.E.

In order to discuss integrating factors, we need first to have some knowledge of exact differentials. The equation,

$$M(x, y)\,dx + N(x, y)\,dy = 0, \tag{S5.4}$$

is exact if

$$\frac{\partial M}{\partial y} = \frac{\partial N}{\partial x}. \tag{S5.5}$$

Then a function, u, exists such that

$$du = \frac{\partial u}{\partial x}\,dx + \frac{\partial u}{\partial y}\,dy = 0.$$

This equation has the obvious solution for u,

$$u = K, \qquad \text{where } K \text{ is a constant.} \tag{S5.6}$$

Thus, if we can determine the function, u, which is itself a function of x and y, we have the solution of the equation (S5.4). Knowing both M and N, it is usually quite easy to determine u by inspection. If this method fails, there are formalized techniques for which the reader is referred to the standard texts previously mentioned.

As examples, take the equations

$$ye^x\,dx + e^x\,dy = 0,$$

and

$$(3x^2 + 4xy)\,dx + (2x^2 + 2y)\,dy = 0.$$

Both are exact differentials as may be confirmed by applying the criterion of Equation (S5.5). In the first case it is quite apparent that, if $u = ye^x$,

$$\frac{\partial u}{\partial x} = ye^x \qquad \text{and} \qquad \frac{\partial u}{\partial y} = e^x,$$

the original equation may be written in the form

$$du = \frac{\partial u}{\partial x}\,dx + \frac{\partial u}{\partial y}\,dy = 0.$$

Therefore, from Equation (S5.6),

$$u = ye^x = K \qquad \text{or} \qquad y = Ke^{-x}.$$

This last result would also be obtained by separating variables and solving the equation,

$$\frac{dy}{dx} + y = 0,$$

to which our first expression is obviously equivalent.

The second expression is somewhat more complicated. Combine coefficients of dx and dy in such a manner that each grouping represents a differential of a product of variables or a differential of a single variable. Rewriting to provide this form

$$(4xy\,dx + 2x^2\,dy) + (3x^2\,dx) + (2y\,dy) = 0.$$

This equation can be rewritten as

$$d(2x^2y) + d(x^3) + d(y^2) = 0 = du \qquad \text{or} \qquad d(2x^2y + x^3 + y^2) = 0 = du.$$

Then,

$$u = 2x^2y + x^3 + y^2 = K,$$

which is the solution to this differential equation. This method is termed the grouping of coefficients.

We are thus able to solve the differential equations when exact differentials are involved. If they are inexact, and may be made exact by multiplication of all terms by a function of x, we may apply the technique of exact differentials for obtaining the solution. In our first example the equation would normally be stated as

$$y' + y = 0.$$

Writing this equation in the form

$$dy + y\,dx = 0,$$

we can readily test to show that it is inexact. But by multiplying both terms by e^x, we have made it exact. Therefore, e^x is an integrating factor. The determination of such factors is then our next operation.

Evaluation of Integrating Factors

First write the first-order, linear O.D.E. in general terms.

$$\frac{dy}{dx} + P(x)y = F(x). \tag{S5.7}$$

Note that this equation not only has variable coefficients but also is nonhomogeneous. Now reorganize the terms,

$$dy + [P(x)y - F(x)]\,dx = 0,$$

so that it is in the form $M\,dy + N\,dx = 0$. Now if the criterion for exactness, Equation S5.5, is applied, we shall find that the equation is inexact. Therefore, we multiply by some undetermined function of x, $\mu(x)$, and then solve for the function so that the criterion of exactness is met.

Multiplying by $\mu(x)$,

$$\mu(x)\,dy + \mu(x)[P(x)y - F(x)]\,dx = 0.$$

Now,

$$\mu(x) = N \qquad \text{and} \qquad \mu(x)[P(x)y - F(x)] = M,$$

and

$$\frac{\partial N}{\partial x} = \frac{\partial \mu(x)}{\partial x} = \mu'(x), \qquad \frac{\partial M}{\partial y} = \mu(x)P(x).$$

Then if $\mu(x)$ is properly chosen,

$$\mu'(x) = \mu(x)P(x).$$

Separating variables,

$$\frac{\mu'(x)}{\mu(x)} = P(x).$$

Therefore,

$$\ln \mu(x) = \int P(x)\,dx \qquad \text{and} \qquad \mu(x) = \exp\left[\int P(x)\,dx\right].$$

We can then make the general statement that $\exp\,[\int P(x)\,dx]$ will always be the correct integrating factor.

Applying this factor to the general Equation S5.7, we obtain

$$\exp\left[\int P(x)\,dx\right]y' + \exp\left[\int P(x)\,dx\right]P(x)y = \exp\left[\int P(x)\,dx\right]F(x).$$

Examination of the left-hand side will indicate that it may be written

$$\frac{d}{dx}\left(ye^{\int P(x)\,dx}\right).$$

Therefore,

$$ye^{\int P(x)\,dx} = \int F(x)e^{\int P(x)\,dx}\,dx + \text{const.}$$

The general solution is, therefore, obtained by multiplying both sides by $e^{-\int P(x)\,dx}$.

$$y = e^{-\int P(x)\,dx}\left[\int F(x)e^{\int P(x)\,dx}\,dx + K\right].$$

In the form usually quoted,

$$y = Ke^{-\int P(x)\,dx} + e^{-\int P(x)\,dx} \int F(x) e^{\int P(x)\,dx}\,dx. \tag{S5.8}$$

It is immediately apparent that the usefulness of this form of the solution depends upon the ease with which the integrals of $P(x)\,dx$ and the more complicated $F(x)e^{\int P(x)\,dx}\,dx$ can be obtained. The former is frequently quite easy while the latter may be exceedingly difficult.

Example S5.1

$$\frac{dy}{dx} + ay = Ke^{-ax}.$$

The integrating factor is obtained from the coefficient of y, and is

$$e^{\int a\,dx} = e^{ax}.$$

Multiplying through by this factor and by dx

$$e^{ax}\,dy + aye^{ax}\,dx = e^{ax}(Ke^{-ax})\,dx = K\,dx.$$

But the left side may be written

$$d(ye^{ax}).$$

Consequently,

$$ye^{ax} = \int K\,dx = Kx + K_1,$$

from which

$$y = Kxe^{-ax} + K_1 e^{-ax} = e^{-ax}(Kx + K_1).$$

Example S5.2

$$x\left(\frac{dy}{dx}\right) + (x + 1)y = x^3$$

or

$$\left(\frac{dy}{dx}\right) + \left(1 + \frac{1}{x}\right)y = x^2. \tag{S5.9}$$

The $P(x) = 1 + 1/x$, from which we easily find that the integrating factor is $\exp(x + \ln x) = xe^x$. Multiplying Equation S5.9 by this factor and simplifying,

$$xe^x\,dy + xye^x\,dx + ye^x\,dx - x^3e^x\,dx = 0. \tag{S5.10}$$

The first three terms will be seen to be $d(xye^x)$. But the last term offers some trouble. It does not contain both terms of $-d\,(x^3e^x)$, for

$$-d(x^3e^x) = -x^3e^x\,dx - 2x^2e^x\,dx.$$

We therefore add and subtract the second term to Equation S5.10. Now we must account for the second term of $d\,(x^2e^x)$ in the same fashion, and so forth for the extra term generated by $d\,(xe^x)$. Thus we can write for Equation S5.10,

$$d(xye^x) - d(x^3e^x) + 3\,d(x^2e^x) - 6\,d(xe^x) + 6\,de^x = 0;$$

integrating, we have

$$xye^x - x^3e^x + 3x^2e^x - 6xe^x + 6e^x - K = 0.$$

Dividing by xe^x and rearranging terms,

$$y = x^2 - 3x + 6 - 6x^{-1} + K(xe^x)^{-1}.$$

This general solution can be substituted into the original differential equation, and will be found to be correct. If a boundary condition $y(1) = 5$ is known, the constant K may be eliminated. The complete solution is then found to be

$$y = x^2 - 3x + 6 - 6x^{-1} + \frac{7e^{(1-x)}}{x}.$$

S5.5 Linear O.D.E. with Constant Coefficients

The limitation to constant coefficients implies, in the tank purging problem, for example, Equation 5.5, a constant mass in the process and a constant effluent rate. Consequently, there is also the implication of constant input rate. These restrictions might frequently be applicable to a real process. It is therefore reasonable to consider solutions to equations of this type. First we treat the case of the first-order equation and second we expand the method to include a differential equation of higher order.

The general first-order equation

$$y' + ay = F(x) \tag{S5.11}$$

may be written as

$$L_1(D)y = F(x).$$

Now we know that $y = \phi(x)$ is the solution to the equation and that performing on y the operations specified in $L_1(D)$ must produce $F(x)$. In other words, if y is known, then by taking the differential of y with respect to x and adding to this differential the value of y multiplied by a, we must obtain $F(x)$. Otherwise, no solution has been obtained. We now propose to show that y may be broken into two parts and we shall subsequently obtain the two parts separately.

If we write $L_1(D)y_c = 0$, disregarding $F(x)$, the equation is known as the *reduced equation*. It obviously has a solution which we call y_c and which is known as the complementary function. If we further define an unknown particular integral, y_p, such that

$$L_1(D)y_p = F(x),$$

it is apparent that

$$L_1(D)y_c + L_1(D)y_p = F(x).$$

But this expression may, according to the usual rules of algebra, be written

$$L_1(D)(y_c + y_p) = F(x);$$

and, since our original differential equation stated that

$$L_1(D)y = F(x),$$
$$y = y_c + y_p.$$

We may then obtain the y_c and the y_p separately, subsequently adding them for the general solution.

It is interesting to note that this same conclusion regarding independence of y_c and y_p could also have been deduced from Equation S5.8. The first term of that equation is the solution to the reduced equation; the second is the particular integral.

The reduced equation, with constant coefficients, is always separable. There is no problem in obtaining y_c. In order to obtain y_p, we shall introduce the Method of Undetermined Coefficients.

The General Method of Undetermined Coefficients

This method has two requirements. First, if two equations of comparable form are equal, then the coefficients of each term shall also be equal. For example,

$$ax^3 + bx^2 + c = Ax^3 + C, \quad \text{only if} \quad a = A, \quad b = 0, \quad \text{and} \quad c = C.$$

Second, it must be possible, by inspection of $F(x)$, to guess a suitable functional form of y_p. For many forms of $F(x)$, this requirement is easily met. We now consider these forms of $F(x)$ and the consequent functional forms of y_p, as shown in Table S5.1.

TABLE S5.1

$F(x)$	Functional Form of y_p
ax^n	$A_n x^n + A_{n-1}x^{n-1} + \cdots + A_1 x + A_0$
be^{mx}	Be^{mx}
$c \cos kx$, $h \sin kx$	$C \cos kx + H \sin kx$

The table might include the possibility, $F(x) =$ const. Since this case can be solved by a simple transformation of the dependent variable, it is not listed for solution by undetermined coefficients. The solution follows.

$$y' + a_1 y = a.$$

Then,

$$y' + a_1 y - a = 0.$$

Now let

$$z = a_1 y - a.$$

Then, $z'/a_1 = y'$, and the original O.D.E. becomes

$$z' + a_1 z = 0.$$

This equation is readily solved to give

$$z = Ke^{-a_1 x},$$

or, in terms of the original variable,

$$a_1 y - a = Ke^{-a_1 x}.$$

Consequently,

$$y = K_1 e^{-a_1 x} + \frac{a}{a_1}\cdot$$

The term a/a_1 is the particular portion of the general solution.

Inspection of the functional forms in Table S5.1 indicates that y_p is chosen to contain the terms of $F(x)$ and all their derivatives. It is then reasonable to conclude that products of the terms listed above under $F(x)$ would be treated similarly. For example, if $F(x) = x^2e^{2x}$, we should assume for y_p a form containing x^2e^{2x} and all its derivatives with respect to x. Therefore we would include xe^{2x} and e^{2x}, thereby assuming

$$y_p = Ax^2e^{2x} + Bxe^{2x} + Ce^{2x}.$$

Similarly, if $F(x) = x \sin kx$, we would choose

$$y_p = Ax \sin kx + Bx \cos kx + C \sin kx + H \cos kx.$$

Once having chosen the form of y_p, the next step is to operate on it according to the terms of $L_1(D)y$. The expression so obtained is necessarily equal to $F(x)$ and this equality is so written. Then all that is necessary is the equating of the coefficients of like terms on both sides of the equation.

Example S5.3

$$y' + 2y = xe^{2x} + 5.$$

The reduced equation is $y' = -2y$; separating variables,

$$\frac{dy}{y} = -2\,dx.$$

Integrating

$$\ln y_c = -2x + \text{const},$$
$$y_c = Ke^{-2x}.$$

This is the complementary function. We now proceed to the particular integral.

From the form of $F(x)$, $xe^{2x} + 5$, assume

$$y_p = Axe^{2x} + Be^{2x} + C.$$

Then,

$$y_p' = Ae^{2x} + 2Axe^{2x} + 2Be^{2x},$$
$$2y_p = 2Axe^{2x} + 2Be^{2x} + 2C.$$

Adding

$$y_p' + 2y_p = 4Axe^{2x} + (A + 4B)e^{2x} + 2C.$$

This expression must be identically equal to $F(x) = xe^{2x} + 5$. Therefore, equating coefficients of like terms,

$$4A = 1;$$
$$A + 4B = 0;$$
$$2C = 5.$$

Therefore,

$$A = \tfrac{1}{4};$$
$$B = -\tfrac{1}{16};$$
$$C = \tfrac{5}{2}.$$

The general solution to the equation is

$$y = y_c + y_p = Ke^{-2x} + (\tfrac{1}{4})xe^{2x} - (\tfrac{1}{16})e^{2x} + \tfrac{5}{2}. \tag{S5.12}$$

We can, and should, prove that this solution is correct by substituting it into the original equation. Thus, obtaining the terms y' and $2y$ as indicated in that equation:

$$y' = -2Ke^{-2x} + (\tfrac{1}{4})e^{2x} + (\tfrac{1}{2})xe^{2x} - (\tfrac{1}{8})e^{2x},$$
$$2y = 2Ke^{-2x} + (\tfrac{1}{2})xe^{2x} - (\tfrac{1}{8})e^{2x} + 5.$$
$$y' + 2y = xe^{2x} + 5 \equiv F(x).$$

Equation S5.12 is thereby proven to be a correct solution.

EXCEPTION One exception to the procedure as listed should be noted. In the case where $F(x)$ contains a term which appears in the complementary function, the functional form chosen for y_p must be modified. The modification is such that the same term found in y_c is not chosen for y_p but rather that term multiplied by x^k, where k is a positive integer of size just sufficient to prevent any term in y_p duplicating one in the complementary function.

If we modify our previous illustration only to the extent of changing the sign on the term $2y$, we have

$$y' - 2y = xe^{2x} + 5.$$

Now $y_c = Ke^{2x}$. This term has the same form as one of the terms which we should normally select for y_p. Therefore, the normal choice of

$$y_p = Axe^{2x} + Be^{2x} + C,$$

duplicates the term e^{2x} found in the complementary function. Multiplication of *the exponential part by x* gives

$$y_p = Ax^2e^{2x} + Bxe^{2x} + C, \tag{S5.13}$$

which contains no term found in the complementary function. Treating Equation S5.13 in the same manner as in the preceding example, we obtain

$$y = Ke^{2x} + (\tfrac{1}{2})x^2e^{2x} - \tfrac{5}{2}.$$

This solution may also be shown to be correct by substitution into the original differential equation. If one had attempted solution, using as the chosen form the ordinary choice

$$y_p = Axe^{2x} + Be^{2x} + C,$$

the xe^{2x} term would have been lost upon adding y' and $-2y$. The solution could not then have been correct.

This particular problem arises in a variation of the tank purging process. If two tanks of identical volume are placed in series, and if the feed to the first tank is pure solvent, the feed to the second tank will be $(C/C_0)e^{-t/\theta}$. The solution to the effluent concentration from the second tank will then involve $e^{-t/\theta}$ in both the complementary function and in the particular integral. The technique of multiplying by t to give $te^{-t/\theta}$ as one term in the choice of y_p will be essential to a correct solution by the method of undetermined coefficients.

Higher-Order Equations

By employing the D-operator and the notation of $L_n(D)y$, we can greatly simplify the discussion of the solution of higher-order linear differential equations. First, we deal with

$$L_n(D)y = F(x). \tag{S5.14}$$

The solution might be written, using the normal rules of algebra,

$$y = \frac{F(x)}{L_n(D)}. \tag{S5.15}$$

The function, $L_n(D)$, is of the form

$$a_n D^n + a_{n-1} D^{n-1} + \cdots + a_1 D + a_0,$$

which is a polynomial with roots $m_1, m_2, \ldots, m_{n-1}, m_n$. For the moment we assume that all the roots are distinct (i.e., no two are equal). Then

$$\frac{1}{L_n(D)} = \frac{1}{(D-m_1)(D-m_2)\cdots(D-m_n)}. \tag{S5.16}$$

The right-hand side of Equation S5.16 can be decomposed by the method of partial fractions into

$$\frac{A_1}{D-m_1} + \frac{A_2}{D-m_2} + \cdots + \frac{A_n}{D-m_n},$$

granted that we properly determine the various A_i. Then

$$y = F(x) \bigg/ \left(\frac{A_1}{D-m_1} + \frac{A_2}{D-m_2} + \cdots + \frac{A_n}{D-m_n} \right). \tag{S5.17}$$

If we now examine the solution of the first-order equation (see Equation S5.8) expressed in the same manner, we find

$$y = F(x) \bigg/ \left(\frac{1}{D-m} \right) = Ke^{mx} + e^{mx} \int e^{-mx} F(x)\, dx.$$

We conclude that the solution of the higher-order equation may be written

$$y = \sum_{i=1}^{n} K_i e^{m_i x} + \sum_{i=1}^{n} A_i e^{m_i x} \int e^{-m_i x} F(x)\, dx. \tag{S5.18}$$

The point of immediate importance is the two summations. The first is a summation of the complementary functions for each root of $L_n(D)$. Thus, for the differential equation

$$L_2(D) = y'' - 3y' + 2y = 2x,$$

we can write $L_2(D) = (D-2)(D-1)$. Then $m_1 = 2$ and $m_2 = 1$. The first summation in Equation S5.18 is $K_1 e^{2x} + K_2 e^{x}$. There is never a basic problem with the first summation since the roots can always be determined even when the form is less simple than in this illustration.

The problem of the second summation can be tackled in two ways. The first method would logically be a determination of the various A_i. The second is the application of the method of undetermined coefficients *when the differential equation contains constant coefficients.*

The A_i may be established by the method of partial fractions. Equation S5.18 may then be integrated, term by term. This method has the same drawbacks as in the first-order case. However, the determination of the A_i is quite simple.

$$A_i = \frac{1}{L_n'(m_i)}.$$

This mathematical statement says to evaluate the derivative of $L_n(D)$ with respect to D, at $D = m_1$, take the reciprocal of the number obtained, and this number is A_1. Obviously we repeat for each value of m at each value of i.

The Particular Integral for Higher-Order Equations

Use the method of undetermined coefficients on the equation

$$L_2(D) = y'' - 3y' + 2y = 2x.$$

We may assume the following form of y_p, since $F(x) = 2x$:

$$y_p = Ax + B.$$

Then,

$$y_p' = A \qquad \text{and} \qquad y_p'' = 0.$$

Therefore,

$$y'' - 3y' + 2y = 2Ax - 3A + 2B = F(x).$$

Since

$$\begin{aligned} F(x) &= 2x, \\ 2A &= 2, \qquad A = 1, \\ -3A + 2B &= 0, \qquad B = \tfrac{3}{2}. \end{aligned}$$

The general solution is

$$y = K_1e^{2x} + K_2e^x + x + \tfrac{3}{2}.$$

If this solution is substituted into the original $L_2(D)y$, $F(x)$ is obtained, proving that the solution is correct.

When one of the terms of the complementary function contains an exponential term which is identical to a term in $F(x)$, it is essential to take the precaution of multiplying the proposed terms in y_p by a power of the independent variable, x, in such fashion that no term in y_p also appears in the complementary function. The precaution is identical to that named for the case of the first-order equation. Several examples will illustrate the procedure.

Example S5.4

$$y'' - 3y' + 2y = xe^x.$$

Therefore,

$$(D - 2)(D - 1)y = xe^x.$$

The complementary function then is $K_1e^{2x} + K_2e^x$. The usual choice for y_p is $Axe^x + Be^x$. But the term e^x appears in the complementary function.

Therefore the normal choice of y_p must be multiplied by x^k, where k is chosen so that *no term* in y_p appears in the complementary function. In this case, k can be unity and

$$y_p = Ax^2e^x + Bxe^x.$$

Using this form, the solution of the differential equation is found to be

$$y = K_1e^{2x} + K_2e^x - \tfrac{1}{2}x^2e^x - xe^x.$$

Note that no two terms in the solution represent the same function of x.

Example S5.5

$$y'' - y' - 2y = 5e^{2x} + \cos 2x.$$

Since

$$(D - 2)(D + 1)y = 5e^{2x} + \cos 2x,$$

the complementary function is $K_1e^{2x}+K_2e^{-x}$. The normal form of y_p is $Ae^{2x} + B\cos 2x + C\sin 2x$. However, e^{2x} appears in the complementary function so that this portion of the chosen y_p must be multiplied by x. Then

$$y_p = Axe^{2x} + B\cos 2x + C\sin 2x.$$

Note here that only that part of the normal y_p which is related to the duplication of functions of x needs to be multiplied by x. The solution of this differential equation is

$$y = K_1e^{2x} + K_2e^{-x} + 5xe^{2x} - \tfrac{3}{20}\cos 2x - \tfrac{1}{20}\sin 2x.$$

Example S5.6

$$y'' - 2y' + y = xe^x,$$

which may be written

$$(D - 1)^2y = xe^x.$$

Then the complementary function is $K_1e^x + K_2xe^x$. The instance of repeated roots is encountered here for the first time. The treatment is similar to that which we have been encountering in the formation of y_p. If no allowance is made for the repeated roots, both terms of the complementary function would be K_ie^x. We should, then, have two identical terms which could be added directly and written as Ke^x. Since all terms must be distinct, the same correction for potential duplication is made in the complementary function which we have previously employed in y_p; i.e., multiplication of one of the terms by x. If the differential equation had been $(D - m)^r$, the complementary function would be

$$(K_1 + K_2x + \cdots + K_{r-1}x^{r-2} + K_rx^{r-1})e^{mx}.$$

Now the normal choice of y_p is $Axe^x + Be^x$, but this function contains xe^x, which also appears in the complementary function. Multiplication by x will not alleviate the situation, since the second term will now be a duplication of another term in the complementary function. We must multiply by x^2 giving

$$y_p = Ax^3e^x + Bx^2e^x.$$

The solution of the differential equation is then found to be

$$y = K_1e^x + K_2ex^x + \tfrac{1}{6}x^3e^x.$$

Example S5.7

$$y''' + 2y'' + y' = xe^{-x} + x.$$
$$D(D + 1)^2 = xe^{-x} + x.$$

Now the complementary function must be of the form

$$K_1e^0 + K_2e^{-x} + K_3xe^{-x}.$$

The first term is obviously just K_1, a constant. Normal choice of y_p would involve selection of the sum of two functions, the first chosen to include the differentials with respect to x of xe^{-x} as well as xe^{-x} itself, and the second chosen to account in the same manner for the term x in $F(x)$. This choice would suggest

$$Axe^{-x} + Be^{-x} + Cx + H.$$

However, the first two terms are not satisfactory until multiplied by x^2 and the last two are not satisfactory until multiplied by x since the complementary function already contains a constant. The final choice of y_p is then

$$y_p = [(Axe^{-x} + Be^{-x})x^2] + [(Cx + H)x].$$

The contributions of the two separate parts in brackets can be determined independently since no term in x^n alone can give one in e^{-x} by differentiation and vice versa. For example, for the terms containing x only

$$y_p' = 2Cx + H,$$
$$y_p'' = 2C,$$
$$y_p''' = 0.$$

Therefore,

$$y''' + 2y'' + y' = 2Cx + (4C + H).$$

By comparison of coefficients,

$$2C = 1,$$
$$4C + H = 0;$$

solving,

$$C = \tfrac{1}{2}, H = -2.$$

The part of y_p containing terms in e^{-x} can be treated in similar fashion and the two results added to give the general solution

$$y = K_1 + K_2e^{-x} + K_3xe^{-x} - \tfrac{1}{18}x^3e^{-x} - \tfrac{1}{9}x^2e^{-x} + \tfrac{1}{2}x^2 - 2x.$$

Differential equations of order higher than unity will not be encountered in problems involving material balances only, but will be found in Chapter 9.

LIST OF SYMBOLS

(Excluding the most common notations of the calculus.)

Letters used to denote arbitrary constants

$A, B, C, H, K; a, b, c, h, k$

Letters used to denote functionality of variables following in parentheses

$F, L, M, N, P; a, f, u; \phi$ (e.g., $M(x, y)$ is a function of both variables, x and y).

Letters used to denote variables

v, x, y

Latin letters

$a_n(x), a_{n-1}(x), \ldots, a_1(x), a_0(x)$	Various functions of x as coefficients in linear form of O.D.E.
D	Differential operator, d/dx
L_n	Linear combination of differential terms of orders zero to n, inclusive
m	Root of a polynominal in D
O.D.E.	Ordinary differential equation
x	Independent variable
y	Dependent variable
y_c	Complementary portion of solution (i.e., solution of reduced equation)
y_p	Particular portion of solution

6 The Principle of Energy Conservation

The principle that in all physical and chemical processes there is something that remains unchanged is repeatedly found as an underlying hypothesis in various areas of science. Since this principle is considered basic, it is incapable of theoretical proof, but it may be experimentally verified, and we are justified in using it as long as we know of no exceptions to it.

In the broadest sense, present knowledge leads us to a belief in the principle of conservation of mass-energy, but for the purpose of this text and most practical problems, it is more convenient to consider mass and energy separately. This chapter will concentrate on the First Law of Thermodynamics, which is essentially a mathematical statement of the principle of energy conservation applied to macroscopic systems. Conservation of mass alone has been discussed in Chapter 4. The use of the First Law of Thermodynamics in the solution of problems will be found in Chapter 9.

6.1 Inherent Energy and Energy in Transition

Analyses of energy are always applied to systems. A system is any body of matter chosen to suit the purpose of the problem. It may be specifically described, such as an air compressor, a steam engine, or a chemical reactor. On the other hand, it may be an abstracted quantity of a substance, such as 1 cu ft of carbon dioxide under a pressure of 10 atm and at 300 °C. Once defined, it is circumscribed by real or imaginary boundaries from the rest of the universe, called the surroundings.

Stored Energy

We are never concerned with the precise quantity of energy inherent in a system, but only with changes occurring in that energy content. The inherent energy may be potential (caused by the mutual attraction of the component parts of the system), kinetic (arising from the relative motion of the component parts of the system), molecular (due to the translation, rotation, or vibration of the molecular species of which the system is composed), chemical (associated with those forces binding the atoms together to form the molecules), atomic (the largest reservoir of all, responsible for keeping the complex atomic structure intact), etc. Some of these forms will be discussed in detail at this point; others will be elaborated in subsequent sections. At the moment it should simply be noted that completely describing the system fixes its energy content, and as long as the system undergoes no change, its capacity for producing a particular effect is specified. Furthermore, if any changes in forms of energy occur wholly *within* the system (e.g., the change of potential into kinetic energy), the surroundings are unaffected.

Potential Energy

Any two masses, regardless of size, exert an attraction on one another. If this force of attraction is multiplied by the distance of separation, the resultant energy is known as potential energy. Thus, there is potential energy in the sun-earth system or in a system composed of gas molecules. Engineers are primarily concerned with that potential energy associated with a mass above an arbitrary datum plane when the force of attraction is that due to the earth's gravitational field. Energy is obviously required to move a mass, m, from the datum plane vertically to the position which it has in the system, or conversely, energy would be released if the mass fell from its existent position to the datum plane, as when a suspended rock falls to the earth's surface. This vertical distance will be given the symbol Z. Z is actually a difference between two positions, but since the datum is arbitrarily assigned a height of zero, we shall speak of Z and not ΔZ. The specification of the datum is important. Consider a piece of chalk held above a desk top. If the datum plane is the surface of the desk, the computed potential energy will be less than if the datum plane were on the floor on which the desk stands, for more energy would be required to raise the piece of chalk from the floor to its original position than would be needed if the chalk were moved only from the top of the desk.

In the English system Z is expressed in feet and mass in pounds (mass). The force involved is $(g/g_c)m$ pounds (force). Thus the potential energy is $(g/g_c)mZ$ foot-pounds (force). In the English system, the ratio g/g_c is essentially unity, so *numerically* potential energy is equivalent to mZ. Although we shall be concerned mostly with the English system, it should be observed that in the metric system, when m is expressed in grams and Z in centimeters, the potential energy will be 980.8 mZ dyne-centimeters or ergs. Since g or g_c are not numerically equal, g_c in this system being unity, grams can never represent force and energy is not expressible in gram-centimeters.

Kinetic Energy

Kinetic energy is defined in terms of the relative motion of two masses. Just as the arbitrary datum plane was assigned a height of zero when we were calculating potential energy, we normally assume that one of the two bodies is at rest and only the other one is moving. Again, the earth is conveniently assumed to be the one at rest. Let u be the velocity of the moving substance with reference to the "stationary" datum. The dimensions of u are L/t; therefore $dL = u\,dt$. Since acceleration, a, has the units L/t^2, $du = a\,dt$ and $dt = du/a$. Then, substituting for dt in the previous expression for dL, it follows that $dL = (u/a)\,du$, and $L = u^2/2a$. From Newton's laws, $lb_f = lb_m(a/g_c)$; therefore, force times distance equals

$$\frac{\text{mass times acceleration}}{g_c} \times \frac{(\text{velocity})^2}{2(\text{acceleration})}.$$

Upon cancellation of the acceleration terms, there results:

$$\text{kinetic energy} = \frac{mu^2}{2g_c}.$$

A glance at the dimensions associated with kinetic energy shows them to be the same as for potential energy, i.e., foot-pounds force or dyne-centimeters.

Internal Energy

For the moment all molecular, chemical, and atomic energy contained within the system will be subsumed under the heading internal energy. It will be demonstrated shortly that the precise nature of internal energy need not be known, and therefore a detailed account of it will not be given here. However, it should be noted that it is usually expressed in thermal units (Btu or calories) rather than in the mechanical energy units used for potential and kinetic energy. However, with the conversion factor 778 ft-lb_f = 1 Btu, one may readily translate one set of units to the other.

6.2 Energy in Transition

It has been stated earlier that as long as energy remains stored within the system, the system has the capacity for producing an effect, but no effect is evident. However, energy may cross the boundaries of the specified system, thus producing changes in the energy content of both the system and its surroundings. Such energy crossing boundaries may take two forms: work (in its most general sense) and heat. It must be emphasized that work and heat are both transitory phenomena; they cease as soon as energy stops crossing the boundary between the system and its surroundings. Work and heat must *never* be thought of as stored quantities, and all terminology that implies storage must be carefully avoided. Thus phrases like "heat content" or "internal work" have no meaning, and if used (as they occasionally are) give rise to misconception.

Work

In its most general form, work is defined as a force operating through a distance. Since the applied force may vary with the distance, one should write $W = \int F\,dL$. In the special case that the force remains constant, $W = F(L_{\text{final}} - L_{\text{initial}})$. There are times when the chemical engineer is faced with problems involving electrical, magnetic, or other force fields. In such cases, the equation for work just given should be used. In most instances, however, he is concerned with work as commonly understood, i.e., the raising of a weight, the turning of a wheel, or some equivalent mechanical operation. Since energy is the capacity for producing an effect, and the effect produced is mechanical, work (in the narrow sense just discussed) is frequently defined as mechanical energy in transition. Such a definition focuses attention on the result; it must be remembered that the source of energy may be other than mechanical, e.g., thermal. In this text, the symbol W will be restricted to energy in transition resulting in a mechanical effect.

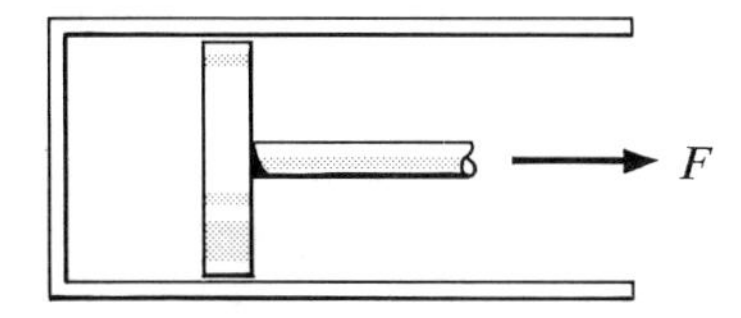

FIGURE 6.1 Simple piston-cylinder device for producing work.

Let us now postulate the simple mechanical device illustrated in Figure 6.1. It consists of a piston within a cylinder. The piston is being driven outward by the expansion of the fluid contained within the cylinder. For convenience, a unit mass of this fluid will be considered. The pressure acting on the piston head, whose area is A, is the force per unit area. Therefore, $F = PA$. As the piston moves, the volume of the enclosed fluid increases by dv for every dL of motion. Since $dv = A\,dL$, $dL = dv/A$ and $\int F\,dL = \int PA(dv/A) = \int P\,dv$.

Thus the units of mechanical work in the English system are $(lb_f/ft^2)(ft^3/lb_m) = ft\text{-}lb_f/lb_m$, which are the same as those previously encountered for potential and kinetic energy. In all three instances, the energy involved could be expressed in thermal units through the use of the conversion factor 778 $ft\text{-}lb_f/Btu$.

Heat

Although the true nature of work was rather early understood, heat was a concept subject to much misunderstanding. Actually, it was not until the beginning of the twentieth century that a clear, unequivocal definition of heat was given. Heat may be defined as the effect which occurs when two bodies at different temperatures are brought into contact with each other. The phrase "in contact" does not necessarily mean physical contact. Heat will flow if the two bodies are in optical contact. In this case (a common example is the sun warming the earth), the heat flow is called radiation. When physical contact is involved, the heat flow is called conduction. It must be emphasized that there can be no heat flow unless the system and its surroundings are at different temperatures, but the mere existence of a temperature difference is no guarantee that such a flow will result. No flow results when the system is insulated from its surroundings. While no practical insulation is perfect, it is often sufficiently effective to enable us to consider that the system is thermally isolated from the surroundings. A process occurring in such a system is known as an adiabatic process.

In this text, the symbol Q will be used to represent energy in transition wholly as a result of a temperature difference between systems and surroundings. It is normally expressed in thermal dimensions, but these units may be converted to their mechanical equivalents.

Sign Conventions

In setting up equations and working problems, we must recognize that sometimes energy flows into the system and sometimes out of the system. Appropriate signs must be used to indicate the different directions of flow. The choice is clearly a matter of convention, but once made, it must be adhered to. Current practice is to regard W as positive when the energy flow is from the system to the surroundings and negative when the surroundings supply energy to the system. On the other hand, Q is positive when the system receives the energy (i.e., when the surroundings have the higher temperature) and negative when the reverse is true. Thus an exothermic chemical reaction has a negative heat. Unfortunately the older literature often reversed either or both of these conventions; thus one must be on guard when consulting any reference material to clarify the choice of convention in each source.

State (or Point) Functions

One further matter must be clarified before we may proceed to set up an equation for the First Law of Thermodynamics. A system is normally described in terms of its properties. The three basic, experimentally determinable, properties normally chosen are the *absolute* pressure, P, the *absolute* temperature, T, and the *specific* volume (inverse of density), v. It will be noted that these properties are independent of the total mass of the system, and are called intensive properties as opposed to extensive properties that depend on mass. It should also be noted that these properties may vary uniformly through as small a change as desired. Thus for an infinitesimally small change in temperature, we write dT, and if we know the total range of temperature over which a process occurs, we may write $\int dT = T_{\text{final}} - T_{\text{initial}}$. All state properties may be treated mathematically in this fashion. Such differentials are called "exact" differentials, and when they occur in equations, they are handled according to the rules of the calculus. Changes in state properties are always expressible as exact differentials. It will be noted that the path by which the system changes from the initial to final state need not be known or specified. A change in a state property is independent of the path by which the change occurs.

Path Functions

On the other hand, work and heat are not properties of a system; they represent effects occurring between a system and its surroundings. While we can speak of an extremely small quantity of heat or work, we cannot speak of a change of Q or W; thus, we cannot write dQ or dW as we wrote dT. Instead, $d'Q$ or $d'W$ are used to denote the fact that the differentials are inexact. The integral of $d'Q$ is simply Q, not $Q_{\text{final}} - Q_{\text{initial}}$, since these latter terms have no meaning. Integration of inexact differentials should be considered as a summation of all the tiny individual items required to give the whole. Furthermore, whereas the integration of dT gives the same result regardless of how the process occurred, obviously the amount of work or heat involved in a process depends on how the process was carried out. It has already been pointed out, for example, that Q will vary depending on whether or not the system has been insulated. It is thus very important to distinguish clearly, both mathematically and physically, between Q and W and such terms as P, v, and T.

6.3 The First Law of Thermodynamics for Static Systems

A large proportion of chemical processes operate on a continuous basis in which matter flows into and out of the system at a steady rate, but there are important instances where batch reactors, autoclaves, and the like are used.

Chemical engineers have occasion to analyze such processes in which there is no flow of mass as well as those in which there is. In these instances only that part of the process occurring while the matter is in a so-called static condition is of concern, not the charging and discharging steps. These nonflow processes will be discussed first in this chapter, and then the steady flow processes will be considered. In both cases the equations which are derived will be based on unit mass; to get total quantities, one must multiply by the amount of mass involved.

There are many approaches to the first law of thermodynamics, but only one will be taken here. The reader is referred to specialized textbooks for a more extensive treatment of the subject. Suppose that a system, such as a unit mass of a gas, undergoes a cyclic process, i.e., several steps which result in a return of the system (although not necessarily the surroundings) to its initial energy state. The first law is a mathematical statement of the experimental observation that in such a cyclic process heat is proportional to work, i.e., $\oint d'W \propto \oint d'Q$ or $\oint d'W = J \oint d'Q$, where J is simply the necessary proportionality factor required to reconcile the difference in units. As has already been noted, the approximate value of this factor is 778 ft-lb$_f$/Btu in the English system. In the equations which follow, this factor will be omitted, since by now the necessity of checking all equations for dimensional consistency is thoroughly appreciated.

Although $\oint d'Q - \oint d'W = 0$ for this special case, most processes are not cyclical, and it is important to get a more general equation. The question that faces us is, what is $d'Q - d'W$ equal to in general? Mathematically, it must be an exact differential, since only exact differentials have a cyclic integral equal to zero. Physically, it follows that it must be a state property of the system, for only state properties may be expressed as exact differentials. Furthermore, it must have the units of energy to satisfy the equation. Thus, if experimental observations of the cyclic process are correct, there must exist for all systems a state property representing energy. If W represents work in the most general sense, the symbol assigned to this state property is E; when W is restricted to mechanical work (as we are doing), the symbol U is often used to note this restriction. Some writers use E for both cases and let the restriction on W be shown in other ways. Summarizing, the First Law of Thermodynamics for static systems may now be written:

$$dE = d'Q - d'W_{\text{general}}, \tag{6.1}$$

$$dU = d'Q - d'W_{\text{mechanical}}, \tag{6.2}$$

with both E and U representing the internal energy of the system. Note that these equations determine the change in the property, not the value of the property itself.

It should be noted that although definitions of W and Q have been given, there has been no need to define U or E descriptively. These functions arise mathematically out of the statement of the first law. It has been previously pointed out that internal energy is considered as the sum of all molecular, chemical, atomic, and other forms of energy inherent in the system, but insofar as classical thermodynamics is concerned this knowledge is unnecessary. We shall, therefore, consider U and E as storage functions representing a reservoir to which energy may be added or from which energy may be withdrawn. Since W and Q represent the only forms of energy crossing boundaries, the first law of thermodynamics may be stated in words as: accounting for energy in all the forms in which it may exist, the total energy entering a system must equal the energy leaving it plus any accumulation within the system. This is, of course, a general statement of the law of conservation of energy. The analogy between this statement and that of the general law of conservation of mass (Section 4.1) is important and should be noted.

Reversible Processes

We may substitute $F\,dL$ for $d'W_{\text{general}}$ or $P\,dv$ for $d'W_{\text{mechanical}}$ in the equations of the previous section, but only under the following additional restriction. Assume that a system undergoes a compression process that can be represented graphically on P-v coordinates as in Figure 6.2. Assume

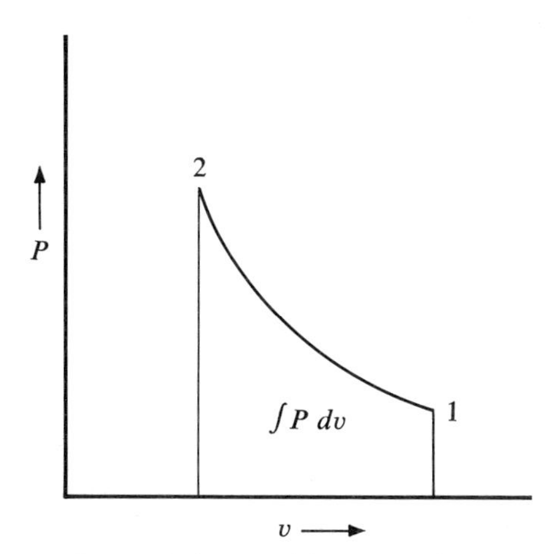

FIGURE 6.2 Work involved in a reversible compression (expansion) process.

further that the process occurs in the complete absence of any mechanical friction, i.e., that each increment of applied force (no matter how small) results in property changes in the system, which when graphed result in the line 1–2. Since no mechanical friction is present, the compressed gas may subsequently be slowly expanded along the path 2–1, returning the system to its original energy state. Of equal importance, the energy state of the surroundings will also have been restored. The energy represented by $\int P\,dv$ is that crossing the boundaries from the surroundings to the system during the process 1–2; this energy was released from the system to the surroundings during the process 2–1 so that at the end there has been no net energy change in the universe in form, amount, or location. Such a process, known as a reversible process, can exist only in the complete absence of friction. Completely reversible processes are hypothetical, of course, but the situation can be approached by very slow processes occurring in well-lubricated equipment. The concept of reversibility is extremely useful, however, and conclusions obtained from such a consideration often have general applications as will be demonstrated shortly.

Irreversible Processes

Return now to the original system and carry out the same process in real equipment. To provide a concrete example for discussion, suppose a gas is being compressed inside a cylinder by the motion of a piston as shown in Figure 6.3. The gas can still be compressed along the path 1–2 of Figure 6.2,

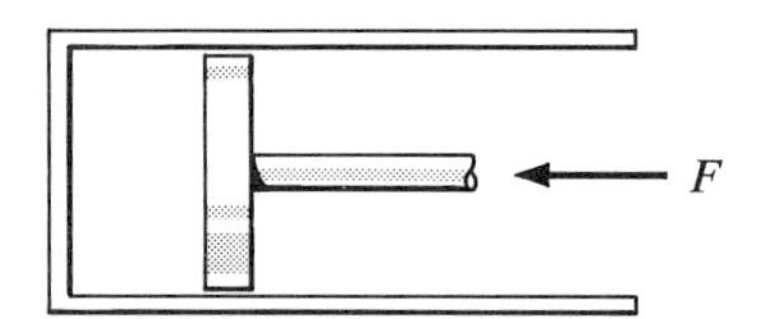

FIGURE 6.3 Simple piston-cylinder compressor.

but in this case the amount of energy which an outside agent (the piston) must expend will be greater than before. It must not only effect the energy change in the system represented by the path 1–2, but it must also supply additional energy to overcome the friction of the piston-cylinder combination. Although the compressed gas could expand along the path 1–2, as it did in the first case, all the energy represented by $\int P\,dv$ is not available to restore the surroundings to their initial state, because some of the energy is required to overcome friction. Thus, the mechanical effect (work) required to compress the gas is greater than $\int_1^2 P\,dv$ and the mechanical effect obtainable from the compressed gas is less than the $\int_2^1 P\,dv$.

Consequently, $d'W$ and $P\,dv$ may not be equated unless the process is reversible. At first, this restriction might seem to be a severe one, but recall that U is a state function and that $U_2 - U_1$ is independent of the path between 1 and 2. Therefore, in computing ΔU, a mechanically reversible process may always be *assumed* for convenience, and for the system we may write $\Delta U = Q - \int P\,dv$. What may *not* be done is to call the integral term the actual mechanical effect produced (or required) unless the process is truly reversible. Since the determination of mechanical friction is largely empirical, thermodynamics usually concerns itself with only the reversible case, the thermodynamicist knowing that his answer represents the limiting situation—the *maximum* amount of work available *from* a process or the *minimum* amount of work *needed* to carry out a process.

6.4 The First Law of Thermodynamics for Systems in Steady Flow

Although chemical engineers do have occasion to analyze systems at rest (e.g., batch reactors), they more frequently encounter flowing systems. When a process is being started up or being shut down, the process variables change with time; this transient condition is discussed elsewhere in this book (Chapter 5 and Section 9.8 of Chapter 9). However, it is desirable to conduct most continuous processes under unchanging conditions called the steady state. This does not mean that process variables such as P, v, and T do not change from point to point along the line of flow; it does mean that at any given point the derivative of a property with respect to time is zero. This situation will now be discussed. All symbols and definitions previously discussed in nonflow processes retain their former significance. For example, W still refers to energy crossing the boundary between the system and its surroundings resulting in an effect equivalent to the raising of a weight; Q is the energy crossing the boundary as a result of a temperature difference, and E and U are storage functions. As noted earlier, the values for each item are for a unit mass of matter. New terms, of no concern to the static case, become important to the steady flow system, and it is the definitions of these quantities that we will consider first.

Flow Energy

When a system is at rest, its potential energy relative to any arbitrary datum plane does not change; but in a flowing system a unit mass of material entering may be at a different distance from the datum than that of the same unit at the exit from the system. Furthermore, this unit mass is likely to have a different velocity at the exit than it had at the entrance due to a difference in pipe sizes or fluid expansion. Thus, one may expect changes in potential and kinetic energies to be encountered.

However, even if there were no such difference (e.g., an incompressible fluid flowing in a horizontal pipe of constant diameter), energy is required to move a unit mass from one portion of the conduit to the next. This energy comes from within the system; it does not come from energy crossing the boundary between the system and the surroundings, and thus it must not be considered work as we have carefully defined it. The basic unit of mass with which we are concerned exists under a pressure, P, in the usual units of force per unit area. When this pressure is exerted on the unit mass, which has a specific volume of L^3/m, the effective result is $F/L^2 \times L^3/m$ or FL/m, i.e., force moving unit mass through a specific distance. Thus the mass is displaced and flow occurs. This Pv product is known as flow energy and will be seen to be a property of the system, since both P and v are state properties. Thus, at any point in the conduit the unit mass will have flow energy characterized by the pressure and the specific volume at that point.

One must carefully distinguish between Pv, which is a product of two state properties and is itself a state property, and $P\,dv$, which is the product of a state property and *change* in a state property. The value of the latter depends on the nature of the change, and thus the $\int P\,dv$ is path dependent. On the other hand, changes in Pv are independent of the path, and the difference in flow energy between two points is simply $P_2v_2 - P_1v_1$, just as the difference in temperature is $T_2 - T_1$. These physical and mathematical distinctions between Pv and $P\,dv$ must be clearly understood.

The Basic Equation for Flow

If the Principle of Conservation of Energy is now applied to a unit mass of a flowing system, the following equation results. The sign convention for Q and W previously established will be followed, and all possible energy terms will be considered.

$$\frac{g}{g_c}Z_1 + \frac{u_1^2}{2g_c} + P_1v_1 + U_1 + Q = W + \frac{g}{g_c}Z_2 + \frac{u_2^2}{2g_c} + P_2v_2 + U_2. \quad (6.3)$$

Expressed in words, this equation states that the sum of the potential, the kinetic, the flow, and the internal energies of the system at entrance together with any heat entering the process must equal any work done by the process plus the sum of the potential, the kinetic, the flow, and the internal energies at the exit of the system. As has previously been observed, Q and U are normally expressed in thermal units, whereas the remainder of the terms are usually given dimensions of mechanical energy. In any application of this equation, it must be made dimensionally consistent by the use of the factor $J = 778$ ft-lb$_f$/Btu. Whether all terms should be expressed in mechanical or thermal units is usually decided by the nature of the unknown in the problem being solved. Since Equation 6.3 is written in terms of a unit mass, it may be converted to total mass by multiplying each term by m, but normally this multiplication is deferred until the end of the problem. A single multiplication of the one term at that time is quicker and less likely to produce arithmetical errors.

Equation 6.3 has been written as an energy balance over a finite system. There are occasions when it is desirable to consider the inlet and outlet separated only by an infinitesimally small distance. The differential form of the steady flow equation then becomes:

$$\frac{g}{g_c}\,dZ + \frac{du^2}{2g_c} + d(Pv) + dU + d'W - d'Q = 0. \tag{6.4}$$

6.5 Comparison of the First Law Expressions for Static and Steady Flow Systems

In recapitulating the equations obtained thus far, the differential form of all equations will be used so that comparisons may more readily be apparent.

$$\text{Static}\begin{cases} dU = d'Q - d'W_{\text{nonflow}} & \text{(reversibility not in question),} \\ dU = d'Q - P\,dv & \text{(process must be mechanically reversible),} \end{cases}$$

$$\text{Flow}\left\{ dU = d'Q - \left[\frac{g}{g_c}\,dZ + \frac{du^2}{2g_c} + d(Pv) + d'W_{\text{flow}}\right] \right. \quad \text{(reversibility not in question).}$$

The term in brackets in the third equation (for flow processes) is often referred to as the mechanical effects of the flow process. We now wish to see how these mechanical effects are related to $P\,dv$.

For convenience, a process occurring in the absence of heat (adiabatic) is considered, although this specification is not essential as long as the quantity of heat is known and specified. Let this process occur so that it will describe the path shown in Figure 6.2, but in one case let it occur under reversible static conditions and in the other case let reversible steady flow conditions prevail. Since U is a state property dependent only on the end states, and since the end states are identical for the two processes, $U_2 - U_1$ (or dU) will be the same for both processes. Since it has been agreed that there is no heat or that the heat will be the same for the two processes, it is evident that the mechanical effects for the flowing system will have a value equal to that for work in the static process, and further, since reversibility has been specified, $P\,dv$ evaluates both. These statements may be summarized in the equation

$$P\,dv = \frac{g}{g_c}\,dZ + \frac{du^2}{2g_c} + d\,(Pv) + d'W_{\text{flow, reversible}} = d'W_{\text{nonflow, reversible}}. \tag{6.5}$$

It is then clear that, in the frequent cases where potential and kinetic changes are negligible,

$$d'W_{\text{flow, reversible}} = P\,dv - d\,(Pv). \tag{6.6}$$

Substituting for $d\,(Pv)$ its equivalent $P\,dv + v\,dP$,

$$d'W_{\text{flow, reversible}} = -v\,dP. \tag{6.7}$$

The various Pv products and integrals are illustrated in Figure 6.4.

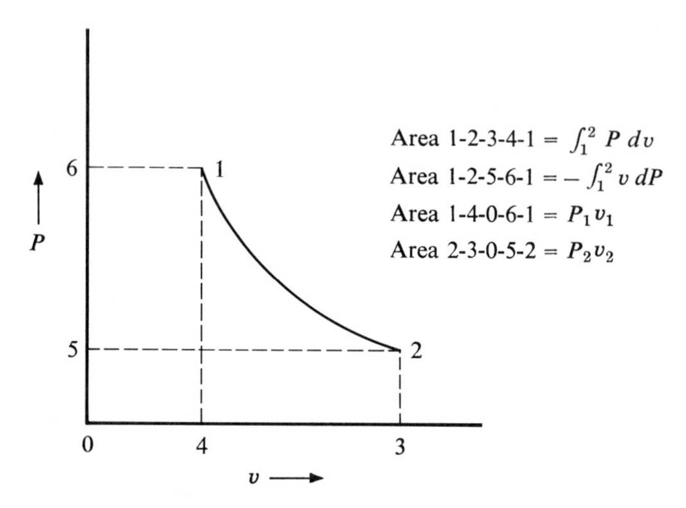

FIGURE 6.4 Graphical representation of certain P–v relations.

6.6 The Concept of Enthalpy

Just as the internal energy function, U, is an inevitable consequence of stating the First Law as $\oint d'Q = \oint d'W$, it may also be shown by formal thermodynamics, beyond the scope of this text, that three other state functions must also exist. Only one of these, enthalpy, is of special concern to us; we shall simply define it here as $H = U + Pv$, that is; the sum of the internal energy and the flow energy. It is normally expressed in thermal units per unit mass, e.g., Btu per lb. Like U, absolute values of H are not known, and we shall be concerned only with changes in H during a process. However, values of H relative to some arbitrary datum reference point are frequently tabulated. A common datum is liquid water at its freezing point, although, for refrigerants, -40 °C (equivalent to -40 °F) is frequently used. Values of relative enthalpies as a function of temperature and pressure are available for air, water, the standard refrigerants, and many other substances in standard references such as Keenan and Keyes, *Thermodynamic Properties of Steam* or the refrigerant tables published by the American Society of Refrigerating Engineers.

The importance of enthalpy (as well as its limitations) can be seen by analyzing the various forms of the first law of thermodynamics presented in the previous section. The most general form of the first law, applicable to all reversible processes

$$U_2 - U_1 = Q - \int_1^2 P\,dv$$

reduces to

$$U_2 - U_1 = Q - P(v_2 - v_1),$$

if constant pressure is specified. Constant pressure processes are very common in chemical engineering; under these conditions it will be seen from the definition of H that

$$Q = H_2 - H_1;$$

i.e., the heat involved in a constant pressure, reversible process is numerically equivalent to the difference of the enthalpies at the end states of the process. This very important relationship constitutes a statement of the heat balance, discussed in detail in Section 9.6.

Now consider the steady state equation when the kinetic energy and potential energy terms are negligible and when no mechanical work is involved. A great many chemical engineering processes do not involve work, and when one considers 778 ft-lb_f = 1 Btu, it is obvious that changes in potential and kinetic energy would have to be very large to have thermal significance. Therefore, the restrictions mentioned are commonly met in chemical engineering practice, and when they are, once again the equation reduces to the heat balance

$$Q = H_2 - H_1.$$

Thus, the difference in the enthalpy of a system at the end states of a process is frequently a measure of the thermal energy crossing the boundaries of the system during the process. This equivalence is used so frequently in this text that the reader may forget that it is true only if the specified restrictions have been met. For this reason, a few specific situations in which $Q \neq H_2 - H_1$ will now be mentioned. First, in a constant volume reversible process (such as one occurring in a sealed calorimeter) $Q = U_2 - U_1$. Second, when a gas obeys the relationship $Pv = RT$, it can be proved that $U = f(T)$ only, and thus, for a constant temperature process $Q = RT \ln (v_2/v_1)$. Third, when pumping or other mechanical work occurs in a process, the term W cannot be omitted. Fourth, there are instances (e.g., high-speed turbines) when kinetic energy changes in a flowing system cannot be ignored. Fifth, the system may undergo an enthalpy change with no heat flow at all. The last situation exists when a gas is compressed in a thoroughly insulated cylinder; energy crosses boundaries as work, and this energy alters the internal energy of the system and also its P-v state. Thus, the system has undergone an enthalpy change, but due to the insulation, no heat is involved. Consequently, it must be reemphasized that $Q = \Delta H$ *only* when the basic equations permit it.

6.7 Summary

The equations and concepts discussed in this chapter are summarized in Table 6.1 for emphasis and ready reference. They will be given in both their differential and integrated form. All equations are based on unit mass.

TABLE 6.1 *Summary of Important First Law Relationships*

Equation	Application and Limitation
$dU = d'Q - d'W_{\text{nonflow}}$ $U_2 - U_1 = Q - W_{\text{nonflow}}$	The basic equations for all static processes involving only mechanical work. They may be applied to reversible and irreversible processes alike.
$dU = d'Q - [(g/g_c)\,dZ + du^2/2g_c + d\,(Pv) + d'W_{\text{flow}}]$ $U_2 - U_1 = Q - [(g/g_c)(Z_2 - Z_1) + (u_2^2 - u_1^2)/2g_c + (P_2v_2 - P_1v_1) + W_{\text{flow}}]$ NOTE The term in brackets is known as the mechanical effects.	The basic equations for all steady flow processes involving only mechanical work. They may be applied to reversible and irreversible processes alike.
$dE = d'Q - F\,dL$ $E_2 - E_1 = Q - \int_1^2 F\,dL$	The only limitation is reversibility. The term $F\,dL$ (or its integral) includes all external effects except heat. The most useful form of these equations appears next.
$dU = d'Q - P\,dv$ $U_2 - U_1 = Q - \int_1^2 P\,dv$	Reversibility is required and, in addition, these equations are limited to cases where the external effects other than heat are mechanical in nature. If the process is static, $\int_1^2 P\,dv = W_{\text{nonflow}}$. If the process is steady flow, $\int_1^2 P\,dv =$ [mechanical effects]$_{\text{flow}}$.
$P\,dv = d'W_{\text{nonflow}}$ $= d'$[mechanical effects]$_{\text{flow}}$ $\int_1^2 P\,dv = W_{\text{nonflow}}$ $=$ [mechanical effects]$_{\text{flow}}$	Reversibility is required, and only mechanical work may be involved but the equations may be used for either the static or steady-flow case as indicated.
$-v\,dP = d'W_{\text{flow}}$ $-\int_1^2 v\,dP = W_{\text{flow}}$	Reversibility is required; only mechanical work is involved; potential and kinetic energy changes must be negligible; and the process must be steady flow.
$dH = d\,(U + Pv) = dU + d\,(Pv)$ $= dU + P\,dv + v\,dP$ $H = U + Pv$	Basic definition of enthalpy.

(*continued*)

TABLE 6.1 (*continued*)

Equation	Application and Limitation
$Q = H_2 - H_1$	Applies to either of the following cases: 1. Constant pressure, reversible processes (static or steady flow) involving only mechanical work or mechanical effects. 2. Steady flow processes (reversible or irreversible) in which mechanical work and potential and kinetic energy changes are negligible.
$W = H_1 - H_2$	Steady flow, adiabatic processes (reversible or irreversible) in which potential and kinetic energy changes are negligible.

NOTES

1. By convention: W is positive when the system delivers mechanical energy to the surroundings. Q is positive when the surroundings deliver thermal energy to the system, i.e., when the surroundings are at the higher temperature.
2. The conversion factor $J = 778$ ft-lb$_f$/Btu must be used to make the equations dimensionally correct, since E, U, H, and Q are expressed in thermal units and the remaining term in mechanical energy units.
3. The Pv terms have not been multiplied by g/g_c, thereby implying that the pressure is expressed correctly in units of force per unit area.

Examples

Since engineers normally desire answers in English units, and most practical problems deal with earth systems where the ratio g/g_c is essentially unity, this ratio is omitted in the following examples.

Example 6.1 Steam Boiler

A steam boiler is essentially a device for converting liquid water into steam. Water is first pumped into the boiler by a feed pump. Then the water flows through tubes heated by the combustion of gas, oil, or solid fuel. The steam produced is finally discharged into pipes that carry it to the point of use. We are concerned with the water and steam only as it exists within the boiler; i.e., the feed pump and the steam pipes are external to the system. Pressure within the boiler is maintained constant by a regulating valve. The small pressure drop in the boiler tubes due to friction is sufficiently small to be ignored. If such friction is ignored, the process is mechanically reversible and the equation $U_2 - U_1 = Q - \int P\,dv$ may be applied. Since P is constant, the equation reduces to $U_2 - U_1 = Q - P(v_2 - v_1)$. Combining the U and Pv terms, there results $Q = H_2 - H_1$. In other words, the heat required by the boiler is the difference between the enthalpy of the discharged steam and that of the inlet water.

This process may also be analyzed by the steady state equation, since the process is constant flow. The dimensions of a boiler make the potential energy difference extremely small, and it can be omitted from further consideration. The kinetic energy change may also be disregarded, since it would require extremely large velocity changes for this term to contribute any important value in terms of thermal units, as can be noted from the relation:

$$\frac{u_2^2 - u_1^2}{2(32.2)(778)} = \text{Btu/lb},$$

when the velocity is expressed in ft per sec. As noted earlier, the feed pump is outside the system; therefore, $W = 0$. Dropping all of these items from the flow equation, we obtain

$$U_2 - U_1 = Q - (P_2v_2 - P_1v_1) \qquad \text{or} \qquad Q = H_2 - H_1,$$

which is the equation previously obtained. Note that the question of reversibility did not arise in the application of the flow equation.

Example 6.2 Steam Engine

One use of steam generated by the boiler just described would be to operate an engine. Although energy would be released in spurts if the steam were to expand in a single cylinder engine, practical engines are so designed that the release of energy as work occurs steadily. Therefore, we are justified in using the steady flow equation. However, mechanical friction is always present in devices of this sort, so we may not use the equation containing the integral of $P\,dv$. This leaves us with but one equation to consider. As before (see steam boiler), we may neglect any changes in potential and kinetic energy, although the latter term might be significant were we considering a high-speed turbine. Let us make the further assumption that the engine is thoroughly insulated. The flow equation then reduces to: $U_2 - U_1 = -[P_2v_2 - P_1v_1 + W]$. Solving for W, the term in which we are interested, there results $W = -(H_2 - H_1)$ or $W = H_1 - H_2$, i.e., the work is the difference between the enthalpies of the steam entering and of that leaving. If heat is lost due to incomplete insulation, the final equation would be $W = H_1 - H_2 + Q$, but note that since heat should be leaving the system, Q will have a negative value and will thus reduce the amount of work the engine can deliver. This is, of course, the reason for insulating the engine.

Example 6.3 Throttling Valve

A throttling valve is a restrictive device placed in a pipe for the purpose of maintaining a constant, high pressure upstream and a constant, lower pressure downstream. This process is clearly both steady flow and irreversible (the restriction in the pipe introduces considerable friction), so the only applicable equation is

$$U_2 - U_1 = Q - \left[(Z_2 - Z_1) + \frac{(u_2^2 - u_1^2)}{2g_c} + (P_2v_2 - P_1v_1) + W\right].$$

Considering the nature of the device, Q, W, and ΔZ will always be zero. The mass rate of flow throughout the pipe is, of course, constant, and if the pipe diameter does not change, the term $\Delta u^2/2g_c$ due to possible changes in the density of the fluid flowing would be negligible. When all of these terms are dropped, there remains only $H_1 = H_2$. A process occurring with no change in enthalpy is called an isenthalpic process. All throttling processes are isenthalpic, but a process may be isenthalpic without involving throttling. For example, the isothermal expansion of a perfect gas in a cylinder to produce work is isenthalpic and yet it is not a throttling process.

Example 6.4 Bernoulli Equation

A very important consequence of the steady flow equation as applied to fluid flow problems is called the Bernoulli Equation. Since it is fully discussed in Chapter 9, it will not be developed here, but the reader should remember that it comes from the First Law of Thermodynamics.

Example 6.5 Heat Transfer in a Jacketed Kettle

Consider a large kettle in which a quantity of a solution is being heated by steam condensing in a jacket surrounding the kettle. For present purposes, we shall consider the entire unit thoroughly insulated. Since the heat required for warming the solution comes from the steam, two systems will have to be analyzed. To distinguish between them, we shall call the contents of the kettle the system and the steam in the jacket the surroundings.

Let us look at the contents of the kettle first. From the point of view of this system, the process is static, reversible (no mechanical friction is involved), and occurs at constant (atmospheric) pressure. Thus,

$$U_2 - U_1 = Q - P(v_2 - v_1) \qquad \text{or} \qquad Q_{\text{system}} = (H_2 - H_1)_{\text{contents}}.$$

From the point of view of the surroundings (the steam in the jacket), the process is steady flow, since the condensed steam constantly escapes through a condensate trap. This trap also maintains a constant pressure within the steam space. There is no work, and potential and kinetic energies undergo no significant change. Thus, the steady flow equation reduces to:

$$Q_{\text{surroundings}} = (H_2 - H_1)_{\text{steam}}.$$

The final step in the analysis is to relate these two equations. This can be done when we realize that the heat required by the contents is supplied by the steam. Remembering the sign conventions already adopted, $Q_{\text{system}} = -Q_{\text{surroundings}}$, since energy is entering the system, but leaving the surroundings. Therefore, the final equation is:

$$(H_2 - H_1)_{\text{contents}} = (H_1 - H_2)_{\text{steam}}.$$

6.8 Conclusions

The previous examples serve to illustrate the general approach to analyzing problems by the First Law of Thermodynamics. First, two questions must be answered: (1) Is the process mechanically reversible? (2) Is the process static or steady flow? Since all real mechanical processes involve some irreversibility, a decision will have to be made whether or not such irreversibility is of significance or is trivial for the purposes of the problem at hand.

The answer to the second question automatically eliminates the first or third of the three general equations listed on p. 279. If the process is mechanically irreversible, the second equation is eliminated and the one remaining equation must be used. If reversibility may be assumed, then we have the choice of either the second equation or one of the others. In this case, the data given may dictate the one to be used, or in other instances either of the two equations may be used (e.g., the steam boiler discussed above).

After one has chosen the general equation properly applicable to the problem at hand, the next step is to determine what terms, if any, may be assumed to be negligible. Finally, any special restraint, such as constant volume or constant pressure, is introduced as dictated by the problem. The equation is then reassembled to give the quantity sought (work, heat, velocity, etc.) in terms of the remaining variables. Since enthalpy is such an important property, U and Pv are usually combined if both occur in the final equation.

These general principles will be found useful in solving those problems in Chapter 9 which deal with the first law of thermodynamics.

LIST OF SYMBOLS

Latin letters

A	Area
a	Acceleration
d	Differential operator for state functions
d'	Differential operator for path functions
E	Internal energy, general
F	Force
f	Force
g	Local acceleration due to gravity

g_c	Acceleration due to gravity at standard sea level; 32.2 ft/sec^2
H	Enthalpy
J	Mechanical equivalent of heat; 778 ft-lb$_f$/Btu
L	Distance or length
m	Mass
P	Absolute pressure, force/area
Q	Heat
R	Universal gas constant
T	Absolute temperature
t	Time
U	Internal energy when work is limited to mechanical work
u	Linear velocity
v	Specific volume, i.e., volume/unit mass
W	Work (limited to mechanical work unless otherwise subscripted)
Z	Height above a datum

Subscripts

$_1$	Initial state
$_2$	Final state

7 Thermophysical Properties

When energy (frequently in the form of heat) enters or leaves a system, it produces changes. These changes may be wholly physical in nature, such as a rise in temperature, a change of state, etc., or they may be accompanied by chemical reactions. This chapter will be concerned only with thermophysics, the heat effects measuring and accompanying physical changes; Chapter 8 will consider thermochemistry, the heat effects measuring and accompanying chemical change.

7.1 Heat Capacity and Specific Heat

It is common knowledge that equal masses of different materials at a single temperature require varying amounts of energy to bring them to a common higher temperature. The property that quantitatively describes this phenomenon is called by some the specific heat and by others the heat capacity. It is important to spend a few moments clarifying this matter of nomenclature.

The property under discussion has been defined as the specified amount of heat required to raise a specific quantity of matter (usually a unit mass) through a specified temperature interval (usually 1°). Thus, it would seem that the term specific heat would be an admirable choice for a property so defined. Indeed many thermodynamicists prefer this designation.

If one were dealing wholly with scientific theory, no problem would arise. However, in the field of engineering another meaning has been attached to the term specific heat; thus the name, heat capacity, has been assigned to the property we have defined above. The phrase specific heat is then used to represent a dimensionless quantity defined as the ratio of the heat capacity of the substance divided by the heat capacity of a reference substance (often water), just as specific gravity is the dimensionless ratio of two densities.

Why does the thermodynamicist object to the term heat capacity? It is a relic of an age that misconstrued the nature of heat and thought of it as a substance that could be stored or released from a body much as one can hold water in a pail or pour water from it. The confusion resulting from this idea of heat was so great that thermodynamicists have been trying to rid scientific language of words and expressions that tend to perpetuate wrong concepts. To a certain extent they have had success. For example, the old term heat content, widely used even as late as the 1930's, has now been almost wholly superseded by the word enthalpy.

Unfortunately, the present case requires a terminology to cover two concepts, one dimensional and one dimensionless. Since only the former is used in scientific work, one definition would suffice, and when specific heat is adopted for this purpose, no confusion results. Engineering and technical work make use of both types of numbers, and until someone can devise an acceptable coined name that will be adopted as enthalpy has been, we are faced with the practical necessity of continuing to use a phrase that is theoretically improper. This being an engineering textbook, specific heat will be used to represent the ratio of two heat capacities, and the latter term will describe the temperature effect of heat upon matter. It cannot be stressed too strongly, however, that heat can neither be stored nor contained; its true nature as energy in transition must ever be kept in mind.

Having decided to adopt the term heat capacity, let us now get a proper definition of it. Earlier in this section there was a descriptive definition which, while correct as far as it goes, is not sufficiently precise. To begin with, it should be noted that heat capacity is susceptible to even small changes in external conditions. Therefore, it should be defined in differential terms. If the symbol C be used temporarily for the general property, heat capacity may be defined mathematically as

$$C = \frac{d'Q}{dx}. \tag{7.1}$$

For the moment, ignore the meaning of x except to note that it is a state function. What this equation states in words is that the heat capacity is equal to the infinitesimal quantity of heat accompanying some differential change of state.

In practice, the kind of process involved in the passage of energy between the system and the surroundings is usually specified. Thus, heat capacity is normally defined as a partial differential

$$C = \left(\frac{\partial' Q}{\partial x}\right)_y. \tag{7.2}$$

In this expression, x commonly refers to one of the three primary state variables, P, v, or T, and y to one of the remaining two. What this equation states is that heat capacity is equal to the infinitesimally small quantity of heat required to effect a differential change of one state variable when another state variable is maintained constant.

Not all of the six possible heat capacities, so defined, are important even for the theoretical thermodynamicist. For our purposes, only two need to be considered. In both cases $x = T$, with the two individual cases resulting when y is either P or v. The names of the remaining parameters and their formal definitions follow:

$$\text{Heat capacity at constant pressure} = C_p = \left(\frac{\partial' Q}{\partial T}\right)_P, \tag{7.3}$$

$$\text{Heat capacity at constant volume} = C_v = \left(\frac{\partial' Q}{\partial T}\right)_V. \tag{7.4}$$

In both of these expressions Q is the heat per unit mass. When the gm, kg, or lb is the chosen unit of mass no special qualification is included. If a mole is the chosen unit of mass, C is called the molar heat capacity. Either a pound mole or a gram mole may be used, depending on the temperature chosen. Since we are dealing with a difference in temperature, we may use degrees Fahrenheit and degrees Rankine interchangeably. Similarly, it makes no difference whether we specify degrees Centigrade or degrees Kelvin. Thus, we may expect to find C in units of Btu/(lb-mole)(°F), Btu/(lb-mole)(°R), cal/(gm-mole)(°C), and cal/(gm-mole)(°K). Further, other units are used in which the lb-mole is replaced by the lb, and the gm-mole by the gm. A check of the units will reveal that cal/(gm)(°C) equals (numerically) Btu/(lb)(°F), and that cal/(gm-mole)(°K) equals (numerically) Btu/(lb-mole)(°R).

Heat capacities may be studied from the standpoint either of general thermodynamics or molecular kinetics or they may be studied empirically. Because of the student's general background at this point in his career, the last approach is used in this book. We shall now look at heat capacities for each of the three states of matter in turn.

7.2 Heat Capacity of Gases

Thermodynamics shows that for any process involving only ideal gases

$$\partial' Q_p = dH = C_p\, dT, \tag{7.5}$$

$$\partial' Q_v = dU = C_v\, dT, \tag{7.6}$$

$$C_p - C_v = R, \tag{7.7}$$

and that both C_p and C_v are functions of temperature only. Although $dH = C_p\, dT$ for any process in which an ideal gas is involved, this relation-

TABLE 7.1a *Heat Capacity Data for Some Inorganic Gases**

T °K	C_p^0, cal/(gm-mole)(°K)													
	O_2	H_2	H_2O	N_2	NO	CO	CO_2	Cl_2	HCl	Br_2	HBr	SO_2	SO_3	H_2S
200	6.961	6.561		6.957	7.278	6.956								
250	6.979	6.769		6.959	7.183	6.958		7.88		8.48				
298.16	7.017	6.892	8.025	6.960	7.137	6.965	8.874	8.11	6.96	8.62	6.96	9.53	12.10	8.19
300	7.019	6.895	8.026	6.961	7.134	6.965	8.894	8.12	6.96	8.62	6.96	9.54	12.13	8.20
400	7.194	6.974	8.185	6.991	7.162	7.013	9.871	8.44	6.97	8.78	6.98	10.39	14.06	8.53
500	7.429	6.993	8.415	7.070	7.289	7.120	10.662	8.62	7.00	8.86	7.04	11.12	15.66	8.93
600	7.670	7.008	8.677	7.197	7.468	7.276	11.311	8.74	7.07	8.91	7.14	11.71	16.90	9.35
700	7.885	7.035	8.959	7.351	7.657	7.451	11.849	8.82	7.17	8.94	7.27	12.17	17.86	9.78
800	8.064	7.078	9.254	7.512	7.833	7.624	12.300	8.88	7.29	8.97	7.42	12.53	18.61	10.21
900	8.212	7.139	9.559	7.671	7.990	7.787	12.678	8.92	7.42	8.99	7.58	12.82	19.23	10.62
1000	8.335	7.217	9.861	7.816	8.126	7.932	12.995	8.96	7.56	9.01	7.72	13.03	19.76	11.00
1100	8.440	7.308	10.145	7.947	8.243	8.058	13.26	8.99	7.69	9.03	7.86	13.20	20.21	11.34
1200	8.530	7.404	10.413	8.063	8.342	8.168	13.49	9.02	7.81	9.04	7.99	13.35	20.61	11.64
1300	8.608	7.505	10.668	8.165	8.426	8.265	13.68	9.04	7.93	9.06	8.10	13.47	20.96	11.92
1400	8.676	7.610	10.909	8.253	8.498	8.349	13.85	9.06	8.04	9.07	8.20	13.57	21.28	12.16
1500	8.739	7.713	11.134	8.330	8.560	8.419	13.99	9.08	8.14	9.08	8.30	13.65	21.58	12.37
1750	8.888	7.953	11.62	8.486	8.682	8.561	14.3							
2000	9.030	8.175	12.01	8.602	8.771	8.665	14.5							
2250	9.168	8.364	12.32	8.690	8.840	8.744	14.6							
2500	9.302	8.526	12.56	8.759	8.895	8.806	14.8							
2750	9.431	8.667	12.8	8.815	8.941	8.856	14.9							
3000	9.552	8.791	12.9	8.861	8.981	8.898	15.0							
3500	9.763	8.993	13.2	8.934	9.049	8.963	15.2							
4000	9.933	9.151	13.3	8.989	9.107	9.015	15.3							
4500	10.063	9.282	13.4	9.035	9.158	9.059	15.5							
5000	10.157	9.389	13.5	9.076	9.208	9.096	15.6							

Source: *Selected Values of Properties of Chemical Compounds*, Manufacturing Chemists Association Research Project, Thermodynamic Research Center, Department of Chemistry, Texas A & M University, June 30, 1966.

* The values of heat capacities in Tables 7.1a, b are not the same as those calculated from the equations using constants from Table 7.1c. A four-term model was used for the values in Tables 7.1a, b. The simpler three-term model was chosen in Table 7.1c to simplify calculations requiring use of these equations.

TABLE 7.1b *Heat Capacity Data for Some Organic gases*

T °K	C_p^0, cal/(gm-mole)(°K)								
	Methane CH_4	Ethane C_2H_6	Propane C_3H_8	*n*-Butane C_4H_{10}	*i*-Butane C_4H_{10}	*n*-Pentane C_5H_{12}	*n*-Hexane C_6H_{14}	Ethene C_2H_4	Propene C_3H_6
250	8.185	11.28	15.37	20.51	20.10	25.29			
298.16	8.536	12.58	17.57	23.29	23.14	28.73	34.20	10.41	15.27
300	8.552	12.64	17.66	23.40	23.25	28.87	34.37	10.45	15.34
400	9.721	15.68	22.54	29.60	29.77	36.53	43.47	12.90	19.10
500	11.13	18.66	27.02	35.34	35.62	43.58	51.83	15.16	22.62
600	12.55	21.35	30.88	40.30	40.62	49.64	58.99	17.10	25.70
700	13.88	23.72	34.20	44.55	44.85	54.83	65.10	18.76	28.37
800	15.10	25.83	37.08	48.23	48.49	59.30	70.36	20.20	30.68
900	16.21	27.69	39.61	51.44	51.65	63.18	74.93	21.46	32.70
1000	17.21	29.33	41.83	54.22	54.40	66.55	78.89	22.57	34.46
1100	18.09	30.77	43.75	56.64	56.81	69.48	82.32	23.54	35.99
1200	18.88	32.02	45.42	58.74	58.89	72.02	85.30	24.39	37.32
1300	19.57	33.11	46.89	60.58	60.71	74.24	87.89	25.14	38.49
1400	20.18	34.07	48.16	62.17	62.29	76.16	90.14	25.79	39.51
1500	20.71	34.90	49.26	63.57	63.67	77.83	92.10	26.36	40.39

TABLE 7.1b (*continued*)

T °K	C_p^0, cal/(gm-mole)(°K)									
	1-Butene C_4H_8	Iso-butene C_4H_8	Acetyl-ene C_2H_2	Benzene C_6H_6	Toluene C_7H_8	Ethyl-Benzene C_8H_{10}	Styrene C_8H_8	Cyclo-Hexane C_6H_{12}	Methanol CH_3OH	Ethanol C_2H_5OH
250										
298.16	20.47	21.30	10.499	19.52	24.80	30.69	29.18	25.40	10.49	15.64
300	20.57	21.39	10.532	19.65	24.95	30.88	29.35	25.58	10.52	15.71
400	26.04	26.57	11.973	26.74	33.25	40.76	38.32	35.82	12.29	19.36
500	30.93	31.24	12.967	32.80	40.54	49.35	45.94	45.47	14.22	22.77
600	35.14	35.30	13.728	37.74	46.58	56.44	52.14	53.83	16.02	25.69
700	38.71	38.81	14.366	41.75	51.57	62.28	57.21	60.87	17.62	28.19
800	41.80	41.86	14.933	45.06	55.72	67.15	61.40	66.76	19.04	30.33
900	44.49	44.53	15.449	47.83	55.22	71.27	64.93	71.68	20.29	32.19
1000	46.82	46.85	15.922	50.16	62.19	74.77	67.92	75.80	21.38	33.83
1100	48.85	48.88	16.353	52.16	64.73	77.77	70.48	79.3		
1200	50.62	50.63	16.744	53.86	66.90	80.35	72.66	82.2		
1300	52.16	52.17	17.099	55.52	68.77	82.57	74.54	84.7		
1400	53.50	53.51	17.418	56.58	70.38	84.49	76.16	86.8		
1500	54.67	54.68	17.704	57.67	71.78	86.16	77.57	88.6		

Source: *Selected Values of Properties of Chemical Compounds*, Manufacturing Chemists Association Research Project, Thermodynamic Research Center, Department of Chemistry, Texas A & M University, June 30, 1966.

ship is true only for constant pressure processes of real gases. Similarly, $dU = C_v\,dT$, while general for ideal gases, is restricted to constant volume processes of real gases. Likewise, the statement that C_p and C_v are functions of temperature only does not apply to real gases.

For these reasons, it is common practice to report heat capacity data for gases in the ideal state (i.e., under zero pressure), and to make any corrections for nonideality by thermodynamic methods beyond the scope of this text. Such data are usually indicated with a superscript ^{0} or * to indicate that they apply only to the ideal gas state. Since many actual gases approach ideal behavior at normal atmospheric pressure, it is normally satisfactory to use such data for processes occurring at low pressures without correction.

Furthermore, it may be shown by the kinetic theory of gases that for all monatomic gases such as helium, $C_v = \frac{3}{2}R$, $C_p = \frac{5}{2}R$, and therefore, that $C_p/C_v = \frac{5}{3}$ or 1.67. The ratio C_p/C_v is variously known as k or γ, and in later courses in thermodynamics it will be seen to be important in problems involving compression of gases. One cannot easily predict k for polyatomic gases, but for diatomic gases such as oxygen it is approximately 1.4. It becomes smaller (approaching unity) as the molecular complexity increases. For all but monatomic gases, k is a function of the temperature as well as the molecular constitution of the gas.

A great deal of experimental heat capacity data exist, and the student should have little difficulty in finding satisfactory information for most of the applications of current interest to him. These data may be expressed in tabular form, as graphs or nomographs, or by equations. These equations are usually of the form:

$$C_P^0 = a + bT + cT^2 + dT^3 + \cdots \tag{7.8}$$

or

$$C_P^0 = A + BT - \frac{C}{T^2}. \tag{7.9}$$

In some instances one may need as many as four terms in the equation to get maximum accuracy, but commonly only a three-term equation is required. Since the temperature in this equation may be expressed in degrees absolute (either Kelvin or Rankine) or as degrees Centigrade or Fahrenheit, the units on the constants a, b, c, d and A, B, C must be carefully noted before the equation is used. Table 7.1a,b lists molal heat capacities for a number of gases, while Table 7.1c lists the constants to be used in Equation 7.8 for several common gases. Note that the equations are valid only over the interval 300 to 1500 °K and that T is expressed in degrees Kelvin. The heat capacity so calculated will be in gm-cal/(gm-mole)(°K), equivalent of course to Btu/(lb-mole)(°R). The constants a, b, c or A, B, C are specific for the equation for which they were determined; thus the constants for Equation 7.8 may not be used for Equation 7.9. Since tables of constants are

TABLE 7.1c *Empirical Constants for Molal Heat Capacities of Gases at Constant Pressure* ($p = 0$)

$C_p = a + bT + cT^2$, where T is in degrees Kelvin; gm-cal/(gm-mole)(°K)
Temperature range 300 to 1500 °K

Gas	a	$b(10^3)$	$c(10^6)$	% Maximum Deviation 300 to 1500 °K
H_2	6.946	−0.196	0.4757	0.5
N_2	6.457	1.389	−0.069	1.4
O_2	6.117	3.167	−1.005	1.5
CO	6.350	1.811	−0.2675	1.5
NO	6.440	2.069	−0.4206	1.6
H_2O	7.136	2.640	0.0459	1.3
CO_2	6.339	10.14	−3.415	2.0
SO_2	6.945	10.01	−3.794	1.5
SO_3	7.454	19.13	−6.628	3.7
HCl	6.734	0.431	0.3613	1.0
C_2H_6	2.322	38.04	−10.97	1.4
CH_4	3.204	18.41	−4.48	3.5
C_2H_4	3.019	28.21	−8.537	1.9
Cl_2	7.653	2.221	−0.8733	1.5
Air	6.386	1.762	−0.2656	1.0
NH_3*	5.92	8.963	−1.764	1.0

Source: O. A. Hougen, K. M. Watson, and R. A. Ragatz, *Chemical Process Principles, Part I* (New York: John Wiley & Sons, Inc., 1954). Reprinted with permission.
* W. M. D. Bryant, *Ind. Eng. Chem.*, *25*, 820 (1933).

usually accompanied by the form of the equation to which they apply, no difficulties should arise.

7.3 Mean Heat Capacities

It has been mentioned previously that, for constant pressure processes, $dH = C_p\,dT$. If the heat capacity were constant, the change in enthalpy could be determined simply by multiplying the change in temperature by the heat capacity. However, if the heat capacity is expressed by an equation such as $a + bT + cT^2$, one is faced with the problem of integrating

$$(a + bT + cT^2)\,dT$$

to determine the change in enthalpy. While this operation certainly is not difficult, repetitive integration can be avoided by the use of mean heat capacities. Mean heat capacity is defined as that value of C_p that will give the same result (when multiplied by the temperature difference) as the formal integration.

This definition may be visualized by reference to Figure 7.1. The curve represents the experimental relationship between C_p and T. The enthalpy change from T_1 to T_2 is the area under the curve between T_1 and T_2, which is

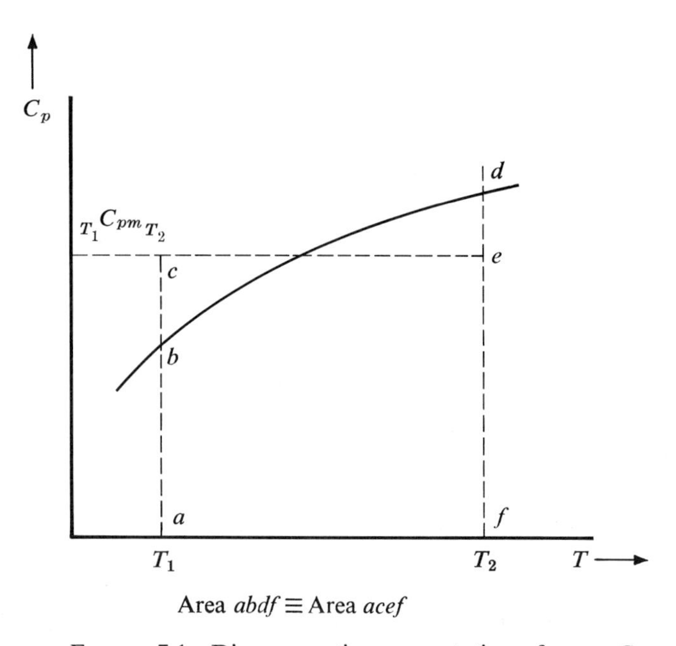

FIGURE 7.1 Diagrammatic representation of mean C_p.

the integral of $C_p\,dT$ between these points. The mean heat capacity, indicated by the dotted line, when multiplied by $T_2 - T_1$ will give the same value; i.e., the rectangular area is the same as the area under the curve. It should be verified that, when the experimental data correspond to the quadratic formula noted above,

$$C_{pm} = a + \frac{b}{2}(T_2 + T_1) + \frac{c}{3}(T_2^2 + T_1T_2 + T_1^2).$$

Obviously the determination of C_{pm} is easiest if T_1 is zero, for then the formula reduces to

$$C_{pm} = a + \frac{b}{2}T_2 + \frac{c}{3}T_2^2.$$

If the temperature were in either degrees Centigrade or degrees Fahrenheit, a zero temperature base might be suitable, but when T represents absolute temperature, a zero base could not be used, since all gases would have solidified before this temperature was reached. As will be noted in Chapter 8, it is common practice to use 25 °C (77 °F, 298 °K, 537 °R) as a standard reference temperature for heats of reaction. Thus, that is the normal base temperature, T_1, chosen for computing C_{pm}. If one chooses this base temperature, he may prepare a table or a chart of mean heat capacities from this temperature to any other. If the intervals are small enough (or enough points are plotted on a graph), one can readily interpolate and get accurately any mean heat capacity desired.

Suppose, however, that in a given problem the lower limit is not 25 °C. A moment's reflection will show that the chart is still readily usable. Let us say that the problem is to find the mean heat capacity of oxygen from 100°–200 °C. The table or chart would give only means from 25–100 °C or 25–200 °C. However, recall that $C_{pm}(T_2 - T_1)$ represents an area under a curve. Therefore, $C_{pm}(200–25)$ is the area under the $C_p - T$ curve between these temperatures, and $C_{pm}(100–25)$ is the corresponding area between these temperatures. Since what we want is the area between 100° and 200°, all we need to do is to subtract the two areas previously determined and we have an area which when divided by (200–100) will give the desired mean heat capacity. While it may be possible to extrapolate prepared charts or tables a short distance below 25 °C, it would be better to compute the mean heat capacity by actual integration. If the temperature range is not large, the mean heat capacity is often very close to the heat capacity at the mean temperature; this, of course, presupposes a linear relationship over the temperature range used.

Figure 7.2 (a,b,c,d) displays mean heat capacity information from 25 °C (77 °F) to t_2 for some common gases.

Example 7.1

▶ STATEMENT: Calculate the molal heat capacity of CO_2 at 500 and 1200 °C.

▶ SOLUTION: From Table 7.1, the equation for the molal heat capacity of this gas is

$$C_p = 6.339 + 10.14 \times 10^{-3}T - 3.415 \times 10^{-6}T^2,$$

with T in °K.

Substituting

$$T = 500 + 273 = 773,$$

$$C_p = 6.339 + 7.8382 - 2.0406,$$

$$= 12.1366 \text{ cal/(gm-mole)(°C)},$$

$$= 12.1366 \text{ Btu/(lb-mole)(°F)}.$$

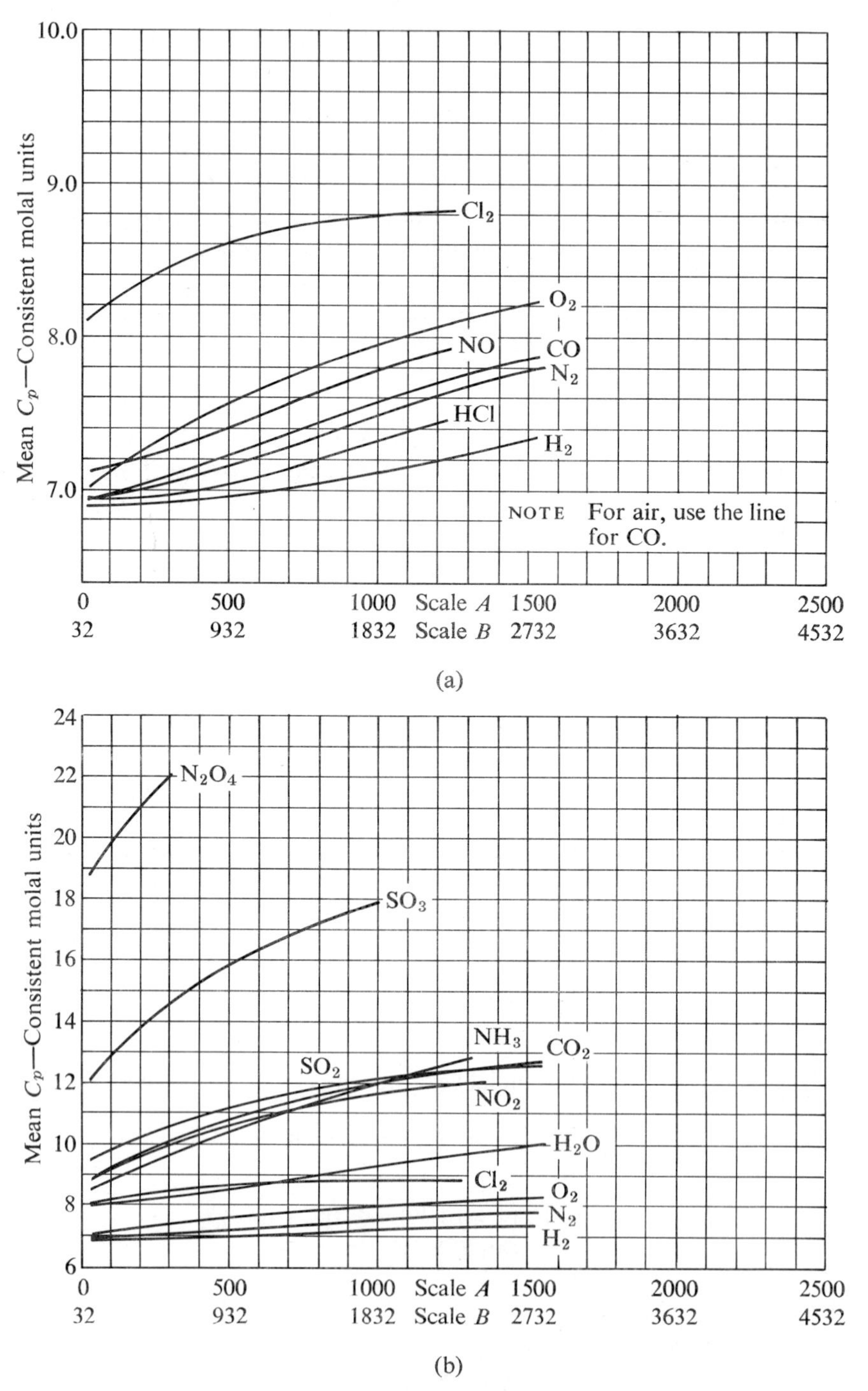

FIGURE 7.2 Mean C_p as a function of temperature. Scale A: 25 to t °C. Scale B: 77 to t °F. [All data from O. A. Hougen, K. M. Watson, and R. A. Ragatz, *Chemical Process Principles, Part I* (New York: John Wiley & Sons, Inc., 1954). Reprinted with permission.]

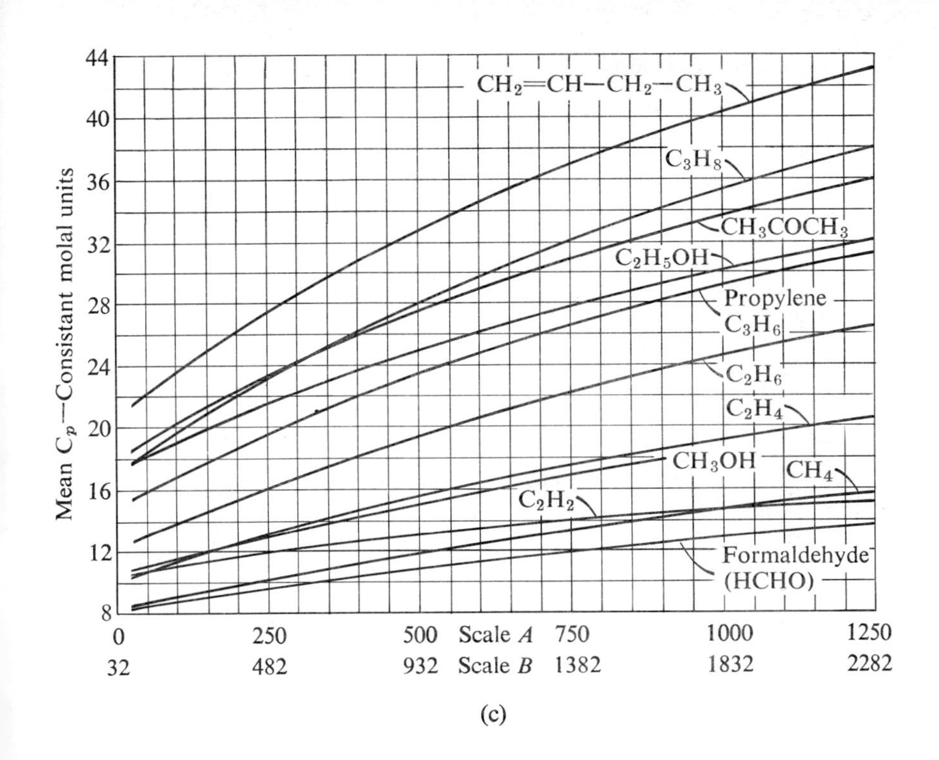

Mean C_p—Consistant molal units
$CH_2{=}CH{-}CH_2{-}CH_3$
C_3H_8
CH_3COCH_3
C_2H_5OH
Propylene
C_3H_6
C_2H_6
C_2H_4
CH_3OH
CH_4
C_2H_2
Formaldehyde
(HCHO)
44
40
36
32
28
24
20
16
12
8
0
250
500
Scale A
750
1000
1250
32
482
932
Scale B
1382
1832
2282

(c)

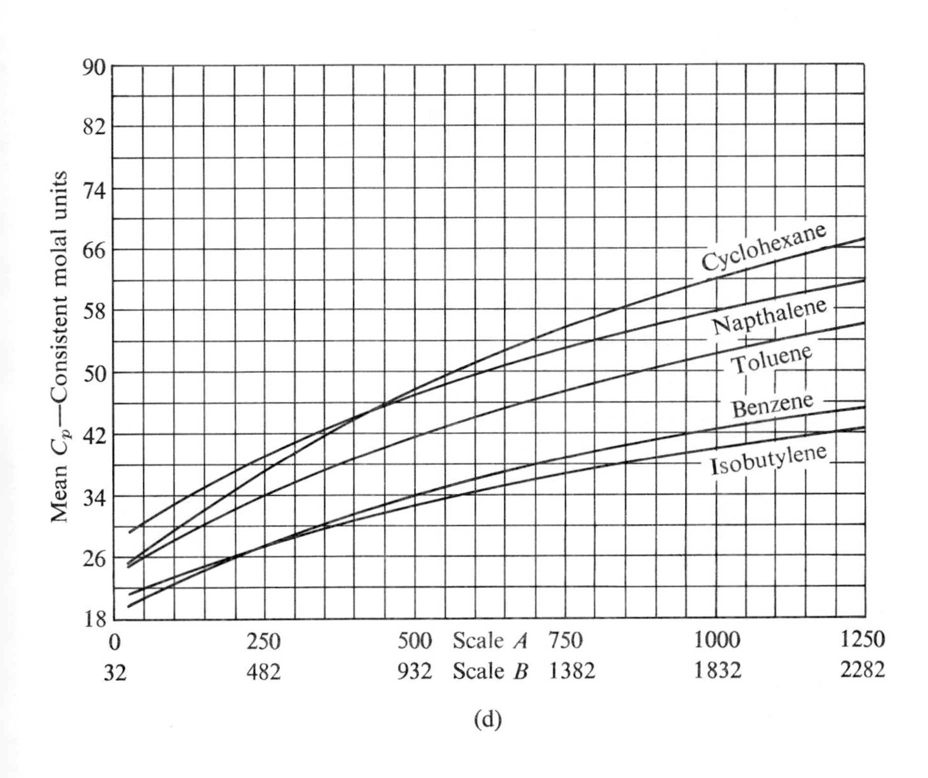

Mean C_p—Consistent molal units
Cyclohexane
Napthalene
Toluene
Benzene
Isobutylene
90
82
74
66
58
50
42
34
26
18
0
250
500
Scale A
750
1000
1250
32
482
932
Scale B
1382
1832
2282

(d)

Substituting

$$\begin{aligned} T &= 1200 + 273 = 1473, \\ C_p &= 6.339 + 14.9362 - 7.4096, \\ &= 13.8656 \text{ cal/(gm-mole)(°C)}, \\ &= 13.8656 \text{ Btu/(lb-mole)(°F)}. \end{aligned}$$

Neither of these values is good to better than 1%. We might then report the values as 12.1 and 13.9, respectively.

Example 7.2

▶ STATEMENT: Calculate the mean molal heat capacity of CO_2 between
a. 25 °C and 1200 °C,
b. 500 °C and 1200 °C.

▶ SOLUTION: From equations in Section 7.3, we see that

$$\begin{aligned} {}_{T_1}C_{pm_{T_2}} &= \frac{\int C_p \, dT}{T_2 - T_1} \\ &= a + \frac{b}{2}(T_2 + T_1) + \frac{c}{3}(T_2^2 + T_1T_2 + T_1^2). \end{aligned}$$

And, obtaining values of a, b, and c from Table 7.1c,

$${}_{T_1}C_{pm_{T_2}} = 6.339 + 5.07 \times 10^{-3}(T_2 + T_1) - 1.138 \times 10^{-6}(T_2^2 + T_2T_1 + T_1^2).$$

Then for Part (a), where

$$T_2 = 1200 + 273 = 1473$$

and

$$T_1 = 25 + 273 = 298.$$

$$\begin{aligned} {}_{298}C_{pm_{1473}} &= 6.339 + 5.07 \times 10^{-3}(1771) - 1.138 \times 10^{-6}(2{,}697{,}487), \\ &= 6.339 + 8.9790 - 3.0697, \\ &= 12.2482. \end{aligned}$$

This value might be reported as 12.2 since no greater precision is justified.
For Part (b):

$$T_2 = 1473 \text{ as in Part (a)}$$

and

$$T_1 = 500 + 273 = 773.$$

Then,

$$\begin{aligned} {}_{773}C_{pm_{1473}} &= 6.339 + 5.07 \times 10^{-3}(2246) - 1.138 \times 10^{-6}(3{,}905{,}887), \\ &= 13.2813, \end{aligned}$$

which we will report as 13.3.

These mean molal heat capacities can obviously be expressed in any of the standard units:

a. cal/(gm-mole)(°C),
b. Btu/(lb-mole)(°F),
c. Pcu/(lb-mole)(°C),
d. Kcal/(Kg-mole)(°C).

Example 7.3

▶ STATEMENT: Calculate the number of Btu required to raise the temperature of 100 lb of CO_2 from 500 to 1200 °C.

▶ SOLUTION: Several courses of action are open to us.

A. Use the value of C_{pm} calculated in Example 7.2. This procedure is the simplest, but it is only through good fortune that C_{pm} over the range of interest is available. Taking advantage of its availability:

$$\text{Moles of } CO_2 = \frac{100}{44} = 2.2727,$$

$$\begin{aligned} C_{pm} \text{ (from Example 7.2)} &= 13.3, \\ Q = (13.3)(2.27)(1200 - 500) &= 21{,}130 \text{ Pcu}, \\ &= 38{,}040 \text{ Btu}. \end{aligned}$$

B. When C_{pm} over the desired temperature range is not available, we might well resort to C_{pm} charts, such as are given in Figure 7.2. Over the range 25 to t °C, we find for CO_2

t	C_{pm}
500	10.76
1200	12.23

Note that the second value agrees with that calculated in Example 7.2 despite disparity in the sources of data. Then for 100 pounds of CO_2

$$\begin{aligned} Q &= 2.27[12.23(1200 - 25) - 10.76(500 - 25)], \\ &= 21{,}020 \text{ Pcu} = 37{,}830 \text{ Btu}. \end{aligned}$$

Obviously the disparity between this answer, and that obtained in Part (A) is due to minor disparities in the data and in reading Figure 7.2.

C. We might have integrated the heat capacity equation directly between 773 and 1473 °K. This method involves more labor than either of the preceding ones and is probably unjustified if information is available for Method B. To complete our illustrations we perform the calculation anyway.

$$
\begin{aligned}
Q &= 2.27 \int_{773}^{1473} C_p \, dT, \\
&= 2.27[6.339T + 5.07 \times 10^{-3}T^2 - 1.138 \times 10^{-6}T^3]_{773}^{1473}, \\
&= 2.27[6.339(1473 - 773) + 5.07 \times 10^{-3}(1473^2 - 773^2) \\
&\qquad - 1.138 \times 10^{-6}(1473^3 - 773^3)], \\
&= 21{,}100 \text{ Pcu} = 37{,}990 \text{ Btu}.
\end{aligned}
$$

7.4 Heat Capacity of Mixtures of Gases

Recall that for ideal gases $dH = C_p \, dT$. Since ideal gases undergo no enthalpy change on mixing, the total enthalpy change of a gaseous mixture (per mole of mixture) is $\Delta H = \sum_{i=1} N_i C_{pmi} \, \Delta T$ where N_i is the mole fraction of the ith component. This is equivalent to saying that the mean molal average heat capacity of the mixture is $\sum_{i=1} N_i C_{pmi}$; thus one simply weights the mean heat capacities of the various components according to the mole fraction of each present.

Example 7.4

▶ STATEMENT: A gaseous mixture of composition

12% CO_2
8% O_2
80% N_2

is cooled from 1200 to 500 °C. Calculate the heat that must be extracted from 1000 cu ft of this mixture, measured at the inlet temperature and 30 in. of mercury.

▶ SOLUTION: Any of the methods in Example 7.3 can be applied to the individual components with the resulting quantities of heat added. We take the simplest approach, that of mean molal heat capacities, using Figure 7.2. The heat capacities, from 25 to t °C are:

	t = 500 °C	t = 1200 °C
CO_2	10.76	12.23
O_2	7.54	8.07
N_2	7.13	7.60

First calculate the number of moles of each component in the measured 1000 cu ft.

$$\frac{1000}{359} \times \frac{273}{1473} \times \frac{30.0}{29.92} = 0.518 \text{ total moles.}$$

$$\begin{aligned} \text{moles } CO_2 &= (0.12)(0.518) = 0.0621, \\ \text{moles } O_2 &= (0.08)(0.518) = 0.0414, \\ \text{moles } N_2 &= (0.80)(0.518) = 0.4140. \end{aligned}$$

Total heat transferred:

$$\begin{aligned} &0.0621[10.76(500 - 25) - 12.23(1200 - 25)] \\ +&0.0414[\ 7.54(500 - 25) - \ 8.07(1200 - 25)] \\ +&0.4140[\ 7.13(500 - 25) - \ 7.60(1200 - 25)] \\ &= -575 - 244 - 2295 = -3115 \text{ Pcu} = -5605 \text{ Btu.} \end{aligned}$$

The answer is first obtained in Pcu units due to the use of the Centigrade scale for temperatures. The minus sign confirms the obvious fact that heat must be extracted to effect the specified temperature change.

7.5 Heat Capacity of Solids

As in the case of gases, one should use actual experimental data for the heat capacity of solids when such information is available. Table 7.2 presents this information in the equation form previously discussed. In some instances, the heat capacity varies almost linearly with temperature over moderate ranges. In this case, interpolation and modest extrapolation are possible with two known values of the heat capacity at different temperatures.

When experimental data are not available, it is helpful to be able to estimate the heat capacity of solids. The law of Petit and Dulong states that all crystalline elements have a heat capacity of 6.2 cal/(gm-atom)(°K), but this is only an approximation that holds best around room temperature for those elements whose atomic weight is above 40. However, this relative constancy of crystalline atomic heat capacities led to Kopp's rule, which is that the heat capacity of a solid compound is equal to the sum of the individual heat capacities of the component atoms. To apply this rule, one must use the atomic values listed in Table 7.3. Note that these values are in cal/(gm-atom) (°K). As an example, if one were interested in estimating the heat capacity of $CaCO_3$, he would add the value of 6.2 for calcium and 1.8 for carbon to three times 4.0, the value for oxygen, thus obtaining 20 cal/(gm-mole $CaCO_3$) (°K). Since the heat capacities of all solids increase with temperature, the value calculated by Kopp's rule could be accurate at only one point. The accuracy is best around room temperature, but even at this condition the rule gives only approximate values, and it should only be used if experimental data are not available.

TABLE 7.2 *Heat Capacity of Selected Inorganic Solids*
C_p = cal/(gm-mole)(°K)

Compound	Equation (T = °K)	Useful Range, °K	Uncertainty %
Al	$4.80 + 0.00322T$	273–931	1
Al_2O_3	$22.08 + 0.008971T - (522500/T^2)$	273–1973	3
$BaCl_2$	$17.0 + 0.00334T$	273–1198	?
$BaSO_4$	$21.35 + 0.0141T$	273–1323	5
$CaCl_2$	$16.9 + 0.00386T$	273–1055	?
$CaCO_3$	$19.68 + 0.01189T - (307600/T^2)$	273–1033	3
CaO	$10.0 + 0.00484T - (108000/T^2)$	273–1173	2
$CaSO_4$	$18.52 + 0.02197T - (156800/T^2)$	273–1373	5
C(graphite)	$2.673 + 0.002617T - (116900/T^2)$	273–1373	2
Cu	$5.44 + 0.001462T$	273–1357	1
CuO	$10.87 + 0.003576T - (150600/T^2)$	273–810	2
Fe(α)	$4.13 + 0.00638T$	273–1041	3
FeO	$12.62 + 0.001492T - (76200/T^2)$	273–1173	2
Fe_2O_3	$24.72 + 0.01604T - (423400/T^2)$	273–1097	2
Fe_3O_4	$41.17 + 0.01882T - (979500/T^2)$	273–1065	2
FeS_2	$10.7 + 0.01336T$	273–773	?
Pb	$5.77 + 0.00202T$	273–600	2
PbO	$10.33 + 0.00318T$	273–544	2
Mg	$6.20 + 0.00133T - (67800/T^2)$	273–923	1
$MgCl_2$	$17.3 + 0.00377T$	273–991	?
MgO	$10.86 + 0.001197T - (208700/T^2)$	273–2073	2
Ni	$4.26 + 0.00640T$	273–626	2
NH_4Cl(α)	$9.80 + 0.0368T$	273–457	5
P(red)	$0.21 + 0.0180T$	273–472	10
K	$5.24 + 0.00555T$	273–336	5
KCl	$10.93 + 0.00376T$	273–1043	2
KNO_3	$6.42 + 0.0530T$	273–401	10
Si	$5.74 + 0.000617T - (101000/T^2)$	273–1174	2
SiC	$8.89 + 0.00291T - (284000/T^2)$	273–1629	2
SiO_2(α quartz)	$10.87 + 0.008712T - (241200/T^2)$	273–848	1
Ag	$5.60 + 0.00150T$	273–1234	1
AgCl	$9.60 + 0.00929T$	273–728	2
$AgNO_3$(α)	$18.83 + 0.0160T$	273–433	2
Na	$5.01 + 0.00536T$	273–371	$1\frac{1}{2}$
NaCl	$10.79 + 0.00420T$	273–1074	2
$NaNO_3$	$4.56 + 0.0580T$	273–583	5
S(rhombic)	$3.63 + 0.00640T$	273–368	3
S(monoclinic)	$4.38 + 0.00440T$	368–392	3
Sn	$5.05 + 0.00480T$	273–504	2
$SnCl_2$	$16.2 + 0.00926T$	273–520	?
Zn	$5.25 + 0.00270T$	273–692	1
$ZnCl_2$	$15.9 + 0.00800T$	273–638	?
ZnO	$11.40 + 0.00145T - (182400/T^2)$	273–1573	1

Source: Perry, *Chemical Engineers' Handbook*, 4th ed. (New York: McGraw-Hill Book Company, 1963).

TABLE 7.3 *Kopp's Rule Values*

Atomic Heat Capacities in cal/(gm-atom)(°K)		
Atom	Number to Be Used for Solids	Number to Be Used for Liquids
C	1.8	2.8
H	2.3	4.3
B	2.7	4.7
Si	3.8	5.8
O	4.0	6.0
F	5.0	7.0
P	5.4	7.4
S	6.2	7.4
All others	6.2	8.0

If one has a simple mixture of solids which do not interact or form solid solutions, the heat capacity of the mixture may be computed by weighting the heat capacity of the individual components according to the mole fractions present in the same way as was done with mixtures of gases.

7.6 Heat Capacity of Liquids

It was noted in the previous section that the heat capacity of solids increases with temperature. It will not be surprising, then, to learn that the heat capacity of a liquid is higher than the corresponding solid. At the melting point, the two heat capacities are virtually the same.

Equations exist to represent the heat capacity of liquids over specific ranges of temperature. Linear equations are often suitable for modest ranges of temperature. Table 7.4 lists some representative data for liquids and solutions.

There is also a Kopp's rule for liquids, but as one would expect, the values for the individual elements are higher than they are for solids. Table 7.3 lists these values, which are in the same units as those for solids. As in the previous case, this approximation rule holds only near room temperature and should be applied only when actual data are unavailable.

There are only a few compounds with a higher heat capacity than water. Therefore, aqueous solutions may be expected to have a lower heat capacity than water, becoming even lower as the concentration of the solute material increases. For dilute solutions, one can nearly ignore the solute; thus, the heat capacity of dilute solutions may be estimated as the heat capacity of water times the fraction of water present. Unlike the cases of gases and solids, the heat capacity of solutions may not be accurately estimated by the additive rule, but such a procedure is better than nothing when data are

TABLE 7.4 *Heat Capacities of Selected Liquids*

Substance	Temperature °C	C_p, cal/(gm)(°C)
Ammonia	−40	1.051
	0	1.098
	60	1.215
	100	1.479
Nitric acid	25	0.417
Sulfuric acid	10–45	$0.339 + 0.00038t$
Sulfur dioxide	10–140	$0.318 + 0.00028t$
Water	15	1.000
	100	1.006
Acetic acid	0–80	$0.468 + 0.000929t$
Acetone	−30–+60	$0.560 + 0.000764t$
Aniline	0	0.478
	50	0.521
	100	0.547
Benzene	5	0.389
	20	0.406
	60	0.444
	90	0.473
n-Butane	−15–+20	$0.550 + 0.00191t$
Carbon disulfide	−100–+150	$0.235 + 0.000246t$
Carbon tetrachloride	0–70	$0.198 + 0.000031t$
Chloroform	−30–+60	$0.221 + 0.000330t$
Ether	0	0.529
	30	0.548
Ethyl acetate	20	0.478
Ethyl alcohol	0	0.535
	25	0.580
	50	0.652
Glycerol	0	0.540
	50	0.598
	100	0.668
n-Heptane	30–80	$0.476 + 0.00142t$
n-Hexane	20–100	0.600
Methyl alcohol	0	0.566
	40	0.616
Nitrobenzene	10	0.538
	50	0.329
	120	0.393
Propane	−30– +20	$0.576 + 0.001505t$

(*continued*)

TABLE 7.4 (*continued*)

Substance	Temperature °C	C_p, cal/(gm)(°C)
Propyl alcohol	0	0.525
	50	0.654
Toluene	0	0.386
	50	0.421
	100	0.470

Source: O. A. Hougen, K. M. Watson, and R. A. Ragatz, *Chemical Process Principles, Part I: Material and Energy Balances*, 2nd ed. (New York: John Wiley & Sons, Inc., 1954).

unavailable. No general rules can be formulated for solutions. Some representative data are provided in Figure 7.3a, b, c, d.

In common with other substances, the heat capacity of solutions and pure liquids increases with temperature, and it is best to use actual data or special correlating equations if they are available.

7.7 Data

Much emphasis has been placed on the desirability of using actual data whenever available. While the tables and charts in this text are sufficient to supply the necessary information for solving the majority of the problems contained in it, one should begin to learn the sources of more extensive data. There are too many to be listed here, but sources of major value are the standard handbooks of physical data such as the *Handbook of Chemistry and Physics* and the *Chemical Engineers' Handbook*. Books of this sort should become part of one's permanent technical library. They should be used frequently so as to become thoroughly familiar with their contents. Since the authors firmly recommend that students provide themselves with references of this sort early in their careers, there has been no attempt to make the tables and charts any more extensive than required for basic problem-solving.

7.8 Vapor Pressure and Latent Heats

Although the forces holding the atoms together as a crystal in the solid state are appreciable, some of the molecules possess sufficient energy to escape into the vapor state. As the temperature rises, more of the solid vaporizes. The pressure these molecules exert is known as the vapor pressure, and it is a unique function of the temperature for each substance. Similarly, every liquid possesses a specific vapor pressure at each temperature. The

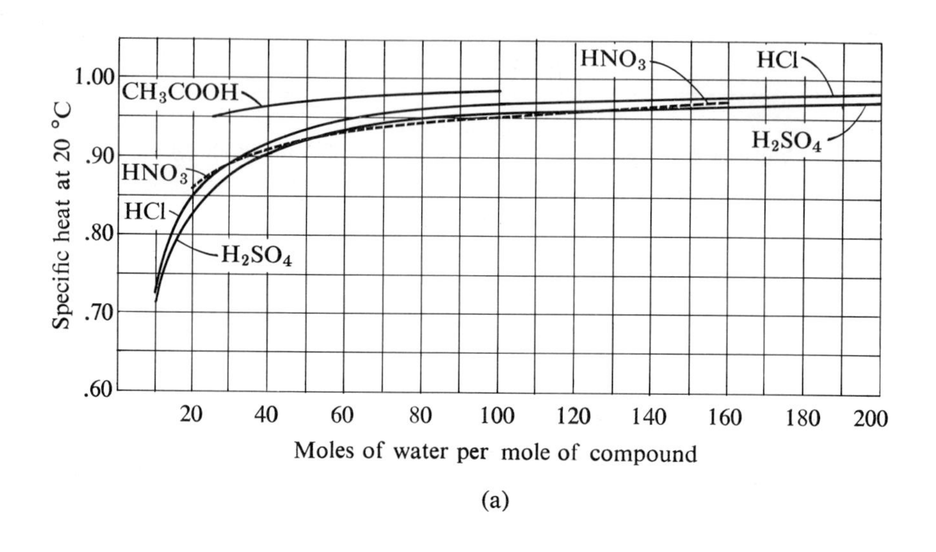

(a)

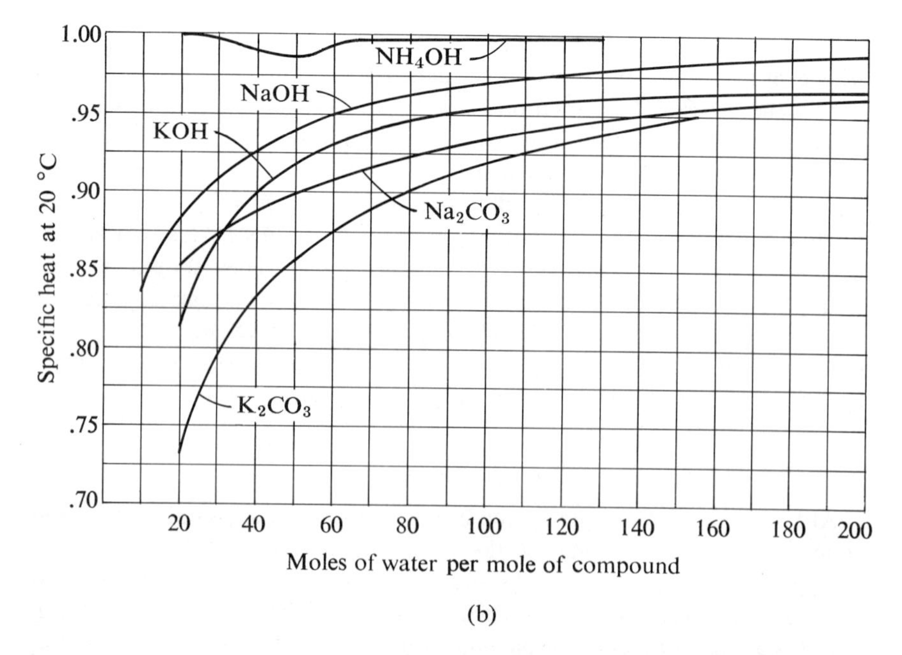

(b)

FIGURE 7.3 Specific heats of aqueous solutions at 20 °C.

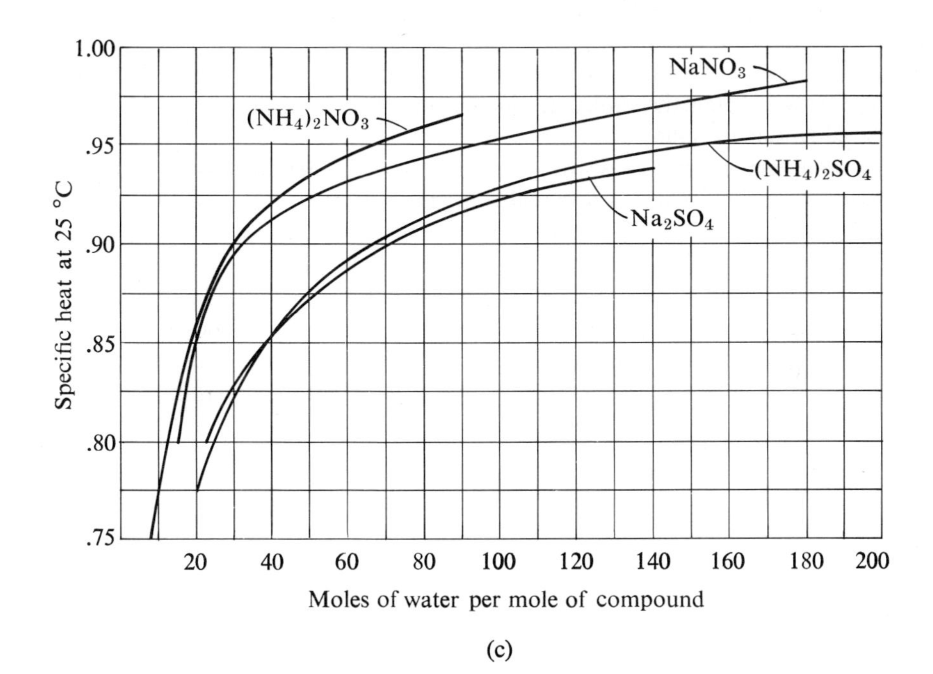
1.00
.95
.90
.85
.80
.75
Specific heat at 25 °C
20
40
60
80
100
120
140
160
180
200
Moles of water per mole of compound
NaNO3
(NH4)2NO3
(NH4)2SO4
Na2SO4

(c)

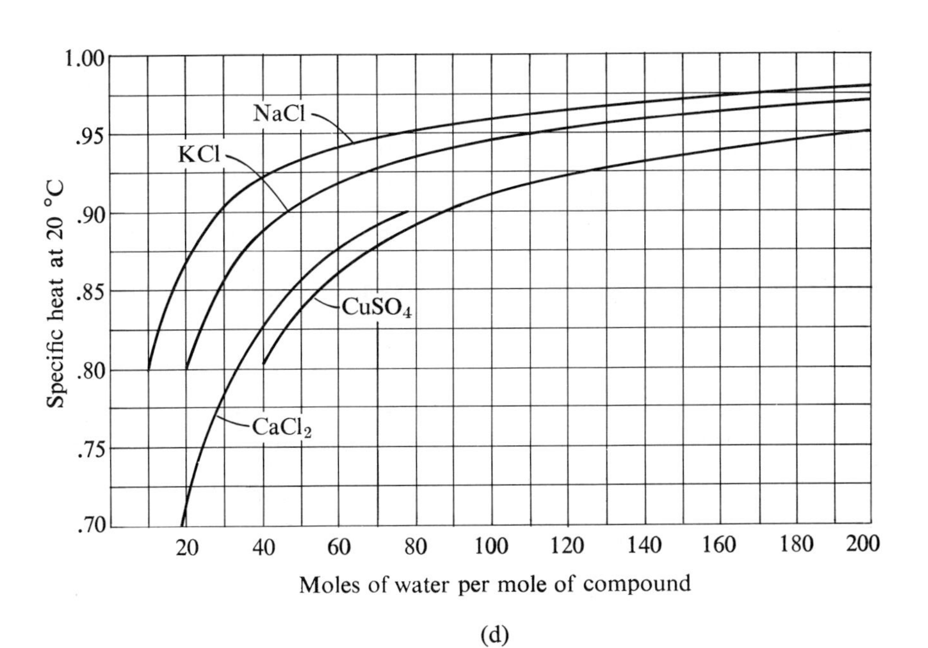
1.00
.95
.90
.85
.80
.75
.70
Specific heat at 20 °C
20
40
60
80
100
120
140
160
180
200
Moles of water per mole of compound
NaCl
KCl
CuSO4
CaCl2

(d)

vapor pressure of a liquid will normally be higher than that of a solid, since the forces keeping the molecules in the liquid state are less than those required to form a solid.

As the temperature rises, the increased energy content of a solid or liquid becomes so great that it is impossible for the original state of aggregation to be maintained. At this temperature, the solid phase begins to liquefy or a liquid starts to boil. In the latter case, boiling occurs when the vapor pressure is exactly equal to the pressure exerted by the surroundings on the liquid. As energy, in the form of heat, continues to flow into the system at constant pressure, it is absorbed wholly in accomplishing the phase change, and the temperature does not change. Constancy of temperature is maintained until

TABLE 7.5a *Heats of Fusion and Vaporization for Some Inorganic Compounds*

	Process					
Substance	Type	Initial State	Final State	Pressure mm Hg	Temperature °C	ΔH kcal/mole
Br_2	fus	c	l	760	−7.3	2.52
	vap	l	g	214	25.00	7.34
Cl_2	fus	c	l		−100.98	1.531
	vap	l	g	760	−34.05	4.878
H_2	vap	l	g	760	−252.77	0.216
HBr	vap	l	g	760	−66.72	4.21
HCl	vap	l	g	760	−85.03	3.86
HNO_3	vap	l	g	48	20.00	9.43
H_2O	fus	c	l	760	0.00	1.4363
	vap	l	g	4.58	0.00	10.767
	vap	l	g	760	100.00	9.717
	vap	l, std	g, std	760	25.00	10.520
H_3PO_4	fus	c	l		42.35	2.52
H_2S	fus	c, I	l	173.9	−85.53	0.5682
	vap	l	g	760	−60.31	4.463
N_2	vap	l	g	760	−195.80	1.333
NH_3	fus	c	l	45.57	−77.74	1.3516
	vap	l	g	760	−33.40	5.581
NO	vap	l	g	760	−151.74	3.2926
O_2	vap	l	g	760	−182.97	1.6299
SO_2	vap	l	g	760	−10.02	5.950
SO_3	sub	c, I	g	760	51.6	15.91
	vap	l	g	760	43.3	9.99

Source: *Selected Values of Properties of Chemical Compounds*, Manufacturing Chemists Association Research Project, Thermodynamic Research Center, Department of Chemistry, Texas A & M University, June 30, 1966.

TABLE 7.5b *Heats of Vaporization for Some Organic Compounds*

Compound	Formula	Normal Boiling Point, °C at 1 atm	Heat of Vaporization ΔH_{vap} kcal/gm-mole	
			At 25 °C	At Normal Boiling Point
Acetaldehyde	CH_3CHO	20.1		6.5
Acetic acid	CH_3COOH	118.2		5.83
Acetylene	C_2H_2	−81.5*		4.2
Aniline	$C_6H_5NH_2$	183.0		9.667
Benzene	C_6H_6	80.10	8.090	7.353
n-Butane	C_4H_{10}	−0.50	5.035	5.352
i-Butane	C_4H_{10}	−11.73	4.570	5.089
n-Butene	C_4H_8	−6.25	4.87	5.238
i-Butene	C_4H_8	−6.90	4.92	5.286
Carbon disulfide	CS_2	46.25		6.40
Carbon monoxide	CO	−191.49		1.444
Carbon tetrachloride	CCl_4	76.7		7.17
Cyclohexane	C_6H_{12}	80.74	7.895	7.19
Dimethyl ether	CH_3OCH_3	−24.82		5.141
Ethane	C_2H_6	−88.63	2.264†	3.517
Ethanol	C_2H_5OH	78.5	10.12	9.22
Ethylbenzene	$C_6H_5C_2H_5$	136.19	10.097	8.60
Ethylene	C_2H_4	−103.71		3.237
Formaldehyde	$HCHO$	−19.3		5.85
Formic acid	$HCOOH$	100.5		5.32
n-Hexane	C_6H_{14}	68.742	7.540	6.896
Hydrogen cyanide	HCN	25.70		6.027
Hydroquinone	$C_6H_4(OH)_2$	285.0		15.8
Methane	CH_4	−161.49		1.955
Methanol	CH_3OH	64.7	8.94	8.41
Nitrobenzene	$C_6H_5NO_2$	210.9		9.739
n-Pentane	C_5H_{12}	36.074	6.316	6.160
Phenol	C_6H_5OH	181.839	13.82	10.920
Propane	C_3H_8	−42.07	3.605	4.487
Propylene	C_3H_6	−47.70		4.402
Resorcinol	$C_6H_4(OH)_2$	276.5		13.7
Toluene	$C_6H_5CH_3$	110.62	9.080	8.00

Source: *Selected Values of Properties of Chemical Compounds*, Manufacturing Chemists Association Research Project, Thermodynamic Research Center, Department of Chemistry, Texas A & M University, June 30, 1966.
* At 900 mm Hg.
† Liquid at saturation pressure.

the last particle of one phase is changed into the other. For a solid, this is called the melting point; for a liquid it is the boiling point. The energy required for a complete change of phase is known as the heat of transition. In this discussion it has been assumed that the total pressure does not change while the transition is occurring. If the pressure were to change, so would the temperature of transition, since, e.g., more energy would be required to overcome the restrictive action of increased pressure.

There are several types of heats of transition: heat of melting, heat of sublimation (transition from solid to vapor without passing through the liquid state), heat of vaporization, and heat of crystalline change (the change of one type of crystal orientation in a solid to another type). These processes are all reversible. Thus, a vapor can condense, a liquid can freeze, etc. The heats involved in the reverse process are numerically the same but opposite in sign to those previously mentioned. If heat is required for the process, it is considered positive. If heat is evolved during the process, it is given a negative sign. These heats of transition are commonly called latent heats, and Table 7.5a,b lists some representative values; the algebraic sign corresponds to the phase change to the more disorganized state, for either solid-liquid or solid-gas transitions.

We shall consider in detail only one phase change, liquid to vapor, and the corresponding latent heat of vaporization. The principle applied here can be extended to other changes of state.

The thermodynamic equation that rigorously relates vapor pressure to the latent heat of vaporization (to be designated here as L_v) is known as the Clapeyron equation:

$$\frac{dP_v}{dT} = \frac{L_v}{T(v_v - v_L)}. \tag{7.10}$$

In this equation, P and T have their usual significance (subscript v indicates that the pressure is vapor pressure); v is the molar specific volume of the vapor and liquid respectively, as noted by the subscripts v and L. Since L_v is normally given in thermal units (it will be seen to be equivalent to an enthalpy change), it will have to be multiplied by an appropriate conversion factor to make the equation dimensionally correct.

7.9 Vapor Pressure Correlation

While Equation 7.10 is thermodynamically rigorous and may be applied to any phase change (with appropriate changes in the subscripts and meanings of symbols), it is easier to use in a modified version. First, we may assume that v_L will usually be very small when compared with v_v, and can, therefore, be ignored. We may also assume that L_v is a constant, but we must recognize that this is not really a fact. L_v is actually a function of the temperature, although it may change slowly. Finally, we may assume that the vapor behaves as an ideal gas; that is, $P_v v_v = RT$. Let us now apply these assumptions to the Clapeyron equation.

$$\frac{dP_v}{dT} = \frac{L_v}{T(v_v - v_L)};$$

dropping v_L:

$$\frac{dP_v}{dT} = \frac{L_v}{Tv_v};$$

applying the gas law:

$$\frac{dP_v}{dT} = \frac{L_v}{T\left(\frac{RT}{P_v}\right)};$$

rearranging:

$$\frac{dP_v}{P_v} = L_v \frac{dT}{RT^2} = d \ln P_v;$$

integrating; assuming L_v constant:

$$\ln P_v = -\frac{L_v}{RT} + A,$$

where A is a constant of integration.
Therefore,

$$\ln P_v = A - \frac{B}{T}, \tag{7.11}$$

since R is constant and L_v has been assumed to be.

Equation 7.11 tells us that if we plot the natural logarithm of the vapor pressure against the reciprocal of the absolute temperature we should obtain a straight line whose slope is $-B$, i.e., $-L_v/R$. Thus we may obtain L_v graphically.

Equation 7.11 is known as the Clausius-Clapeyron equation. Unfortunately, it contains so many assumptions that when $\ln P_v$ is actually plotted against $1/T$, the resulting correlation is a curved line rather than straight. The advantage of a straight line is that it may be determined with only two experimental points and can then be used both for extrapolation and interpolation.

Equation 7.11 can be modified, however, so that linear results are obtained with actual data. If we insert an arbitrary constant, θ, in the equation as follows:

$$\ln P_v = A - \frac{B}{T - \theta}, \tag{7.11a}$$

we obtain what is known as the Antoine equation. If θ is made 43 °K (which means that the equation becomes dimensional), excellent straight lines result.

TABLE 7.6 *Vapor Pressure—Antoine Equation Coefficients*

$\log_{10} P = A - B/(\theta + t);\ t = B/(A - \log_{10} P) - \theta$

P in mm Hg, t in °C

Compound	Formula	A	B	θ
Acetaldehyde	CH_3CHO	7.0565	1070.6	236.0
Acetic acid	CH_3COOH	7.29963	1479.02	216.81
Acetone	CH_3COCH_3	7.23157	1277.03	237.23
Acetylene	C_2H_2	7.0949	709.1	253.2
Ammonia	NH_3	7.36050	926.132	240.17
Benzene	C_6H_6	6.90565	1211.033	220.790
1,2-Butadiene	C_4H_6	7.1619	1121.0	251.00
1,3-Butadiene	C_4H_6	6.85941	935.531	239.554
n-Butane	C_4H_{10}	6.83029	945.90	240.00
i-Butane	C_4H_{10}	6.74808	882.80	240.00
n-Butene	C_4H_8	6.84290	926.10	240.00
i-Butene	C_4H_8	6.84134	923.200	240.00
Cyclohexane	C_6H_{12}	6.84498	1203.526	222.863
Ethane	C_2H_6	6.80266	656.40	256.00
Ethanol	CH_3CH_2OH	8.16290	1623.22	228.98
Ethene (Ethylene)	C_2H_4	6.74756	585.00	255.00
Ethylbenzene	$C_6H_5C_2H_5$	6.95719	1424.255	213.206
Formaldehyde	$HCHO$	7.1561	957.24	243.0
Formic acid	$HCOOH$	7.37790	1563.28	247.06
n-Heptane	C_7H_{16}	6.90240	1268.115	216.900
n-Hexane	C_6H_{14}	6.87776	1171.530	224.366
Hydrogen Bromide	HBr	6.28370	539.62	225.30
Hydrogen Chloride	HCl	7.167160	744.4906	258.704
Hydrogen Sulfide	H_2S	6.99392	768.1302	247.093
Iodine (c)	I_2	9.8109	2901.0	256.00
Methane	CH_4	6.61184	389.93	266.00
Methanol	CH_3OH	8.07246	1574.99	238.86
Naphthalene	$C_{10}H_8$	6.84577	1606.529	187.227
Nitric Oxide	NO	8.74300	682.938	268.27
Nitrogen Tetroxide	N_2O_4	7.38499	1185.722	234.18
Nitrous Oxide	N_2O	7.00394	654.260	247.16
n-Pentane	C_5H_{12}	6.85221	1064.63	232.000
Phenol	C_6H_5OH	7.13457	1516.072	174.569
Phosphorus Trichloride	PCl_3	6.8267	1196.	227.0
Phosphine	PH_3	6.71559	645.512	256.066
Propane	C_3H_8	6.82973	813.20	248.00
1-Propanol	$CH_3CH_2CH_2OH$	6.79498	969.27	150.42
Propene (Propylene)	C_3H_6	6.81960	785.00	247.00
n-Propionic Acid	CH_3CH_2COOH	7.54760	1617.06	205.67
Sulfur Dioxide	SO_2	7.28228	999.900	237.190
Sulfur Trioxide	SO_3	9.05085	1735.31	236.50
Toluene	$C_6H_5CH_3$	6.95464	1344.800	219.482
o-Xylene	$C_6H_5(CH_3)_2$	6.99891	1474.679	213.686
p-Xylene	$C_6H_5(CH_3)_2$	6.99052	1453.430	215.307

Source: *Selected Values of Properties of Chemical Compounds*, Manufacturing Chemists Association Research Project, Thermodynamic Research Center, Department of Chemistry, Texas A & M University, June 30, 1966.

The plot of Equation 7.11a is usually called a Cox chart. While 43 °K is an arbitrary value, it holds well if the normal boiling point of the liquid is above 250 °K. If sufficient experimental data are available, θ may be determined for the substance in question. In fact, three values of P_v *vs* T permit the evaluation of the three constants A, B, and θ, although one would have considerable concern regarding their precision with so few data.

In passing it may be noted that when a Cox chart is prepared for a homologous series of organic compounds, the lines tend to converge to a common point. Thus, once this convergence point is found, only one bit of data is required for the correlation of vapor pressure for other members of the same series.

The Antoine equation, and the resulting Cox chart, is only one of several methods for correlating vapor pressure data, but other methods will not be listed here. Representative data appear in Table 7.6.

7.10 Latent Heat of Vaporization

The simplest method of estimating the latent heat of vaporization is by Trouton's rule, which states that

$$\frac{L_v}{T_b} = K \cong 21.$$

L_v is the latent heat of vaporization in cal per gm-mole at the normal boiling point, T_b °K. The "constant," K, which is approximately 21 for many nonpolar liquids actually varies with T_b, becoming higher as T_b increases. For polar liquids $K \gg 21$.

An improvement on Trouton's rule, which takes into account the variation of K with temperature, has been suggested by Kistiakowsky:

$$\frac{L_v}{T_b} = 8.75 + 4.571 \log_{10} T_b,$$

where the variables have the same dimensions as before. While this equation gives excellent results for nonpolar liquids, it may not be used to estimate L_v for polar liquids.

Section 7.9 has shown how L_v is related to a Cox chart. If such a chart is available L_v may be obtained from the slope of the line. Note that values of L_v were not required in the preparation of this type of chart.

Two other methods have been suggested for obtaining L_v by using the known L_v of a reference substance. One of these is the Othmer method and the other the Dühring method. If one writes the Clausius-Clapeyron equation in the form

$$d \ln P_v = \frac{L_v}{R} \frac{dT}{T^2}$$

for the substance in question (see development of Equation 7.11) and again for the reference substance:

$$d \ln P_{v,R} = \frac{L_{v,R}}{R} \frac{dT}{T^2},$$

and divides the first by the second, there results,

$$\frac{d \ln P_v}{d \ln P_{v,R}} = \frac{L_v}{L_{v,R}} \left(\frac{T_R}{T}\right)^2.$$

When T_R is chosen equal to T

$$d \ln P_v = \frac{L_v}{L_{v,R}} d \ln P_{v,R},$$

which upon integration becomes

$$\ln P_v = \frac{L_v}{L_{v,R}} \ln P_{v,R} + \text{const.}$$

Obviously $\log_{10}$ may be used in place of natural logarithms since only the value of the constant will be affected. Therefore, if $\log_{10} P_v$ is plotted against $\log_{10} P_{v,R}$ (each pair of values at the same temperature), a straight line results whose slope is $L_v/L_{v,R}$. From this slope and the known value of $L_{v,R}$, L_v can be readily computed. Note that all the assumptions inherent in the Clausius-Clapeyron equation apply to this solution, which is known as the Othmer method. However, if the temperature dependencies are comparable for the substance in question and for the reference, the line will be quite straight over appreciable temperature intervals. The value of this method lies in the applicability of this condition, particularly when the two substances are structurally related.

In a similar fashion, under the condition $P_v = P_{v,R}$, the equation can be reduced to

$$L_v = L_{v,R} \left(\frac{T}{T_R}\right)^2 \frac{dT_R}{dT},$$

so that if T_R is plotted *vs.* T, the slope is

$$\frac{L_v}{L_{v,R}} \left(\frac{T_R}{T}\right)^2,$$

and again L_v can be computed. This approach is the Dühring method. The Othmer method is generally superior to that of Dühring.

The following examples illustrate the use of all these methods.

Example 7.5

▶ STATEMENT: Calculate L_v for *n*-Hexane, C_6H_{14}, at the normal boiling point.

DATA *Chemical Engineers' Handbook*

t °C	p, C_6H_6 mm Hg		p, C_6H_{14} mm Hg	$1/T$ °K $\times 10^3$
32	—	†	200	3.280
*43	200*		309*	3.165
50.1	—		400	3.095
56.6	—	†	500	3.035
*61	400*		590*	2.995
61.9	—	†	600	2.985
*67.6	500*	†	725*	2.935
*73.2	600*		1000*	2.887

†Pairs for Dühring } 2 only needed
*Pairs for Othmer }

L_v Data, C_6H_6, cal/gm		
t °C	L_v	
60	97.47	
68.7	—	Hexane boiling point
80	94.17	Benzene boiling point
100	90.57	

▶ SOLUTIONS:

1. Trouton:

$$L_v/T_b \doteq 21,$$
$$L_v = 21(68.7 + 273) = 7176,$$
$$L_v/M = 83.4 \text{ cal/gm}.$$

2. Kistiakowsky: $T_b = 273 + 68.7 = 341.7$.

$$L_v = 8.75(341.7) + 4.571(341.7) \log_{10} 341.7,$$
$$= 6946 \text{ cal/gm-mole},$$
$$= \frac{6946}{86.2} = 80.7 \text{ cal/gm}.$$

3. Clausius:

$$\ln p = 17.58 - \frac{3750}{T},$$

$$L_v/R = 3750,$$

$$\frac{L_v}{M} = \frac{(3750)(1.988)}{86.2} = 86.8 \text{ cal/gm}.$$

4. Othmer:

$$\text{Slope} = 0.941,$$
$$L_{v,R} = 96 \text{ cal/gm at } 69\ ^\circ\text{C},$$
$$L_v = 0.941(96)(78),$$
$$\frac{L_v}{M} = \frac{0.941(96)(78)}{86.2} = 81.8 \text{ cal/gm}.$$

5. Dühring:

$$\text{Slope} = 1.027,$$
$$L_v = L_{v,R}(1.027)\left(\frac{342}{353.2}\right)^2,$$
$$\text{where} \quad L_{v,R} = (94.2)(78) \text{ at } 80\ ^\circ\text{C}.$$
$$L_v = 82.1 \text{ cal/gm}.$$

True value $\doteq$ 80.1 cal/gm, Table 7.5b.

7.11 Concluding Remarks

The number of correlation methods existing for the subjects discussed in this chapter is very large. Only the simpler ones have been given here. For the student who wishes a reasonably complete survey of this subject with literally hundreds of references to the original literature, the following two references are listed:

W. R. Gambill, "How to Estimate Engineering Properties," a series of articles appearing from February 1947 through January 11, 1960 in *Chemical Engineering*.

R. C. Reid, and T. K. Sherwood, *The Properties of Gases and Liquids—Their Estimation and Correlation* (New York: McGraw-Hill Book Co., 1966).

LIST OF SYMBOLS

Latin letters

$a, b, c, d, \ldots$ / $A, B, C, \ldots$	Constants in empirical heat capacity equation
A, B	Constants in Antoine equation
C	Molar heat capacity—subscript $_p$ indicates "at constant pressure"; subscript $_v$ indicates "at constant volume"; superscript ^{0} or * indicates "for an ideal gas"; subscript $_m$ indicates "integral mean"
H	Enthalpy
i	Any component of a mixture
k	Ratio of C_p/C_v

L_v	Latent heat of vaporization, usually in cal/gm-mole
N	Mole fraction, subscripted to show the component involved
P	Pressure; subscript $_v$ indicates vapor pressure
Q	Heat
R	Universal gas constant
T	Absolute temperature, °K or °R; subscript $_b$ indicates normal boiling point in °K
t	Ordinary temperature, °C or °F
v	Molar specific volume; subscript $_v$ indicates vapor; subscript $_L$ indicates liquid

Greek letter

θ	Constant in Antoine equation, often taken as 43 °K

PROBLEMS

7.1 Convert the following quantities:

a. a rate of heat flow of 5000 Btu/(hr)(ft^2) to cal/(sec)(cm^2).
b. a thermal conductivity of 10 Btu-ft/(hr)(ft^2)(°F) to cal-cm/(sec)(cm^2)(°C) and to cal/(sec)(cm)(°C).
c. a heat capacity of 0.3 Btu/(lb)(°F) to cal/(gm)(°C).
d. a specific heat of 0.21 to heat capacity in Btu/(lb-mole)(°F) for the material $CaCO_3$.
e. a latent heat of 5.3 kcal/gm-mole to Btu/lb for formic acid, HCOOH.

7.2 If $C_p = a + bt$, prove that C_{pm} is the average of the heat capacities at the two terminal temperatures.

7.3 Numerous equations have been proposed to fit curves of heat capacities as a function of temperature. One form, the *descending* power series, is

$$C_p = a + \frac{b}{t} + \frac{c}{t^2} + \cdots.$$

For the three-term form, calculate:

a. the amount of heat necessary to raise the temperature of a material from t_1 to t_2.
b. the mean heat capacity between t_1 and t_2.

7.4 Heat capacity relationships exhibit a general monotonic increase with temperature. If these relationships are approximated as follows,

$$C_p = a + bt, \qquad 0 \leq t \leq t_1,$$

and

$$C_p = a_1 = a + bt_1, \qquad t > t_1,$$

calculate

a. a general form of C_{pm} from 0 to $t > t_1$.
b. the maximum C_{pm}, granted that one exists.

7.5 Several general forms have been used for expressing the dependence of heat capacities on temperature. Two of the more common are:

$$C_p = a + bt + ct^2$$

and

$$C_p = a + bt - \frac{c}{t^2}.$$

If the constants b and c are positive and negative, respectively, in these two expressions, can a maximum occur in C_{pm} over the temperature range t_1 to t, where t_1 is nonzero. If so, locate this point in terms of the constants b, c, and t_1.

7.6 Mean heat capacity data are available (e.g., see Figure 7.2) for the specific ranges of temperature 25 to t °C and 77 to t °F. Show how to calculate the mean heat capacity for any general range, t_1 to t_2.

7.7 The heat capacities of the following materials are not readily available:

acetanilid	$C_6H_5NHCOCH_3$,
acrylic acid	$CH_2{=}CH{-}COOH$.

Estimate values by Kopp's rule, reporting results as Btu/(lb-mole)(°F).

7.8 Representative molar heat capacity equations for NO and CO_2 are:

$$\text{NO:} \quad 6.440 + 2.069 \times 10^{-3}T - 0.4206 \times 10^{-6}T^2;$$

$$\text{CO}_2\text{:} \quad 18.036 + 0.4474 \times 10^{-4}T - 1.5808 \times \frac{10^{-6}}{T^{1/2}}.$$

In both equations T is expressed in °K. Calculate:

a. mean molar heat capacity of NO from 77 to 300 °F.
b. mean molar heat capacity of CO_2 from 0 to 300 °C.

7.9 The molal heat capacity of SO_3 is reported to be

$$C_p = 7.454 + 19.13 \times 10^{-3}T - 6.628 \times 10^{-6}T^2,$$

with T in °K. Calculate:

a. the heat capacity relationship per unit weight.
b. the molal heat capacity at 123 °C.
c. the molal heat capacity at 253 °F.
d. the mean molal heat capacity between 25 and 1000 °C. Compare the value with that obtained from Figure 7.2.
e. the number of Btu required to heat 100 lb of SO_3 from 25 to 1000 °C.

7.10 A molal heat capacity equation which has been proposed for CO_2 is

$$C_p = 9.00 + 2.71 \times 10^{-3}t - 0.256 \times 10^{-6}t^2,$$

where t is expressed in °F.

a. calculate the form of the equation where temperature is expressed in °K, and compare the result with the relationship presented in Table 7.1c.

b. calculate the heat required to raise the temperature of one pound of CO_2 from 60 to 400 °F.
c. calculate the integral mean heat capacities for the ranges 0 to 60 °F to 400 °F and show that their use produces the same result obtained in part (b).

7.11 Molar heat capacity data for CO_2, are:

T °K	C_p
300	8.894
400	9.871
500	10.662
600	11.311
700	11.894
800	12.300
900	12.678
1000	12.995

Use these data to calculate the amount of heat required to raise the temperature of one pound of CO_2 from 60 to 400 °F.

7.12 A mixture of combustion gases leaves the furnace at 1100 °F and is subsequently cooled to 400 °F by passage over a waste heat boiler. This boiler is fed water at 125 °C and produces saturated steam at 25 psig with all the heat abstracted from the cooling of the combustion gases being used for this purpose.

The analysis of the gases is 11% CO_2, 6% O_2, 7% H_2O, and 76% N_2. The pressure of the combustion gases is approximately atmospheric throughout the process.

It is estimated that 8000 cu ft of combustion gas (at 1100 °F) is produced per pound of fuel burned.

Calculate:

a. Btu recovered by cooling the flue gases over the waste heat boiler, per pound of fuel burned.
b. pounds of steam produced in waste heat boiler per pound of fuel burned.

7.13 A gaseous mixture containing 80% N_2, 12% CO_2, and 8% O_2 is cooled from 500 to 300 °F, both conditions being substantially at one atmosphere. The gas carried soot (i.e., graphitic carbon) amounting to 50 grains per cu ft, measured at inlet conditions. This carbon must, of course, also be cooled. Calculate the enthalpy change in the cooling, per 10,000 cu ft of the original mixture.

7.14 Ten pounds of wet steam at 100 psia is found to have an enthalpy of 9000 Btu, relative to liquid water at 32 °F. Calculate the quality of the steam, where quality is defined as the fraction of "dry" steam in a mixture of steam and saturated water. Note the desirability of using steam tables.

7.15 Water is fed to a steam boiler at 75 °F. In the boiler this water is heated and converted to steam at 510 psig. This steam is not quite dry, its quality being 99.5 (see Problem 7.14 for definition of quality). This nearly saturated steam is next passed to a superheater where it is given 100 °F superheat at constant pressure. Per 1000 lb of steam calculate the Btu required:

a. to produce the 99.5 quality steam.
b. to provide the superheat.

7.16 A mixture of benzene and chlorobenzene contains 80% of the latter. It is to be cooled from 70 to 20 °C. Assume that the enthalpy of the mixture can be treated as the sum of the enthalpies of the components. Then, using the data supplied below, estimate the heat, in Btu, which must be extracted per 100 lb of the mixture.

DATA

	Sp Ht	°C
Benzene	0.419	6 to 60
	0.340	10
	0.482	65
Chlorobenzene	0.273	0
	0.298	10
	0.308	20

7.17 The mixture in Problem 7.16 is distilled so as to produce essentially pure benzene vapor overhead and pure chlorobenzene liquid at the bottom. The feed is at 20 °C, the benzene vapor leaves at 80 °C, and the chlorobenzene leaves at 130 °C. Calculate the heat required, in Btu and in Pcu, per 100 lb of the mixture.

7.18 The following data for *n*-butane are obtained from the literature:

t °F	Vapor Pressure, Atm
260	24.662
280	29.785
290	32.624

Estimate the latent heat of vaporization by two methods.

7.19 The normal boiling point of *n*-butane is 31.1 °F. The reported latent heat of vaporization is 165.8 Btu/lb. Estimate the value of the latent heat using:

a. Trouton's rule.
b. the Kistiakowsky equation.

7.20 Using the following data, and latent heats of vaporization for ammonia from the literature, estimate the latent heat of vaporization of HCl at −100 °C:

NH_3		HCl	
p, mm Hg	°C	*p*, mm Hg	°C
1	−109.1	100	−114.0
5	−97.5	200	−105.2
10	−91.9	400	−95.3
20	−85.8	760	−84.8
40	−79.2		
Melting Point	−77.7		

7.21 Vapor pressure data for water are as follows:

	Vapor Pressures, mm Hg	
t °C	Over Liquid Water	Over Ice
−5	3.163*	3.013
0	4.579	4.579
20	17.535	—
40	55.324	—

* Supercooled.

The latent heat of vaporization of water at 0 °C = 1073.4 Btu per lb. Use an Othmer Chart (see Section 7.10) to estimate the latent heat of fusion, expressing the result in units of cal per gm, kcal per gm-mole, and Btu per lb.

7.22 For each of the materials listed in Problem 3.30, obtain vapor pressure data for a suitable reference substance and draw a line on Othmer coordinates (see Section 7.10).
Using these lines,

a. obtain an estimate of the vapor pressure at the average temperature.
b. estimate the latent heat of vaporization at the higher of the two listed temperatures, expressed in units of cal per gm and Btu per lb-mole.

7.23 A vat contains 50 gal of a material whose thermal properties closely approximate those of water. The liquid is initially at 210 °F. Assuming perfect thermal insulation from the surroundings and constant latent heats of vaporization, calculate the quantity of the liquid vaporized when the temperature of the liquid has dropped to 100 °F.

7.24 Based on the data listed below, estimate the energy in Btu required to

a. raise temperature of 10 lb of stearic acid ($C_{18}H_{32}O_2$) from 77 to 266 °F, the melting point being 155.8 °F.
b. raise temperature of 50 lb of diphenyl ($C_{12}H_{10}$) from 40 to 150 °C, the melting point being 52.9 °C.

DATA

Latent Heats of Fusion		
Stearic acid	47.54 cal/gm at 68.8 °C	
Diphenyl	No information	
Sp ht stearic acid		
Liquid	0.550	75–137 °C
Solid	0.399	15 °C
Sp ht diphenyl		
Liquid	0.425	80 °C
	0.439	100 °C
	0.452	120 °C
	0.471	150 °C
Solid	0.385	40 °C

8 Thermochemical Relations

8.1 Q and ΔH

In this chapter attention will be focused on the heat released by (or required for) the interaction of chemical species. Most often this interaction can be expressed by a chemical equation, hence the term thermochemistry. The principles presented in Chapter 6 apply to chemical systems as well as to mechanical ones; therefore, $Q = \Delta H$ if the process (reaction) occurs either (1) under steady state conditions (reversibly or irreversibly) when kinetic and potential energy changes may be ignored, or (2) reversibly at constant pressure. Since the emphasis in this chapter will be on reactions meeting one of these requirements, we shall take the enthalpy change during the reaction as equivalent to the heat involved. However, this relationship is not universal. For example, $Q = \Delta U$ for a reversible reaction occurring in a sealed calorimeter (i.e., at constant volume).

If the system loses energy, ΔH will be negative and the heat flow will be from the system to its surroundings. Such a reaction is called exothermic. Stated the other way around, Q is negative for an exothermic reaction. Conversely, when heat is required for the reaction, ΔH as well as Q will be positive. Such a reaction is spoken of as being endothermic. Some older textbooks and some reference works define an exothermic reaction as one in which Q is positive, i.e., $Q = -\Delta H$. When consulting reference books to obtain data it is important to verify the convention used.

8.2 The Standard State

The standard or reference state is a purely arbitrary one that is used simply for convenience and uniformity. For gas calculations the standard state has been specified at 273 °K and 1 atm (Chapter 3). In thermochemical calculations a different reference is adopted. One atmosphere pressure will be used, but the temperature datum will be 25 °C (or 298 °K), and we shall refer to any substance which is present in its normal state of aggregation (solid, liquid, or gas) under these conditions as being in its standard state.

8.3 Special Designations

Sometimes we may wish to write an equation in which some of the reactants or products are not in their normal (i.e., standard) states. Thus, although water is a liquid at 25 °C and 1 atm, we may wish to consider it in its gaseous form, i.e., water vapor. To remove ambiguity under such conditions, the following designations in parentheses after the chemical symbol are appropriate:

s = solid state, e.g., H_2O(s) refers to ice.
c = crystalline; this symbol is used to indicate that the substance exists in the crystalline state.

Since some crystalline substances have two or more modifications, the specific name of the modification, or a Greek letter denoting it, may have to be used, e.g., S(rhombic) or C(β), the latter designating graphite.

l = pure liquid.
aq = aqueous, i.e., an infinitely dilute water solution.

The exact concentration may be further specified by molality (e.g., 0.5 m) or mole fraction (e.g., $x = 0.1$). By itself, (aq) means infinitely dilute (dil is sometimes used). The symbol ∞ is sometimes also used to signify infinite dilution. In this text (aq) will signify any dilution of such magnitude that further addition of solvent will result in no appreciable heat effect.

g = gas; if the pressure and temperature differ from 1 atm and 25 °C, they must also be specified.

If no qualifying symbol is used, it may be assumed that the substance is in its normal state of aggregation at 25 °C and 1 atm.

8.4 Standard Heat of Reaction

If a reaction is assumed to occur at 25 °C and 1 atm with all the reactants entering and all the products leaving at these conditions, the heat evolved (or required) for the reaction going to completion from left to right under stoichiometric conditions is called the standard heat of reaction. The word "stoichiometric" means that the reaction involves the relative amounts of reactants and products specified by a balanced chemical equation.

Enthalpy is a state function like pressure or temperature. Just as the temperature difference of a system (T_{final}-$T_{initial}$) is independent of how the system undergoes that temperature change, so the change of enthalpy of a substance from any given set of conditions to any other set of conditions is independent of the path followed. Therefore, if one knows the enthalpy of each product and each reactant at a particular pressure and temperature, one can readily determine the enthalpy change that has occurred during the reaction without knowing the mechanism of the reaction. Consequently, within the limitations set forth in Section 8.1, the standard heat of a reaction is the difference between the sum of the enthalpies of all products (each at its specified state) and the sum of the enthalpies of all reactants (each at its specified state), i.e.,

$$\Delta H_R^0 = \sum H^0 \text{ (products)} - \sum H^0 \text{ (reactants)}. \qquad (8.1)$$

The zero superscript (^{0}) refers to the standard state. The superscript would be omitted if ΔH_R referred to other conditions.

8.5 Heats of Formation and Their Use

Enthalpy is a relative term, i.e., unlike pressure, it cannot be expressed as an absolute value but only so much more or so much less than the enthalpy at some set of datum conditions. Since we have already chosen a standard state, we shall take 25 °C and 1 atm as the state where $H = 0$; i.e., this will be the datum plane from which differences will always be computed. Let us further specify that all elements in their normal state at 25 °C and 1 atm (e.g., O_2, N_2, C, etc.) will have an enthalpy of zero.

Now suppose that these elements react to form compounds; e.g.,

$$\begin{aligned} C(\beta) + O_2 &\rightarrow CO_2, \\ C(\beta) + \tfrac{1}{2}O_2 &\rightarrow CO, \\ H_2 + \tfrac{1}{2}O_2 &\rightarrow H_2O(l), \\ H_2 + \tfrac{1}{2}O_2 &\rightarrow H_2O(g), \\ \tfrac{1}{2}N_2 + \tfrac{3}{2}H_2 &\rightarrow NH_3(g), \\ \tfrac{1}{2}H_2 + \tfrac{1}{2}Cl_2 &\rightarrow HCl(g), \\ &\text{etc.} \end{aligned}$$

Note that each of these reactions is written as a balanced chemical equation with the quantities of reacting elements so adjusted that exactly 1 mole of *product* is formed. If all reactants are in their standard states and if each product is in its specified state and also at 25 °C and 1 atm, then the heat of each reaction would be the heat of formation of the product formed (i.e., $\Delta H_F^0 = \Delta H_R^0$) (see Equation 8.1).

Since all of the reactants are elements in their standard state, the ΔH_F^0 provides an enthalpy of the product relative to 25 °C and 1 atm. For example, experiments show that the ΔH_F^0 of NH_3 is $-11{,}040$ cal per gm-mole. This negative number signifies that 1 gm-mole of ammonia has an enthalpy 11,040 calories less than that of the elements from which it was formed. In this way, any chemical compound can be given a value of enthalpy. Since all of these values are relative to the same datum, they may be added and subtracted as if they were absolute values in order to get changes of enthalpy. This procedure is analogous to measuring the height of two persons with the same measuring scale from some arbitrary reference position. If the first person has an *indicated* height of 475 in. above the reference position and the other an *indicated* height of 480 in. above the same reference, we know that the second person is 5 in. taller than the first, even though the absolute numbers do not give us the exact height of either individual.

Heats of Reaction from Heats of Formation

The importance of this approach is that standard heats for any reaction can be computed, given a table of heats of formation such as Table 8.1. Consider, for example, the following reaction:

$$CaCO_3 \rightarrow CaO + CO_2. \tag{8.2}$$

One could immediately say that limestone does not readily decompose at room temperature, and he would be right. However, the standard state is not necessarily the state at which reactions actually occur. It is, as has been stated, simply a convenient reference state. We can calculate a heat of reaction appropriate to this reference state, even if no spontaneous reaction in fact occurs. This number can subsequently be converted into a meaningful value for the reaction as it can actually take place (see Section 9.6).

To obtain the standard heat of reaction for Equation 8.2, first consult Tables 8.1 and 8.2 and obtain the appropriate heats of formation. For the substances under discussion, they are:

$$\begin{aligned}
\Delta H_F^0,\ CaCO_3 &= -288{,}450 \text{ cal/gm-mole},\\
\Delta H_F^0,\ CaO &= -151{,}900 \text{ cal/gm-mole},\\
\Delta H_F^0,\ CO_2 &= -94{,}052 \text{ cal/gm-mole}.
\end{aligned}$$

We now make use of Hess's law of constant heat summation, which states that the enthalpy change in a reaction is dependent only on the substances present at the beginning and end of a reaction and the thermodynamic states in which they exist (i.e., phase, temperature, and pressure), but is independent of the reaction or reactions relating them. Hess's law is then only a restatement of the fact that enthalpy is a state property, independent of the path taken in producing a change.

To use this law, rewrite the information on standard heats of formation in the form of chemical equations:

$$Ca + C + \tfrac{3}{2}O_2 \rightarrow CaCO_3 - 288{,}450 \text{ cal/gm-mole,} \tag{8.3}$$

$$Ca + \tfrac{1}{2}O_2 \rightarrow CaO - 151{,}900 \text{ cal/gm-mole,} \tag{8.4}$$

$$C + O_2 \rightarrow CO_2 - 94{,}052 \text{ cal/gm-mole.} \tag{8.5}$$

Now subtract Equation 8.3 from the sum of Equations 8.4 and 8.5 obtaining:

$$Ca + \tfrac{1}{2}O_2 + C + O_2 - Ca - C - \tfrac{3}{2}O_2 = CaO + CO_2 - CaCO_3 + (-151{,}900 - 94{,}052 + 288{,}450) \text{ cal.} \tag{8.6}$$

Treating Equation 8.6 algebraically, canceling like terms and rearranging so that all chemical compounds will have the plus sign associated with them,

$$CaCO_3 \rightarrow CaO + CO_2 + (-151{,}900 - 94{,}052 + 288{,}450) \text{ cal.} \tag{8.7}$$

The standard heat for this reaction is then computed to be

$$-151{,}900 - 94{,}052 + 288{,}450 = 42{,}498 \text{ cal.}$$

The sign on the answer is important. Whenever a positive number is obtained an endothermic reaction is indicated (i.e., heat is required for the reaction). A negative answer would mean that heat was evolved during the process.

If Equation 8.3 is balanced, the elements used in writing the equations of formation must cancel as they did in Equation 8.7, and the sum of the heats of formation of the reactants must always be subtracted from the sum of the standard heats of formation of the products as in Equation 8.7. A general rule may then be formulated to eliminate the lengthy procedure that was involved in writing Equations 8.3 through 8.7. This general rule states that the standard heat of any reaction may be computed by subtracting the sum of the standard heats of formation of the reactants from the sum of the standard heats of formation of the products. Stated symbolically, this rule becomes

$$\Delta H_R^0 = \sum_i n_i \, \Delta H_{F,i}^0 - \sum_j n_j \, \Delta H_{F,j}^0 . \tag{8.8}$$

It will be seen that Equation 8.8 conforms to the defining Equation 8.1, since ΔH_F^0 for each species is the H^0 for that species relative to the specified datum. The subscript $_R$ indicates the reaction, the subscript $_F$ indicates formation, and the superscript ^{0} indicates the standard state (i.e., 25 °C and 1 atm). In the summations, the subscript $_i$ refers to products and the subscript $_j$ refers to reactants. For both reactants and products the symbol n refers to the stoichiometric number of moles of any chemical species. The summations will be made more clear by the next example.

Heat Units in Heats of Reaction

This is a good place in our discussion to take note of the units in which heats of reaction are expressed. Standard heats of formation are commonly

TABLE 8.1 *Standard Heats of Formation of Some Inorganic Compounds*

Compound	Formula	State	Heat of Formation, 25 °C kcal/gm-mole
Aluminum chloride	$AlCl_3$	c	−166.2
Aluminum oxide	Al_2O_3	c, α, corundum	−399.09
Ammonia	NH_3	g	−11.04
Ammonium chloride	NH_4Cl	c, II	−75.38
Ammonium hydroxide	$NH_4(OH)$	aq, 4 moles, H_2O	−87.64
Ammonium nitrate	NH_4NO_3	c, IV	−87.27
Ammonium sulfate	$(NH_4)_2SO_4$	c	−281.86
Barium chloride	$BaCl_2$	c	−205.6
Barium sulfate	$BaSO_4$	c	−350.2
Boron trifluoride	BF_3	g	−265.4
Calcium carbide	CaC_2	c	−15.0
Calcium carbonate	$CaCO_3$	Calcite	−288.45
Calcium chloride	$CaCl_2$	c	−190.0
Calcium hydroxide	$Ca(OH)_2$	c	−236.0
Calcium oxide	CaO	c	−151.90
Calcium phosphate	$Ca_3(PO_4)_2$	c	−986.2
Calcium silicate	$CaSiO_3$	c	−378.65
Calcium sulfate	$CaSO_4$	Anhydrite	−342.42
Copper sulfate	$CuSO_4$	aq	−184.0
Ferric chloride	$FeCl_3$	c	−96.8
Ferric nitrate	$Fe(NO_3)_3$	aq	−160.4
Ferric oxide	Fe_2O_3	c	−196.5
Ferrous chloride	$FeCl_2$	c	−81.5
		aq	−99.7
Ferrous oxide	FeO	c	−64.3
Hydrogen bromide	HBr	g	−8.66

(*continued on facing page*)

reported in gram-calories per gram-mole of each compound, as indicated in Equations 8.3 through 8.5, or quite frequently, in kilogram-calories per gram-mole. In the latter units, the heats of formation in Equations 8.3 through 8.5 would be listed as +288.45, −151.9, and −94.052, respectively. The precision of values reported in kilocalories is more evident than for those reported in calories, since the number of significant figures after the decimal point is clear. With calories, ambiguous zeros may be included.

In Equations 8.6 and 8.7 the calculated standard heats of reaction are reported in gram-calories without specific reference to the quantity of mass. The molal quantities are clearly defined by the stoichiometric equation;

TABLE 8.1 (*continued*)

Compound	Formula	State	Heat of Formation, 25 °C kcal/gm-mole
Hydrogen chloride	HCl	g	−22.063
		aq	−39.56
Hydrogen sulfide	H_2S	g	−4.815
Lithium chloride	$LiCl$	c	−97.70
Magnesium carbonate	$MgCO_3$	c	−266.0
Magnesium chloride	$MgCl_2$	c	−153.40
Magnesium oxide	MgO	c	−143.84
Nitric acid	HNO_3	l	−41.404
Nitric oxide	NO	g	21.600
Nitrogen dioxide	NO_2	g	8.091
Nitrogen tetroxide	N_2O_4	g	2.309
Phosphoric acid	H_3PO_4	c	−306.2
Potassium hydroxide	KOH	c	−101.78
Silicon dioxide	SiO_2	c, II, quartz	−205.4
Sodium carbonate	Na_2CO_3	c	−270.3
Sodium chloride	$NaCl$	c	−98.232
Sodium hydroxide	$NaOH$	c, II	−101.99
Sodium nitrate	$NaNO_3$	c, II	−101.54
Sodium sulfate	Na_2SO_4	c, II	−330.90
Sulfur dioxide	SO_2	g	−70.96
Sulfur trioxide	SO_3	g	−94.45
Sulfuric acid	H_2SO_4	l	−193.91
Water	H_2O	g	−57.7979
		l	−68.3174
Zinc chloride	$ZnCl_2$	c	−99.40
Zinc sulfate	$ZnSO_4$	c	−233.88

Source: *Selected Values of Chemical Thermodynamic Properties*, U.S. Department of Commerce, National Bureau of Standards.

consequently, any indication of quantities of mass in the calculated heats of reaction would be redundant. It would also, in many instances, be confusing. Such instances will always arise when the stoichiometric number of moles of a compound differs from unity. For example, in the reaction,

$$CO_2 + C \rightarrow 2\,CO;\ \Delta H_R^0 = +41.221 \text{ kcal},$$

the 41.221 kcal are required per mole of CO_2 and per mole of carbon, but per 2 moles of CO. If the equation had been written

$$\tfrac{1}{2}CO_2 + \tfrac{1}{2}C \rightarrow CO,$$

TABLE 8.2 *Standard Heats of Formation and Combustion of Some Organic Compounds*

Compound	Formula	State	Heat of Combustion, 25 °C kcal gm-mole		Heat of Formation, 25 °C kcal gm-mole
			CO_2(g), H_2O(l), N_2(g)	CO_2(g), H_2O(g)	
Acetic acid	CH_3COOH	l			−116.4
Acetaldehyde	CH_3CHO	g			−39.76
Acetyl chloride	CH_3COCl	l			−65.8
Acetylene	C_2H_2	g	−310.62	−300.10	+54.194
Acetyl nitrile	CH_3CN	g			+19.81
Adiponitrile	$C_5H_8CN_2$	g			+33.34
		l			+19.19
Amyl acetate	$C_7H_{14}O_2$	l	−1040.		−96.58
Amyl alcohol	$C_5H_{12}O$	l	−786.7		−93.46
Aniline	$C_6N_5NH_2$	l	−812.†		+20.28*
Benzene	C_6H_6	g	−798.08	−757.52	+19.820
		l	−780.98	−749.42	+11.718
i-Butane	C_4H_{10}	g	−685.64	−633.05	−32.15
n-Butane	C_4H_{10}	g	−687.64	−635.05	−30.15
i-Butene	C_4H_8	g	−645.43	−603.36	−4.04
n-Butene	C_4H_8	g	−649.45	−607.37	−0.03
Carbon	C	[solid graphite]	−94.0518	−94.0518	0.00
Carbon dioxide	CO_2	g	0.00	0.00	−94.0518
Carbon disulfide	CS_2	l	−256.97		+21.0
Carbon monoxide	CO	g	−67.6361	−67.6361	−26.4157
Cumene	C_9H_{12}	g	−1257.31	−1194.19	+0.940
		l	−1246.52	−1183.40	−9.848
Cyanamid	NH_2CN	l			+11.18
		c			+9.15
Cyclohexane	C_6H_{12}	g	−944.79	−881.67	−29.43
		l	−936.88	−873.76	−37.34
Diethyl ether	$C_4H_{10}O$	l	−652.59		−65.20

Compound	Formula	State	Heat of Combustion, 25 °C kcal/gm-mole		Heat of Formation, 25 °C kcal/gm-mole
			CO_2(g), H_2O(l), N_2(g)	CO_2(g), H_2O(g)	
Dimethyl ether	CH_3OCH_3	g			−44.3
Ethane	C_2H_6	g	−372.82	−341.26	−20.236
Ethanol	C_2H_5OH	g			−56.24
		l			−66.356
Ethene	C_2H_4	g	−337.23	−316.20	+12.496
Ethyl amine	$C_2H_5NH_2$	g			−12.24
Ethylbenzene	C_8H_{10}	g	−1101.13	−1048.53	+7.120
		l	−1091.03	−1038.43	−2.977
Formaldehyde	HCHO	g			−27.7
Formic acid	HCOOH	g			−86.67
		l			−97.8
Hexamethylene diamine	$(CH_2)_6(NH_2)_2$	g			−30.57
n-Hexane	C_6H_{14}	g	−1002.57	−928.93	−39.96
		l	−995.01	−921.37	−47.52
Hydrogen	H_2	g	−68.3174	−57.7979	0.00
Methane	CH_4	g	−212.80	−191.76	−17.889
Methanol	CH_3OH	g			−48.10
		l			−57.036
Methyl amine	CH_3NH_2	g			−6.7
Nitrobenzene	$C_6H_5NO_2$	l	−739.†		
n-Pentane	C_5H_{12}	g	−845.16	−782.05	−35.00
Phenol	C_6H_5OH	c			−39.44
Propane	C_3H_8	g	−530.60	−488.53	−24.820
Propene	C_3H_6	g	−491.99	−460.43	+4.879
Styrene	C_8H_8	g	−1060.90	−1081.83	+35.22
		l	−1050.51	−1008.44	+24.83
Toluene	C_7H_8	g	−943.58	−901.50	+11.950
		l	−934.50	−892.42	+2.867

Values for all compounds (except those marked * or †) have been obtained from *Selected Values of Properties of Chemical Compounds*. Those marked * are from *IEC Fundamentals*, Vol. 4 (November 1965), 389, and those marked † are from the *International Critical Tables* (1929).

the standard heat of reaction would be 20.610 kcal and this figure would be per mole of CO but per $\frac{1}{2}$ mole of CO_2 or carbon. Therefore no units of mass are ascribed to a standard heat of reaction, the quantity calculated applying to the chemical equation as written.

The units of gram- and kilo-calories are frequently inconvenient for the engineer who is customarily working with mass in pounds. No great problem results, however, for we know from Section 2.7 that gram-calories on a basis of mass in grams and Pcu on a basis of mass in pounds will produce identical numerical values. In other words, the standard heat of formation of CaO is either $-151{,}900$ gm-cal per gm-mole or $-151{,}900$ Pcu per lb-mole. Consequently, the standard heat of dissociation of $CaCO_3$, for the equation written as 8.2, can be expressed as $+42{,}498$ gm-cal or as 42,498 Pcu. The former will be used when quantities are to be expressed in grams and the latter when the quantities are to be expressed in pounds. When heat units are to be expressed in Btu, the remarks in Section 2.7 show that the number of Pcu (or gram-calories) need only be multiplied by the factor 1.8, thereby accounting for the change in temperature scale. The standard heat of reaction for the dissociation of 1 lb-mole of $CaCO_3$, will, therefore, be $+76{,}496$ Btu. The chemical engineer will find that flexibility in the use of the various heat units is an absolute essential. He should be thoroughly conversant with their interrelation and with the underlying reasons.

In the previous example, only 1 mole of reactant was involved, and only 1 mole of each product species was formed. There are cases in which more than 1 mole of a given species is present; and there are also cases where a substance will not be in its normal state of aggregation at 25 °C and 1 atm, as in the following example:

$$CH_4 + 2H_2O(g) \rightarrow CO_2 + 4H_2. \tag{8.9}$$

To obtain the ΔH_R^0 for the above reaction, add the enthalpies for each mole of product and subtract from the total the sum of the enthalpies for each mole of reactants. As before, the enthalpies for reactants and products are the ΔH_F^0 values. Therefore,

$$\Delta H_R^0 = \Delta H_{F,CO_2}^0 + 4\,\Delta H_{F,H_2}^0 - \Delta H_{F,CH_4}^0 - 2\,\Delta H_{F,H_2O(g)}^0.$$

Tables will ordinarily not contain a value for $\Delta H_{F,H_2}^0$ since, as previously noted, all elements at 25 °C and 1 atm have an enthalpy of zero relative to the standard state. Note especially that ΔH_F^0 values must be for the state of aggregation as expressed in the chemical equation being analyzed.

If, in the preceding example, the equation had been written

$$CH_4 + 2H_2O(l) \rightarrow CO_2 + 4H_2, \tag{8.10}$$

a different value would have been obtained from the standard heat of reaction. The difference would, naturally, result from the difference in the states of aggregation of water in the two equations. Subtracting the first writing from the second, there results

$$2H_2O(l) \rightarrow 2H_2O(g). \qquad (8.11)$$

The difference in the standard heats of reaction is then the latent heat of vaporization of 2 moles of water at the standard state of 25 °C and 1 atm. This latent heat may be obtained by treating Equation 8.11 as a chemical equation and applying the rule stated in Equation 8.8 to determine the standard heat of formation of gaseous water from liquid water. Referring to Table 8.1, we find the following values of standard heats of formation:

$$H_2O(l): \quad -68{,}317 \text{ cal/gm-mole};$$
$$H_2O(g): \quad -57{,}798 \text{ cal/gm-mole}.$$

Subtracting the standard heat of formation of the reactant from that of the product, we obtain

$$-57{,}798 + 68{,}317 = +10{,}519 \text{ cal/gm-mole},$$

as the latent heat of 1 mole of water at 25 °C and 1 atm. Inspection of the sign of this quantity provides the information that the reaction as written is endothermic, a fact with which we are already familiar in this particular case. The heat of reaction of Equation 8.10 will then be less than that of Equation 8.9 by 2 times 10,519 cal per gm-mole.

8.6 Heats of Combustion and Their Use

It is not always possible to determine experimentally the heats of formation for some compounds, especially those of organic character. For example, although one could write the following equation, it would be quite impossible to carry out the indicated reaction so as to obtain the ΔH_F^0 of acetic acid:

$$2C + 2H_2 + O_2 \rightarrow CH_3COOH.$$

Fortunately such compounds are usually capable of being oxidized (i.e., burned) with the formation of CO_2 and H_2O. If the compound contains other elements in addition to C and H_2 (e.g., S) these elements will ordinarily appear as oxides also (e.g., SO_2). When we write an oxidation equation involving 1 mole of the compound under study, the ΔH_R^0 associated with this equation is known as the standard heat of combustion of the compound in question. For example, the heat of combustion of acetic acid would involve the equation

$$CH_3COOH + 2O_2 \rightarrow 2CO_2 + 2H_2O.$$

Since water is almost inevitably a product of organic combustion reactions, should we treat it as $H_2O(l)$, which is its proper standard state, or as $H_2O(g)$, which is the form it will actually have in a combustion process? There is no single answer to this question for it is convenient to have ΔH_c^0 values for both situations. If the ΔH_c^0 has been determined with water in its liquid phase, we speak of this value as the gross heat of combustion, or gross heating value (HV). If water is to be designated in the vapor phase, the value of ΔH_c^0 is known as the net heat of combustion, or net heating value. The net HV is always a smaller absolute number than the gross HV, since additional heat is evolved when water condenses.

Incidentally, all ΔH_c^0 values must be negative, since all combustion reactions are exothermic. This observation permits one to check the convention being used in an unfamiliar reference book. If all ΔH_c^0 values are listed as negative, then the convention of the reference is that of this book. If they are recorded as positive numbers, then the older convention, $Q = -\Delta H$, positive for exothermic reactions, is being employed, and the signs should be changed before the numbers are used according to the rules stated here.

Heats of reaction may be determined from heats of combustion as readily as from heats of formation, as can be shown by an example. Again we make use of Hess's law. Suppose we wish to compute the ΔH_R^0 of the following synthesis:

$$CH_3COOH + C_2H_5OH \rightarrow CH_3COOC_2H_5 + H_2O(l). \tag{8.12}$$

First observe that water is in the liquid form, so we must obtain values for gross heats of combustion. If water did not appear as a product (or reactant), it would make no difference whether gross or net values were used as long as we were consistent. Secondly, we know that water will not burn, but is itself a product of combustion. The combustion reactions for the three remaining substances are:

$$\begin{aligned}
&\text{(A)} \quad CH_3COOH + 2O_2 \rightarrow 2CO_2 + 2H_2O(l) + \Delta H_{c,\text{acid}}^0,\\
&\text{(B)} \quad C_2H_5OH + 3O_2 \rightarrow 2CO_2 + 3H_2O(l) + \Delta H_{c,\text{alcohol}}^0,\\
&\text{(C)} \quad CH_3COOC_2H_5 + 5O_2 \rightarrow 4CO_2 + 4H_2O(l) + \Delta H_{c,\text{ester}}^0.
\end{aligned}$$

Now, add Equations A and B and subtract Equation C from their sum:

$$\begin{aligned}
&CH_3COOH + 2O_2 + C_2H_5OH + 3O_2 - CH_3COOC_2H_5 - 5O_2\\
&\quad \rightarrow 2CO_2 + 2H_2O(l) + 2CO_2 + 3H_2O(l) - 4CO_2 - 4H_2O(l)\\
&\quad + \Delta H_{c,\text{acid}}^0 + \Delta H_{c,\text{alcohol}}^0 - \Delta H_{c,\text{ester}}^0.
\end{aligned}$$

Again treating chemical symbols as mathematical symbols

$$\begin{aligned}
&CH_3COOH + C_2H_5OH \rightarrow CH_3COOC_2H_5 + H_2O(l)\\
&\qquad\qquad + \Delta H_{c,\text{acid}}^0 + \Delta H_{c,\text{alcohol}}^0 - \Delta H_{c,\text{ester}}^0
\end{aligned}$$

or

$$\Delta H_R^0 = \Delta H_{c,\text{acid}}^0 + \Delta H_{c,\text{alcohol}}^0 - \Delta H_{c,\text{ester}}^0.$$

The water present in the balanced chemical equation will always appear in the right amount when the combustion equations are properly added and subtracted. Naturally, if incorrect or inconsistent states of aggregation are used, the quantities will not cancel as they did above. In this example, only one mole of each species appeared in the desired equation. If n moles of any species is present, the entire combustion equation for that species, including the ΔH_c^0, must be multiplied by n.

We are, therefore, in a position to state a general principle:

$$\Delta H_R^0 = \sum_j n_j \Delta H_{c,j}^0 - \sum_i n_i \Delta H_{c,i}^0,$$

$$= -\left(\sum_i n_i \Delta H_{c,i}^0 - \sum_j n_j \Delta H_{c,j}^0\right), \tag{8.13}$$

where, as in Equation 8.8, subscript $_i$ refers to products and subscript $_j$ to reactants. Note that products are subtracted from reactants in this case, whereas when we were dealing with heats of formation, reactants were subtracted from products. Equation 8.13 is written so that the form will be consistent with Equation 8.8; the minus sign preceding the summations is of the utmost importance.

There is another important point to be made. Values of ΔH_c^0 are large numbers, since combustion reactions are highly exothermic. Heats of reaction, however, are frequently comparatively small. Referring to Equation 8.12 and using Table 8.2 to provide values of the standard heats of combustion,

$$\Delta H_{c,CH_3COOH}^0 = -208{,}340 \text{ cal/gm-mole};$$
$$\Delta H_{c,C_2H_5OH}^0 = -326{,}700 \text{ cal/gm-mole};$$
$$\Delta H_{c,CH_3COOC_2H_5}^0 = -538{,}760 \text{ cal/gm-mole}.$$

Then applying Equation 8.13, we obtain $\Delta H_R^0 = +3{,}720$ cal. When one subtracts one large number from another and gets a small number, the number of significant figures in the final answer is less (often much less) than the significant figures in the numbers from which the answer came. As a consequence, the percentage error in the answer (due to uncertainty in the final significant figure has been increased [see Chapter 2]). Note how much difference would be introduced in the answer to the above problem if $\Delta H_{c,CH_3COOC_2H_5}^0$ had been changed by 0.1%. This observation leads to two conclusions:

1. Never round values of ΔH_F^0 (or ΔH_c^0) to be used in calculating ΔH_R^0.
2. After calculating ΔH_R^0, round the result according to the least precise value of ΔH_F^0 (or ΔH_c^0).

For example, in the calculation of ΔH_R^0 for the formation of ethyl acetate from ΔH_c^0, the least precise of the various ΔH_c^0 values was that for C_2H_5OH, $-326{,}700$ cal per gm-mole, with a value good only to hundreds. The answer should also be reported to hundreds only, or $+3700$ cal.

In any particular problem one may find ΔH_F^0 values for some of the substances present and ΔH_c^0 values for others. Obviously, such data cannot be mixed as they stand. However, since ΔH_R^0 may be determined from ΔH_c^0 data, and since ΔH_F^0 is a particular form of ΔH_R^0, one may use ΔH_c^0 data to get ΔH_F^0. For example, if one wishes to get ΔH_F^0 for the reaction

$$C(\beta) + 2H_2 \rightarrow CH_4$$

from ΔH_c^0 data, he would use Equation 8.13.

$$\Delta H_R^0 = \Delta H_F^0 = -[\Delta H_{c,CH_4}^0 - \Delta H_{c,C(\beta)}^0 - 2\,\Delta H_{c,H_2}^0]. \tag{8.14}$$

Then with all information as ΔH_F^0, Equation 8.8 may be used.

In performing calculations of the type indicated in Equation 8.14, it must be clearly understood that the various ΔH_F^0 values shall all be listed in consistent phases of the various oxidation products. For hydrocarbons (C_aH_b) and for carbohydrates ($C_aH_bO_c$) the products will be CO_2(g) and H_2O, where the latter may be in either the liquid or gas phase (i.e., the gross or net heating values, respectively). The various ΔH_c^0 values used to determine a ΔH_R^0 must all be gross, or all net. They cannot be mixed in any single calculation.

In the case of organic compounds which contain other elements, such as sulfur, halides, and nitrogen, the oxidized forms reported may also be chosen variously. For instance, the oxidized form of sulfur reported in one value of ΔH_c^0 might be SO_2(g) and in another H_2SO_4(aq). The oxidized forms of nitrogen might be N_2(g) or HNO_3(aq); those for chlorine might be Cl_2(g) or HCl(aq), etc. In using values of compounds containing these elements great care must be taken to ascertain the oxidized forms for which the heat of combustion is reported. Fortunately heats of formation are now more commonly reported for organic compounds than was the case a few years ago so that this problem arises less frequently, but the infrequent use of the technique of converting ΔH_c^0 to ΔH_F^0 must not allow us to forget some of these precautions.

8.7 Sources of Data

Tables 8.1 and 8.2 provide information on ΔH_F^0 and ΔH_c^0 for a limited number of compounds. More extensive tables are available in standard handbooks; those most commonly available to the chemical engineer are the *Chemical Engineers' Handbook* (McGraw-Hill Book Co.) and the *Handbook of Chemistry and Physics* (Chemical Rubber Publishing Co.). The vast potential array of organic compounds leads one to believe that standard heats of formation (or combustion) may not be found for every compound of interest. When such a situation exists we must resort to estimation methods. Fortunately, such methods have been quite well developed and refined. The reader is referred to an article by Verma and Doraiswamy, *Industrial and Engineering Chemistry Fundamentals*, *4*, 389–396 (1965).

8.8 Effect of Temperature on Heat of Reaction

Before introducing other thermochemical quantities, we shall consider what happens to the heat of reaction when a synthesis occurs at a temperature other than the standard 25 °C, 1 atm. Ammonia, for example, is commercially produced at a temperature of about 500 °C and at pressures of 200–1000 atm, conditions appreciably removed from the standard state. In this section we shall observe the effect of changing temperatures, maintaining, however, all reactants and products at a single temperature, while keeping the pressure at 1 atm. In the next section we shall say something about the effects of pressure on ΔH_R. In Chapter 9 we shall make calculations where temperatures of reactants and products need not be identical.

The fact that enthalpy is a state function and that ΔH is independent of how we go from one state to another must constantly be kept in mind. Consider any general reaction $aA + bB \rightarrow rR + sS$. We have already seen how to calculate the ΔH_R^0 of this reaction from the ΔH_F^0 or ΔH_c^0 of each species. Suppose first that we took each reactant and changed its temperature from 25 to t °C. What would be the enthalpy change involved?

We shall assume initially that neither reactant undergoes a phase change during the process; i.e., if A is a solid it remains a solid, and if B is a liquid it remains a liquid. Since, for a constant pressure process, $\Delta H = \int C_p\, dT$, the total enthalpy change of the reactants for the proposed Δt is

$$a \int_{25^\circ}^{t^\circ} C_{p,A}\, dT + b \int_{25^\circ}^{t^\circ} C_{p,B}\, dT.$$

If, however, a phase change occurs, it must be taken into consideration. For example, suppose A is a solid that melts at t', a temperature between 25 and t °. For such a substance, the total enthalpy change will be

$$\Delta H_A = a \int_{25^\circ}^{t'} C_{p,A(s)}\, dT + a(\Delta H_{\text{melting at } t'}) + a \int_{t'}^{t} C_{p,A(l)}\, dT.$$

If both melting and evaporation take place, the first at t' and the second at t'', add the appropriate terms so that

$$\Delta H_A = a \int_{25^\circ}^{t'} C_{p,A(s)}\, dT + a(\Delta H_{\text{melting at } t'}) + a \int_{t'}^{t''} C_{p,A(l)}\, dT$$
$$+ a(\Delta H_{\text{vap at } t''}) + a \int_{t''}^{t} C_{p,A(g)}\, dT.$$

Other possible phase changes are transitions (e.g., monoclinic to rhombic) and sublimation. Regardless of the complexity of phase changes, the total ΔH for all reactants between 25 °C and 1 atm and t °C and 1 atm can readily be determined. Since 1 atm is usually close enough to the ideal state, values of C_p, uncorrected for pressure changes can justifiably be used in the calculation.

The enthalpy change involved when each mole of *product* goes from 25 to t° is identical to that described for each reactant.

We now have the relative enthalpies of all products and all reactants at t and 1 atm. The difference is, by definition, ΔH_R at t and 1 atm. We may summarize this fact as follows:

$$\Delta H_R \text{ at } t \text{ and } 1 \text{ atm} = \Delta H_R^0 + \sum_i \Delta H_{\substack{(\text{products})\\ 25-t^\circ}} - \sum_j \Delta H_{\substack{(\text{reactants})\\ 25-t^\circ}}. \quad (8.15)$$

This equation is called Kirchhoff's law.

Kirchhoff's equation will now be rewritten in a slightly different form, that most commonly reported. In deriving it, two assumptions will be made which must not be forgotten when it is used: (1) there is no change of state of aggregation of any reactant or product with temperature; (2) all C_p data are represented by the equation, $a + bT + cT^2$.

In Equation 8.15 take t_1 as different from t by only dT, pressure remaining constant at 1 atm. Then

$$\lim_{t_1 \to t} (\Delta H_{R,t_1} - \Delta H_{R,t}) = d\,\Delta H_R = d\left(\sum_i \Delta H_{\text{products}} - \sum_j \Delta H_{\text{reactants}}\right)_t. \quad (8.15a)$$

Each term in the right-hand portion of the equation may be represented by $C_p\,dT$ or $(a + bT + cT^2)\,dT$. If the reaction under consideration is of the general form,

$$aA + bB \rightleftarrows rR + sS,$$

then

$$d\sum \Delta H_{\text{products}} = r(a_R + b_R T + c_R T^2)\,dT + s(a_S + b_S T + c_S T^2)\,dT$$

and

$$d\sum \Delta H_{\text{reactants}} = a(a_A + b_A T + c_A T^2)\,dT + b(a_B + b_B T + c_B T^2)\,dT.$$

Subtracting $\sum \Delta H_{\text{reactants}}$ from $\sum \Delta H_{\text{products}}$ as required by Equation 8.15a, we may write

$$d\,\Delta H_R = (\Delta a + \Delta b T + \Delta c T^2)\,dT,$$

where Δa, Δb, and Δc are obtained from the heat capacity equations; i.e.,

$$\begin{aligned}\Delta a &= r a_R + s a_S - a a_A - b a_B,\\ \Delta b &= r b_R + s b_S - a b_A - b b_B,\\ \Delta c &= r c_R + s c_S - a c_A - b c_B.\end{aligned}$$

The coefficient preceding each constant taken from the appropriate C_p equation represents the number of moles of that chemical species present in the balanced chemical equation.

If the preceding equation is integrated,

$$\Delta H_R = \Delta aT + \tfrac{1}{2}\Delta bT^2 + \tfrac{1}{3}\Delta cT^3 + \Delta H_0, \tag{8.15b}$$

where ΔH_0 is the constant of integration. Equation 8.15b is the useful form of Equation 8.15. The integration constant, ΔH_0, may be obtained from any known value of ΔH_R at any given T. Since we usually know ΔH_R (at 25 °C = 298 °K), we can readily obtain an equation for ΔH_R at any temperature, providing that no phase changes occur between the temperature used to calculate ΔH_0 and that used for ΔH_R, and that C_p data are in the proper form. An equation analogous to 8.15b may be derived for C_p data in the form $a + bT - c/T^2$; this derivation is suggested as an exercise.

It is important to realize that ΔH_R computed in the above manner is valid only if all reactants and products are at the temperature t, at which the reaction occurs. In actual practice, this requirement is not frequently fulfilled. Kirchhoff's equation has, then, little real practical application. Further remarks on this subject can be made after heat balances have been introduced. Such remarks will be found in Section 9.7.

8.9 Effect of Pressure on Heat of Reaction

A complete discussion of this topic is beyond the scope of this book, but can be found in texts on thermodynamics, where it is shown that

$$dH = C_p\,dT + \left[v - T\left(\frac{\partial v}{\partial T}\right)_P\right] dP. \tag{8.16}$$

When pressure is constant this equation reduces to the form used in the preceding section, i.e., $dH = C_p\,dT$. When pressure is not constant, the quantity in brackets must be determined and integrated over the pressure range involved. It so happens that, with liquids, pressure has little effect on the enthalpy unless the pressure is very large or conditions are near the critical pressure and temperature. The effect of pressure on the enthalpy of solids is even less. Therefore, from a practical point of view we may ignore the small enthalpy effects due to pressure change when dealing with solids or liquids.

With gases, $Pv = RT$ when the ideal gas law is applicable; then $P\,dv + v\,dP = R\,dT$ and

$$\left(\frac{\partial v}{\partial T}\right)_P = \frac{R}{P} = \frac{v}{T}.$$

Therefore, the term in brackets in Equation 8.16 reduces to $v - T(v/T) = 0$, and disappears. Consequently, if the gas is ideal, pressure will have no effect on its enthalpy.

The only remaining case is that of a gas which is not ideal. Under these circumstances the pressure effect on enthalpy may be appreciable. It is not easy to evaluate this effect analytically because nearly all equations of state fail to be explicit in v. There are techniques (usually graphical) that make the determination possible, but to discuss them here would take us too far afield, and we shall ignore the correction. In other words, the assumption introduced in Chapter 3, the applicability of the ideal gas law, is here reaffirmed. When this assumption is clearly untenable, any standard thermodynamics text should be consulted.

On occasion little error is introduced by ignoring the pressure effect on ΔH due to nonideal gas behavior. This situation holds when gases exist on both sides of the chemical equation in such quantities that the ΔH correction for pressure on the reactants is nearly balanced by the ΔH correction for pressure on the products; then ΔH_R is nearly the same at high pressure as it is at 1 atm. This circumstance is more likely to occur if the number of moles of gaseous reactants is nearly equal to the number of moles of gaseous products.

8.10 Incomplete Reactions

The ΔH_R computed by the methods previously given assumes a reaction going completely to the right. Commercial syntheses are seldom so complete. Therefore, to get the actual heat released (or required), one needs to know the extent of the actual reaction. This information may be determined experimentally, or it may be calculated by the methods of thermodynamics and kinetics. The heat released (or required) is limited by the number of moles which react, independent of the number fed.

8.11 Adiabatic Reactions

Up to this point we have assumed that the reaction under consideration is proceeding isothermally, i.e., at a constant temperature of 25 or t °C. It is, however, possible that the reaction is occurring in an insulated vessel that prevents either the escape of heat or its admission. Under these conditions the temperature will rise if the reaction is exothermic or fall if it is endothermic. In this case it is desirable to know what the final temperature will be. This problem is easily attacked through heat balances and will be discussed in detail in Section 9.6.

8.12 Thermochemistry of Solutions

Early contacts with chemistry elicit the fact that considerable heat is evolved when sulfuric acid is added to water. This process is but one example of heats of solution. In the general case this heat effect may be either positive or negative and may result from a variety of causes. First, the substance may ionize, e.g., form H^+, OH^- or corresponding cations and anions like Cl^- or Na^+. This ionization is, of course, a chemical reaction with a corresponding ΔH_R. Secondly, a hydrate may be formed, analogous to the reaction

$$CaSO_4 + 2H_2O \rightarrow CaSO_4 \cdot 2H_2O.$$

This, also, is a chemical reaction with its own characteristic ΔH_R. There may also be other forms of solute-solvent interaction involving other, and less powerful, forces of association, such as hydrogen bonding, or van der Waals' forces. Some heat effects also accompany these associations. The net result of all of these effects (and there can be more than one occurring simultaneously) is to make the enthalpy of the solution different from that of its components. If the solution is effected isothermally at 25 °C and 1 atm, this enthalpy change is called the standard heat of solution. As in any chemical reaction, the value of the heat of solution will depend on the amount of the reactants, but in this case there is no fixed or stoichiometric ratio to be specified. Therefore, it will be necessary either to state the concentration of the solution being prepared or the number of moles of solvent used per mole of solute. Beyond a certain ratio of solvent to solute additional solvent will have no measurable thermal effect; such a solution has been defined (Section 8.3) as infinitely dilute even though there is a finite amount of solute present.

We shall first mention briefly the case where a very small drop (theoretically an infinitesimal amount) of the solute is added to an existing solution (which may already contain some solute or none at all, i.e., pure solvent) the resulting differential enthalpy change may be expressed mathematically as

$$\left(\frac{\partial H}{\partial n_A}\right)_{n_B, P, T},$$

which is simply a statement that the number of moles of B (solvent) are maintained constant as we measure at fixed t and P the extremely small change in enthalpy due to the infinitesimal change in the amount of A (solute). This quantity is called the partial molal enthalpy and is but one example of partial molal quantities of value in thermodynamics. It is clearly dependent upon the concentration of the solution undergoing change as well as upon the specific levels of P and T applicable to each circumstance studied.

The partial molal enthalpy is a very refined method of dealing with heats of solution. The necessary data for practical calculations are rarely available. It does not, therefore, concern us here. We are much more interested in the integral enthalpy change, i.e., the sum of all the effects when we go from one concentration to another. Data for integral heats of solutions are more commonly available and may be presented in graphical or tabular form, e.g., Table 8.3. Use of these data is very easy. Once the concentration of the final solution that is being prepared from pure components is known (expressed as moles of solvent per mole of solute), the heat evolved (a minus number) or

TABLE 8.3 *Standard in Kcal/gm-mole solute*

Compound	In Moles of Water As Indicated	ΔH°_{soln}, 25 °C
Acetic acid CH_3COOH	0.5 H_2O	+0.13
	1 H_2O	+0.15
	2 H_2O	+0.16
	3 H_2O	+0.14
	5 H_2O	+0.09
	10 H_2O	−0.05
	25 H_2O	−0.205
	50 H_2O	−0.270
	100 H_2O	−0.305
	500 H_2O	−0.335
	1000 H_2O	−0.339
	5000 H_2O	−0.342
	∞ H_2O	−0.343
Ammonia NH_3	1 H_2O	−7.06
	2 H_2O	−7.66
	3 H_2O	−7.83
	4 H_2O	−7.95
	5 H_2O	−8.03
	10 H_2O	−8.19
	20 H_2O	−8.23
	30 H_2O	−8.24
	40 H_2O	−8.24
	50 H_2O	−8.25
	100 H_2O	−8.26
	200 H_2O	−8.28
	∞ H_2O	−8.28
Ammonium chloride NH_4Cl	10 H_2O	+3.810
	25 H_2O	+3.783
	50 H_2O	+3.756
	100 H_2O	+3.752
	200 H_2O	+3.734
	300 H_2O	+3.724
	400 H_2O	+3.716
	500 H_2O	+3.709
	1000 H_2O	+3.689
	5000 H_2O	+3.652
	10,000 H_2O	+3.643
	∞ H_2O	+3.620

Compound	In Moles of Water As Indicated	ΔH°_{soln}, 25 °C
Ethanol	0.05 H_2O	−0.278
CH_3CH_2OH	0.1 H_2O	−0.23
	0.2 H_2O	−0.41
	0.5 H_2O	−0.92
	1 H_2O	−0.194
	2 H_2O	−0.412
	3 H_2O	−0.666
	4 H_2O	−0.898
	5 H_2O	−1.122
	7 H_2O	−1.466
	10 H_2O	−1.760
	25 H_2O	−2.224
	50 H_2O	−2.384
	100 H_2O	−2.454
	200 H_2O	−2.294
Hydrochloric acid	1 H_2O	−6.268
HCl	2 H_2O	−11.668
	5 H_2O	−15.308
	10 H_2O	−16.608
	20 H_2O	−17.155
	40 H_2O	−17.448
	50 H_2O	−17.509
	100 H_2O	−17.650
	200 H_2O	−17.735
	400 H_2O	−17.796
	1000 H_2O	−17.850
	5000 H_2O	−17.909
	10,000 H_2O	−17.924
	100,000 H_2O	−17.949
	∞ H_2O	−17.960
Sodium chloride	10 H_2O	+0.464
NaCl	25 H_2O	+0.726
	50 H_2O	+0.892
	100 H_2O	+0.982
	400 H_2O	+1.020
	1000 H_2O	+1.004
	5000 H_2O	+0.972
	10,000 H_2O	+0.961
	100,000 H_2O	+0.940
	∞ H_2O	+0.930

(*continued*)

TABLE 8.3 (*continued*)

Compound	In Moles of Water As Indicated	ΔH°_{soln}, 25 °C
Sodium hydroxide	3 H_2O	−6.904
NaOH	4 H_2O	−8.229
	5 H_2O	−9.025
	7 H_2O	−9.846
	10 H_2O	−10.158
	15 H_2O	−10.238
	25 H_2O	−10.231
	50 H_2O	−10.164
	100 H_2O	−10.118
	500 H_2O	−10.127
	1000 H_2O	−10.149
	5000 H_2O	−10.196
	10,000 H_2O	−10.211
	∞ H_2O	−10.246
Sulfuric acid	0.5 H_2O	−3.76
H_2SO_4	1.0 H_2O	−6.71
	2 H_2O	−10.02
	5 H_2O	−13.87
	10 H_2O	−16.02
	25 H_2O	−17.28
	50 H_2O	−17.53
	75 H_2O	−17.61
	100 H_2O	−17.68
	200 H_2O	−17.91
	500 H_2O	−18.34
	1000 H_2O	−18.78
	5000 H_2O	−20.18
	10,000 H_2O	−20.81
	100,000 H_2O	−22.38
	∞ H_2O	−22.99

Source: F. O. Rossini, *et al.*, *Selected Values of Chemical Thermodynamic Properties*, National Bureau of Standards Circular 500, U.S. Government Printing Office, 1952.

heat required (a positive number) to maintain the temperature constant at 25° can be read from the table. For dilution or concentration from one known concentration to another, two values are read and their difference represents the appropriate heat effect. If the reaction is not isothermal, but adiabatic, then the temperature rise or fall can be obtained by straightforward heat balances, Section 9.6.

Example 8.1

▶ STATEMENT: 10 lb of calcium chloride is to be dissolved in sufficient water to make a 20% solution. If the process is fed both water and calcium chloride at 25 °C, calculate the quantity of heat, in Btu, which must be removed in order for the solution to be at 25 °C.

▶ SOLUTION: The molecular weights of $CaCl_2$ and H_2O are 111.0 and 18.0, respectively. Therefore,

$$CaCl_2 = 0.0901 \text{ lb-moles.}$$

The required quantity of water is

$$\left(\frac{10}{0.20}\right) 0.80 = 40 \text{ lb.}$$

The quantity corresponds to 2.22 lb-moles of water. Expressed in moles of water per mole of solute, the concentration is

$$\frac{2.22}{0.0901} = 24.64.$$

The heat of solution at 25°C is −17,000 cal per gm-mole of solute. In English units, the figure is −30,600 Btu per lb-mole. Since there are only 0.0901 lb-moles of $CaCl_2$, the solute in this problem, the required quantity is

$$(-30{,}600)(0.0901) = -2760 \text{ Btu.}$$

The minus sign indicates, of course, that heat is evolved.

Example 8.2

▶ STATEMENT: 100 lb of 78.4% sulfuric acid is to be concentrated to 98%. Calculate the heat in Btu, required in this concentration, process at 25 °C.

▶ SOLUTION: Basis: 100 lb of the 78.4% acid

$$78.4 \text{ lb } H_2SO_4 = 0.8 \text{ lb-moles,}$$
$$21.6 \text{ lb } H_2O = 1.2 \text{ lb-moles.}$$

Concentration, expressed in ratio of moles = 1.5. From Table 8.3, the heat of solution per mole is

$$-9000 \text{ cal/gm-mole.}$$

After concentration there is still 78.4 lb of acid and the amount of water is

$$\frac{78.4}{0.98} 0.02 = 1.6 \text{ lb} = 0.0888 \text{ lb-moles.}$$

The concentration is 0.111, expressed as the ratio of water to acid. Then, from Table 8.3 the heat of solution per mole is -1000 cal per gm-mole. It is, therefore, apparent that the heat of concentration is

$$-1000 - (-9000) = 8000 \text{ cal/gm-mole}, \\ = 14{,}400 \text{ Btu/lb-mole}.$$

For the 0.8 lb-moles involved the heat required is 11,500 Btu, the positive sign indicating that the heat is required in the process.

Another and very effective method for representing enthalpy data of solutions is by means of enthalpy-concentration diagrams. Since the number of systems for which these charts are readily available is very limited we shall not illustrate enthalpy concentration charts or their use. The reader is referred to Hougen, Watson, and Ragatz, *Chemical Process Principles, Part I*, 2nd ed., for further information on this subject.

SUMMARY We have observed that the solution of any thermochemical problem is based on a few simple facts:

1. Enthalpy is a state function. To get ΔH we need know only the initial and final states; the path between them is unimportant.

2. If we know a value of H for *each* substance present at the beginning of a reaction and a value for each substance present at the end of a reaction *relative to a common base*, the ΔH for that reaction is

$$\sum \Delta H_{\text{products}} - \sum \Delta H_{\text{reactants}}.$$

3. A convenient way of getting the relative ΔH for any substance is to calculate its ΔH_F^0 from the constituent elements assigned a value of $\Delta H = 0$ at 25 °C, the base temperature commonly chosen.

4. The change in the relative ΔH of any substance with temperature is $\int_{25}^{t'} C_p\, dT$ unless a phase change is involved in which case it is

$$\int_{25}^{t'} C_{p_{\phi,2}}\, dT + \Delta H_\phi + \int_{t'}^{t} C_{p_{\phi,1}}\, dT.$$

More than one phase change is handled in a similar manner.

5. $\Delta H = Q$ for the reaction if the reaction is reversible and at constant pressure or occurs in a system under steady flow conditions when kinetic and potential energy changes may be ignored.

LIST OF SYMBOLS

Latin letters

A, B, C	Specific chemical substances in a general equation; may also be used as subscripts
a, b, c	Empirical constants in heat capacity equations

aq	Designates an infinitely dilute solution (equivalent to ∞)
Btu	British thermal unit
C	Molal heat capacity (subscripted when referring to a specific substance), consistent units
c	When used with a substance, indicates the crystalline form
g	When used with a substance, indicates the gaseous state
H	Enthalpy, in general, heat units/mole
H^0	Enthalpy, in general, at the standard state, heat units/mole
ΔH	A change of enthalpy, in general
ΔH_c	Heat of combustion; superscript ^{0} indicates at the standard state
ΔH_F	Heat of formation; superscript ^{0} indicates at the standard state
ΔH_R	Heat of reaction; superscript ^{0} indicates at the standard state
ΔH^0	A change of enthalpy, in general, at the standard state
i	A subscript indicating the ith product in a chemical reaction
j	A subscript indicating the jth reactant in a chemical state
l	When used with a substance, indicates the liquid state
m	Molarity
n	Number of moles (often subscripted to refer to specific items)
P	Pressure, force/unit area
Pcu	Pound centigrade unit
Q	Heat transferred to or from a system by or into the surroundings
R	Universal gas constant in consistent units
s	When used with a substance, indicates the solid state
T	Absolute temperature
t	Temperature in general
t', t'', t_1	Specific temperatures
ΔU	Change in internal energy, heat units per mole
v	Molal volume, volume units per mole
x	Mole fraction

Greek letters

β	Graphitic form of carbon
ϕ	A subscript indicating a phase change

PROBLEMS

8.1 From heat of formation data, calculate the standard heats of reaction of the following.

a. $SO_2(g) + \frac{1}{2}O_2(g) + H_2O(l) \rightarrow H_2SO_4(l)$.
b. $CaCO_3(s) \rightarrow CaO(s) + CO_2(g)$.
c. $Zn + 2HCl(aq) \rightarrow ZnCl_2(aq) + H_2$.
d. $SO_2 + \frac{1}{2}O_2 \rightarrow SO_3$.
e. $Cu + H_2SO_4(aq) \rightarrow CuSO_4(aq) + H_2$.
f. $N_2(g) + 3H_2(g) \rightarrow 2NH_3(g)$.
g. $N_2(g) + O_2(g) \rightarrow 2NO(g)$.

h. $CaCl_2(s) + BaSO_4(s) \rightarrow BaCl_2(s) + CaSO_4(s)$.
i. $2FeO + \frac{1}{2}O_2 \rightarrow Fe_2O_3$.
j. $Ca_3(PO_4)_2(s) + 3SiO_2(s) + 5C(s) \rightarrow$

$$3[CaO \cdot SiO_2](s) + 5CO(g) + 2P(s).$$

k. $4NH_3(g) + 5O_2(g) \rightarrow 4NO(g) + 6H_2O(g)$.
l. $3NO_2(g) + H_2O(l) \rightarrow 2HNO_3(aq) + NO(g)$.
m. $P_2O_5(s) + 3H_2O(l) \rightarrow 2H_3PO_4(aq)$.

8.2 Calculate the heats of formation of the following compounds from the standard heat of combustion data:

a. benzene ($C_6H_6(l)$).
b. ethyl alcohol ($C_2H_5OH(l)$).
c. diethyl ether ($C_2H_5OC_2H_5(l)$).
d. propane (g).
e. phenol (s).

8.3 Calculate the standard heats of reaction of the following reactions, expressed in calories.

a. $C_2H_5OH(l) + O_2(g) = CH_3COOH(l) + H_2O(l)$.
ethyl alcohol — acetic acid

b. $3C_2H_2(g) = C_6H_6(l)$.
acetylene — benzene

c. $CH_3OH(l) + \frac{1}{2}O_2(g) = HCHO(g) + H_2O(l)$.
methyl alcohol — formaldehyde

d. $C_2H_2(g) + H_2O(l) = CH_3CHO(g)$.
acetylene — acetaldehyde

e. $CH_3CHO(g) + \frac{1}{2}O_2 \rightarrow CH_3COOH(l)$.
acetaldehyde — acetic acid

f. $C_6H_5NO_2(l) + 3Fe(s) + 6HCl(aq) =$
nitrobenzene

$$C_6H_5NH_2(l) + 3FeCl_2(aq) + 2H_2O(l).$$
aniline

8.4 In the contact process for sulfuric acid synthesis pure sulfur is first oxidized to sulfur dioxide, and then in a separate reactor further oxidation to sulfur trioxide occurs. Finally, sulfuric acid is formed from the SO_3 by hydration with water.

a. what is the ΔH_R^0 for the primary oxidation step?
b. what is the ΔH_R^0 for the secondary oxidation step?
c. what is the ΔH_R^0 for the hydration step?
d. what is the overall ΔH_R^0 for the synthesis of sulfuric acid from sulfur?

8.5 Ammonium sulfate, a fertilizer, is made by absorbing ammonia gas in a sulfuric acid solution. Calculate the ΔH_R^0 for this synthesis, assuming

a. $2NH_3 + H_2SO_4(aq) \rightarrow (NH_4)_2SO_4(aq)$.
b. $2NH_3 + H_2SO_4(l) \rightarrow (NH_4)_2SO_4(s)$.

8.6 Acetylene is generated when water is allowed to drip onto CaC_2. The basic equation is

$$CaC_2(s) + 2H_2O(l) \rightarrow Ca(OH)_2(s) + C_2H_2(g).$$

What is the ΔH_R^0 for this reaction?

8.7 Acids neutralize bases to give salts and water. Calculate the heats of neutralization per mole of water produced for the following cases. In each case consider that all acids and bases used and all salts produced are present in very dilute solutions.

a. sulfuric acid plus sodium hydroxide.
b. hydrochloric acid plus sodium hydroxide.
c. nitric acid plus sodium hydroxide.
d. sulfuric acid plus potassium hydroxide.
e. hydrochloric acid plus potassium hydroxide.
f. nitric acid plus potassium hydroxide.

8.8 From the results of Problem 8.7, draw conclusions regarding the heat of ionization of water at 25 °C and 1 atm.

8.9 Two methods for the manufacture of aniline from nitrobenzene are (a) iron-acid reduction in the liquid phase and (b) hydrogen reduction in the vapor phase. These reactions may be represented as follows:

a. $C_6H_5NO_2(l) + 3Fe + 6HCl(aq) \rightarrow C_6H_5NH_2(l) + 3FeCl_2(aq) + 2H_2O(l)$.
b. $C_6H_5NO_2(g) + 3H_2 \rightarrow C_6H_5NH_2(g) + 2H_2O(g)$.

Per mole of aniline formed, which involves the greater heat?

8.10 Calculate the ΔH_F^0 for each of the following hydrocarbons.

a. CH_4.
b. C_2H_6.
c. C_2H_4.
d. C_2H_2.
e. C_3H_8.
f. C_3H_6.
g. C_6H_6.

8.11 Derive a general equation for the ΔH_F^0 of any hydrocarbon C_xH_y from its heat of combustion and check your equation against the answers obtained in Problem 8.10.

8.12 When an organic compound of the general formula $C_rH_sCl_tN_uO_vS_w$ is burned, the products are

$$CO_2(g),\ H_2O(g),\ Cl_2(g),\ N_2(g),\ \text{and}\ SO_2(g).$$

Derive an equation of the form $\Delta H_F^0 = A\,\Delta H_c^0 + Br + Cs + Eu + Fv + Gw$ by determining the numerical values of A, B, C, D, E, F, and G.

8.13 Use the equation developed in Problem 8.12 to determine the ΔH_F^0 of the following compounds:

CH_3OH,	C_6H_5OH,
C_2H_5SH,	$C_2H_5OC_2H_5$.
CH_3NH_2,	

8.14 When amyl alcohol reacts with acetic acid, amyl acetate, and water (l) are formed. What is the ΔH_R^0?

a. calculate the ΔH_F^0 of each species first and use these values to get ΔH_R^0.
b. calculate ΔH_R^0 directly from heat of combustion data.

8.15 The first step in the manufacture of synthetic nitric acid is the catalytic oxidation of NH_3 to NO according to the following reaction:

$$4NH_3(g) + 5O_2(g) \rightarrow 4NO(g) + 6H_2O(g).$$

a. what is the ΔH_R^0 for this reaction?
b. what is the ΔH_R at 920 °C, the usual reaction temperature?

Commercial procedure is to feed the ammonia vapor at room temperature into a stream of air preheated to 750 °C before it passes over the catalyst. Sufficient air is used to provide a 20% excess of oxygen over that theoretically required to oxidize the ammonia completely to nitric acid (additional oxygen is required in successive steps). Approximately 90% of the NH_3 is converted to NO in the first step, and the gases leave at the reaction temperature.

c. What is the actual heat involved in the primary converter under commercial operating conditions per mole of ammonia fed?

8.16 Sulfur dioxide may be catalytically oxidized to sulfur trioxide over a V_2O_5 catalyst at 425 °C and 1 atm. Assuming this reaction is 97% complete, how much heat (in Btu per day) is liberated in a converter for a plant making 500 tons of 100% sulfuric acid per day?

8.17 The electrolytic process for making sodium hydroxide and chlorine may be represented by the overall reaction

$$2NaCl + 2H_2O \rightarrow 2NaOH + Cl_2 + H_2.$$

If a saturated salt solution is fed to the electrolytic cell and 75% of the salt fed is electrolyzed, how much heat is liberated in a cell producing 25 tons of sodium hydroxide per day at 25 °C?

8.18 If equal weights of the following materials are mixed, how much heat is evolved or required to maintain the final solution at 25 °C? Express your answers in Btu per pound of solution produced.

a. 100% H_2SO_4 and water.
b. 50% H_2SO_4 and water.
c. NaOH(s) and water.
d. 100% HNO_3 and water.
e. HCl(g) and water.
f. 20% HCl solution and water.
g. NH_4Cl(s) and water.
h. NaCl(s) and water.
i. NH_3(g) and water.

8.19 How much heat is required to concentrate 100 tons of 55% H_2SO_4 to 93% H_2SO_4 by removal of excess water? The process is conducted at 25 °C.

8.20 Write an equation (defining all symbols used) that may be used to calculate heats of dilution or energy required for concentration.

9 *Energy Balances—Applications*

This chapter will deal with application to practical problems involving various forms of heat and mechanical energy of material from the last three chapters. There will be two major categories with some subdivision in each class. The first will contain situations involving energy changes that have mechanical connotations: potential energy, kinetic energy, pressure-volume changes, and mechanical work. Emphasis will be on flow processes although the nonflow process will be covered briefly. The flow process in pipes without addition of external mechanical work provides important applications for the engineer. Processes in this category include friction loss in straight pipes and a number of flow meters.

The second category will exclude all forms of mechanical energy changes and mechanical work, thereby treating the heat balance only. Again the applications may be logically subdivided. The two classifications here will be purely physical processes as opposed to those which include chemical reaction.

9.1 Establishing the System and Surroundings

The terms *system* and *surroundings* have been defined in Chapter 6. In brief summary, the system may be batch or continuous. If batch, the system is contained and closely defined by physical boundaries. If continuous the portion of the entire system that is under consideration must be designated in some manner, the use of the balance boundary as employed in material balances being suitable. For example, consider Figure 9.1. Any of the balance boundaries indicated might be of interest. The system differs according to the choice. If balance boundary I is chosen, the energy input to the system by the pump is the main consideration. If balance boundary II is

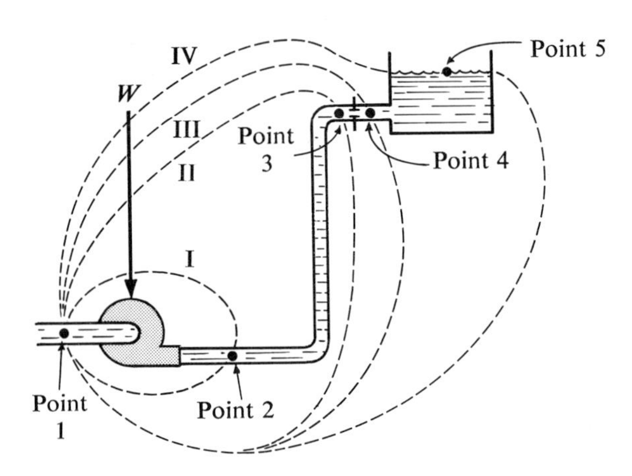

FIGURE 9.1 Diagram showing system and surroundings according to choice of terminal points.

chosen, this same energy input is included, but a portion of the input is used to raise the fluid to Point 3. At boundary III any permanent losses involved in forcing the fluid through the orifice are included. At boundary IV an additional increase in height is required, plus losses incurred in slowing the flowing stream on entrance to the tank. Note that none of the boundaries includes a line terminus except IV. The other boundaries were chosen at convenient points in relation to the gain or consumption of energy in some specific part of the process. Regardless of the choice, anything outside the boundary chosen constitutes *surroundings*.

9.2 The Mechanical Batch Process

The demands of large scale production dictate the use of continuous processes wherever possible. However, this emphasis does not mean that the batch process is obsolete. There are many specialty materials produced in small quantities where batch equipment is the only reasonable design. There are other situations where continuous processes would be desirable but the conversion of batch to continuous operation has not been successful. Therefore, it is wise to acknowledge the batch process in a study of energy balances. While we include only the following single illustration, we do not thereby imply that we minimize the importance of batch operations.

Example 9.1

▶ STATEMENT: 5000 cu ft of air at 1 atm and 60 °F is to be compressed to 5 atm. The final temperature desired is 60 °F. Two *reversible* processes are contemplated:

a. cooling at constant pressure followed by heating at constant volume;
b. heating at constant volume followed by cooling at constant pressure.

For each process calculate the heat requirements, the work requirements, and the change in internal energy and enthalpy. Assume that air acts as an ideal gas at all points in either process and that the heat capacities are constant with respect to temperature and pressure with molal values of

$$C_v = 5,$$
$$C_p = 7,$$

both values in consistent units, such as calories per (gram-mole)(°C) or Btu per (pound-mole)(°F).

▶ SOLUTION: The processes defined are batch, devoid of kinetic energy or potential energy changes. Therefore, Equation 6.2 applies in the form

$$dU = d'Q - \int P\,dv.$$

The constant g/g_c has been assumed to be nearly unity since there is no specification of any unusual location of the process.

BASIS The initial 5000 cu ft appear to be a less reasonable basis than the pound-mole since the actual number of pound moles would have to be used in every term in the energy balance. The basis will, therefore, be the pound-mole and all final quantities will be multiplied by the actual number of pound-moles involved. Pound-moles in 5000 cu ft as specified:

$$5000 \times \frac{492}{520} = 4730 \text{ ft}^3 \text{ at standard conditions}$$

and,

$$\frac{4730}{359} = 13.18 \text{ moles.}$$

Final molal volume, at 5 atm, 60 °F, per mole:

$$359 \times \frac{1}{5} \times \frac{520}{492} = 75.8 \text{ ft}^3.$$

Initial molal volume:

$$359 \times \frac{520}{492} = 379 \text{ ft}^3.$$

PROCESS (a) At constant pressure, $v/T = \text{const}$; then, cooling to temperature T, at which $v = 75.8 \text{ ft}^3$,

$$T = 520 \frac{75.8}{379} = 104 \text{ °R} = -356 \text{ °F}.$$

Also, cooling at constant pressure,

$$\begin{aligned} Q &= \Delta H = H_2 - H_1, \\ &= C_p \Delta T = 7(104 - 520) = -2910 \text{ Btu/lb-mole}. \end{aligned}$$

By basic definition,

$$\begin{aligned} \Delta U &= \Delta H - \Delta(Pv) = \Delta H - P\,\Delta v \\ &= -2910 - \frac{(14.7)(144)(75.8 - 379)}{778} = -2080 \text{ Btu/lb-mole}. \end{aligned}$$

The second step is heating at constant volume. Then there is no (Pv) work, since $dv = 0$, and $\Delta U = Q = C_v \Delta T$,

$$\Delta U = Q = 5(520 - 104) = 2080 \text{ Btu}.$$

The complete process is the sum of the two parts:

$$\begin{aligned} Q &= -2910 + 2080 = -830 \text{ Btu}. \\ \Delta U &= -2080 + 2080 = 0 \text{ Btu}. \end{aligned}$$

For a first-law process,

$$\begin{aligned} \Delta U &= Q - W, \\ 0 &= -830 - W, \\ W &= -830 \text{ Btu} = \text{work performed } on \text{ the system}. \end{aligned}$$

Moreover, since $T_1 = T_2$, $\Delta H = 0$. Then from the definition of enthalpy, $\Delta(Pv) = 0$, which is necessarily true at constant temperature.

Finally, for the 13.18 moles of air being processed,

$$\begin{aligned} mQ &= 13.18(-830) = -10{,}930 \text{ Btu}. \\ mW &= 13.18(-830) = -10{,}930 \text{ Btu}. \end{aligned}$$

PROCESS (b) Heating at constant volume, to give a final pressure of 5 atm,

$$T = 520 \times \frac{5}{1} = 2600 \text{ °R} = 2140 \text{ °F},$$

$$Q = C_v \Delta T = 5(2600 - 520) = 10{,}400 \text{ Btu/mole}.$$

For the second step, the cooling at constant pressure,

$$Q = C_p \Delta T = 7(520 - 2600) = -14{,}500 \text{ Btu/mole}.$$

Again applying the basic definition,

$$\Delta U = \Delta H - \Delta(Pv) = -14{,}500 - \frac{5(14.7)(144)(75.8 - 379)}{778},$$
$$= -10{,}400 \text{ Btu/mole.}$$

Then, since the complete process is the sum of the two parts:

$$\begin{aligned} Q &= 10{,}400 - 14{,}500 = -4100 \text{ Btu/mole,} \\ \Delta U &= 10{,}400 - 10{,}400 = 0, \\ \Delta H &= 0 \text{ (since } T_1 = T_2), \\ W &= Q - \Delta U = -4100 - 0 = -4100 \text{ Btu/mole.} \end{aligned}$$

Consequently, for the 13.18 moles of air being processed

$$\begin{aligned} \Delta U &= 0, \\ \Delta H &= 0, \\ mQ &= 13.18(-4100) = -54{,}100 \text{ Btu,} \\ mW &= 13.18(-4100) = -54{,}100 \text{ Btu.} \end{aligned}$$

For either process the point functions ΔU and ΔH have the same values, being dependent only upon the initial and final conditions which are unchanged by the manner in which the process is conducted. But Q and W, which are path functions, give different quantities depending on the process.

9.3 Mechanical Flow Processes

Before illustrating mechanical energy balances in flow systems we shall first review the basic form of equation already written for this balance (Equation 6.4) and second, develop some other useful forms. In Chapter 6, Equation 6.4 was developed and noted to apply to any flow system, regardless of reversibility. It may be written

$$dU = d'Q - dZ - u\,du/g_c - d(Pv) - d'W_{\text{flow}}. \quad (6.4)$$

First consider the units of the various terms, keeping in mind the conventions: (a) dimensional factors J and g/g_c are usually omitted; and (b) the equation as written refers to unit mass of fluid flowing. Then the units of the last three terms can be expressed as foot-pounds force per pound-mass, where

W_{flow} is in ft-lb$_f$/lb of mass flowing,
P is lb$_f$/unit area,
v is specific volume in unit volume/lb-mass,
u is velocity, and
g_c is the gravitational constant in units of acceleration $\times$ (lb/lb$_f$).

If Z is multiplied by g/g_c, the same units apply to this potential energy term. But U and Q are normally expressed in heat units and generally refer to unit mass if no notation of mass is noted. In Equation 6.4 we would then expect these symbols to be so expressed. If the heat unit chosen is the Btu, multiplication by J, previously noted to be equal to 778 ft-lb$_f$ per Btu, will make all terms in the equation dimensionally consistent. Each term in the equation is then expressed in units of mechanical energy per unit mass.

When each term of Equation 6.4, expressed in these units, is multiplied by the mass rate of flow, the dimensions of each term are changed to foot-pounds force per unit time, the rate of doing work. Since power is the rate of doing work, this modification of the equation expresses all terms in units of power. Rather than multiplying term by term, it may be simpler, in specific cases, to perform the balance on a unit-rate *basis*, solving for the unknown in terms of mechanical energy per unit mass, and subsequently multiplying this term only by mass rate. This treatment is often convenient when power requirements of a pump are to be specified (see Example 9.3).

Still another useful form can be obtained by multiplying each term of Equation 6.4, expressed in units of foot-pounds force per pound-mass, by the ratio g_c/g. Since the dimensions of this ratio are pound-mass per pound-force, the resulting units of each term are solely those of feet, expressed in terms of the fluid flowing in the system. Equation 6.4, treated in this manner, then becomes

$$J\frac{g_c}{g}\,dU = J\frac{g_c}{g}\,d'Q - dZ - u\frac{du}{g} - \frac{g_c}{g}\,d(Pv) - d'W_{\text{flow}}, \qquad (6.4a)$$

where W' is expressed in feet of fluid flowing. Although the units of this equation are not expressed in terms of traditional energies or power, they may be easily converted in any one of a number of ways. The first would be the obvious reversal of the procedure by which we obtained 6.4a from 6.4, multiplying by g/g_c, obtaining energy per unit mass. A second would be by multiplication by mass rate to obtain foot-pounds mass per unit time, followed by multiplication by g/g_c. Still a third would be multiplication by the density and g/g_c to give pressure in pounds force per unit area followed by multiplication by the so-called mass velocity, in units of mass/(unit time) (unit area). These conversions will again produce units of power, as can be readily verified. (Reference to the supplement to Chapter 2 will confirm the implications of the procedure just noted, i.e., that pressure and head of fluid are equally useful methods for expressing the same intensive property of a system, as long as the density of the fluid is known at the position of measurement.)

The choice of units in the energy balance is, then, at the discretion of the engineer for the particular problem being undertaken. Examples 9.2, 9.3, and 9.4 will illustrate this point. The first will deal in foot-pounds force for the entire mass being handled, the second in power units, and the third in feet.

Returning to the problem of developing other useful forms of the energy balance in flow situations, equate the expressions for dU from Equations 6.2 and 6.4 to obtain, in linear units

$$P\,dv = dZ + u\frac{du}{g} + d(Pv) + d'W_{\text{flow}}.$$

Rearranging, while noting that

$$d(Pv) = P\,dv + v\,dP,$$

we obtain

$$dZ + u\frac{du}{g} + v\,dP + d'W_{\text{flow}} = 0.$$

If no mechanical work has been exchanged between the system and the surroundings,

$$dZ + u\frac{du}{g} + v\,dP = 0. \tag{9.1}$$

Equation 9.1 is known as the *Classical Bernoulli Equation.* According to previous remarks regarding units and symbolic representations, the following equivalent forms are equally suitable:

$$\frac{g}{g_c}\,dZ + u\frac{du}{g_c} + v\,dP = 0; \qquad \text{units of each term are ft-lb}_f\text{/lb};$$

and, designating mass rate of flow by the symbol, w,

$$w\frac{g}{g_c}\,dZ + wu\frac{du}{g_c} + wv\,dP = 0; \qquad \text{units of each term are ft-lb}_f\text{/unit time}.$$

For cases where work is exchanged between the system and the surroundings, *and* where frictional losses are incurred in the system, terms may be added to Equation 9.1 to include these effects. The resulting equation is known as the *Extended Bernoulli Equation* and is written, in simplest symbolic form,

$$dZ + u\frac{du}{g} + v\,dP + d'W_{\text{flow}} + d'F = 0. \tag{9.2}$$

The frictional term is a path function, as noted by the form of the operator.

Equations 9.1 and 9.2 have now been added to the list of tools available for dealing with mechanical energy balances. Let us add one more equation before considering the general problem of integrated forms and illustrative uses.

If the original form of Equation 6.4 is retained (i.e., including dU), and it is noted from Chapter 6 that change in enthalpy is defined as

$$dH = d(U + Pv),$$

Equation 6.4, which is not restricted with respect to reversibility, can be rewritten as

$$dH + dZ + u\frac{du}{g} = d'Q - d'W_{\text{flow}}. \tag{9.3}$$

Equations 6.4, 9.2, and 9.3 are all representative of the energy balance in flow systems and the most useful form may be chosen for each individual situation. Equation 9.1 is also available, but it will be recognized to be only a special case of Equation 9.2.

For integrated forms, we shall examine Equation 9.2 in detail and then obtain a comparable form of Equation 9.3 by analogy. The integrated form of Equation 9.1 is obviously identical to that of Equation 9.2, except that some terms are omitted.

Consider that the process operates between two points, numbered 1 and 2. Point 1 designates the beginning of the system and point 2 the end. We may then write an energy balance, in terms of input and output items.

POINT 1 Input kinetic and potential energies =

$$Z_1 + \frac{u_1^2}{2g}.$$

POINT 2 Output kinetic and potential energies =

$$Z_2 + \frac{u_2^2}{2g}.$$

Output energy in the form of friction $= {}_1F_2$, where the subscripting indicates output between the points 1 and 2.

Output energy in the form of mechanical work $= {}_1W_2$, with subscripting having the same meaning and work output being defined as positive, according to the conventions defined in Chapter 6.

Output in the form of work of self-expansion; from the $v\,dP$ term, =

$$\int_1^2 v\,dP.$$

Now combining all these forms of energy in an energy balance, and combining like terms,

$$(Z_2 - Z_1) + \frac{(u_2^2 - u_1^2)}{2g} + \int_1^2 v\,dP + {}_1F_2 = -{}_1W_2. \tag{9.4}$$

When incompressible fluids are involved, v is a constant and the integral term becomes

$$\frac{(P_2 - P_1)}{\rho}.$$

By analogy, the integrated form of Equation 9.3 is

$$(H_2 - H_1) + (Z_2 - Z_1) + \frac{(u_2^2 - u_1^2)}{2g} = {}_1Q_2 - {}_1W_2. \tag{9.5}$$

Equations 9.4 and 9.5 will frequently be expressed in Δ-form. These optional forms follow, with 9.4a expressing the situation for incompressible fluids only.

$$\Delta Z + \Delta\left(\frac{u^2}{2g}\right) + \frac{\Delta P}{\rho} + {}_1F_2 = -{}_1W_2. \tag{9.4a}$$

$$\Delta H + \Delta Z + \Delta\left(\frac{u^2}{2g}\right) = {}_1Q_2 - {}_1W_2. \tag{9.5a}$$

By comparing Equations 9.2 and 6.4, we can draw some interesting conclusions regarding friction, heat effects, and internal energy. First, subtract Equation 9.2 from 6.4:

$$dU + d(Pv) + dZ + \frac{u\,du}{g} - d'Q = -d'W \tag{6.4}$$

$$v\,dP + dZ + \frac{u\,du}{g} + d'F = -d'W \tag{9.2}$$

$$dU + d(Pv) - v\,dP - d'Q - d'F = 0.$$

Now, when we limit the comparison to incompressible fluids, the terms $d(Pv)$ and $v\,dP$ are identical and cancel. Then we are left with

$$d'F = dU - d'Q.$$

If the process is also isothermal, $dU = 0$, and $d'F = -d'Q$, which indicates that all frictional losses leave the system (leave due to the minus sign on $d'Q$) in the form of heat. On the other hand, if the process is adiabatic, $d'Q = 0$ and $d'F = dU$. Under these conditions, the friction is all utilized to increase the internal energy, the effect corresponding to a rise in temperature.

The remarks just made regarding the isothermal process might mislead one into believing that U is solely a function of T and that dU must always be zero whenever temperature is constant. Nothing could be further from the truth. In general,

$$U = \phi(P, v, T),$$

a relationship which reduces to

$$U = \psi(T)$$

only under two very special circumstances. One is a process in which all material involved may be treated as a perfect gas. The other is the situation in which all material may be treated as a perfectly incompressible fluid. It is the latter special case which allowed the conclusions just drawn.

There may be occasions when we are interested in enthalpy changes and also in pressure effects. Since the flow equation (Equation 9.3), containing enthalpy terms, has eliminated pressure terms, we shall need another relationship. For simplicity in establishing the concepts consider the frictionless case with incompressible fluids. Then, from Equation 6.7, including the potential and kinetic energy terms,

$$d'W_{\text{flow}} = -v\,dP - dZ - \frac{du^2}{2g}. \tag{6.7a}$$

We also know, from Equation 6.4, that

$$d'W_{\text{flow}} = -dU + d'Q - d(Pv) - dZ - \frac{du^2}{2g}. \tag{6.4}$$

Equating the right-hand sides of Equations 6.7a and 6.4 and rearranging terms,

$$d'Q = dU + d(Pv) - v\,dP = dH - v\,dP.$$

This last expression provides some interesting information. First, consider the case of no pressure difference at the two ends of the system (e.g., a pumping problem in which the fluid surfaces at each end are exposed to atmospheric pressure). Then the enthalpy increase is exactly equal to $d'Q$, the net heat added. Second, if the pressure at outlet is greater than that at inlet (as in Problem 9.110),

$$\Delta H > Q,$$

since

$$Q + \frac{(P_2 - P_1)}{\rho} = H_2 - H_1.$$

Increase in enthalpy then equals the sum of the increases in thermal energy due to compression plus the net heat added. Third, if isothermal conditions prevail, there is no increase in enthalpy* and, when $P_2 > P_1$, the net Q must be negative, removing the increase in energy due to compression. Finally, under adiabatic conditions,

$$Q = 0,$$

and

$$H_2 - H_1 = \frac{(P_2 - P_1)}{\rho}.$$

• • • • •

* For limitations on this statement see discussion on p. 290.

If we now decide to include the effects of friction, $d'F$ should appear as an output term in Equation 6.7a, which is to say that it should have the same sign as that of the $d'W$ term. Then

$$d'F + d'W_{\text{flow}} = -v\,dP - dZ - \frac{du^2}{2g_c}. \tag{6.7b}$$

Since reversibility is not in question in Equation 6.4, elimination of potential and kinetic energy terms between Equations 6.7b and 6.4 gives

$$d'F + d'Q = dH - v\,dP.$$

Then, if there is no change in pressure, the increase in enthalpy is due to the sum of the two thermal effects, Q and F, as would be intuitively expected. Under isothermal conditions, where there is no enthalpy change, and for the case of $P_2 > P_1$, Q must be negative to indicate the removal of the energy of friction and of the compressive effects. If adiabatic, also with $P_2 > P_1$, the enthalpy increase will equal both compressive and frictional effects.

In considering these remarks, the relative size of thermal and mechanical effects is important, particularly when dealing with incompressible fluids. In general, mechanical effects will be small. For example, with water, the number of Btu per pound accompanying unit increase in pounds per square *inch*, is only 3×10^{-3}. The $-v\,dP$ term is then essentially negligible in most instances. Thus, we can say that, where appreciable heat effects are present, it makes little difference to the required work whether or not we change the pressure.

The *datum plane* is never a matter of concern in mechanical balances. The equations are always written as differences between two states. Therefore, the datum plane is implied to be that of the initial state. Note that this situation does not exist in the case of heat balances (Section 9.6) where both temperature and phase must be carefully specified.

Example 9.2

▶ STATEMENT: Water is being pumped from a well into the bottom of a storage tank at a rate of 500 gal per hr. The well is 200 ft deep and the level of water in the tank is 60 ft above ground level. The pump is driven by a 5-horsepower motor; only 45% of the power available is used in pumping. In winter the well water temperature is 40 °F. To prevent freezing in the lines leaving the tank, the tank temperature is to be retained at 45 °F when the water rate into the tank exactly equals that withdrawn (i.e., the level remains the same and steady state conditions are maintained).

Calculate the net heat input required from the heater in the inlet line.

▶ SOLUTION: The flow system consists of the well, piping, pump, and tank as shown in Figure 9.2. The process is irreversible, due to friction losses.

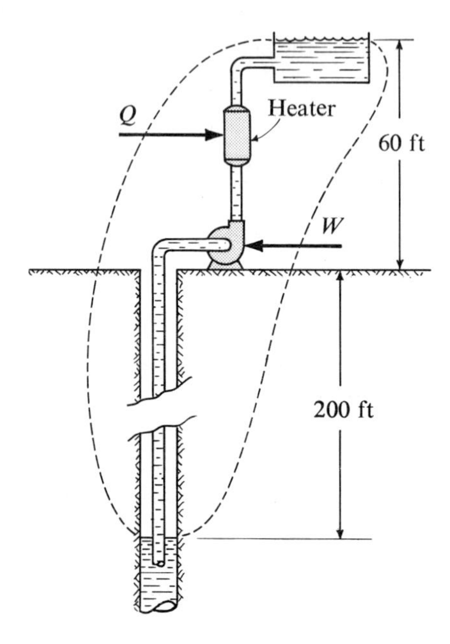

FIGURE 9.2 Mechanical pumping system containing a heater.

Net heat input will be interpreted to mean total heat input minus losses from the entire system since the latter figure is not given.

Equation 9.5a will be used on a per pound basis.

BASIS One hour of operation, equivalent to 500 gal of water pumped. The pumping rate is

$$500\,\frac{\text{gal}}{\text{hr}} \times \frac{1}{7.48}\,\frac{\text{ft}^3}{\text{gal}} = 66.8\,\frac{\text{ft}^3}{\text{hr}}.$$

The density of water is almost exactly 62.4 lb/ft^3 at any temperature encountered in the system. The pumping rate is then

$$66.8\,\frac{\text{ft}^3}{\text{hr}} \times 62.4\,\frac{\text{lb}}{\text{ft}^3} = 4170\,\frac{\text{lb}}{\text{hr}}.$$

The total head change is 260 ft. Kinetic energy changes are zero since the velocities of water in the well and in the tank are both zero.

Potential energy per 500 gal:

$$\left(\frac{g}{g_c}\right)\left(\frac{\text{lb-force}}{\text{lb-mass}}\right)(4170\text{ lb-mass})(260\text{ ft}) = 1{,}082{,}000\text{ ft-lb force.}$$

$$\frac{1{,}082{,}000}{778} = 1392\text{ Btu.}$$

Mechanical work: done *on* system; negative work

$$
\begin{aligned}
{}_1W_2 &= -(5\text{ hp})(0.45)\,\frac{(33{,}000\text{ ft-lb force})}{(\text{min})(\text{hp})}\left(60\,\frac{\text{min}}{\text{hr}}\right),\\
&= -4{,}460{,}000\text{ ft-lb force/hr},\\
&= -5730\,\frac{\text{Btu}}{\text{hr}}.
\end{aligned}
$$

Enthalpy change: heat capacity over this range is essentially constant with value of 1.0 Btu/(lb)(°F).

$$
\Delta H = \left(4170\,\frac{\text{lb}}{\text{hr}}\right)\left(1.0\,\frac{\text{Btu}}{(\text{lb})(^\circ\text{F})}\right)(45 - 40)(^\circ\text{F}) = 20{,}850\,\frac{\text{Btu}}{\text{hr}}\ (20.85 \times 10^3).
$$

Balance:

$$
\begin{aligned}
Q &= 1392 - 5730 + 20{,}850,\\
&= 16{,}510 \text{ net Btu required per hr by the system.}
\end{aligned}
$$

Example 9.3

▶ STATEMENT: A pump draws solution of specific gravity $1.84^{20/4}$ from one tank and transfers it to another through a standard 2-in. pipe (diameter = 2.067 in.) at a rate of 130 gal per min. The pump motor is rated at 5 hp and 65% of this power may be utilized in pumping. The end of the discharge pipe is 50 ft above the level of liquid in the storage tank.

Calculate the friction loss and express this in terms of a head, in feet of fluid flowing.

Calculate the pressure which must be developed by the pump, in pounds force per square inch, if the pump inlet is of 3-in. standard pipe (diameter = 3.068 in.).

▶ SOLUTION: The minute seems to be as convenient a basis as any.

PART (a) Calculation of frictional loss. The process, illustrated in Figure 9.3, is assumed to be isothermal within Balance boundary I.

The volume of flow in 1 min is

$$
130\,\frac{\text{gal}}{\text{min}} \times \frac{\text{ft}^3}{7.48\text{ gal}} = 17.38\text{ ft}^3.
$$

The velocity in a standard 2-in. pipe is

$$
\begin{aligned}
\frac{17.38\text{ ft}^3}{\text{min}} \times \frac{4}{\pi(2.068)^2\text{ sq in.}} \times \frac{144\text{ sq in.}}{\text{sq ft}} &= 744\text{ ft/min},\\
&= 12.4\text{ ft/sec}.
\end{aligned}
$$

The mass flow per minute is

$$
\left(17.38\,\frac{\text{ft}^3}{\text{min}}\text{ of water}\right) \times 1.84\left(\frac{\text{density of fluid}}{\text{density of water}}\right) \times 62.42\,\frac{\text{lb of water}}{\text{ft}^3}
$$

$$
= 1990 \text{ lb of the fluid being pumped } (19.90 \times 10^2).
$$

The pump delivers to the system

$$-[(5)(0.65)\text{ hp}]\,\frac{33{,}000\text{ ft-lb}_f/\text{min}}{\text{hp}} = -107{,}200\text{ ft-lb}_f/\text{min}\ (107.2\times 10^3).$$

The potential energy required is

$$\frac{g}{g_c}(50\text{ ft})\times 1990\text{ lb/min} = 99{,}500\text{ ft-lb}_f/\text{min}\ (99.50\times 10^3).$$

Note especially that the flow becomes discontinuous (no longer contained in the system) at the pipe outlet. The balance may be drawn at that point but cannot be drawn through the surface of the liquid in the tank.

There is kinetic energy at outlet since the balance boundary includes the end of the pipe. There is no kinetic energy at inlet since the balance boundary passes through the quiescent level of liquid in the tank. The kinetic energy change required is then

$$\frac{(1990)(12.4)^2}{2g_c} = 4760\text{ ft-lb}_f/\text{min}\ (4.760\times 10^3).$$

Pressures at inlet and outlet are both atmospheric. Therefore the work term for self-expansion, is zero. The balance, as indicated by Equation 9.4a, can then be written in terms of foot-pounds force per minute:

$$\begin{aligned} 99.5\times 10^3 + 4.760\times 10^3 + {}_1F_2 &= 107.2\times 10^3, \\ {}_1F_2 &= 2.9\times 10^3\text{ ft-lb}_f/\text{min}. \end{aligned}$$

To express power consumption in terms of head, convert to foot-pounds mass per minute and divide by mass rate of flow. Since g/g_c is essentially unity, this frictional head is

$$\frac{2.9\times 10^3}{1.99\times 10^3} = 1.46\text{ ft.}$$

PART (b) Calculation of pressure developed by the pump.

Pressure developed is the difference between pressures at inlet and outlet. It may be calculated by writing the extended Bernoulli equation across the pump only (boundary line II in Figure 9.3). Assume that the difference in level from inlet to outlet is zero and that friction is accounted for in the efficiency factor (i.e., the 0.65 figure) relating to the pump. Then,

$$\frac{\Delta P}{\rho} = -{}_1W_2 - \frac{\Delta u^2}{2g}, \qquad \text{in consistent units.}$$

Velocity at pump inlet is

$$12.4\times\left(\frac{2.068}{3.068}\right)^2 = 5.64\text{ ft/sec.}$$

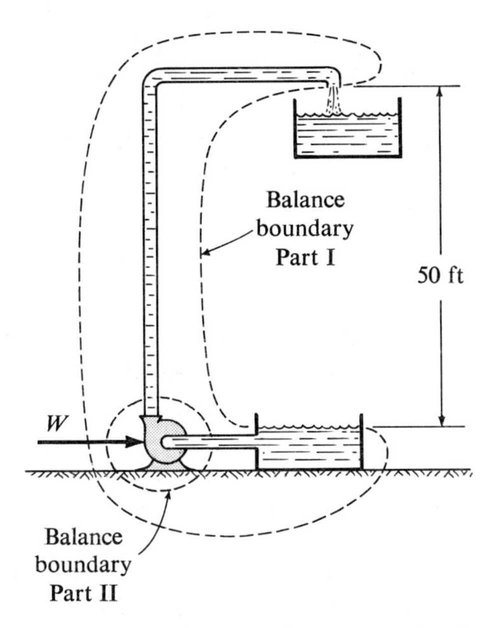

FIGURE 9.3 Mechanical pumping system—isothermal.

The kinetic energy loss is then

$$\frac{12.4^2 - 5.64^2}{2g} = 1.90 \text{ ft of fluid.}$$

When g/g_c = unity, the work in feet is

$$\frac{g_c}{g}\, 107.2 \times 10^3 \frac{\text{ft-lb}_f}{\text{min}} \times \frac{\text{min}}{1.990 \times 10^3 \text{ lb}} = 53.9 \text{ ft of fluid}$$

and

$$\Delta P = \frac{g}{g_c}(1.84)(62.4)(53.9 - 1.90) = 5960 \text{ lb}_f/\text{ft}^2\ (59.6 \times 10^2),$$

$$\Delta P = 41.5 \text{ lb}_f/\text{sq in.}$$

Two items in this example are worth special note. First any units may be used in the balance as long as each term is expressed in the same manner. To emphasize this point, part (a) used a balance with terms expressed as foot-pounds force per minute, while part (b), using the same data, drew the balance in terms of feet of fluid flowing in the system. The former is obtained from the latter by multiplying by g/g_c and by mass rate of flow as shown. The second item of importance is the relatively small contribution of kinetic energy unless velocities are extremely large. In part (a), where the velocity was greater than 12 ft per sec, the contribution of the kinetic energy was only about 5%

of the total energy requirement. We can correctly conclude that kinetic energy terms in general can be neglected unless the linear velocity is large. We cannot establish a minimum value of velocity which must be considered since the relative importance of the kinetic energy term must be assessed against the magnitude of the other terms in the balance. Certainly we can conclude, however, that there may be many occasions when kinetic energy can be neglected legitimately.

Example 9.4

▶ STATEMENT: Water at 75 °F is flowing at a velocity of 20 ft per sec in a straight horizontal insulated pipe whose inside diameter is suddenly increased to 2.00 in. Calculate: (a) the change in enthalpy of the water if there are no contributions from external work; (b) *maximum* enthalpy change for sudden enlargement from a 1.00 in. pipe; (c) maximum enthalpy change if the velocity in the inlet section (i.e., the velocity of approach) were 2 ft per sec rather than 20 ft per sec?

▶ SOLUTION: The process of enlargement from 1.00 to 2.00 in. is represented in Figure 9.4. The distance from inlet boundary to outlet boundary is, of course, negligible. This diagram will serve for all three questions posed. The basis is best chosen as 1 sec.

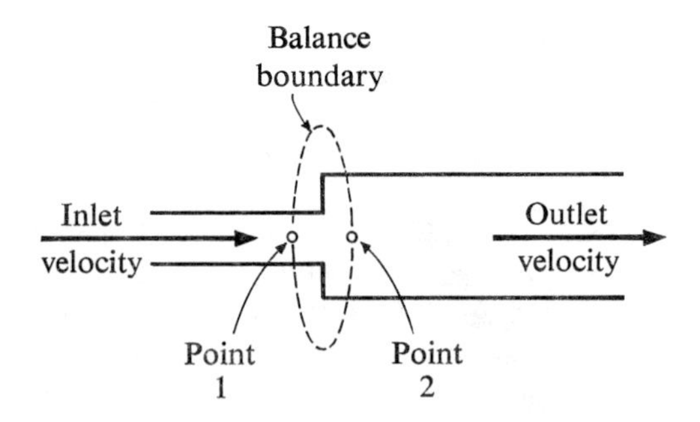

FIGURE 9.4 Diagram of a sudden enlargement.

PART (a) At the boundary shown, ΔZ is zero at the center of gravity of the system. Work, W, is zero. The insulation will be assumed to be sufficient so that Q may be assumed to be zero. Then Equation 9.5 is applicable. In differential form,

$$\frac{u\,du}{g} = -dH, \qquad \text{in consistent units.}$$

Integrating from Point 1 to Point 2

$$\frac{u_2^2 - u_1^2}{2g} = H_1 - H_2 = -\Delta H.$$

A positive value of ΔH would indicate an increase in water temperature, and vice versa.

The velocity at Point 2 is related to that at Point 1 through the square of the ratio of the diameters, since $W = u_1 A_1 \rho_1 = u_2 A_2 \rho_1$. Therefore,

$$\left(\frac{1}{2}\right)^2 20 = 5 \text{ ft/sec.}$$

The kinetic energy in feet, or in foot-pounds force per pound, is, with g/g_c equal to unity:

$$\frac{5^2 - 20^2}{64.34} = -5.83 \text{ ft.}$$

The mass rate of flow is

$$\frac{20 \text{ ft}}{\text{sec}} \times \pi\left(\frac{1}{2}\right)^2 (\text{in.})^2 \times \left(\frac{\text{ft}}{12 \text{ in.}}\right)^2 \times \frac{62.2 \text{ lb}}{\text{ft}^3} = 6.78 \text{ lb/sec.}$$

The change in kinetic energy expressed in foot-pounds force per second is $(-5.85)(6.78) = -39.5$, and expressed in Btu per second, is

$$-\frac{39.5}{778} = -0.051.$$

Therefore ΔH per sec $= 0.051$ Btu per sec. The temperature rise expected from this energy exchange is less than 0.01 °F.

PART (b) For the maximum enthalpy change due to expansion, the cross section at Point 2 should be infinite and the velocity would be zero. The kinetic energy terms would then be $-400/64.34 = -6.22$ ft. The enthalpy change could then be calculated by multiplying the ΔH per sec for the finite expansion by the ratio of the head losses:

$$(0.051)\left(\frac{-6.22}{-5.83}\right) = 0.054 \text{ Btu/sec.}$$

PART (c) The maximum enthalpy change with a 2 ft per sec velocity of approach can be readily seen to be $\frac{1}{100}$ of the preceding figure.

These enthalpy changes (and attendant temperature changes) are quite obviously negligible. This type of process appears to be both isothermal (i.e., constant temperature) and adiabatic (no external heat added or removed from the system). Of course, both conditions cannot be met at the same time, as has just been shown, but the small actual temperature change makes the assumption of both conditions very close to the fact. Let us then assume a frictionless process involving no exchange of external work and inquire regarding the pressure drop (or increase) over the section. The Bernoulli equation 9.1 then becomes

$$(u_1^2 - u_2^2)/2g_c = \int_1^2 v\, dP, \qquad \text{in consistent units,} \tag{9.6}$$

and, for liquids, the right-hand side becomes $(P_2 - P_1)/\rho$. For the first situation indicated in Example 9.4, the pressure increase from Point 1 to Point 2 must equal the calculated kinetic energy head loss between the same two points, multiplied by the fluid density. The latter is constant since the temperature is essentially constant. The decrease in kinetic energy is 5.83 ft. The pressure increase is, therefore,

$$(5.83)(62.2) = 363 \text{ lb-force/sq ft}$$

or

$$2.52 \text{ lb-force/sq in.}$$

These relatively large pressure changes and concomitant small temperature changes suggest that measuring the pressure drop across a sudden change in cross section might well provide an excellent flow meter. This idea will now be expanded and some of the attendant difficulties will be briefly noted.

9.4 Venturi and Orifice Meters

Meters of these types are based upon the principle that a constriction placed in a line will completely convert kinetic energy to Pv-energy and therefore provide a means for the measurement of rate of flow. The present discussion will be restricted to the same conditions which existed in the last example, frictionless flow, isothermal and adiabatic conditions, and for the time being, incompressible fluids (i.e., liquids).

Under these limiting conditions any restriction placed in a smooth pipe will result in the following energy balance, based on Equation 9.6:

$$\frac{u_2^2 - u_1^2}{2g_c} = \frac{\Delta P}{\rho} = \Delta h, \qquad \text{in consistent units,} \tag{9.7}$$

where Δh indicates the difference in head, expressed as the height of a column of the fluid flowing in the line, usually in units of feet. Then, if the difference in head is measured, as with a manometer (see Supplement to Chapter 2), the difference between the squares of the average velocities can be easily calculated. Now if the cross sections of the streams at Points 1 and 2 are known, the relationship between u_1 and u_2 is also known so that either one of these velocities can be calculated. The mass rate of flow can be readily calculated, knowing the cross section of the stream at this point and the density of the fluid.

Suppose that we investigate the principle of this instrument without worrying about the complications of the form of the stream (i.e., whether section is circular, what form the constricting disturbance takes, and the frictional losses which must occur). If the cross sections at Points 1 and 2 are 4 and 1 sq in., respectively, then the velocity at Point 2 with relation to that at Point 1 must be 4:1. Equation 9.7 can be rewritten in terms of either velocity using the relationship

$$u_2 = 4u_1.$$

Therefore,

$$\Delta h = (16 - 1)\frac{u_1^2}{2g_c} = 15\frac{u_1^2}{64.34}.$$

Now if the measured head, Δh, is 10 ft of the flowing fluid,

$$\frac{(64.34)(10)}{15} = u_1^2.$$

Then

$$u_1 = 6.5 \text{ ft/sec}.$$

Suppose further that the fluid flowing had a density of 1.23 gm per cu cm. Then the mass rate of flow is:

$$\left(6.5\frac{\text{ft}}{\text{sec}}\right) \times \left(\frac{4}{144}\right) \text{sq ft} \times (1.23)(62.42)\frac{\text{lb}}{\text{ft}^3} = 13.8\frac{\text{lb}}{\text{sec}}.$$

The situation is not quite this simple. While a lump of solder, or of weld, could effect the constriction, it would be difficult to measure the cross section at the constriction. The constriction must be carefully made in order to relate the high and low velocity cross sections. Two basic forms, the orifice and the Venturi, and one hybrid, the nozzle, will be briefly discussed, including the major features of each. Fine distinctions within each class will be deliberately ignored. For these points, which provide a maximum accuracy, the reader is referred to texts which discuss these instruments in detail.*

The Orifice Meter

This form of flow meter is illustrated in Figure 9.5. The orifice consists of a hole drilled in a flat plate with the opening beveled to provide a sharp edge; the bevel faces downstream. The plate containing the orifice is fastened in place by any suitable means, often between two flanges. Pressure taps *d* and *e* are placed above and below the orifice. Pressure readings could be taken from these taps by any suitable devices, such as bourdon pressure gauges; alternately the difference in pressure between the two points may be read directly on a manometer.

The pressure exerted by the flowing fluid varies as a function of the length of the pipe-and-orifice combination, indicating that the exact location of these taps may be of considerable importance. (See Figure 9.5.) The downstream pressure particularly varies markedly with position. The upstream tap will not need to be precisely located since the upstream pressure varies for a very short distance only. It is always located at a position of known cross section.

• • • • •

* E.g., W. L. McCabe and J. C. Smith, *Unit Operations of Chemical Engineering* (New York: McGraw-Hill Book Co., 1966).

In contrast, it is almost impossible to locate the downstream pressure tap where the stream cross section is accurately known. The orifice itself is the only position of this kind, and location of a tap at this point is impossible except in very large installations. We must, therefore, investigate the effect of alternate locations.

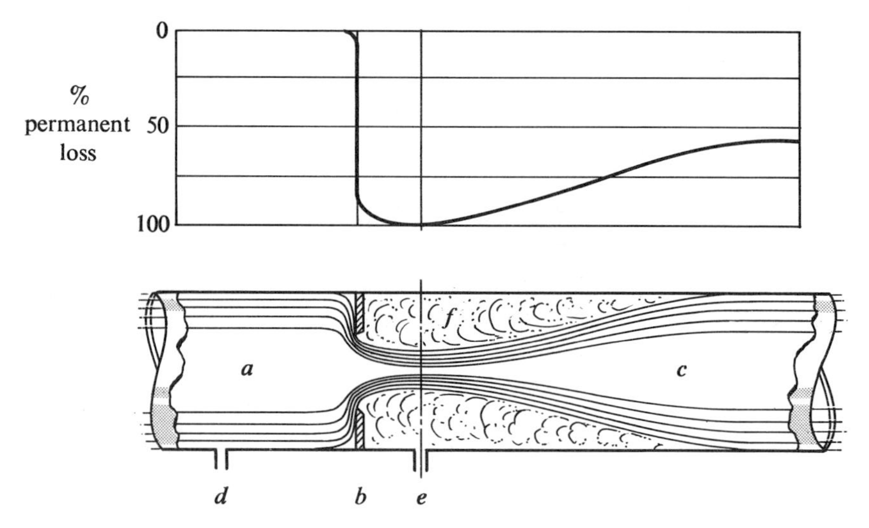

FIGURE 9.5 Diagram of sharp-edged orifice-type flow meter: a. upstream section; b. orifice; c. downstream, pressure recovery, section; d. upstream, high pressure, tap; e. downstream tap located at *vena contracta*; f. region of turbulent eddies.

As the stream leaves the orifice it has a strong forward component along the axis of the pipe. It also has an appreciable inward component caused by forcing a large part of the upstream flow to move inward to pass through the orifice. The resultant of these two forces provides a continuation in the reduction of the main stream cross section for some distance after the orifice plate is passed. Eventually, a point of minimum cross section (known as the *vena contracta*) is reached. Subsequently, the main stream tends to expand until it again fills the entire cross section at some distance down the pipe. Around the jet in the region where the main stream does not fill the cross section there is fluid in aimless flow, known as eddies. Energy is consumed to keep these (useless) eddies in motion and this expenditure of energy is wasted so that the recovery of full pressure is never attained downstream. In other words the instrument consumes an appreciable amount of power in wasted motion. It is inefficient in this regard, and also in relation to the Venturi (as we shall see), but it is inexpensive to build. Thus the traditional problem in engineering is encountered—low cost, high operating expense *vs.*

high cost, low operating expense; which one should be chosen? This problem is considered in detail in Example 9.5.

The existence of a *vena contracta* at a point downstream from the actual orifice also indicates that the simple energy balance will not be correct without some modification. The standard procedure here is to introduce an arbitrary constant, correcting for the fact that the orifice velocity and downstream pressure are not measured at the same cross section. If the *vena contracta* position is the location of the downstream tap—and it can be so located quite well as long as the ratio of orifice to pipe diameters is small—then the pressure reduction is greater than that predicted from the kinetic energy change based upon velocities calculated at the two known cross sections. We shall then insert an orifice discharge coefficient, C_0, with value $0 < C_0 < 1$, to modify the previous equation by accounting for the difficulty of measuring downstream conditions. Then

$$\sqrt{u_2^2 - u_1^2} = C_0\sqrt{2g_c\,\Delta h}. \tag{9.8}$$

If the orifice and pipe diameters are d_0 and d_p, respectively, then

$$u_1 = \left(\frac{d_0}{d_p}\right)^2 u_2 = \beta^2 u_2.$$

Therefore,

$$\sqrt{u_2^2 - u_1^2} = \sqrt{u_2^2 - \beta^4 u_2^2} = u_2\sqrt{1 - \beta^4}$$

and

$$u_2 = \frac{C_0}{\sqrt{1 - \beta^4}}\sqrt{2g_c\,\Delta h}. \tag{9.9}$$

When the ratio, $d_0/d_p \leq 0.25$, $\beta^4 \leq 0.004$ and the term $1 - \beta^4$ can be considered to be unity with an error $\leq 0.4\%$. In terms of velocities, if the ratio of the diameters is 4:1, the ratio of the velocities is 16:1 and, neglecting u_1^2 can introduce an error in the term $1 - \beta^4$ no greater than one part in 256, with approximately one-half as great an error in $\sqrt{1 - \beta^4}$. Under these circumstances, Equation 9.9 is often reported simply as

$$u_2 = C_0\sqrt{2g_c\,\Delta h}.$$

The empirical coefficient C_0 is dependent upon the ratio of diameters, the diameter of the pipe (due to the decrease in the contribution of roughness to the frictional losses as the pipe-size increases) and to the placement of the taps (particularly the downstream tap with relation to the position of the *vena contracta*). For taps with the downstream position at the *vena contracta* and the upstream position one pipe diameter above the orifice, the orifice coefficient may be taken to be 0.61. This value will vary with the factors noted above. Unless very low flow rates are encountered,* the variation of this constant will not exceed $\pm 1\%$. For detailed information on more exact values of C_0 the reader is referred to the *Chemical Engineers' Handbook* and the sources there noted.

Since the economics of any piece of equipment are important to the engineer, the permanent pressure (or head) lost in a flow instrument is important. This loss represents a power requirement solely due to the presence of the orifice in the line. The permanent loss is represented by the upper diagram in Figure 9.5; the line tracing the percentage of permanent pressure loss eventually levels off at a value of about 60% at some appreciable distance downstream from the plate. In the case represented, the permanent loss is then 40% of the temporary loss in pressure induced by the orifice. The temporary loss refers to the pressure reduction measured by the manometer. The permanent loss is reported as a fraction of the manometer reading. It is highly dependent upon the ratio of diameters as indicated in Figure 9.6.

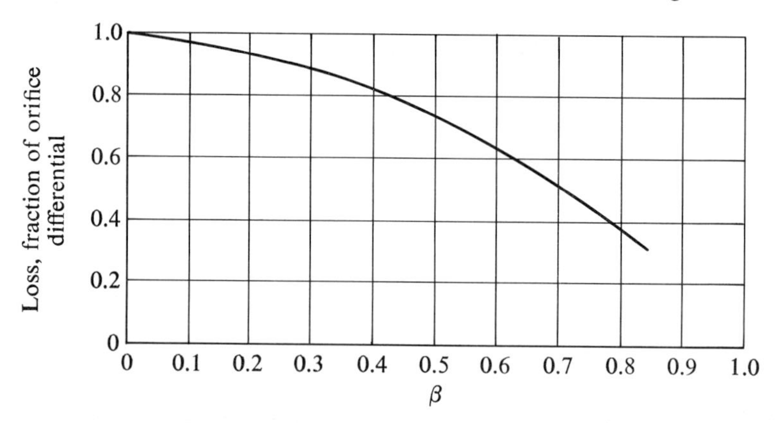

FIGURE 9.6 Overall pressure loss in orifice meters. [From *Fluid Meters: Their Theory and Applications*, 4th ed. (New York: American Society of Mechanical Engineers, 1937), p. 35. Reprinted with permission.]

The Venturi Meter

The Venturi meter operates on the same principle as the orifice; the difference between the two is in the manner in which the restriction is formed.

• • • • •

* If flow becomes sufficiently slow, the form of the flow pattern in a pipe changes. At these very low flows, all particles move along the pipe in a fixed position with relation to the pipe axis. As might be expected, this type of flow is known as "streamline flow." (It is also known as "viscous flow.") At high velocities the flow pattern is disturbed, the motion becomes random, and the advance of one particular element cannot be predicted. This type of flow is termed "turbulent flow." It is in the latter region that the coefficient, C_0, has values closely approaching 0.61.

The Venturi is an instrument of high initial cost and low operating cost, these characteristics being exactly the reverse of those of the orifice. In the Venturi there are three carefully machined parts illustrated in Figure 9.7: first, an approach section with precisely specified taper decreasing the diameter to the throat; second, a throat of constant cross section *where the downstream tap is located*; third, another precisely tapered section enlarging the cross section gradually to the full pipe size.

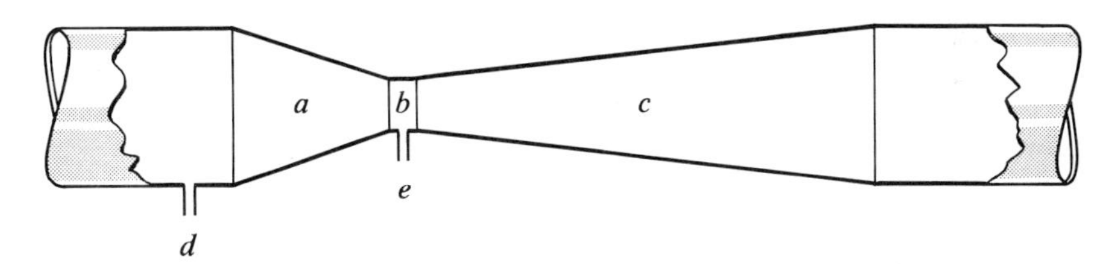

FIGURE 9.7 Diagram of Venturi-type flow meter: a. converging, upstream section; b. throat; c. diverging, downstream section; d. upstream, high pressure, tap; e. downstream tap.

The taper in the first section reduces the magnitude of the radial component of flow in comparison to that of the axial component, preventing appreciable further contraction in the throat (the minimum cross section). Therefore, there is no great disparity between the maximum calculated velocity and the point of minimum pressure. Granted that this objective is attained we can conclude that a discharge coefficient approximately equal to unity would be expected with such an instrument. This conclusion is entirely correct, an average coefficient for well-constructed Venturis being 0.98.

The tapered expansion section is intended to form boundaries within which the fluid pressure can be recaptured by gradual reduction of velocity without attendant, useless, power-consuming eddies. Again granted the attainment of the objective, we should conclude that the power recovery would be 100%. This conclusion is *nearly* correct in a carefully constructed instrument, actual losses of head usually being in the neighborhood of 10%, but occasionally as low as 2% in large installations. Equation 9.9 applies, with C_V substituted for C_0. (Note: C_V here is a Venturi coefficient not to be confused with heat capacity at constant volume; thus the capital V.)

Nozzles

Nozzles have the general characteristic of a Venturi on the upstream side and of the orifice on the downstream side as illustrated in Figure 9.8. From the previous analysis of the orifice and the Venturi we can conclude that the nozzle coefficient should closely approximate unity and the permanent loss of

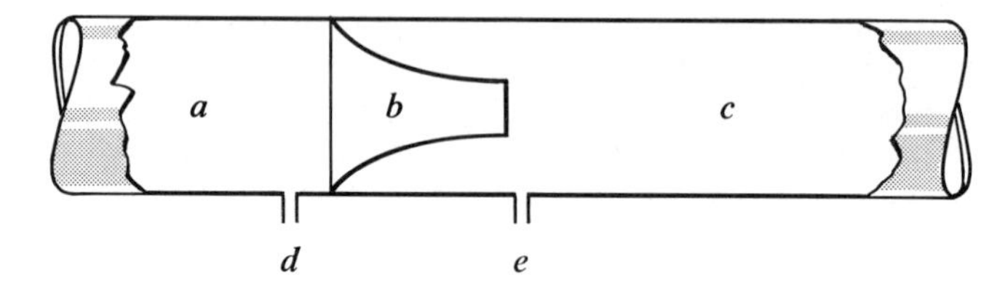

FIGURE 9.8 Diagram of nozzle-type flow meter: a. converging, upstream section; b. nozzle; c. pressure recovery section; d. upstream, high pressure, tap; e. downstream tap.

power over the instrument should be similar to that of the orifice. These conclusions are correct. Actual nozzle coefficients, C_N, vary from 0.98 to 1.00 depending upon the form of the converging section. Those with a standard conical converging section similar to that illustrated for the Venturi have values of C_N similar to the C_V reported for the Venturi, 0.98. Those with bell-shaped converging sections similar to that in Figure 9.8 can have coefficients as high as unity.

Example 9.5

▶ STATEMENT: A 3-in. standard pipe (inside diameter = 3.068 in.) is carrying a solution of alcohol and water, 12% alcohol by weight, at a rate of 40 gal per min, at 25 °C. A flow-measuring instrument is to be inserted in the line. The diameter of the constricted section is to be 1.00 in.

Calculate the reading, in inches, that will be obtained on an ordinary differential manometer located in a room at 25 °C, with mercury as the heavier fluid when the instrument is:

a. an orifice.
b. a Venturi.
c. a nozzle.

Also calculate, for each instrument, the horsepower requirement.

▶ SOLUTION: The velocities in the pipe and in the restricted sections will be identical in all instruments.

$$u_2 = \frac{40 \text{ gal}}{\text{min}} \times \frac{\text{ft}^3}{7.48 \text{ gal}} \times \frac{(4)(144)}{(\pi)(1.0)^2 \text{ ft}^2} \times \frac{\text{min}}{60 \text{ sec}} = 16.32 \text{ ft/sec},$$

$$\beta = \frac{1.00}{3.068} = 0.326,$$

$$\beta^4 = 0.0113; 1 - \beta^4 = 0.9887; \sqrt{1 - \beta^4} = 0.994.$$

Then for any one of the instruments, with a proper subscript on C_i, using Equation 9.9,

$$16.32 = \frac{C_i}{(0.994)}\sqrt{64.34\,\Delta h}; \qquad i = o,\ V,\ \text{or}\ N,$$

$$\sqrt{\Delta h} = \frac{2.02}{C_i},$$

$$\Delta h = \frac{4.08}{C_i^2},$$

in feet of the alcohol solution flowing.

In general, for the ordinary differential manometer (see Equation S2.1), the reading, R_m, in feet, will be

$$R_m = \frac{\rho_1 \Delta h}{\rho_2 - \rho_1}.$$

In view of the fact that densities enter as a ratio, specific gravities of the fluids in question, referred to water at the same temperature, will be substituted:

$$\begin{aligned} \text{Sp gr, 12\% alcohol} &= 0.9664^{25/4}; \\ \text{Sp gr, mercury} &= 13.53^{25/4}. \\ \hline \text{Difference in sp gr} &= 12.56. \end{aligned}$$

Then the reading will be

$$R_i = \left(\frac{4.08}{C_i^2}\right)(12)\left(\frac{0.9664}{12.56}\right) = \frac{3.77}{C_i^2}\ \text{in. of mercury.}$$

For the Venturi and for the nozzle, $C_i = 0.98$, $C_i^2 = 0.96$, and $R_V = R_N = 3.94$ in. of mercury. For the orifice, $C_0 = 0.61$, $C_0^2 = 0.372$, and $R_0 = 10.15$ in. of mercury.

Losses: We shall assume that 10% of the drop in head across the constriction is permanently lost in the case of the Venturi and that the losses in the orifice and nozzle are identical and obtainable from Figure 9.6. From this figure, with $\beta = 0.326$, this permanent loss is estimated to be 88% of the orifice or nozzle differential. The actual permanent losses in head are then estimated to be:

$$\text{Venturi:}\quad (0.10)\left(\frac{4.08}{0.98^2}\right) = 0.426\ \text{ft of alcohol;}$$

$$\text{Orifice and nozzle:}\quad (0.88)\left(\frac{4.08}{C_i^2}\right), \qquad i = o\ \text{or}\ N;$$

Orifice: 9.66 ft of alcohol;

Nozzle: 3.75 ft of alcohol.

If these heads are multiplied by the mass rate of flow and by g/g_c, which later we shall consider to have a value of unity, the losses will be obtained in units of foot-pounds force per unit of time. The time unit will be whatever was used in the mass flow rate.

The mass flow rate is the volume rate times the density:

$$\left(\frac{40}{7.48}\right)(0.9664)(62.4) = 328 \text{ lb/min.}$$

Horsepower requirements of the various instruments:

$$\text{Orifice:} \quad \frac{(9.66)(328)}{33{,}000} = 0.096 \text{ hp;}$$

$$\text{Nozzle:} \quad \frac{(3.75)(328)}{33{,}000} = 0.0372 \text{ hp;}$$

$$\text{Venturi:} \quad \frac{(0.426)(328)}{33{,}000} = 0.0042 \text{ hp.}$$

For power at 1¢ per kilowatt-hour and for a year of 300 24-hr operating days, the power costs for the orifice amount to $5.32 per yr. With fixed charges (i.e., interest, taxes, maintenance, depreciation, etc.) figured at 25% of the equipment cost, and an orifice cost of $20.00, installed, fixed charges per year are $5.00. The total yearly cost of the orifice is $10.32. To decide whether to use an orifice or a Venturi, calculate the allowable cost of manufacturing and constructing the specified Venturi. The power costs per year are only 23¢. This leaves $10.32 − 0.23 = $10.09 for fixed charges. At the same fixed charge rate (25%), the Venturi would be competitive with the orifice at $10.09/0.25 = $40.36 and would be preferred at any cost < $40.26. It is inconceivable that the Venturi could be manufactured and installed so cheaply. The orifice would have to be chosen for economic operation. Venturis become more favorable as flow rate, and consequently permanent loss in the orifice, increases. Comparison of the orifice with the nozzle can be made in similar fashion.

9.5 Compressible Fluids

All applications of energy balances have, to this point, been limited to liquids. Consequently, the densities have been constant and integration of the term $v\,dP$ has caused no difficulty. Let us examine the problems encountered in dealing with gases (compressible) and attempt some solutions even if rather restrictive assumptions are needed. Consider first the case of flow meters. Instead of using Equation 9.4, take the form of the balance that involves enthalpy change in differential form, Equation 9.3,

$$dH = \frac{u\,du}{g_c}. \tag{9.10}$$

In writing Equation 9.3 in this form we have assumed that $dZ = 0$, $d'W = 0$ and $d'Q = 0$ as in the earlier discussion of flow meters. It is worth noting that dZ is not necessarily zero since these flow meters do not have to be placed in horizontal lines; if the line is vertical or sloped, the elimination of dZ is not possible and the equations derived to this point, and those to be derived in this section, will not hold without modification. This modification will be left to the reader since it involves a complication but does not involve any new theoretical problems.

Integrating Equation 9.10:

$$2g_c(H_2 - H_1) = u_2^2 - u_1^2.$$

Be very careful not to confuse H and h at this point. The lower case symbol h is used to refer to head while the upper case symbol H refers to enthalpy.

We look closely at the elimination of one of the velocity terms. Previously we wrote $u_1 = (A_{c2}/A_{c1})u_2$. This expression was correct only because of the constant density of an incompressible fluid. Such a relationship can be true only if it represents a material balance over the two positions of the balance boundary; i.e., it must express the statement that mass rate in equals mass rate out. Since mass rate equals volume rate times density

$$u_1 A_{c1} \rho_1 = u_2 A_{c2} \rho_2. \tag{9.11}$$

This relationship is commonly known as the *continuity equation*. When dealing with compressible fluids, ρ_1 and ρ_2 are not equal and must be retained in the continuity equation. It is convenient to replace the ρ_i by specific volumes, v_i. Then

$$u_1^2 = u_2^2 \left(\frac{A_{c2}v_1}{A_{c1}v_2}\right)^2 = u_2^2\, \beta^4 \left(\frac{v_1}{v_2}\right)^2.$$

The energy balance over the flow meter, with the empirical constant C_i included, becomes

$$u_2 = C_i \left\{\frac{2g_c(H_1 - H_2)}{[1 - \beta^4(v_1/v_2)^2]}\right\}^{1/2}, \qquad i = o,\ V, \text{ or } N. \tag{9.12}$$

Equation 9.12 is not at all convenient since it contains enthalpy terms, not directly measurable, as well as the ratio of specific volumes which is dependent upon pressures and temperatures. We can eliminate the enthalpy difference by substitution of

$$H_1 - H_2 = \frac{C_p(T_1 - T_2)}{M} = \frac{C_pT_1[1 - (T_2/T_1)]}{M},$$

where the molecular weight has been introduced so as to maintain the heat capacity on a molal basis. If we further assume that the ideal gas law applies, we may substitute

$$\frac{P_2v_2}{P_1v_1} = \frac{T_2}{T_1}.$$

Substituting these expressions into Equation 9.12, we have,

$$u_2 = C_i \left\{ \frac{2g_c C_p T_1 \left[1 - \left(\frac{P_2 v_2}{P_1 v_1}\right)\right]}{M\left[1 - \beta^4 \left(\frac{v_1}{v_2}\right)^2\right]} \right\}^{1/2}. \tag{9.12a}$$

In order to eliminate the specific volume terms, we must remember that the expansion followed an adiabatic path. The pressure-volume relationships are dependent upon this path (i.e., the work of self-expansion is a path-dependent process). For ideal gases, these relationships are frequently represented in general by the statement

$$Pv^{\alpha} = \text{a constant},$$

where α varies with the process. If the expansion is isothermal, α equals unity and the usual ideal gas law applies. If the process is adiabatic, as in the present case, $\alpha = C_p/C_v = \kappa$, where the symbol, κ, is assigned to the ratio of the molal heat capacities. If frictional losses are appreciable, $C_p/C_v < \kappa$.

In the present instance our assumption of adiabatic conditions is equivalent to implying reversibility. The value of α is then established for any known gas flowing through the measuring instrument, and

$$\frac{v_1}{v_2} = \left(\frac{P_2}{P_1}\right)^{1/\kappa}.$$

Substituting this expression into both numerator and denominator of Equation 9.12a produces, after a little manipulation,

$$u_2 = C_i \left\{ \frac{2g_c C_p T_1 \left[1 - \left(\frac{P_2}{P_1}\right)^{(\kappa - 1)/\kappa}\right]}{M\left[1 - \beta^4 \left(\frac{P_2}{P_1}\right)^{2/\kappa}\right]} \right\}^{1/2}. \tag{9.13}$$

The heat capacity can also be eliminated in favor of the constants, κ and R. Continuing to assume the ideal gas law,

$$C_v = C_p - R.$$

Substitution of this expression into the relationship,

$$\frac{C_p}{C_v} = \kappa,$$

results in

$$C_p = \frac{R\kappa}{\kappa - 1}.$$

Finally substituting for C_p in Equation 9.13 we obtain

$$u_2 = C_i \left[\frac{2g_c \dfrac{\kappa}{\kappa - 1} \dfrac{RT_1}{M} [1 - (P_2/P_1)^{(\kappa - 1)/\kappa}]}{1 - \beta^4 (P_2/P_1)^{2/\kappa}} \right]^{1/2} \tag{9.14}$$

Note that a reading of ΔP or Δh is no longer sufficient to establish the velocity through the orifice. Now both the temperature at inlet and one of the absolute pressures are required. Ratios of pressures are more fundamental to solution of Equation 9.14 than are differences in pressures. The necessary pressure data may be obtained by adding an open-ended manometer to the differential manometer as shown in Figure 9.9 or by measuring the pressures individually at the two taps.

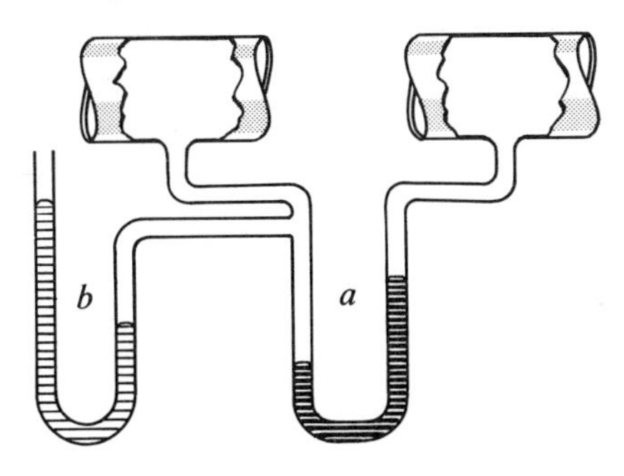

FIGURE 9.9 Differential manometer a. with total pressure manometer b. on high pressure side.

The reader may wonder at this point why manometers should be used in such an instrument when one is primarily interested in the two absolute pressures. The reason lies in the precision of the readings. If the pressure drop is small, as it will frequently be for instruments measuring the flow of gases, the ratio P_2/P_1 will be nearly unity. After the power is taken and the result is subtracted from unity, the difference is extremely small. Any small errors in either P_1 or P_2 will be greatly amplified in the numerator of Equation 9.14 as a result of the subtraction of numbers of comparable magnitude. In contrast, if a manometer is used to measure pressure drop, this type of error can be minimized. By adjustment of manometer fluids or by proper design (see Supplement to Chapter 2) small pressure drops can be measured quite precisely. Then regardless of the precision in measuring directly one of the pressures, the ratio will be quite precise.

As an example of the errors which might be introduced by direct reading when pressure differences are small, take the case of an upstream pressure of 15.3 psig and a true difference of 10 in. of water. This difference is equivalent to 0.352 psig and the downstream pressure is truly 14.948 psig. Then

$(P_2/P_1)^{(\kappa-1/\kappa)}$ is 0.9936 when air is the fluid and the factor $1 - (P_2/P_1)^{(\kappa-1)/\kappa}$ is 0.0064. If the downstream pressure had been read by a gauge, it could hardly be read better than 15.0, which would have made the factor 0.0056, in error by almost 10%. If, as seems likely, one of the gauges had been read differently from the true reading by as little as 0.1 psig the error would be in the neighborhood of 50%. But, if the upstream pressure had been erroneously read as 15.2 psig and the differences were still 10 in. of water, the ratio of the pressures is not changed by 0.1%.

Compressible Fluids in Straight Pipes

Consider straight horizontal pipes of constant cross section. As a result of pressure losses down the pipe due to friction, the gas will expand and the velocity will increase. The energy balance, written in differential form, will take the form:

$$\frac{u\,du}{g} + v\,dP + d'F = 0. \tag{9.15}$$

In Equation 9.15 all terms are expressed as heads, in feet of fluid flowing in the line, with v in units of volume per unit of *mass* and P in units of pounds-mass per square foot. Because of the dependence of velocity on pressure and of the latter on the extent of the frictional losses per unit of length, the form of Equation 9.15 is inconvenient. Introduce a new quantity, the mass velocity, designated G. The quantity is simply the mass rate of flow divided by the area. Thus,

$$G = \frac{w}{A}.$$

Now the linear velocity, u, will be the product of the mass velocity in units of mass/(area)(time) and the specific volume, v:

$$u = Gv.$$

Then,

$$du = G\,dv,$$

since G must be a constant by the Continuity Equation. Substituting for u and du in the energy balance,

$$\frac{G^2 v\,dv}{g} + v\,dP + d'F = 0.$$

Dividing by v^2,

$$\frac{G^2}{g}\frac{dv}{v} + \frac{dP}{v} + \frac{d'F}{v^2} = 0. \tag{9.16}$$

The first term may now be directly integrated, since G^2/g is a constant. The second term may be integrated if the perfect gas law is assumed. Then

$v = RT/PM$ and the term becomes $(M/RT)P\,dP$. This term can be integrated if the process is isothermal or if T is used as the average temperature when conditions of the process are such that the temperature varies only slightly. In the latter case the average temperature, $\overline{T}$, is used in place of T.

In order to deal with the third term, we need information regarding frictional losses. Without resort to lengthy discussion of the fluid dynamics involved, let us make use of the *Fanning equation* (see the *Chemical Engineers' Handbook*),

$$d'F = \frac{2fu^2}{gd_p}\,dx. \tag{9.17}$$

In Equation 9.17, dx is the differential distance along the pipe, d_p is the internal diameter of the pipe and f is the Fanning friction factor, a dimensionless constant for any given mass rate of flow. Values of f are obtained from empirical correlations of this dimensionless friction factor as an arbitrary function of the dimensionless *Reynolds' number*. The Reynolds' number may be written in several ways, two of the most useful being $d_p u\rho/\mu$ and $4w/\pi\mu\,d_p$. The latter is readily seen to be a constant for a given mass rate of flow. The relationship of f to Reynolds' number is represented graphically in Figure 9.10.

If Gv is substituted for u in Equation 9.17 and the result is substituted for $d'F$ in the energy balance, the third term of Equation 9.17 becomes $(2fG^2/g\,d_p)\,dx$. All the factors in parentheses are constant so that this term can be easily integrated. The final form of the energy balance may then be written

$$\frac{G^2}{g}\ln\frac{v_2}{v_1} + \left(\frac{M}{2R\overline{T}}\right)(P_1^2 - P_2^2) + \frac{2fG^2L}{gd_p} = 0.^* \tag{9.18}$$

A more convenient form can be obtained by multiplying through by g/G^2 and by converting the first term to pressure units by substituting the simple gas law. Then

$$-\ln\left(\frac{P_2}{P_1}\right) + \frac{gMP_1^2}{2G^2R\overline{T}}\left[\left(\frac{P_2}{P_1}\right)^2 - 1\right] + \frac{2fL}{d_p} = 0. \tag{9.18a}$$

The only unknown in this form of the equation is P_2. An iterative solution is suggested due to the transcendental form of the equation.

The problem most likely to be encountered is the calculation of the pressure drop to be expected for a specific length of pipe and for given mass flow rate, as described above. Another question could be asked, the length of pipe that

• • • • •

* The reader should assure himself that all the terms of this equation have consistent units, each with dimensions of lb_m/ft^5.

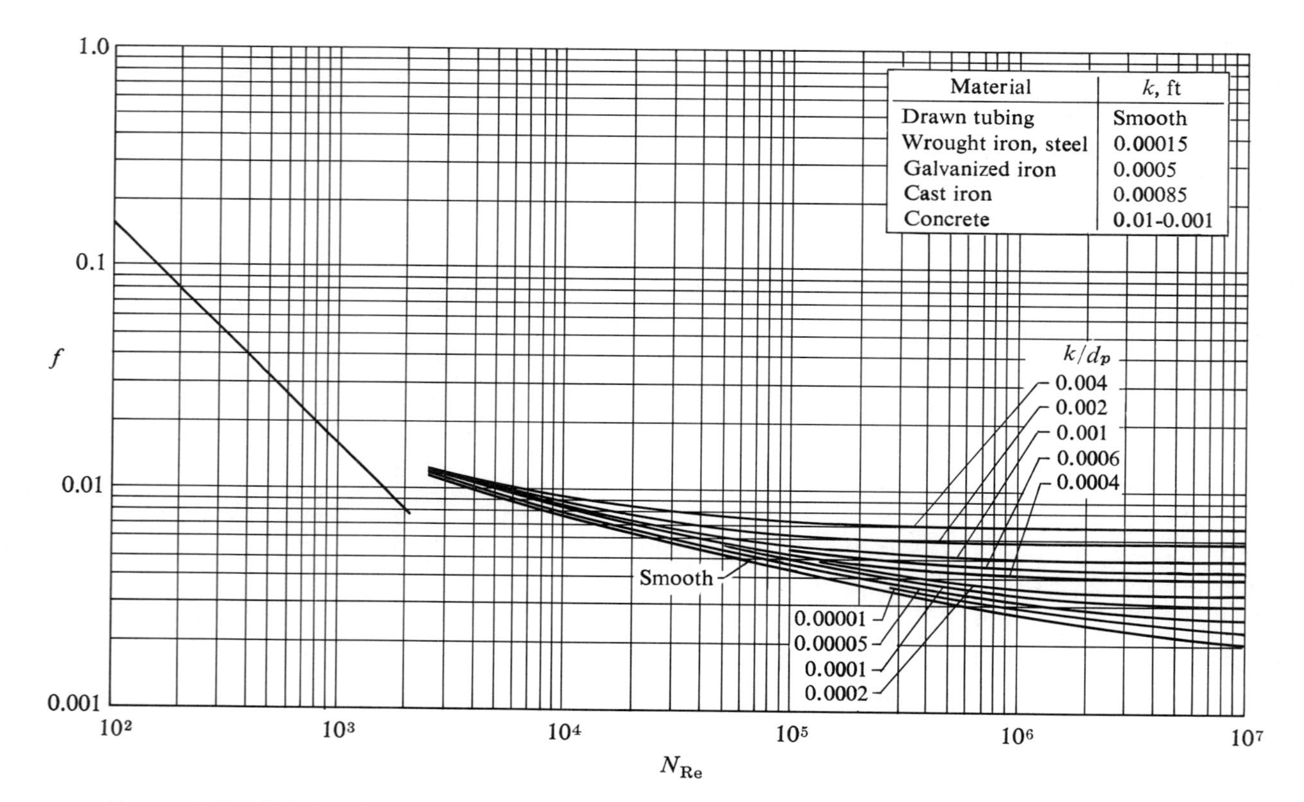

FIGURE 9.10 Friction factor chart, N_{Re} vs. f. [From L. F. Moody, *Trans. A.S.M.E.*, *66* (1944), 671. Reprinted by permission.]

would result in a specific pressure drop for a given mass rate of flow, G. Then L is the only unknown and may be calculated directly.

Nonhorizontal Pipes

In this instance, a potential energy term dZ, will also appear in the energy balance. If the angle that x linear feet of straight pipe makes with the horizontal is ζ, we can write

$$x \sin \zeta = Z$$

and

$$\sin \zeta \, dx = dZ.$$

After dividing by v^2, or multiplying by ρ^2 which is the same thing, the term which is added to our previous differential equation, 9.16, is

$$\sin \zeta \rho^2 \, dx.$$

Unless an average value of ρ is used, this term is difficult to handle. When the average density, $\bar{\rho}$, is used, the integration gives

$$\sin \zeta \bar{\rho}^2 L.$$

When the total change in gas density is small, this procedure is reasonable. When it is large, the problem could be handled in exactly the same fashion by considering, successively, small segments of length.

Example 9.6

▶ STATEMENT: Air at 75 °F enters 95 ft of straight horizontal standard 1-in. wrought iron pipe (1.049 in. inside diameter). The process is isothermal. The inlet pressure is 20 psig. The mass rate of flow is 100 lb per hr. Calculate the outlet pressure.

▶ SOLUTION: Equation 9.18a is applicable to this case. G, v_1, P_1, $\overline{T}$, and f can be calculated from the data

$$G = \frac{w}{A} = 100 \frac{\text{lb}}{\text{hr}} \times \frac{1}{3600} \frac{\text{hr}}{\text{sec}} \times \frac{4}{\pi} \times \frac{144}{(1.049)^2} \frac{1}{\text{ft}^2},$$

$$= 4.63 \text{ lb/(ft)}^2\text{(sec)}.$$

$$P_1 = 20.00 + 14.7 = 34.7 \text{ psia}.$$

$$\overline{T} = 75 + 460 = 535 \text{ °R}.$$

$$v_1 = \left(\frac{359}{29}\right)\left(\frac{535}{492}\right)\left(\frac{14.7}{34.7}\right) = 5.69 \text{ ft}^3\text{/lb}.$$

The Reynolds' number is most conveniently expressed, for the present purposes, as $4w/\pi\mu d_p$. The viscosity of air is required in this number. It is an intrinsic property, which is listed in standard textbooks. The value at 1 atm and 75 °F is 0.0175 centipoises (cP); the viscosity is nearly independent of pressure. The poise (P) is a unit in the metric system, expressed as grams per centimeter-second. The reader may confirm that 0.0175 cP is converted to units of pounds per (hour)(foot) by multiplying centipoises by 2.42.

$$\mu = 0.0175 \times 2.42 = 0.0424 \text{ lb/(hr)(ft)}.$$

Therefore,

$$\frac{4w}{\pi\mu d_p} = \frac{(4)(100)}{(3.1416)(0.0424)} \frac{1.049}{12},$$
$$= 3.44 \times 10^4.$$

The value of f may be obtained from Figure 9.10. Since k is in feet and k/d_p is dimensionless, k/d_p is 0.0018. Therefore, $f = 0.007$. The gas law constant, R, should be expressed in units of ft-lb$_f$ per (lb-mole)(°R). Since a pound-mole of air occupies 359 cu ft at standard conditions,

$$R = \frac{Pv}{T} = \frac{(14.7)(144)(359)}{492} = 1545 \text{ ft-lb}_f\text{/(lb-mole)(°R)}.$$

Substituting these quantities into Equation 9.18a,

$$\ln\frac{P_1}{P_2} + \frac{(32.2)(29)}{2(1545)(535)} \frac{(34.7 \times 144)^2}{(4.63)^2} \left[\left(\frac{P_2}{P_1}\right)^2 - 1\right] + \frac{2(0.007)(95)(12)}{1.049} = 0;$$

$$\ln\left(\frac{P_1}{P_2}\right) - 6.59 \times 10^2 \left[1 - \left(\frac{P_2}{P_1}\right)^2\right] + 15.22 = 0.$$

If the pressure drop is small, the first term of this equation will be nearly zero and may be ignored as a first approximation. It can readily be seen that this is the case in the present example, since $\frac{15.22}{659} = 0.0231$. Consequently, $(P_2/P_1)^2 = 0.9816$ and, to a first approximation $P_2/P_1 = 0.991$. This answer may be sufficiently accurate. If not, other values of the pressure ratio between 0.99 and 1.00 may be substituted into the equation, writing the right-hand side as ε instead of zero. A plot of ε *vs.* the trial pressure ratio will provide a more precise solution, the desired value of P_2/P_1 being found at the point where $\varepsilon = 0$. For the present purposes we shall assume that the value of 0.991 is satisfactory, since the friction plot cannot be expected to be good to better than 1%.

We can calculate the frictional loss by integrating Equation 9.15,* now that P_2 is known:

$$\int_1^2 \frac{u\,du}{g} + \int_1^2 v'\,dP = -\int_1^2 d'F;$$

$$\frac{u_2^2 - u_1^2}{2g} + \frac{RT}{M}\int_1^2 \frac{dP}{P} = -{}_1F_2;$$

$$\frac{u_2^2 - u_1^2}{2g} + \frac{RT}{M}\ln\frac{P_2}{P_1} = -{}_1F_2.$$

We now see that u_2 and u_1 are related through the continuity equation:

$$u_1 A_{c1}\rho_1 = u_2 A_{c2}\rho_2, \qquad \text{where} \qquad A_{c1} = A_{c2}.$$

Then,

$$u_2 = \frac{u_1 v_2}{v_1}.$$

Since

$$u_1 = 4.63\,\frac{\text{lb}}{(\text{ft})^2(\text{sec})} \times 5.69\,\frac{\text{ft}^3}{\text{lb}} = 26.4 \text{ ft/sec},$$

and

$$\frac{v_2}{v_1} = \frac{1}{0.991};$$

$$u_2 = \frac{26.4}{0.985} = 26.7 \text{ ft/sec};$$

$$\therefore \frac{u_2^2 - u_1^2}{2g} = \frac{(26.7)^2 - (26.4)^2}{62.37} = \frac{718 - 698}{62.37} = 0.204 \text{ ft-lb},$$

which will be seen to be negligible in comparison to the work of self-expansion. The self-expansion term shows that

$$\frac{RT}{M}\ln\frac{P_2}{P_1} = \frac{(1545)(535)}{29}\,2.303\log_{10} 0.985,$$

$$= -426 \text{ ft-lb}.$$

$${}_1F_2 = 426 \text{ ft-lb},$$

$$\equiv 0.55 \text{ Btu}.$$

• • • • •

* It would seem that ${}_1F_2$ might be more easily calculated from the definition,

$$d'F = \frac{2fu^2}{gd_p}\,dx,$$

or its equivalent

$$d'F = \frac{2fG(v)^2}{gd_p}\,dx.$$

In either of these expressions we find a variable which, inconveniently, is a function of length, u in the first expression and v in the second. An average value of v could be used in the second or calculation can be made by returning directly to the energy balance as we do here.

Example 9.7

▶ STATEMENT: Carbon dioxide gas at 100 °F is flowing through a standard 2-in. line (inside diameter = 2.06 in.). The inlet pressure is 20 psig. A nozzle, with throat diameter of 0.5 in., is placed in the line. The pressure drop across the nozzle, as measured on a differential manometer filled with mercury, is 10 in. of mercury.

Calculate the mass rate of flow of carbon dioxide.

▶ SOLUTION: Equation 9.14 is applicable if the gas is assumed to be ideal at all points in the nozzle. Values of β and P_2 may be calculated from the data. A value of κ requires C_p and C_v data.

$$\beta = \frac{0.5}{2.06} = 0.243, \qquad \beta^4 = 3.46 \times 10^{-3}.$$

Taking $C_p = 9.2$ and $C_v = 7.2$ as average values for CO_2 (*Chemical Engineers' Handbook*),

$$\kappa = \frac{C_p}{C_v} = 1.28;$$

$$\frac{\kappa}{\kappa - 1} = 4.57; \qquad \frac{\kappa - 1}{\kappa} = 0.219; \qquad \frac{2}{\kappa} = 1.563;$$

$$P_2 = P_1 - \Delta P.$$

ΔP may be calculated from the manometer reading, $\Delta P = R(\rho_2 - \rho_1)$, if the density of gas in both arms of the manometer is the same. It seems unlikely that the gas density can be appreciable in relation to the (13.53)(62.4) = 845 lb per cu ft of mercury. Certainly neglecting ρ_1 could not cause an error of more than 1 in 8000. It will be then treated as negligible in this calculation.

$$\Delta P = \left(\frac{g}{g_c}\right)\left(\frac{10}{12}\right)\frac{845}{144} = 4.89 \text{ lb}_f/\text{sq in.}$$

$$P_2 = 34.7 - 4.89 = 29.8 \text{ lb}_f/\text{sq in.}$$

Calculate the ratio P_2/P_1 and the various powers required:

$$\frac{P_2}{P_1} = \frac{29.8}{34.7} = 0.859;$$

$$\left(\frac{P_2}{P_1}\right)^{0.219} = 0.9673; \qquad 1 - \left(\frac{P_2}{P_1}\right)^{0.219} = 0.0327;$$

$$\left(\frac{P_2}{P_1}\right)^{1.563} = 0.789; \qquad \beta^4\left(\frac{P_2}{P_1}\right)^{1.563} = 2.74 \times 10^{-3}.$$

Therefore $[1 - \beta^4(P_2/P_1)^{2/\kappa}]$, is substantially equal to unity. Equation 9.14 is thereby simplified to

$$u_2 = C_N \left[2g_c \frac{\kappa}{\kappa - 1} \frac{RT_1}{M} \left(1 - \frac{P_2}{P_1}\right)^{(\kappa-1)/\kappa}\right]^{1/2},$$

$$= 0.98 \left[(64.34)(4.57)\left(\frac{1545}{44}\right)(560)(0.0327)\right]^{1/2},$$

$$= 426 \text{ ft/sec.}$$

Calculation of the mass rate of flow requires the density at the nozzle or the velocity in the main pipe. Choosing the former, and using the adiabatic relationship, Pv^κ = constant, to obtain the throat temperature,

$$T_2 = T_1 \left(\frac{P_2}{P_1}\right)^{(\kappa-1)/\kappa} = (560)(0.9673) = 541\ °\text{K},$$

$$= 81\ °\text{F}.$$

Then,

$$v_2 = \frac{359}{44}\left(\frac{541}{492}\right)\left(\frac{14.7}{29.8}\right) = 4.42 \text{ ft}^3/\text{lb}.$$

Now the mass velocity G, at the throat, is known, for

$$G_2 = \frac{u_2}{v_2} = \left(\frac{426 \text{ ft}}{\text{sec}}\right)\left(\frac{\text{lb}}{4.42 \text{ ft}^3}\right),$$

$$= 96.4 \text{ lb/(ft)}^2\text{(sec)}.$$

Mass velocity is *not* constant throughout the nozzle as in Example 9.6, for the cross section is no longer constant. The mass rate of flow is

$$w = G_2 A_2,$$

$$= (96.4)\left(\frac{\pi}{4}\right)\left(\frac{1}{2}\right)^2 \frac{1}{144} \times 3600 = 473.0 \text{ lb/hr.}$$

9.6 Physical and Chemical Processes

In this group of processes we exclude mechanical work. Many such situations exist. All physical separation processes and most chemical reactors would fall in this category. This statement does *not* imply that mechanical work is never involved in these processes. It merely means that there are many useful energy balance positions around parts of these processes that do not include pumps and fans. For example, a distillation unit will have pumps to feed the column, and others to transport the distillate and bottoms streams to other parts of the plant. The energy requirements of the column itself can be calculated with the balance boundary *inside* these mechanical elements; mechanical energy demands will not then enter such an energy balance.

On this basis of no external work added or removed, the energy balance, Equation 9.3, can be written in the form

$$dH + dZ + \frac{u\,du}{g} = d'Q, \qquad \text{in consistent units.}$$

When any appreciable heat loads exist (i.e., Q), the potential and kinetic energy tends to be relatively unimportant since a very large factor (778 ft-lb$_f$/Btu) relates mechanical energy to heat energy. An exception might occur if the difference in heights between entering and leaving streams is enormous and/or if the velocities are tremendous. Assuming then that there are many instances in which these terms are truly negligible, Equation 9.3 reduces to

$$dH = d'Q$$

or

$$H_2 - H_1 = {}_1Q_2. \tag{9.19}$$

In this type of situation, the energy balance devolves into a balance of terms expressed entirely in terms of heat units. Situations where Equation 9.19 is applicable are so numerous that special consideration will be given to processes of this type; balances of this type will be called *heat balances.*

Heat Balances

The heat balance is sometimes erroneously referred to as the *enthalpy balance*. The inclusion of the Q term makes such a terminology fundamentally incorrect. An enthalpy balance must, by definition, involve only enthalpy terms, thereby excluding any exchange of transient heat in a physical process and also excluding a process in which heat of reaction is involved.

Physical processes such as pure gas throttling do exist under circumstances such that $\Delta H = 0$. Commonly encountered situations that embrace the majority of important industrial cases will, nevertheless, involve transient heat quantities and consequently must be classed as *heat balances.* These processes and situations will be discussed in this section.

The heat balance states that a positive difference between the enthalpy of the products and the enthalpy of the feed streams is equal to the heat added to the process. Two points should be emphasized. First, the enthalpy change referred to is the *sum* of all output enthalpies minus the *sum* of all input enthalpies. Second, the source of the heat added is not specified and the term, ${}_1Q_2$, must be taken to mean a *sum* of heats added from all possible sources. The predominant sources are heaters (frequently steam but not exclusively so), coolers extracting rather than adding heat, chemical reactions, and enthalpy changes accompanying physical processes such as dilution, absorption, adsorption, wetting, etc.

In the establishment of heat balances, especial pains should be taken in specifying the basis. Not only is the mass, or time, important as in the straight material balance, but also the reference state and the base temperature, for enthalpy terms are relative to both the state of aggregation and the temperature, as has been noted (see Chapters 6, 7, and 8). An inconsistency between the datum conditions for the various streams would obviously invalidate the calculations.

It is necessary to include in the datum both the reference state of aggregation *and* the temperature since temperature alone would not be sufficiently specific, particularly in the case of a substance that might exist in either of two states of aggregation at the chosen temperature. This situation is encountered in steam tables where the reference temperature is 32 °F and 1 atm *and* the state of aggregation is liquid water rather than the equally possible solid state (i.e., ice). (See abbreviated steam tables in Appendix C.)

It is also extremely important when chemical reaction is involved that the heats of reaction be established at some datum temperature and that the state of aggregation of the reactants and products be established according to convenience, without regard to their actual state in reaction. For example, in the reaction

$$CO + H_2O = CO_2 + H_2,$$

the water *reacts* in the vapor state, but the heat of reaction may be calculated for water in the liquid phase. The datum has then arbitrarily been established as liquid water. An account will then have to be made in the heat balance of the discrepancy in the water states between the datum condition and the condition as charged by including the latent heat of vaporization as an input item. Very few difficulties will be encountered in heat balances if the state of aggregation of reactants and products are properly reconciled with the datum conditions. Proper choice of the datum conditions can, moreover, simplify the calculations tremendously.

Heat Balance in Physical Processes

The general situation for a heat balance, as defined above and including ${}_1Q_2$, is represented diagrammatically in Figure 9.11. One or more streams enter, represented in the diagram by streams A and B and not limited in number to two. Since they are feed streams to the system we may expect that both temperature and phase will frequently be specified by the conditions of the problem. Similarly, one or more streams leave, represented in the diagram by streams R and S, and again not limited in number to two. The amount of information regarding their temperatures and phases will depend upon the information to be calculated. Heat inputs and heat losses will be noted. The commonest loss is a radiation loss. If the system is perfectly

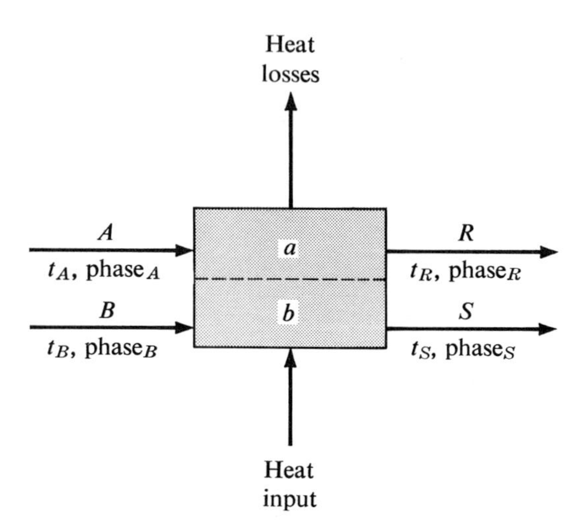

FIGURE 9.11 General diagram for heat balance in a physical process.

insulated these losses are zero and may be so considered in many practical cases even though perfection of lagging (i.e., the insulation) cannot be truly attained. This assumption merely states that the losses are negligible in relation to the other heat quantities involved in the process.

Two general types of problems exist. In one, complete information regarding the temperatures and phases of streams $A, B, \ldots, R, S, \ldots$, is given, the heat input is given and the heat losses are to be calculated. Alternately in the *same* type of problem, the heat losses are given (or assumed equal to zero) and the heat input is required. In the *other* type of problem, the temperatures and phases of the input streams, $A, B, \ldots$, are given, the heat input and losses are specified in some fashion, and the temperatures of the output streams are required, their phases being known.

If there is a partition within the system, as indicated by the dotted line in Figure 9.11, then the two "sides" of the system, marked *a* and *b* in the figure, can be treated as completely separate elements; balances may be performed on each element separately, and the energy absorbed on one side of the partition is evolved on the other side.

Without a physical separation all effluent streams may emerge from the same chamber, as with flash chambers, simple continuous stills, and evaporators, for example. These streams must, then, be at the same temperature and pressure (see similar remarks and inherent restrictions, Supplement to Chapter 4, p. 189). This restriction does not imply, however, that the compositions and phases of the effluent streams must also be identical. If such were the case, the physical separations performed by these units would be impossible. On the other hand, the lack of a complete physical partition is no guarantee that the effluents will leave at the same temperature and pressure. In units such as the absorption tower or the distillation column (Figures 4.4

and 4.1a, respectively), at least two streams leave the unit, but at different geographical locations within the unit, so that temperatures and pressures cannot be identical. The separation in this type of unit is effected by phase differences, with gases and vapors entering at the bottom and leaving at the top, or vice versa for liquids.

The following examples illustrate some of the situations cited.

Example 9.8

▶ STATEMENT: 100 lb of air per min, at 75 °F and 5 psig, enter a heater. The heating element is a steam coil to which 1 lb per min of steam, saturated at 5 psig, is charged. The steam is condensed in the coil and leaves through a steam trap at 190 °F, still at 5 psig. The heater is sufficiently well insulated so that an assumption of negligible radiation loss is justified.

Calculate the temperature of the air leaving the heater.

▶ SOLUTION: The system is shown schematically in Figure 9.12. This problem is of the second type listed above, with the outlet temperature to be calculated. It is also of the type where input streams are kept separate by

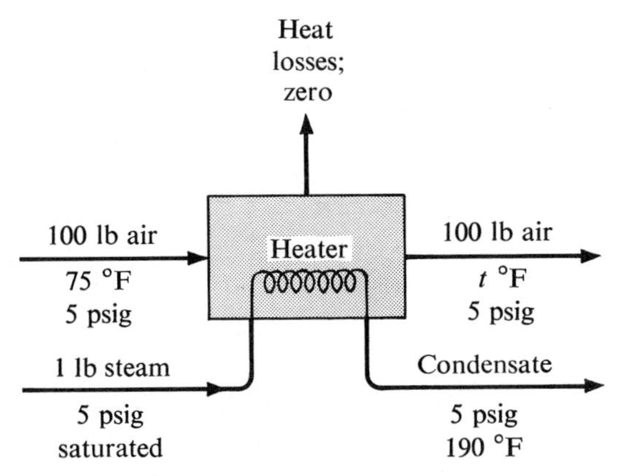

FIGURE 9.12 Element for heating air with stream condensing in a coil.

physical partitions. This latter condition allows a choice of two separate datum conditions, one for the air and one for the steam. Since the steam tables are the most convenient means of dealing with enthalpies of water in liquid and/or vapor form, there is no free *choice* of datum on the steam side, the 32 °F, liquid water condition being established by the tables. Certainly the retention of 32 °F as datum temperature for air would be foolish, for we can simplify the work considerably by choosing 75 °F (and, of course, the gas phase) and thereby eliminate enthalpy in the inlet air stream. We shall, therefore, handle the datum specification in this fashion, keeping the two sets of datum conditions separate.

BASIS The choice of 1 min will establish the calculations on the basis of inlet quantities.

Material Balances Heat balances are frequently preceded by material balances to establish quantities unspecified in the problem statement. While situations will be found where this consecutive attack will not work (see Chapter 10), no such difficulty is encountered in the present instance. In fact, it almost appears that material balances need not be mentioned since it is obvious that the 100 lb of air that entered will constitute the mass of one exit stream and the 1 lb of steam that entered must constitute the mass of the condensate stream. These conclusions come from material balances, no matter how simple. The first step in dealing with any problem involving a heat balance is, then, the establishment of material balances. We have done so here with obvious results. In more complex problems the material balances will be equally helpful and more necessary.

Heat Balance

Steam side: Heat in: 1156.0 Btu (steam tables).
Heat out: 157.95 Btu (steam tables).

Heat lost by steam: 998.0 Btu.

Air side: Heat in: zero due to choice of 75 °F as datum.

Heat out: $\frac{100}{29}\int_{75}^{t} C_p\,dt$, with C_p in molal units;

$\frac{100}{29}$ = number of moles of air fed.

Balance:

$$\int_{75}^{t} C_p\,dt = (0.29)(998.0) = 289.42. \tag{9.20}$$

There are now several choices open for the solution of this equation. Given C_p *vs.* t data we can integrate analytically if C_p is given as a function of t, or we can use mean heat capacities if these are available. Let us assume that C_p is actually independent of pressure over the very small range from 14.7 to 19.7 psia involved, an independence which is theoretically correct with ideal gases. Then choose the last method, that of the mean heat capacities.

Mean heat capacities are given in Figure 7.2, from 77 °F to t °F. Then, per mole of air, Equation 9.20 may be written in the form

$$_{77}C_{pm_t}(t - 77) - {}_{77}C_{pm75}(75 - 77) = 289.42. \tag{9.21}$$

$_{70}C_{pm75} = 7.0$ (From Figure 7.2.).

Then,

$$_{77}C_{pm_{75}}(75 - 77) = (7.0)(-2) = -14.0 \text{ Btu}.$$

Therefore, from Equation 9.21,

$$_{77}C_{pm_t}(t - 77) = 275.4. \tag{9.21a}$$

Since the value of the mean heat capacity will vary with the value of t, the solution of this equation requires an iterative approach. First, guess a value of t in the range of 77 to 227 °F (i.e., the final temperature cannot be below the inlet temperature nor above the steam temperature). Obtain the value of C_{pm} at this temperature. Then compare the calculated value of the left-hand side of Equation 9.21a with the known value of 275.4. If Equation 9.21a is written in the form

$$_{77}C_{pm_{t^{(i)}}}(t^{(i)} - 77) - 275.4 = \varepsilon,$$

where $t^{(i)}$ is the ith guess of the outlet temperature, the solution is obtained when $\varepsilon = 0$. Since it is unlikely that we shall hit the temperature exactly in a reasonable number of guesses, a plot of ε *vs.* $t^{(i)}$ is a useful means of interpolating.

In the specific problem at hand, first choice of t will be 50 °C (122 °F). From Figure 7.2, C_{pm} is found to be 7.0. Then $\varepsilon = 37.33$ Btu. If we choose 40 °C (104 °F) we find that we cannot read any real difference in the mean heat capacity. Thus for this particular problem we can write

$$7.0(t - 77) = 275.4.$$

The solution is at $t = 116.6$ °F with somewhat more precision than is justified by the data. Certainly we are not justified in reporting anything more precise than 117° and 120 °F is probably good enough.

As a matter of curiosity let us use an empirical correlation of C_p *vs.* T and integrate the equation. The following value of C_p for air is obtained from Table 7.1c:

$$C_p = 6.386 + 1.762 \times 10^{-3}T - 0.2656 \times 10^{-6}T^2.$$

Then, over the interval 297 °K (75 °F) to T °K:

$$\int_{297}^{T} C_p\,dT = 6.386(T - 297) + 0.881(T^2 - 297^2) \times 10^{-3} - 0.0885(T^3 - 297^3) \times 10^{-6}. \tag{9.22}$$

Since the heat capacity is in calories per (gram-mole)(°K), which is equivalent to Pcu per (pound-mole)(°K), the integral will be evaluated in units of Pcu per pound-mole. We previously had the value to which the integral is equated in Btu, 289.42. This value must be converted to Pcu for the present calculation. This new value is $\frac{289.42}{1.8} = 160.78$. Therefore,

$$160.78 = 6.386T - 1896.6 + 0.881 \times 10^{-3}T^2 - 77.71 - 0.885 \times 10^{-6}T^3 + 2.32,$$

$$2132.8 = 6.386T + 0.881 \times 10^{-3}T^2 - 0.885 \times 10^{-6}T^3. \qquad (9.23)$$

There are many ways to solve a cubic equation. An excellent computer method is the Newton–Raphson procedure (see Supplement to this chapter in which Equation 9.23 is so solved). It may also be solved by direct iteration. Let us proceed in the latter manner. Suppose that we choose a temperature of 47 °C for the first trial. Then $T = 320$ °K and the right-hand side of Equation 9.23 equals 2105, which is quite close to the correct value of 2132. As a second trial value, choose 52 °C, so that $T = 325$ °K. Then the right-hand side of the equation equals 2138. Although this $f(T)$ is certainly nonlinear, no great harm will result in assuming linearity over such a small temperature interval. Linear interpolation then indicates that the correct solution is approximately 51 °C or 124 °F. The value obtained previously was 117 °F. Even for a substance as well known as air, then, discrepancies will not be surprising when two different sources of C_p-data are involved, particularly when we work, as in this case, in the extreme ends of empirically fitted equations.

Example 9.9

▶ STATEMENT: Change the previous example only slightly. Instead of using the pound of steam in a coil, and thereby effecting a physical separation of the air and steam, blow the steam directly into the air stream so that the process becomes a mixer rather than a heater. Figure 9.13 represents this process. Now calculate the temperature of the outlet gas.

▶ SOLUTION: The problem is of the same type with respect to information desired. It requires prediction of a temperature and consequently is in the nature of a design problem; if the outlet temperature were given and the amount of heat lost were required, it would be an evaluation of a performance. The problems differ sharply in that in this case the two feed streams are not kept separate and *one set of datum conditions* is required. The choice seems to be quite obvious for this simple case, 75 °F and the gas phase. Certainly we should choose *one* of the inlet temperatures. The choice of the air temperature stems from the main emphasis being centered on air and on the fact that, given a choice of two temperatures for a datum, the lower is usually somewhat less likely to cause trouble; all enthalpies at higher temperatures will be positive with respect to this datum.

BASIS One minute, as before.

Material balances The outlet stream is 101 lb of air-water mixture.

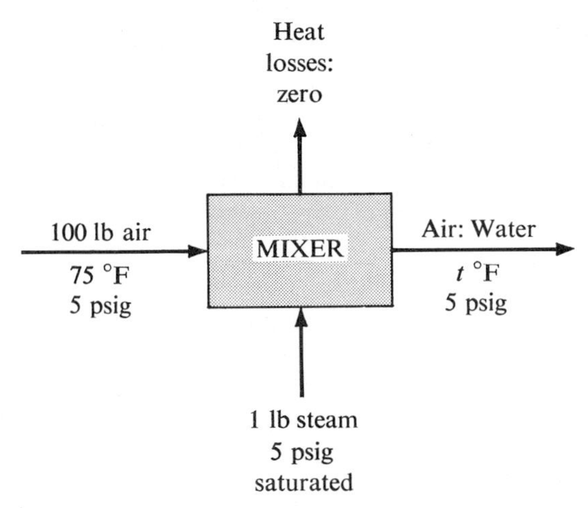

FIGURE 9.13 Element for mixing air and live steam.

Heat Balance

Heat in: Air: None.

Steam: The steam temperature is 227 °F. Again assume that the heat capacities at atmospheric pressure apply at 5 psig, this time for both air and the $\frac{1}{18}$ mole of steam.

$$_{77}C_{pm_{227}} = 8.1,$$

$$_{75}C_{pm_{77}} = 8.0,$$

$$\left(\frac{1}{18}\right)[8.1(227 - 77) - 8.0(75 - 77)] = 68.4 \text{ Btu.}$$

Heat out: With gases at low pressure it is reasonable to assume that the gas mixture is ideal so that the enthalpy is the sum of the individual enthalpies of the components, as if these components existed pure under the same conditions as in the mixture.

If $t < 100$ °F the heat capacities will be nearly constant at 6.90 for air and 8.0 for water vapor. Then,

$$\text{Air:} \quad \left(\frac{100}{29}\right)(6.9)(t - 75) = 23.8t - 1787$$

$$\text{Steam:} \quad \left(\frac{1}{18}\right)(8.0)(t - 75) = 0.445t - 33.3$$

$$\text{Total:} \quad 24.2t - 1820,$$

expressed in Btu.

Balance

$$24.2t - 1820 = 68.4,$$

$$t = 78.05 \text{ °F.}$$

Certainly, although we can calculate three or more significant digits, there is no likelihood of the answer being this precise. The answer might be reported as 78 °F.

This example illustrates the difference between the heating and mixing processes, the primary point of importance being the failure in the second case to condense the steam. Since no latent heat was obtained, the increase in temperature of the air was minor.

Example 9.10

▶ STATEMENT: In the manufacture of sulfuric acid, SO_2 is oxidized to SO_3 in a catalytic converter and the SO_3 is absorbed in dilute sulfuric acid to form strong acid. The catalytic conversion is a two-step process. Gases entering the first converter analyze 8.7% SO_2, 9.8% O_2 and the remainder nitrogen; 80% of the SO_2 is converted. The heat generated by this oxidation raises the temperature of the gas stream to 625 °C while the entering temperature to the second converter should be 400 °C. Per pound of sulfur, calculate the heat which must be extracted from the gas stream in a cooler between the first and second converters.

▶ SOLUTION: The system is shown schematically in Figure 9.14. Obviously, this problem is of the second type, information being available on enthalpies of all streams, with the transient heat from the process being required.

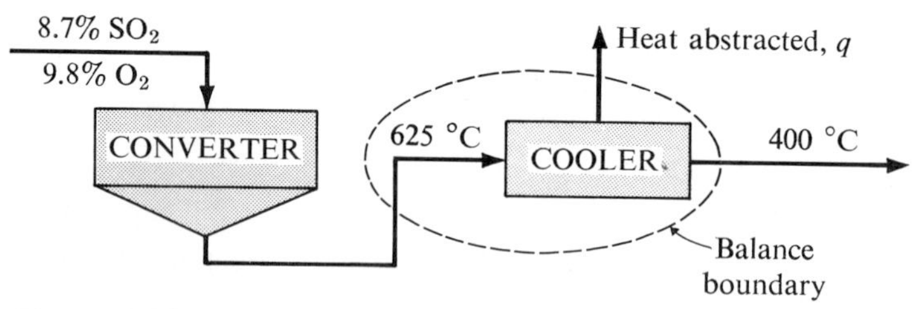

FIGURE 9.14 Contact sulfuric acid process: catalytic converter and cooler.

Material balances One hundred pound-moles of the gas to the converter appears to be the *best basis*. The number of moles of each component is then given by the data. The alternate basis of 1 lb of sulfur has no obvious advantage.

8.7 moles of SO_2; 8.7 moles of S; 278.4 lb S
9.8 moles of O_2
81.5 moles of N_2

100.0 moles, total

The reaction is: $SO_2 + \frac{1}{2}O_2 = SO_3$.

80% of SO_2 converted $= (0.80)(8.7) = 6.96$ moles of SO_3.
Remaining $SO_2 = 8.7 - 7.0 = 1.7$ moles of SO_2.

Remaining $O_2 = 9.8 - \dfrac{7.0}{2} = 6.3$ moles of O_2.

Nitrogen: constant $= 81.5$ moles of N_2.

Heat balance

Datum conditions: the phase is obviously gaseous; the 77 °F base used in mean molal heat capacities appears to be logical as a temperature datum. The enthalpies of inlet and outlet streams can then easily be calculated using values of C_{pm}.

The first step is the tabulation of the mean molal heat capacities, as obtained from Figure 7.2.

Temperature	SO_3	SO_2	O_2	N_2
(625 °C) 1157 °F	16.45	11.1	7.66	7.25
(400 °C) 752 °F	15.25	10.4	7.43	7.11

Heat Input at 625 °C (1157 °F)	
SO_3:	7.0 (16.45)(1157 − 77)
SO_2:	1.7 (11.10)(1157 − 77)
O_2:	6.3 (7.66)(1157 − 77)
N_2:	81.5 (7.25)(1157 − 77)
Total:	(773.2)(1157 − 77) = 835,060 Btu (835.06 × 10³)

Heat Output at 400 °C (752 °F)	
SO_3:	7.0 (15.25)(752 − 77)
SO_2:	1.7 (10.40)(752 − 77)
O_2:	6.3 (7.43)(752 − 77)
N_2:	81.5 (7.11)(752 − 77)
Total:	(750.70)(752 − 77) = 506,720 Btu (506.72 × 10³)

Heat abstracted in cooler: $-Q$.
Total heat output $= 506.72 \times 10^3 - Q$, in Btu.
Balance $835.06 \times 10^3 = 506.72 \times 10^3 - Q$.

$$-Q = 328.34 \times 10^3 \text{ Btu/8.7 moles of S.}$$
$$= 1180 \text{ Btu per lb of sulfur charged.}$$

Heat Balances in the Presence of Chemical Reactions

For any reaction in which the standard heats of formation ($\Delta H^0_{F,298}$), or the standard heats of combustion ($\Delta H^0_{c,298}$), are known, a standard heat of reaction (ΔH^0_{298}) can be calculated by methods described in Chapter 8. It is unlikely, however, that this temperature will be the one at which the reaction is conducted, or, for that matter, that any constant temperature will be maintained throughout the reaction. Some system is needed to effect a general heat balance around a process with reaction progressing at some indefinite, and not necessarily constant, temperature. Fortunately, enthalpies are state functions. There is no concern as to the path followed in arriving at the inlet and outlet temperatures. It is then possible to prescribe any convenient hypothetical path which will allow a simple means of evaluating the total enthalpy change from inlet to outlet. Such a path is indicated schematically in Figure 9.15.

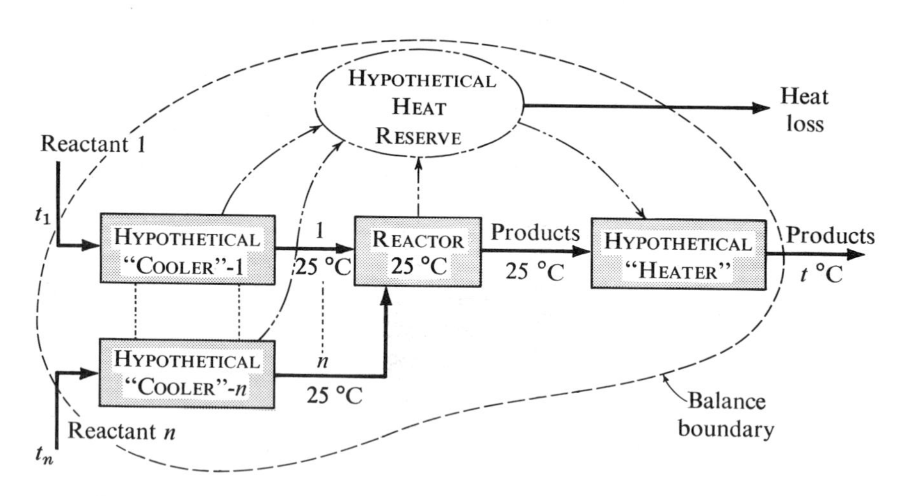

FIGURE 9.15 Hypothetical system for heat balances involving chemical reaction.

Any number of reactant streams may be considered. Thus the diagram indicates n streams; each stream passes through an individual hypothetical cooler in which its temperature is reduced to 25 °C. During the reaction, heat is evolved or consumed, and this heat must be removed or supplied as fast as required so that all materials in the reactor maintain the temperature of 25 °C at all times. The products then emerge from the reactor at 25 °C. Thus we have presumed conditions whereby the heat evolved or consumed may be calculated by the procedures established for *standard* heats of reaction. In doing so we have removed or added quantities of heat from or to the cooler and from or to the reactor. These transient quantities of heat may be held in some nonexistent hypothetical reserve. Thus, in our hypothetical path we end with products at 25 °C and an external quantity of heat held in reserve. We now add the heat to the products, bringing the latter to the final temperature. Note that the balance boundary never cuts one of the hypothetical streams so that the path chosen to describe the action does not affect the results.

It is easy to see that the datum conditions are established and are no longer a choice of the engineer. The datum conditions are fixed by the calculational method at 25 °C (298 °K), with the phases of reactants and products being fixed by the phases used in the stoichiometric equation. Thus the hypothetical coolers must not only adjust the temperature of the reactants to 25 °C but also must adjust the phases of these reactants to those written in the chemical equation. At the end of the process, where the heat hypothetically held in reserve is added back to the products, this amount of heat is used to accomplish two objectives, the first to adjust the phases to those which exist in the actual process, and second to use any remaining portion to raise the temperature of the products above 25 °C.

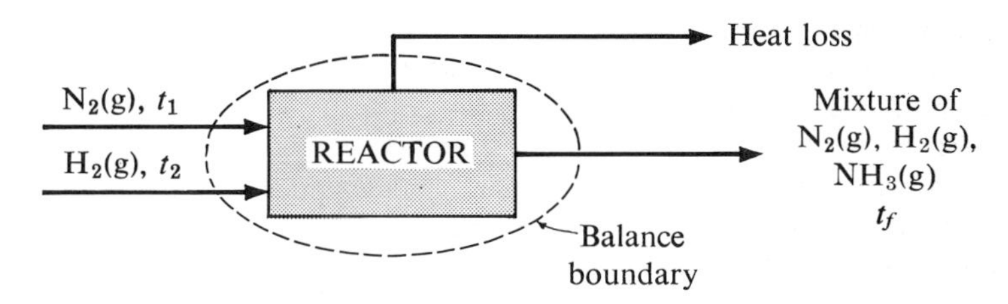

FIGURE 9.16 Reactor for ammonia synthesis.

Consider the process shown in Figure 9.16. The reaction in question is

$$\tfrac{1}{2}N_2(g) + \tfrac{3}{2}H_2(g) = NH_3(g).$$

Now if the heat of reaction is calculated using standard heats of formation of gaseous materials only (N_2, H_2, and NH_3), then all heat from the hypothetical reserve will be used to raise the temperature of the products, no adjustment of phase being required. On the other hand, if the equation is written

$$\tfrac{1}{2}N_2(g) + \tfrac{3}{2}H_2(g) = NH_3(l),$$

the hypothetical process produces liquid ammonia at 25 °C; part of the heat from the hypothetical reserve must be used to change this product to vapor; the remainder will raise the temperature of the mixture of ammonia vapor and unreacted gases. If correctly performed, both of these procedures produce exactly the same final temperature, for the heats of reaction will differ by the latent heat of vaporization of NH_3 at 25 °C, 1 atm. Obviously the first is simpler.

As a second example consider the catalytic water gas reaction. The standard heat of reaction could be calculated for either of the following statements of the chemical reaction:

$$\text{(a) } CO(g) + H_2O(g) = CO_2(g) + H_2(g) \qquad \Delta H_{298} = -9838 \text{ Pcu.}$$
$$\text{(b) } CO(g) + H_2O(l) = CO_2(g) + H_2(g) \qquad \Delta H_{298} = 681 \text{ Pcu.}$$

The first equation is written in the manner in which the reaction is actually conducted, with steam (i.e., water vapor) rather than with liquid water being one of the reactants. Then, the enthalpies of CO and H_2O would be calculated at the inlet temperatures, relative to those of the gaseous states of each at 25 °C (77 °F). These are the enthalpies that would be held in reserve. On the other hand, if the standard heat of reaction is obtained from Equation (b), the procedure must be changed to meet the datum conditions specified in this equation. Then the enthalpy of CO will be calculated as before, but the enthalpy of the steam will be that obtained by cooling the steam in the vapor state from its inlet condition to 25 °C in the vapor state *and* also from condensing it to liquid water at this temperature (i.e., a summation of the sensible *and* latent heats). These two enthalpy changes for CO and $H_2O(l)$ are added and placed in reserve; the quantity in reserve in this instance is greater than that where Equation (a) was employed. The increase exactly balances the reduction in ΔH^0_{298}, Equation (a) being appreciably exothermic while Equation (b) is mildly endothermic. This example points out the role of the phase in the specification of the datum conditions.

Since occasions may arise when it is difficult to analyze whether a heat of reaction constitutes an input or an output, it is well to define a standard manner of indicating, diagrammatically, this stream. The proper direction for the arrow indicating this stream can best be rationalized by considering some general process conducted at 25 °C. Let all reactants enter at this temperature and all products similarly leave at this temperature. Further indicate any heat added from the surroundings, ${}_1Q_2$, by an arrow inward according to the convention that heat added is positive. Then from the heat balance

$$\Delta H^0_R = {}_1Q_2,$$

it is obvious that positive heat of reaction must be indicated by an arrow outward, as indicated in Figure 9.17, since there is no sensible heat in any of the reactant and product streams and the heat of reaction must exactly balance the heat added from the surroundings. Thus an endothermic reaction (ΔH positive) constitutes a heat output and an exothermic reaction (ΔH negative) reverses the direction of the arrow and constitutes a heat input.

Types of Problems

Heat balances in physical and chemical systems differ fundamentally only in the sources of heat. In the physical system, only Q is encountered. In the

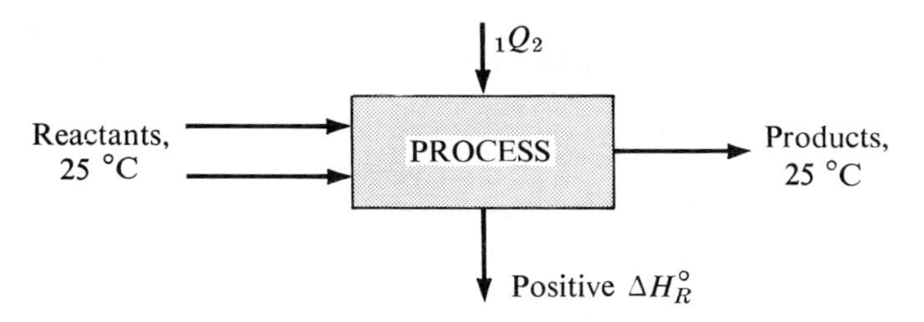

FIGURE 9.17 General diagrammatic representation of Q and ΔH_R°.

chemical system, both Q and ΔH_{298} appear. The overall aspects of heat balance problems are, however, not changed by this difference. The basic types of problems encountered in physical systems are duplicated in chemical systems. In the first type, all net heat streams are known, with the temperature of the products unknown. In the second, all information about temperatures is known and transient heat from system to surroundings (or vice versa) is to be calculated. Both types are illustrated in Examples 9.11 and 9.12.

Example 9.11

▶ STATEMENT: In Example 9.10, the temperature of the gases from the first converter (SO_2 to SO_3) was stated to be 625 °C. We shall now ask the question: Will this temperature be attained if the gases enter the converter at 425 °C?

▶ SOLUTION: The system was shown schematically in Figure 9.14. The problem is of the temperature-unknown type. The material balances have been solved (Example 9.10) on a basis of 100 lb-moles of gas fed to the converter. This basis is entirely satisfactory for the present problem. The quantities, to and from the converter, as previously calculated, are:

Moles Entering		Moles Leaving
—	SO_3	7.0
8.7	SO_2	1.7
9.8	O_2	6.3
81.5	N_2	81.5

Heat Balance

The reaction is:

$$SO_2(g) + \tfrac{1}{2}O_2(g) = SO_3(g).$$

The datum conditions are 25 °C (77 °F) and gaseous phases throughout. The standard heats of formation, at 25 °C, are

SO_2(g)	−70.94 kcal/gm-mole,
SO_3(g)	−94.39 kcal/gm-mole,
O_2(g)	zero.

The standard heat of reaction, at 25 °C, is

$$\begin{aligned}\Delta H^0_{298} &= -94.39 - (-70.94) = -23.45 \text{ kcal/gm-mole,}\\ &= -23{,}450 \text{ Pcu/lb-mole,}\\ &= -164{,}150 \text{ Pcu/7.0 moles converted } (-164.2 \times 10^3).\end{aligned}$$

At the inlet temperature of 425 °C = 797 °F, the pertinent heat capacities are as follows.
Mean heat capacities per mole, range of 77 to 797 °F:

SO_2	10.95,
O_2	7.48,
N_2	7.12.

Heat capacities as a function of T, °K, per mole:

$$\begin{aligned}SO_3&: 7.454 + 19.13 \times 10^{-3}T - 6.628 \times 10^{-6}T^2,\\ SO_2&: 6.945 + 10.01 \times 10^{-3}T - 3.794 \times 10^{-6}T^2,\\ O_2&: 6.117 + 3.167 \times 10^{-3}T - 1.005 \times 10^{-6}T^2,\\ N_2&: 6.457 + 1.389 \times 10^{-3}T - 0.069 \times 10^{-6}T^2.\end{aligned}$$

Heat capacities as a function of T, °K, in reacting quantities, listing coefficients only and omitting powers of ten which are consistent throughout:

SO_3:	52.178	133.910	46.396
SO_2:	11.806	17.017	6.500
O_2:	38.357	19.952	6.332
N_2:	526.246	113.204	5.624
Total gas:	628.587	284.083	64.852
Integrated:	628.587	142.042	21.617

Heat balance

Input: 1. Reaction: 164,200 Pcu.
2. Inlet gas:

SO_2:	(8.7)(10.95)(720) =	68,591 Btu (68.6 × 10^3),
O_2	(9.8)(7.48)(720) =	52,780 Btu (52.8 × 10^3),
N_2:	(81.5)(7.12)(720) =	417,800 Btu (417.8 × 10^3),
	Total	539,170 Btu (539.2 × 10^3)
	=	299,540 Pcu (299.5 × 10^3).

Input total: 463,740 Pcu (4.637 × 10^5).

Output:

$$\begin{aligned}628.587T + 142.042 \times 10^{-3}T^2 - 21.617 \times 10^{-6}T^3 \\ - (628.587)(298) - (142.042 \times 10^{-3})(298)^2 \\ + (21.617 \times 10^{-6})(298)^3.\end{aligned}$$

The numerical portion of this expression amounts to

$$-199 \times 10^3 \text{ Pcu}.$$

Balance

$$21.617 \times 10^{-6}T^3 - 142.042 \times 10^{-3}T^2 - 628.587T + 663{,}100 = 0.$$

For the Newton-Raphson Method,* the left-hand side is $f(T)$.

$$f'(T) = 64.852 \times 10^{-6}T^2 - 284.083 \times 10^{-3}T - 628.587.$$

Using 900 °K as the value of $T^{(1)}$ for the first iteration (i.e., 627 °C)

$$\begin{aligned}f(900) &= -2823.5,\\ f'(900) &= -831.7.\end{aligned}$$

The value of $T^{(2)}$ for the second iteration is

$$T^{(2)} = T^{(1)} - \frac{-2823.5}{-831.7} = 900 - 3.39 = 896.61.$$

Using $T^{(2)} = 896.6$ in the next iteration, we obtain
$T^{(3)} = 897.6$ °K,

and a final iteration produces $T^{(4)} = 897.7$ °K. Since 624 °C = 898 °K, it would appear that the required temperature cannot quite be obtained. The heat capacity equations are good only to about 1.5%. Therefore, the 1.0 °C difference from the desired temperature is not significant. We cannot expect to obtain an answer which is good to better than 12 or 13 °C. The second iteration was not, in fact, justifiable.

Example 9.12

▶ STATEMENT: The fuel fed to a furnace is a producer gas at 25 °C analyzing 35% CO and 65% N_2. The air for combustion is also fed at 25 °C, 25% excess air being used. The products of combustion leave the furnace at 1400 °C. Calculate the radiation loss from the furnace per ton of carbon in the producer gas.

• • • • •

* See Supplement to Chapter 9.

▶ SOLUTION: The system is shown schematically in Figure 9.18. This problem is of the type where all temperatures are known, all mass quantities may be calculated by material balances, and a heat stream (radiation loss) is the only unknown.

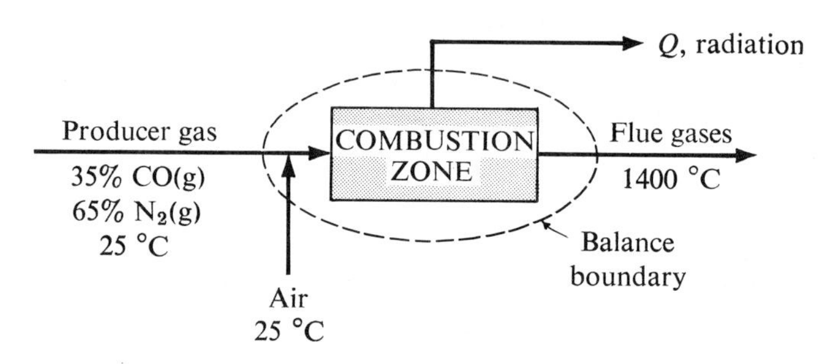

FIGURE 9.18 Combustion of producer gas.

The datum conditions are specified by the presence of chemical reaction, the temperature being 25 °C with all components in the gaseous phase. The last statement is made without the presence of an equation for the reaction only because of the obvious fact that the entire process deals only with gases and any other specification of phase would be ridiculous.

BASIS There are a number of reasonable choices: 1 ton of carbon in the inlet gas; 1, 10, or 100 moles of producer gas; 1, 10, or 100 moles of CO. Each has some advantage. The first provides calculations directly in the units requested. The second uses the only stream with complete composition data. The third would be the simplest with relation to the chemical reaction. We have here chosen the second on the principle that a completely analyzed stream often provides the simplest calculation. Basis: 1 mole of producer gas at entrance.

Material balances

Input: 0.35 moles of CO,
0.65 moles of N_2,
$CO(g) + \frac{1}{2} O_2(g) = CO_2(g)$.

O_2 required: $\frac{1}{2}(0.35) = 0.175$ moles.
O_2 supplied: $(1.25)(0.175) = 0.219$ moles.
N_2 supplied: $(0.219 \text{ moles } O_2)(79 \text{ moles } N_2/21 \text{ moles } O_2) = 0.823$ moles.

Combustion products:
CO_2: 0.35 moles,
O_2: $0.219 - 0.175 = 0.044$ moles,
N_2: $0.65 + 0.823 = 1.473$ moles.

Heat balance

The datum conditions have been specified above.

Standard heat of reaction:

$$\Delta H^0_{298} = \Delta H^0_{F,CO_2} - \Delta H^0_{F,CO}$$
$$= -94{,}052 - (-26{,}416) = -67{,}636 \text{ Pcu/lb-mole CO.}$$
$$(0.35)(\Delta H^0_{298}) = (0.35)(-67{,}636) = -23{,}673 \text{ Pcu for CO charged.}$$

Heat *input* in feed streams: zero, since both are at datum temperature. Heat *output*: enthalpies of the various components will be calculated separately and summed.

1400 °C = 2552 °F.

	O_2	CO_2	N_2
${}_{2552}C_{pm_{77}}$	8.18	12.50	7.74*

Enthalpy, O_2 = (0.044)(8.18)(2552 − 77) = 891 Btu (0.891 × 10^3)
Enthalpy, N_2 = (1.473)(7.74)(2552 − 77) = 28,220 Btu (28.22 × 10^3)
Enthalpy, CO_2 = (0.35)(12.50)(2552 − 77) = 10,830 Btu (10.83 × 10^3)
Total = 39,940 Btu (39.94 × 10^3)

*Data from Figure 7.2.

Heat of reaction acts as input (exothermic). Therefore, converting the heat of reaction to Btu, the balance reads:

$$(1.8)(23{,}673) = 39{,}940 - Q_{\text{radiation}};$$
$$-Q_{\text{radiation}} = 2{,}671 \text{ Btu/0.35 moles of carbon } (2.67 \times 10^3).$$

Now, 0.35 moles of carbon = 4.2 lb. Therefore, the radiation losses per ton of carbon fed will be

$$(2671)\left(\frac{2000}{4.2}\right) = 1{,}271{,}900 \text{ Btu } (1.272 \times 10^6).$$

9.7 The Kirchhoff Equation

The Kirchhoff equation has been presented in Chapter 8. Now that energy balances have been thoroughly discussed, it is interesting to re-evaluate this equation. It quite apparently is simply a heat balance since the only terms involved in the derivation of the equation are: (a) the standard heat of reaction; and (b) enthalpies of all products and reactants, relative to their enthalpies at the temperature of the standard heat of reaction. The arbitrary constant, ΔH_0, which is evaluated through knowledge of any $\Delta H_{R,t}$, is merely an integration constant. It has no physical significance as has been noted in Chapter 8.

Since the Kirchhoff equation is merely a heat balance, it has no special value. Although it determines a value of ΔH_t at any chosen value of t, the same can be done by a heat balance in which reactants and products enter and leave the system at the chosen t and the value of ΔH_t is calculated as the amount of heat required to maintain this temperature of products. If the process is exothermic, $-Q(= -\Delta H_t)$ must be removed from the system (as in a cooling coil); if endothermic, Q is the amount of heat which must be supplied to the system to maintain temperature t.

Not only can we, by the principles of heat balances, obtain the same information as provided by the Kirchhoff equation, but also we must doubt that we should ever want such information. The completely isothermal reaction is rare and could, as noted, be solved for any practical problem by applying heat balances.

The one practical requirement of a $\Delta H_{R,t}$ would appear to be the calculation of a standard heat of formation or a standard heat of combustion from data supplied for some temperature other than the usual 25 °C. This situation *can* exist, for the standard 298 has had universal acceptance only in recent decades. Formerly information at 288 °K, 291 °K, and 293 °K was also found and different sources used different standards. Under those circumstances, it was possible to find one item of required information in one source, another in another, and so forth. Then conversion to one unique temperature was required to obtain the desired heat of reaction. Even in this circumstance, it is obvious that heat capacity data were required to convert from one temperature system to another; with such data the required information at a standard temperature could be obtained by heat balance methods.

We conclude that the Kirchhoff equation adds nothing to our fund of information, nor does it provide a unique method of solution of any practical problem.

9.8 Unsteady State Energy Balances

Problems involving time- and position-dependent temperatures can arise in a manner similar to that discussed in Chapter 5 for mass quantities. In this section we offer a few examples of these situations. In principle the establishment of the differential equations which describe the processes will be identical to that presented in Chapter 5 and the solutions will generally be similar to ones presented there. In a few instances the solutions go beyond Chapter 5, primarily in the involvement first, of equations of order higher than one, and second, of simultaneous differential equations. Methods for the former were detailed in the Supplement to Chapter 5, but no applications were noted. A method for simultaneous equations will be developed here when needed.

In Chapter 5 we took great pains to write true mass balances as a first step. Then we divided through by Δ(time) and wrote the equations in the limit as

Δ(time) approached zero, these last two steps producing the differential equations from the mass balances. It was apparent that we then ended with a rate balance rather than a mass balance. The differential equations always stated:

$$\text{Mass rate in} = \text{mass rate out} + \text{rate of accumulation of mass.}$$

For a system with n components there were n independent equations of this type.

In this chapter we propose, wherever no confusion is apt to occur, to write the rate balances directly. Note that there is a single energy balance which is available around any process. The two forms of the balance are:

$$\text{Total energy in} = \text{total energy out} + \text{accumulation of energy.} \quad (9.24)$$

$$\text{Rate of energy in} = \text{rate of energy out} + \text{rate of accumulation of energy.} \quad (9.25)$$

The Heated Tank

Liquid in a tank is to be heated. The time required for the liquid temperature to reach any given value is desired. Obviously an equation is required that will express liquid temperature, t, as a function of time, θ.*

The system to be considered has a number of known characteristics. It consists of a kettle, or tank, so perfectly stirred that temperature throughout the liquid does not vary with location. (See Figure 9.19.) The heating medium will be saturated steam. Condensate will be withdrawn at the saturation temperature. The steam is condensed in a jacket or a coil. In either case the total available heating surface, or area A, will be completely covered at all times. The tank or kettle is charged initially with m pounds of fluid of density

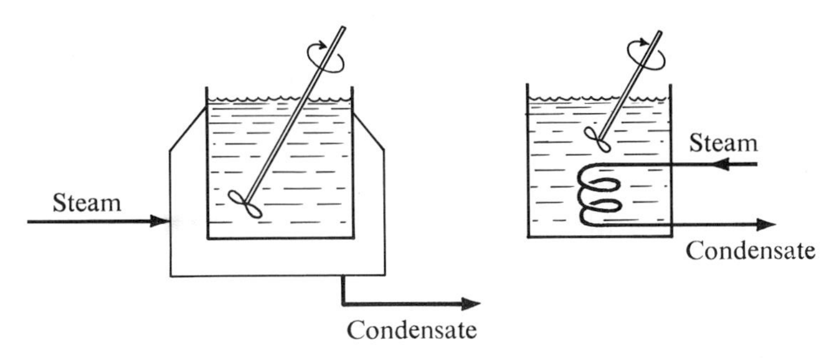

FIGURE 9.19 Perfectly stirred batch heaters.

• • • • •

* θ, rather than t, will be used as the symbol for time, so that both t and T are available to represent temperatures.

ρ and this density varies little over the range of temperature under consideration. The heat capacity of the liquid is known as a function of temperature and will be written $C(t)$. Heat losses to the surroundings may be assumed to be zero.

There is no heat output from this system according to the conditions stated. The heat balance needs to equate only input from the steam to the accumulation in the liquid. The accumulation is an enthalpy and will afford no difficulty. The transfer of heat from the hot to the cold side requires an independent rate expression, reminiscent of the transfer of mass across the phase boundary in Section 5.4. It seems reasonable to think that this rate will be proportional to the difference in temperatures of the two streams. Certainly an ice cube melts much faster in hot water than in cold, the temperature difference being larger in the former case. We shall then write the rate as proportional to $T - t$, where we use T as the temperature of the hot side and t as the temperature on the cool side, both at the same, but indefinite time. It is also reasonable to think that the rate of heat transfer is proportional to the area of pipe acting as the boundary between the two fluids. This area is designated as A. It is the submerged heating surface of the coil or the jacket.

The heat input rate may then be written in the form

$$UA(T - t),$$

where U is a proportionality constant known as a heat transfer coefficient and expressed in units of

$$\frac{\text{quantity of heat}}{(\text{unit time})(\text{unit area})(\text{unit } \Delta t)}.$$

Obviously a consistent system of units should be used. The most common is Btu/(hr)(sq ft)(°F). This coefficient is dependent upon the rate of agitation but will temporarily be written as independent of temperature.

Consider a time period, θ. Choose as a datum some temperature t_d. The reference phase must be liquid, that being the only phase of the mass being heated.

Heat input from the heating medium in time interval θ to $\theta + \Delta\theta$:

$$\tfrac{1}{2}UA\{[T - t(\theta)] + [T - t(\theta + \Delta\theta)]\}\Delta\theta.$$

Accumulation of enthalpy

$$mC(t)\{[t(\theta + \Delta\theta) - t_d] - [t(\theta) - t_d]\}.$$

Form the balance, Equation 9.24, by equating these two, divide by $\Delta\theta$, and take the limit as $\Delta\theta$ approaches zero. The datum temperatures obviously cancel and we obtain:

$$m\, C(t)\frac{dt(\theta)}{d\theta} \tag{9.26}$$

Separating variables,

$$\frac{C(t)\,dt(\theta)}{T - t(\theta)} \tag{9.27}$$

Equation 9.27 may be integrated if $C(t)$ is expressed in polynomial form, $C = a + bt + ct^2$. Then Equation 9.27 may be rewritten in the form

$$a\frac{dt}{T - t} + b\frac{t\,dt}{T - t} + c\frac{t^2\,dt}{T - t} = \frac{UA}{m}\,d\theta. \tag{9.28}$$

Each of the terms may be integrated as follows:

First term: $-a[\ln(T - t)]$.
Second term: $-b[T\ln(T - t) - (T - t)]$.
Third term: $-c[T^2\ln(T - t) - 2T(T - t) + \frac{1}{2}(T - t)^2]$.
Right-hand side: $\frac{UA\theta}{m} + \text{const.}$

The constant of integration may be obtained from the initial condition, $t = t_1$ when $\theta = 0$.

The additional accuracy introduced by using a temperature-dependent form of $C(t)$ is rarely needed since values of U are frequently no better than ± 15 to 25%. If an average value of C is used over the temperature range of the process, the final form of the $t - \theta$ relationship is greatly simplified, probably without sacrifice of any justifiable precision. The left side of Equation 9.27 will now contain the first term only, and the simplified equation is

$$-\frac{d(T - t)}{T - t} = \frac{UA}{mC}\,d\theta. \tag{9.29}$$

Integrating,

$$-\ln(T - t) = \frac{UA}{mC}\theta + \text{const.}$$

Applying the initial condition previously noted,

$$\ln\frac{T - t_1}{T - t} = \frac{UA}{mC}\theta. \tag{9.30}$$

During the entire period of operation (i.e., $\theta = 0$ to $\theta = \theta$), an amount of heat is transferred equal to Q. This quantity is obviously equal to the gain in enthalpy of the liquid in the tank or kettle,

$$mC(t_2 - t_1),$$

which may be rewritten as

$$mC[(T - t_1) - (T - t_2)].$$

If this expression is equated to Q, solved for mC, and the result substituted into Equation 9.30, we obtain

$$Q = UA\theta \frac{(T - t_1) - (T - t_2)}{\ln[(T - t_1)/(T - t_2)]} = UA\theta\, \Delta t_{lm}. \tag{9.31}$$

The expression Δt_{lm} is read "the logarithmic mean temperature difference." It occurs frequently in calculations involved in both steady state and unsteady state heating processes.

Varying U • We have seen fit to use $C(t)$ simply as C in many practical instances, and we noted that variation of U with t had been disregarded. The latter assumption may not always be justified, since the rate of heat transfer depends upon fluid properties (among other things) and the viscosity is one property which may vary widely with change in temperature. We should not leave this type of process then without considering the effect of writing $U(t)$ in place of U. If we do so, Equation 9.29 would have to be revised to read

$$\frac{dt}{U(t)(T - t)} = \frac{A}{mC}\, d\theta. \tag{9.32}$$

The $U(t)$ is rarely expressed analytically. The left-hand side of the equation may be integrated graphically or numerically when this condition pertains. Obviously for graphical integration t is plotted *vs.* $1/U(t)(T - t)$ and integrated between the limits $t = t_1$ and $t = t$. If the time is known rather than the final temperature, an iterative procedure will be necessary.

Loss of Heat from Surface • A simple modification of the previous case will be introduced. Heat loss from all parts of the apparatus except the liquid surface will be considered to be negligible. Heat loss from the surface will be presumed to take place according to the relationship $q = U_1S(t - t_a)$, in which S is the area of the surface of the liquid and t_a is the external air temperature, assumed henceforth to be a constant. Both U and C will be assumed to be constant, as will U_1. Now the terms of the heat balance are:

$$\text{Input:} \quad UA(T - t),$$
$$\text{Output:} \quad U_1S(t - t_a),$$
$$\text{Accumulation:} \quad mC\frac{dt}{d\theta}.$$

The balance may be written in the form:

$$UAT + U_1St_a - (UA + U_1S)t = mC\frac{dt}{d\theta}.$$

Let $a = UAT + U_1St_a$, and let $b = UA + U_1S$.

Then

$$\frac{d\theta}{mC} = \frac{dt}{a - bt}.$$

Integrating

$$\frac{\theta}{mC} = -\frac{1}{b}\ln(a - bt) + \text{const.}$$

As usual, the constant may be evaluated from initial conditions, where $\theta = 0$ and $t = t_1$. Then,

$$\theta = \frac{mC}{b}\ln\frac{a - bt_1}{a - bt} = \frac{mC}{UA + U_1S}\ln\frac{UA(T - t_1) + U_1S(t_a - t_1)}{UA(T - t) + U_1S(t_a - t)}. \quad (9.33)$$

Position-Dependent, Steady State Balances

The previous part of Section 9.8 dealt with the traditionally transient process in the literal (i.e., time) sense. We now turn to the analysis of temperature profiles within heaters and interchangers in the steady state. Heaters and interchangers are exactly what the names imply. In the heater the temperature of a cold* fluid is raised by passing the fluid through a tube (or tubes) outside which there is a hot fluid. In a heater the hot fluid will be a vapor at its saturation temperature and the heat passed to the cold fluid will be in the form of latent heat. In an interchanger the two fluids, one hot and one cold, are both exchanging sensible heat, the one which is fed at the higher temperature being cooled during its passage through the unit, and vice versa.

The immediate objective is the development of a temperature-length profile for the fluid being heated in the heater and for each stream in the interchanger. These relationships provide a means for calculating the length of pipe needed to effect a given temperature change, or the temperature that may be expected at outlet when the fluids are passed through existing equipment. The approach is analogous to that used for mass transfer towers in Section 5.7.

All diagrams will represent double pipe units, as in Figure 9.20. The choice of this type of diagram does not preclude the application of the relationships to units containing multiple tubes in parallel as long as the flow patterns in those units duplicate those in the double pipe units. Multiple pipe units increase the capacity (i.e., the rate at which fluid may be handled) without affecting the temperature-length relationships. For example, two double pipe interchangers in parallel will obviously heat twice as much material as one. If the two pipes are placed in a common shell we have a multiple tube unit acting exactly the same as the two double pipe units in parallel.

• • • • •

* The use of the words "cold" and "hot" implies a relative state. A "cold fluid" is at a lower temperature than the temperature of a "hot fluid" and vice versa.

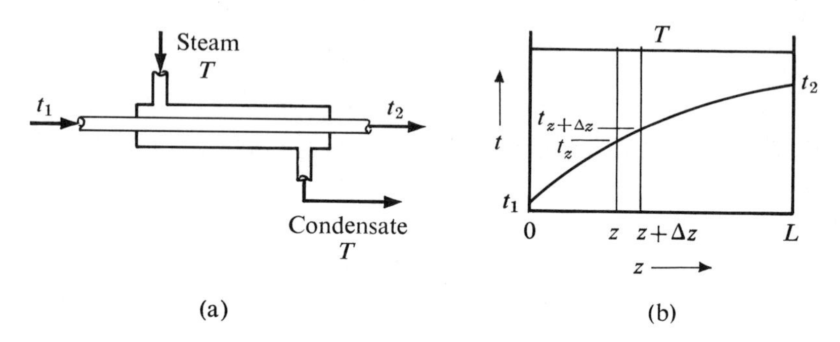

FIGURE 9.20 Diagrams of double pipe steam heaters.

A double pipe heater is illustrated in Figure 9.20 along with the generalized temperature profiles. The constant hot temperature of the fluid is due to the restriction to heating by means of condensing a vapor; e.g., as in a steam heater. The cold fluid initially at temperature t_1, is pumped through the inner pipe at constant rate, w, a symbol chosen to distinguish it from the fixed mass, m, in the tanks of the previous part of this section. Saturated steam is fed to the outer, or annular section of the heater (i.e., the jacket) at a rate sufficient to maintain the pressure and therefore the constant temperature. We assume no pressure drop of appreciable proportions through the jacket so that the saturation temperature is unchanged throughout. Condensate, also at the saturation temperature, is withdrawn from the bottom of the jacket.

Choose any section of the interchanger for analysis, starting at a length z from the inlet position for the cold fluid and continuing to a length $z + \Delta z$. Write a heat balance around this section. (Truly the balance should be an energy balance but the changes in all forms of mechanical energy will be negligible in all common applications.) This procedure is reminiscent of that used in Section 5.7 on mass transfer towers. It is, in fact, exactly analogous, but without one complication encountered there. In the present instance we are interested in temperatures in the streams on both sides of the tube and these temperatures are obviously expressed in the same units. In mass transfer the concentrations on the two sides of the phase boundary were expressed in different units and had to be reconciled before they could be subtracted. The heat balance, is, therefore, much simpler to handle. We can concentrate much more closely on the temperature-time relationships without being diverted by the reconciliation of concentration units.

The heat balance across this arbitrarily located section involves three streams. There is enthalpy entering in the cold stream. There is enthalpy leaving in the same stream after its temperature has been raised by passage through the section. There is heat transferred from the fluid on the hot side, a second input stream. We need some independent expression of the rate of heat transfer from the hot to the cold fluid. We shall again assume proportionality to $T - t$ and to the area through which the heat transferred is flowing. This area is a cylindrical surface which, over a length of pipe Δz, may be written as $\pi\, d_p\, \Delta z$. Do not confuse this area with the cross section of the pipe. The fluid flows through this cross section, but the heat flows through the cylindrical surface. We now have for q, the rate at which heat is transferred,

$$q\big|_{z}^{z+\Delta z} = U\pi\, d_p\, \Delta z(T - t),$$

where U, the proportionality constant, is again a heat transfer coefficient.

We may now write all the necessary terms in the heat balance. In this instance we shall write each term as a rate, rather than as a quantity of heat. The datum temperature will be t_d and the phase will be liquid.

Rate of enthalpy input: $wC(t)[t(z) - t_d]$.

Rate of enthalpy output: $wC(t)[t(z + \Delta z) - t_d]$.

Rate of heat added: $\frac{1}{2}U\pi\, d_p\, \Delta z\{[T - t(z)] + [T - t(z + \Delta z)]\}$.

Equating inputs to output, dividing by Δz, taking the limit as Δz approaches zero and recognizing that t_d terms cancel,

$$w\frac{dC(t)t(z)}{dz} = U\pi\, d_p[T - t(z)]. \tag{9.34}$$

Note the similarity of Equations 9.26 and 9.34. Integration problems encountered in Equation 9.26 are identical to those potentially present with Equation 9.34. The assumption of $C(t) = C$ is usually justified. Making this assumption and separating variables,

$$\frac{dt}{T - t} = \frac{U\pi\, d_p}{wC}\, dz. \tag{9.35}$$

Noting that T is a constant and integrating,

$$-\ln(T - t) = \frac{U\pi\, d_p z}{wC} + \text{const.}$$

The constant of integration can be evaluated from the initial condition $t = t_1$ when $z = 0$. Therefore,

$$\ln\frac{T - t_1}{T - t} = \frac{U\pi\, d_p z}{wC}. \tag{9.36}$$

When $z = L$, $t = t_2$, giving the form of the integrated expression usually reported,

$$\ln\frac{T - t_1}{T - t_2} = \frac{U\pi\, d_p L}{wC}. \tag{9.37}$$

It should be quite apparent that the procedure used with tanks can be repeated to obtain

$$q = U\pi\, d_p L\, \Delta t_{lm} = UA\, \Delta t_{lm}. \tag{9.38}$$

Parallel and Countercurrent Flow Interchangers • Interchangers, in contrast to heaters, do not have constant temperature in either stream. For the countercurrent case the flow in the inner pipe of Figure 9.20 is unchanged. In the jacket, the second fluid enters where the condensate left the heater, and leaves where the stream entered. The temperature profiles are represented in Figure 9.21. We retain the symbols, T and t, the former for the fluid which enters hot and the latter for the fluid which enters cold.

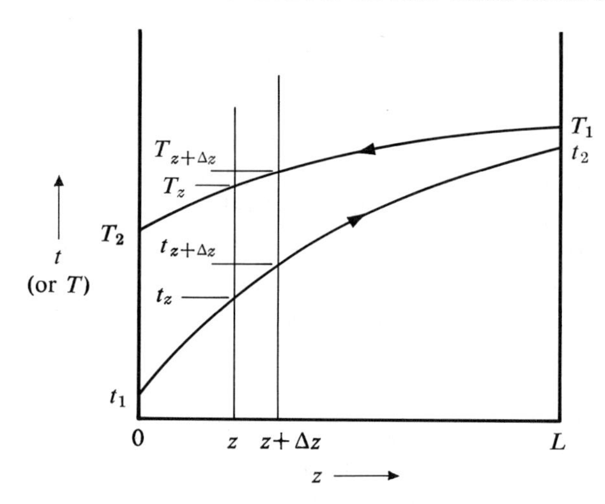

FIGURE 9.21 Temperature-length relationships in a countercurrent interchanger.

The heat balance may be developed in exactly the same manner employed for the steam heater. The only difference lies in the necessity of designating the T as a function of length. Making this modification in Equation 9.34 and separating variables,

$$\frac{dt(z)}{T(z) - t(z)} = \frac{U\pi\, d_p}{wC}\, dz. \tag{9.34a}$$

The left-hand side of the equation cannot be integrated until the two temperatures are related (See Section 5.7 when the concentration difference is expressed in terms of both y and x). The necessary relationship can be established by a heat balance from either end of the unit to some general position within the unit where the temperatures of the two streams are $T(z)$ and $t(z)$. In order to write the balance we need symbols to represent the weight rate and heat capacity of the jacket stream. Designate these as w_j and C_j, respectively. Then, writing the balance around the section of the interchanger from $z = 0$ to $z = z$,

Sum of the two input streams $= wC(t_1 - t_d) + w_jC_j[T(z) - t_d]$;
Sum of the two output streams $= wC[t(z) - t_d] + w_jC_j(T_2 - t_d)$.

Equating these two in a heat balance, and noting that the datum temperatures must cancel,

$$wC[t_1 - t(z)] = w_jC_j[T_2 - T(z)].$$

Solving for $T(z)$,

$$T(z) = \frac{wC}{w_jC_j}[t(z) - t_1] + T_2. \tag{9.39}$$

Substituting Equation 9.39 into Equation 9.34a,

$$\frac{dt(z)}{(wC/w_jC_j)t(z) - (wC/w_jC_j)t_1 + T_2 - t(z)} = \frac{U\pi\, d_p}{wC}\, dz.$$

The left side is of the form

$$\frac{dt(z)}{(a - 1)t(z) + b},$$

where

$$a = \frac{wC}{w_jC_j}$$

and

$$b = -at_1 + T_2.$$

Integrating,

$$\frac{1}{a - 1}\ln[(a - 1)t(z) + b] = \frac{U\pi\, d_p z}{wC} + \text{const.}$$

From the initial condition,

$$t(z) = t_1;\ z = 0,$$

$$\text{const} = \frac{1}{a - 1}\ln[T_2 - t_1].$$

Therefore the final integrated form is

$$\frac{w_jC_j}{wC - w_jC_j}\ln\frac{(wC/w_jC_j)(t - t_1) + (T_2 - t)}{T_2 - t_1} = \frac{U\pi\, d_p z}{wC}. \tag{9.40}$$

It is obvious that when $t = t_2$, $z = L$, the total length of the interchanger.

Although it is certainly *not* obvious it is true that, by a technique similar to that used previously we can show that the logarithmic mean temperature difference, and Equation 9.38 also apply to this countercurrent interchanger.

By analogous arguments Equation 9.38 also applies to the parallel current case, the profiles of which are shown in Figure 9.22. Equation 9.40 does *not* apply. A similar but not identical equation can be derived.

Types of Problems in Heaters and Interchangers • As noted on page 413, there are two main types of problems involving interchangers. If the unit

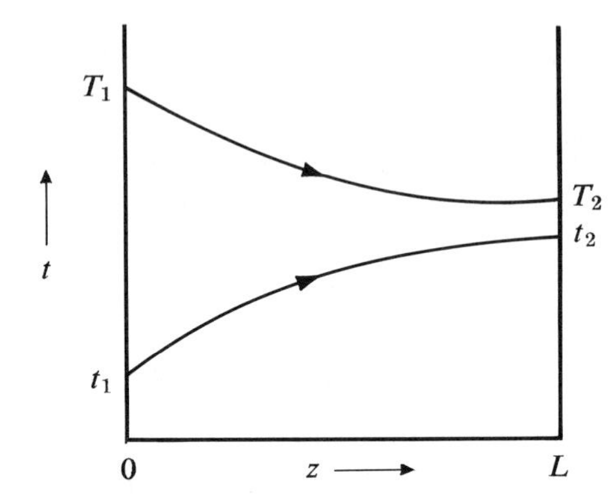

FIGURE 9.22 Temperature-length relationships in a cocurrent interchanger.

is being designed, at least three of the terminal temperatures are known, input and output desired for one and available input of the other. Both flow rates are known and the degrees of freedom are not sufficient to allow specification of all terminal temperatures. Then the problem is the amount of heat transfer surface to provide in order to effect the required temperature change in this stream. The normal procedure would be to complete the terminal temperatures by an overall heat balance, calculate all t's, and use Equation 9.38 to calculate the area.

If the interchanger unit exists and it is proposed to effect heat transfer between two streams of known input rates, properties and temperatures, neither terminal t is known. It is then more convenient to use Equation 9.37, 9.40, or the counterpart if the process is parallel current.

The Heated Rod, Surroundings at Constant Temperature • An uninsulated rod, heated at one end only, is not in itself of much practical importance but it provides a background for the more geometrically complicated finned heaters and coolers. The objective in studying this system is the determination of the *steady state* temperature-length profile, for any fixed temperature at the heated end. In order to establish this profile, two types of auxiliary information are required. First, we must have some measure of the characteristics of the rod for conducting heat, based upon its dimensions and upon its inherent ability to allow heat flow within the material of which the rod is constructed. Second, the characteristics of the surroundings to which heat is continually being lost from the surface of the rod must also be known. We shall examine these two in the order named.

The "rod" may be of any shape and cross section. For the present consider only a cylindrical constant cross section of radius, r. The ability of a solid to transmit heat through itself by conduction is measured in terms of a quantity called the thermal conductivity, k. Materials such as metals which transmit heat readily in conduction have high thermal conductivities. Good insulating materials are at the other end of the scale, with low thermal conductivities. In the English system, thermal conductivities are usually reported in (Btu)(feet) per (hour)(square feet)(°F). This unit is exactly the same as a heat transfer coefficient multiplied by length. The thermal conductivity is, in fact, a heat transfer coefficient operating through a given length.

The flow of heat through conduction only is stated by Fourier's law of the conduction of heat. If we assume that there is no difference in temperature from axis to surface in any given cross section, we can use this law in its one-dimensional form,

$$q = -kA_c \frac{dt}{dz}, \tag{9.41}$$

where, as usual, q is *rate* of passage of heat, A_c is cross section of the rod, the unsubscripted symbol A being reserved for cylindrical surface, and z is distance down the rod.

The surroundings in this instance will be a fluid at constant temperature, t_f. Under these circumstances, and by arguments similar to those used previously (p. 410)

$$q = hA(t - t_f), \tag{9.42}$$

where A is the cylindrical surface through which heat is flowing from the rod to the surrounding fluid and h is a heat transfer coefficient.*

The process, as now defined, can be represented at any section of the rod by the diagram shown in Figure 9.23. Steady state temperatures at every z are assumed to have been attained. Then the heat balance states,

$$\underset{\text{input}}{q_1} = \underset{\text{output}}{q_2 + q_3}. \tag{9.43}$$

Each of these q can be written in terms of the characteristics of the system, as just discussed:

$$q_1 = -kA_c \left.\frac{dt}{dz}\right|_z = -k\pi r^2 \left.\frac{dt}{dz}\right|_z,$$

$$q_2 = -kA_c \left.\frac{dt}{dz}\right|_{z+\Delta z} = -k\pi r^2 \left.\frac{dt}{dz}\right|_{z+\Delta z},$$

$$q_3 = hA(t - t_f) = h(2\pi r\,\Delta z)(t - t_f).$$

• • • • •

* The change from U to h is made to account for the fact that heat must flow only through the resistance offered by the single fluid. In the cases where U was used, heat flowed through a composite resistance consisting of one fluid, the retaining wall between phases, and then through another fluid.

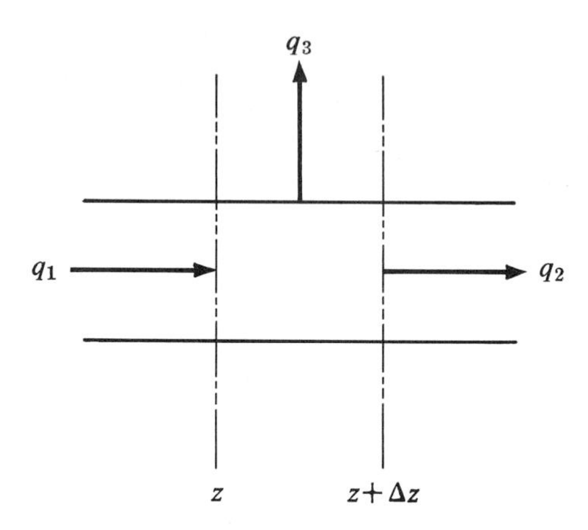

FIGURE 9.23 Section of rod heated at left end and cooled by air flowing around the external surface of the rod.

Writing the balance, canceling π's and pertinent r's and dividing by Δz, we obtain,

$$\frac{\left.\dfrac{dt}{dz}\right|_{z+\Delta z} - \left.\dfrac{dt}{dz}\right|_{z}}{\Delta z} = \frac{2h}{kr}(t - t_f). \tag{9.44}$$

The left-hand side of Equation 9.44 is the definition of the second derivative, so that Equation 9.44 becomes,

$$\frac{d^2t}{dz^2} - \frac{2h}{kr}(t - t_f) = 0.$$

Since t_f is a constant, this equation may be rewritten,

$$\frac{d^2T}{dz^2} - \frac{2h}{kr}T = 0, \qquad \text{where } T = t - t_f. \tag{9.45}$$

Equation 9.45 is an ordinary, linear, second-order, homogeneous, differential equation. It can be easily solved by methods discussed in the Supplement to Chapter 5. Writing in operator form,

$$\left(D^2 - \frac{2h}{kr}\right)T = 0.$$

Therefore,

$$(D + \sqrt{2h/kr})(D - \sqrt{2h/kr})T = 0.$$

The solution is, therefore,

$$T = K_1e^{sz} + K_2e^{-sz}, \qquad \text{where } s = \sqrt{2h/kr}.$$

Since the differential equation was of second order, there are two arbitrary constants, K_1 and K_2, in this general solution. These constants may be evaluated if we can find two sets of boundary conditions. It will always be true that $t = t_1$ when $z = 0$. If the rod were infinitely long, it would be necessary for the temperature to be that of the surroundings; then substituting this boundary condition at $z = \infty$

$$0 = K_1 e^{\infty} + K_2 e^{-\infty} = K_1 e^{\infty}.$$

Therefore, K_1 must equal zero, since e^{∞} cannot be zero. Now substituting the initial conditions

$$t_1 - t_f = K_1 + K_2 = K_2.$$

The complete solution is

$$\frac{t - t_f}{t_1 - t_f} = e^{-z\sqrt{2h/kr}}. \tag{9.46}$$

The Heated Rod, Surroundings in Countercurrent Flow • Change the conditions specified on p. 419 only to the extent that the surroundings are not at constant temperature but are governed by the rate of flow of a fluid past the rod, flowing axially and entering cold at the colder end of the rod. The heat balance on the section of the rod is unchanged except that t_f is no longer a constant. A second heat balance can be made on the fluid flowing through the same section of length Δz. It enters at temperature t_f and leaves at temperature $t_f + \Delta t_f$, picking up heat from the rod. Then the heat balance provides another differential equation,

$$wC\, dt_f = 2\pi rh\, dz(t - t_f),$$

which may be written in the form

$$\frac{dt_f}{dz} = \frac{2\pi rh}{wC}(t - t_f). \tag{9.47}$$

Now there are two ordinary differential equations to describe the two temperature profiles, one for the rod and one for the fluid stream. They must be solved simultaneously. Since both have constant coefficients, the solution can be easily effected through the use of D-operators. Writing Equations 9.45 and 9.47 in operator form,

$$(D^2 - \alpha)t = \alpha t_f, \qquad \text{where } \alpha = \frac{2h}{kr},$$

$$(D + \beta)t_f = +\beta t, \qquad \text{where } \beta = \frac{2\pi rh}{wC}.$$

Solving the second of these equations for t_f,

$$t_f = \frac{\beta t}{D + \beta}.$$

Substituting this value of t_f into the first of the two equations,

$$(D^2 - \alpha)t = -\frac{\alpha\beta t}{D + \beta}.$$

Multiplying both sides by $(D + \beta)$,

$$(D^3 + \beta D^2 - \alpha D - \alpha\beta)t = -\alpha\beta t.$$

Since the $\alpha\beta t$ term cancels, we are left with a third-order homogeneous differential equation with constant coefficients.

$$(D^3 + \beta D^2 - \alpha D)t = 0.$$

The roots of the polynomial in D are

$$s_1 = 0; \qquad s_2, s_3 = \frac{-\beta \pm \sqrt{\beta^2 + 4\alpha}}{2} = -\frac{\pi r h}{wC} \pm \sqrt{\left(\frac{\pi r h}{wC}\right)^2 + \frac{2h}{kr}}. \tag{9.48}$$

The general solution is, therefore,

$$t = K_1 + K_2 e^{s_2 z} + K_3 e^{s_3 z}. \tag{9.49}$$

For the complete solution we need three sets of boundary conditions so as to eliminate the three K's. We can use the initial condition and the boundary condition employed on p. 421 and add the fact that at $z = \infty$, the first derivative must be zero since the temperature of the rod could no longer be changing at this position. We then have:

$$z = 0, \qquad z = \infty, \qquad z = \infty,$$

$$t = t_1, \qquad t = t_{fl}, \qquad \frac{dt}{dz} = 0.$$

Taking the derivative of Equation 9.49,

$$\frac{dt}{dz} = s_2 K_2 e^{s_2 z} + s_3 K_3 e^{s_3 z}. \tag{9.50}$$

Now examination of Equation 9.48 indicates that either s_2 or s_3 must be negative, while the other one is positive. Thus, for the negative one, $e^{s_i z}$ is zero when $z = \infty$. Let us assign the negative sign to s_3. Then using the third set of boundary conditions, in Equation 9.50,

$$0 = s_2 K_2(\infty) + 0.$$

Since s_2 cannot equal zero, K_2 must, so that this term drops out as it did in Equation 9.46. Now we can use the other boundary condition and the initial condition to obtain:

$$t_1 = K_1 + K_3,$$
$$t_{fl} = K_1 + 0.$$

The complete solution is, therefore,

$$\frac{t - t_{fl}}{t_1 - t_{fl}} = \exp(s_3 z).$$

where s_3 has already been designated as the negative root.

The equation for t_f can be developed in the same manner, or we can substitute the value just obtained to obtain the ordinary, first-order equation

$$\frac{dt_f}{dz} + \frac{2\pi rh}{wC} t_f = -\frac{2\pi rh}{wC} f(z).$$

This equation may be solved by the methods suggested in the Supplement to Chapter 5.

The Heated Fin • Instead of a rod, consider a fin extending from some heated surface into cooler surroundings. In the present instance we shall choose surroundings at constant temperature, t_f. The fin will be flat, of constant thickness, w_o, but of equilateral triangular shape attached to a flat wall. At the wall it is of height H_f and its length measured at right angles to the wall is L. Such a fin is illustrated in Figure 9.24.

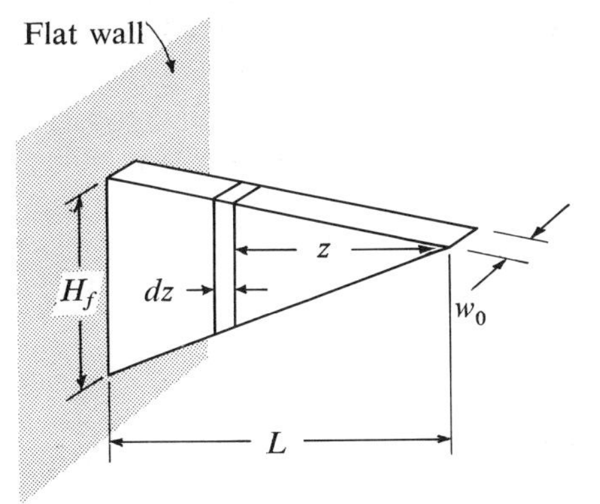

FIGURE 9.24 Triangular thin fin set in flat wall.

The general distance, z, and the differential distance, dz, are measured from the tip, contrary to the flow of heat in the z direction. This choice is a matter of convenience in expressing the necessary cross sectional and surface areas; care will subsequently be taken to see that signs are properly adjusted to account for this selection.

The height at the position z will be, by similar triangles, $H_f z/L$. Thus the cross section through which heat flows within the fin in the z direction* is

• • • • •

* A completely rigorous analysis of this situation requires inclusion of heat flow in the plane of the fin at right angles to the z direction. According to Mickley, Sherwood, and Reed, *loc. cit.*, most fins are designed so that this added complication is not justified.

$H_f w_o z/L$. The *surface area* in the differential length, through which heat flows to the surroundings, is $[(2H_f z/L) + 2w_o]\,dz$. We shall assume that the fin is very thin, as is frequently the case, and neglect the contribution of the width, $2w_o$. The surface area is then $2H_f z\,dz/L$.

Following previous procedures, consider the heat flow into the differential section at z, the flow out at $z + dz$, and the flow out to the surroundings. Since temperatures will enter these three expressions only as differences, specification of a datum temperature is not required. Now because of the choice of z direction the flow in at z is negative. It shall then be written

$$kA_c \frac{dt}{dz}\bigg|_z = kH_f w_o \left(\frac{z}{L}\right) \frac{dt}{dz}.$$

The output at $z + dz$ is also negative and shall be written

$$kA_c \frac{dt}{dz}\bigg|_{z+dz} = kH_f w_o \frac{z}{L}\frac{dt}{dz} + \frac{d}{dz}\left(kH_f w_o \frac{z}{L}\frac{dt}{dz}\right) dz.$$

Note particularly that the increment applies to the entire heat conduction rate, not just the temperature gradient. The second term is the incremental increase in q at a position dz farther along the z axis. In dealing with the heated rod we were not aware of this situation since the conduction equation involved the variable, z, only in the differential term.

The output to the surroundings involves a heat transfer coefficient, which we continue to identify as h. This output is

$$-hA(t - t_f) = -\left(\frac{2H_f hz}{L}\right)(t - t_f)\,dz.$$

The heat balance then states

$$\frac{d}{dz}\left(kH_f w_o \frac{z}{L}\frac{dt}{dz}\right) dz - \left(\frac{2H_f hz}{L}\right)(t - t_f)\,dz = 0.$$

Dividing by dz and multiplying by $L/kH_f w_o$,

$$\frac{d}{dz}\left(z\frac{dt}{dz}\right) - \left(\frac{2h}{kw_o}\right) z(t - t_f) = 0.$$

Now expanding the first term,

$$z\frac{d^2t}{dz^2} + \frac{dt}{dz} - \left(\frac{2h}{kw_o}\right) z(t - t_f) = 0.$$

Since t_f is a constant, we can transform the variable, using $y = t - t_f$, giving

$$z\frac{d^2y}{dz^2} + \frac{dy}{dz} - \left(\frac{2h}{kw_o}\right) zy = 0. \tag{9.51}$$

Equation 9.51 is a second-order, differential equation with variable coefficients. It is a form of Bessel's equation. The solution is known but its attainment is beyond the scope of the solutions reported in the Supplement to Chapter 5. The main point of importance, knowing that we can solve, analytically or numerically, *any* differential equation, is that we have been able without much difficulty to establish the equation governing this situation.

*Heater in a Control Loop** • As a final transient process, we consider a heater in a reservoir tank, the circuit being designed so that the heater will come on when the temperature in the tank falls below a set value. The rate of heat input from the coil will be proportional to the difference in temperature between the maximum for which the circuit is designed, t_m, and the temperature in the tank, t. This input rate to the coil may be written

$$b(t_m - t),$$

where b is an arbitrary constant dependent upon the design of the heater. We assume that $t < t_m$ under all circumstances. The rate at which heat is supplied to the liquid in the tank is

$$hA(t_c - t),$$

where A is the surface area of the coil, t_c is the temperature of the coil at any instant and may be assumed to be constant throughout the coil, and where h is the heat transfer coefficient, as used in previous illustrations.

The mass of liquid in the tank will be assumed to vary so little with temperature that it may be taken to be constant and equal to m pounds. The mass of coil is m_c. The weight of flow through the tank is w. Heat capacities are taken to be constant over the small temperature ranges contemplated and represented as C for liquid and C_c for the coil.

We wish to predict the temperature-time profile for liquid in the tank. The liquid is perfectly stirred so that no appreciable gradients exist throughout the tank when changes in inlet temperature occur. The tank has been operating at temperature, t_o. The inlet temperature is suddenly increased to t_i.

▶ SOLUTION: Establish first the heat balance on the system consisting of liquid in the tank. There are two rates of heat input, that from the coil and that from the inlet liquid. The former is a function of the difference in temperature, $t_c - t$, as already noted. The latter is a function of the difference in temperature, $t_i - t_d$, where t_d is some arbitrarily chosen datum temperature. This input can be written

$$wC(t_i - t_d).$$

• • • • •

* Adapted from a problem in R. B. Bird, W. E. Stewart, and E. N. Lightfoot, *Transport Phenomena* (New York: John Wiley & Sons, Inc., 1960).

The rate of output is expressed as a function of the difference between the tank temperature and the datum temperature, $t - t_d$, and may be written

$$wC(t - t_d).$$

We see immediately that there is a net input rate of

$$wC(t_i - t),$$

since we assumed that the inlet temperature has suddenly risen. This net input is the same as would be obtained if the tank temperature were chosen as the datum temperature.

The accumulation in the tank is $mC\, dt/d\theta$. The heat balance is, therefore,

$$\frac{dt}{d\theta} = \frac{hA}{mC}(t_c - t) + \frac{wC}{mC}(t_i - t) = -\left(\frac{hA}{mC} + \frac{w}{m}\right)t + \frac{hA}{mC}t_c + \frac{w}{m}t_i. \quad (9.52)$$

This equation includes two variable temperatures, t and t_c. In order to relate the two another heat balance must be written. The system considered for the second will be the heater alone.

The input rate to the heater has been defined as $b(t_m - t)$. The output rate is the same as the net input from the heater to the liquid $hA(t_c - t)$. The accumulation rate is the change in coil temperature per unit time, $dt_c/d\theta$, multiplied by mass and heat capacity of the heater. The heat balance over the heater system is then

$$\begin{aligned}\frac{dt_c}{d\theta} &= -\frac{hA}{m_cC_c}(t_c - t) + \frac{b}{m_cC_c}(t_m - t),\\ &= \left(\frac{hA}{m_cC_c} - \frac{b}{m_cC_c}\right)t - \frac{hA}{m_cC_c}t_c + \frac{b}{m_cC_c}t_m.\end{aligned} \quad (9.53)$$

In order to simplify the algebra, some new constants will be defined:

$$a = \frac{hA}{mC} + \frac{w}{m}; \quad e = \frac{hA}{mC}; \quad f = \frac{w}{m}, \quad \text{for use in Equation 9.52.}$$

$$g = \frac{hA}{m_cC_c}; \quad p = \frac{hA}{m_cC_c} - \frac{b}{m_cC_c}; \quad j = \frac{b}{m_cC_c}, \quad \text{for use in Equation 9.53.}$$

Writing the two equations in D-operator form, using these definitions,

$$(D + a)t = et_c + f, \quad (9.54)$$

$$(D + g)t_c = pt + j. \quad (9.55)$$

Solving Equation 9.55 for t_c,

$$t_c = \frac{pt + j}{D + g},$$

and substituting this expression into Equation 9.54,

$$\begin{aligned}(D + a)(D + g)t &= e(pt + j) + Df + fg,\\ &= ept + ej + fg, \qquad \text{since } Df = 0.\end{aligned}$$

Rearranging,

$$[D^2 + (a + g)D + (ag - ep)]t = ej + fg.$$

Since all the lower case letters are constants, this equation is a linear, second-order, differential equation. It may be made homogeneous by a transformation of variable,

$$y = t - \frac{ej + fg}{ag - ep},$$

Then,

$$[D^2 + (a + g)D + (ag - ep)]y = 0.$$

The solution is, obviously,

$$y = K_1 e^{s_1\theta} + K_2 e^{s_2\theta},$$

where s_1 and s_2 are the roots of the quadratic in D. If we write these roots in the form,

$$s_1, s_2 = \frac{-\eta \pm \sqrt{\eta^2 - 4\nu}}{2},$$

$$\eta = a + g \qquad \text{and} \qquad \nu = ag - e\eta.$$

We can now make some comments on the stability of the system. If $\eta^2 > 4\nu$, the system is said to be overdamped and the temperature in the tank moves slowly to a new stable value. If $\eta^2 = 4\nu$, the system is said to be critically damped and moves monotonically to the new stable value as quickly as possible. If $\eta^2 < 4\nu$, the system is said to be underdamped and oscillates about the new final position, the amplitude of the oscillations becoming gradually smaller. It is then possible to design the most favorable control system of this type by designing the coil so that $\eta^2 = 4\nu$.

Obviously we can proceed in the same manner to obtain a $t_c - \theta$ relationship. It is, however, of vastly less interest than the $t - \theta$ relationship just developed.

LIST OF SYMBOLS

(Omitting obvious designations; i.e., temperature scales, abbreviations of units and dimensions, stream designations in diagrams, chemical symbols, heat unit abbreviations.)

Latin letters

A	Surface area
A_c	Cross-sectional area
a, b, c	Arbitrary constants

a, e, f, g, j, p	Arbitrarily defined symbols, on p. 426 only
C	Heat capacity
C_c	Heat capacity of a heating coil
C_N, C_O, C_V	Discharge coefficients for nozzles, orifices, and Venturis, respectively
$C(t)$	Heat capacity, to be considered as a function of temperature
${}_{t_1}C_{pm_{t_2}}$	Mean molal heat capacity at constant pressure, over the temperature range, t_1 to t_2
C_v, C_p, C_{pm}	Molal heat capacities, at constant volume, at constant pressure, and mean value at constant pressure, respectively
D	Differential operator, d/dx
d	Usual exact differential operator
d'	Path-dependent differential operator
d_o, d_p	Diameters, of orifice opening and inside pipe, respectively
E	Internal energy, in general
e	Exponential operator
F	Loss due to friction, expressed as a head of fluid; also written as ${}_1F_2$ to indicate the loss from point 1 to point 2.
f	Friction factor, dimensionless
G	Mass velocity, expressed as mass/(unit time)(unit area)
g	Local acceleration due to gravity
g_c	Gravitational constant
H	Enthalpy
H_f	Fin height, used only on pp. 423-24
ΔH_t	Heat of reaction at temperature t
ΔH^0_{298}	Heat of reaction at 298 °K
$\Delta H^0_{F,298}$	Standard heat of formation, at 298 °K
$\Delta H^0_{c,298}$	Standard heat of combustion, at 298 °K
h	Head, in feet of flowing fluid; also used as film coefficient of heat transfer, unsteady state section only
J	Mechanical equivalent of heat, 778 ft-lb$_f$/Btu
K_i	Arbitrary constants of integration, i = an integer
k	Thermal conductivity, expressed as heat units/(unit time)(unit length)(unit temperature difference)
L	Distance or length
M	Molecular weight
m	Mass in general
m_c	Mass of a heating coil, used only on p. 426
n	Number of components
P	Pressure, expressed as force/unit area
Q	Heat energy in transit; also written ${}_1Q_2$ to emphasize heat in transit between points 1 and 2
q	Rate of transfer of heat
R	Gas law constant, consistent units
r	Radius
R_i	Manometer reading; $i = N, o,$ or V
R_m	Manometer reading

R_N, R_O, R_V	Manometer readings for nozzle, orifice, and Venturi applications, respectively
S	Liquid surface area
s	Roots of polynomials in D-operator format
T	Absolute temperature
T	Higher of two temperatures, in unsteady state section only
$\bar{T}$	Average temperature
$T(z)$	Temperature as a function of distance, z
t	Temperature
$t^{(i)}$	ith iterated value of temperature; i = integer
$t(z)$	Temperature as a function of distance, z
$t(\theta)$	Temperature as a function of time, θ
t_a	Air temperature
t_c	Coil temperature
t_d	Datum temperature
t_f	Fluid temperature
t_1	Inlet temperature
t_o	Initial temperature
Δt_{lm}	logarithmic mean temperature difference, defined as $\dfrac{\Delta t_1 - \Delta t_2}{\ln(\Delta t_1/\Delta t_2)}$
U	Internal energy, mechanical; also used as overall heat transfer coefficient, unsteady state section only
u	Average velocity over the cross section in which fluid is flowing
v	Specific volume, expressed as volume per pound
W	External mechanical work, positive when done by the system; also written ${}_1W_2$ to indicate the amount of external work between the points 1 and 2
w	Mass rate of flow, expressed as mass per unit time
w_o	Width of a fin
z	Distance in general

Greek letters

α	Constant in Pv^{α}
α	Arbitrarily defined function in unsteady state section only
β	Equal to d_o/d_p, in Section 9.4
β	Arbitrarily defined function in unsteady state section only
Δ	Usual difference symbol
ε	A value which becomes zero when a correct trial value of the root is obtained
ζ	Designation of an angle
η	Function defined and used on p. 427 only
θ	Time
κ	Equal to C_p/C_v

ν	Function defined and used on p. 427 only
ρ	Density
ϕ	Function
ψ	Function

PROBLEMS

The organization of problems in this chapter is as follows:

Series 1–99: Heat balances
Series 100–199: Mechanical energy balances
Series 200–299: Unsteady state and length-dependent energy balances

In the specifications of the following problems there is no guarantee either that all the necessary data are supplied, or that there is no unnecessary information. Missing data essential to the solutions of the problems will be largely in the category of physical or thermodynamic properties, obtainable from tables in the text or from handbooks. Atmospheric pressure may be assumed for any stream for which a specific pressure is not noted.

▶ HEAT BALANCES

9.1 Bromine is heated from −10 °C to 50 °C, the latter temperature being below the boiling point. The data below may be used to estimate the latent heat of fusion. No information exists on heat capacities of bromine in the liquid and solid states. Kopp's rule may be used to estimate these quantities.

	Vapor Pressure, mm	
t °C	Br_2	H_2O
−28.4	7.6	—
−7.3*	44.4	2.65
0	65.9	4.58
3.7	76.0	5.97
22.5	190.0	20.44

Calculate the heat required, expressed in Pcu per pound-mole bromine.

9.2 Glacial acetic acid at 80 °C is to be cooled by water in a heat interchanger. The final temperature of the acid is to be 25 °C. Water is available at 17.5 °C and may be heated to 52 °C. The interchanger is to cool 2000 gal of the acid per day. The overall heat transfer coefficient in the cooler is 120 Btu per (hr)(ft^2)(°F). Calculate the quantity of water to be used.

• • • • •

* Melting point of bromine.

9.3 Toluene, C_7H_8, has a heat of vaporization of 86.5 cal per gm at 100 °C. The C_p of the liquid is 0.44; that of the vapor is 0.32. Calculate:

a. heat of vaporization in calories per gram-mole at 40 °C.
b. heat of vaporization in Btu per pound-mole at 40 °C.
c. enthalpy of toluene vapor at 150 °C relative to the liquid at 18 °C.

9.4 Flue gases in passing at constant pressure over a superheating unit are cooled from 900 to 750 °F. The pressure is 29.2 in. Hg. The gases analyze CO_2, 12%; CO, 5%; O_2, 2%; N_2, 81%. One thousand pounds of saturated steam per hour is fed to the superheater at 67 lb (gauge), steam emerging at the same pressure containing 100 °F of superheat. Calculate the cubic feet per hour, entrance conditions, of flue gas required.

9.5 Flue gases, containing, on a dry basis, 12% CO_2, 1.5% CO, 6% O_2, and 80.5% N_2 leave a furnace at 800 °F and 29.8 in. Hg. They are cooled to 300 °F under a waste heat boiler generating 320 lb per hr of 50 lb gauge dry, saturated steam, and leave the flue at this temperature and a pressure of 29.6 in. Hg. Further analysis shows that, on a wet basis, the flue gases contain 8.7% water vapor. Inlet water to waste heat boiler is at 75 °F. Calculate the cubic feet of gas leaving the flues per minute.

9.6 Fifty thousand cubic feet (dry, 29.2 in. Hg, 68 °F) of flue gas per hour, analyzing 15% CO_2, 81% N_2, and 4% O_2, leave a furnace at 800 °F. This gas is cooled to 300 °F under a waste heat boiler, generating process steam, dry, saturated, at 10 lb gauge pressure. Feed water enters at 70 °F. Calculate the pounds of steam which may be generated per hour.

9.7 In an ammonia oxidation unit, the air is preheated in the interchanger by the hot gases leaving the converter. The latter analyze 12% NO, 18% H_2O, 2% O_2, and the remainder N_2. They enter the exchanger at 820 °C, the rate of flow being 1.24 cu ft per min. This rate of flow is measured at the outlet conditions, 145 °C and 29.2 in. Hg. Calculate the Btu extracted per hour from the converter gases.

9.8 A flue gas analyzing 20% CO_2, 10% $H_2O(g)$, and 70% N_2 is cooled from 1200 °C to 200 °C. Calculate the heat extracted from each cubic foot of gas (measured at 30 in. of Hg, 200 °C).

9.9 The following data are available for the integral heat of solution of NaOH at 25 °C:

Mole H_2O/mole NaOH	ΔH_{soln}, cal/gm-mole NaOH
0	−0
2	−4,600
4	−8,300
6	−9,900
8	−10,200
10	−10,300
16	−10,300
20	−10,200
50	−10,100

Calculate the number of Btu necessary to concentrate in a long tube, forced circulation evaporator, 100 lb of 18.2% solution NaOH to 52.6% under a pressure of 1.94 psia.

DATA The boiling point of the 52.6% solution at this pressure is about 195 °F. Assume products at this temperature and feed at 75 °F. Room temperature is 68 °F and heat losses from the evaporator are negligible.

9.10 An evaporator is fed 1000 lb per hr of 10% caustic solution at 30 °C. Operating at 20 in. of vacuum gauge, with a barometer at 754 mm Hg, 30% caustic is produced. The average specific heat of the feed is 0.88. Heat is supplied by steam at 5.3 lb per sq in. gauge and condensate leaves at 202 °F. The boiling point of 30% caustic under these conditions is 86 °C. Calculate the pounds of steam required to operate the evaporator, assuming that radiation may be considered to be negligible.

9.11 Ten tons per day of 25° Baumé hydrochloric acid is to be produced by absorption of HCl in water. The gas fed to the tower is an air-HCl mixture of 25% HCl. The exit gas is to contain 0.2% HCl. Calculate the heat that must be dissipated if the temperature of the liquid is to be maintained at 20 °C throughout the tower and the temperature of the gas remains substantially unchanged.

9.12 The grainer was, at one time, a common form of concentrator and crystallizer for manufacturing common salt from brine. The following data were obtained in a performance test on such a unit.

Heating surface of the steam coils = 1700 sq ft.
Average temperature, entering brine = 120 °F.
Average temperature, brine and crystals in grainer = 180 °F.
Average steam temperature in coils = 225 °F.
Average temperature of condensate from coils = 220 °F.
Quality of entering steam = 100%.
Weight of condensate, 24 hr = 80,000 lb.
Pounds salt produced, plus adhering brine, 24 hr = 30,000 lb
Percentage of water in wet salt = 21.0% (based on total yield of salt and brine).
Percentage of NaCl in entering brine = 26.0%.

Calculate:

a. heat required to bring the brine to the temperature of the grainer.
b. heat required for evaporation of water.
c. total heat supplied to grainer.

9.13 In the manufacture of urea from ammonia and carbon dioxide, the first reaction product is ammonium carbamate ($NH_4CO_2NH_2$). Ammonia and CO_2 are fed to a 400-atm reactor at 200 °C; the reactor is maintained at this temperature so that the carbamate is formed at 200 °C. The heat that must be removed to maintain this temperature is found to be 67,000 Btu per lb-mole of carbamate formed. The reaction is:

$$CO_2 + 2\,NH_3 = NH_4CO_2NH_2.$$

Calculate the standard heat of combustion of carbamate at 25 °C for combustion products $H_2O(g)$, $CO_2(g)$, $N_2(g)$.

9.14 Butane (C_4H_{10}) is being dehydrogenated in a process shown in the accompanying flow sheet, with off-gas analysis as indicated. Only $2\frac{1}{2}\%$ of the butane is converted to butene (C_4H_8); a total of 5% reacts, thereby accounting for the generation of the off-gas and the carbon. Unconverted butane is recycled to the reactor from the separation plant.

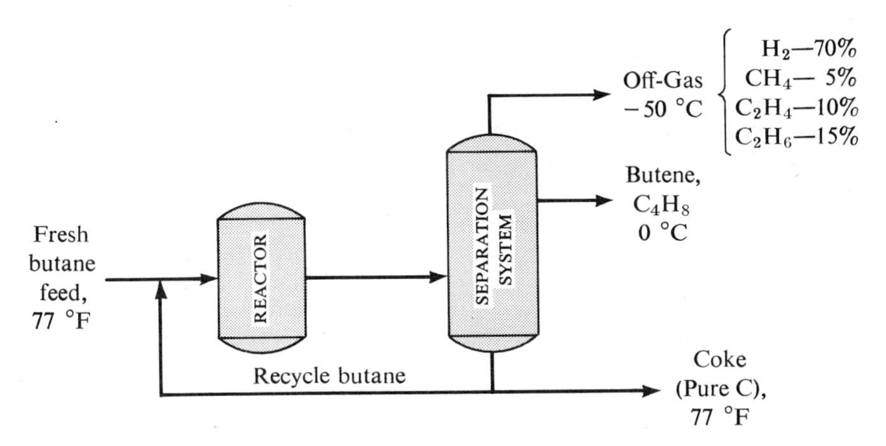

All streams are at atmospheric pressure. All temperatures are as indicated on the flow sheet. Calculate the amount of heat required to be added to the system (reactor plus separation system) per ton of fresh butane feed.

DATA AND ASSUMPTIONS Heat capacities may be considered to be constants over the small temperature ranges involved. They are reported below as Pcu per (pound-mole)(°C). Standard heats of formation are reported as kilocalories per gram-mole.

Substance	$\Delta H^0_{F,298}$	C_p
C(s)	0	3.45
H_2(g)	0	6.5
CH_4(g)	−17.889	8.0
C_2H_4(g)	+12.496	9.5
C_2H_6(g)	−20.236	11.5
C_4H_8(g)	+0.280	19.5
C_4H_{10}(g)	−29.812	22.0

9.15 The concentration of a dilute acid containing 40 weight % H_2SO_4 at 70 °F is to be raised to 60 weight % by the addition of pure liquid H_2SO_4 which is available at 150 °F. The final temperature must not be allowed to exceed 150 °F. Calculate:

a. heat removed per pound of final solution.
b. temperature if no cooling were used and no heat loss took place during mixing.

9.16 The statement has been made that regardless of the fuel used, the enthalpy and heat capacity of flue gas, *per standard cubic foot*, is essentially independent of the fuel composition. This statement is supposed to hold only for combustion with *air*, and with equal excesses. Demonstrate the correctness of this statement by calculation of heat capacity and enthalpy for two extreme fuels, pure carbon and pure hydrogen, each burned completely with 50% excess air.

9.17 In the hypothetical reaction

$$A + B = C,$$

the heat of reaction at 20 °C is −10,000 cal per gm-mole of C. The heat capacities, in calories per (gram-mole)(°C), are:

$$A: \quad 6.0 + 0.05t,$$
$$B: \quad 6.5 + 0.20t,$$
$$C: \quad 15.0 + 0.60t.$$

In these quantities, temperature, t, is expressed in °C. Calculate the temperature at which the reaction changes from exothermic to endothermic.

9.18 Pyrofax (pure propane) at $11 per 100 lb and natural gas at $1.25 per 1000 standard ft^3 are both available. The natural gas has a *gross* heating value of 950 Btu per standard ft^3 and contains 67.0% ethane (C_2H_6), 32.3% methane (CH_4), and 0.7% N_2. Calculate the proper choice of fuels, based on *net* heating values.

9.19 A form of pyrites ore contains 84% FeS_2, 2.0% water, and 14.0% inert inorganic gangue. Twenty tons of this ore are burned per day to provide SO_2 for the manufacture of sulfuric acid. The burner gas contains 8.1% SO_2, 10.5% oxygen, and 81.4% nitrogen, and leaves the roaster at a temperature of 800 °F. The cinder contains no residual sulfur, contains all of the original iron as Fe_2O_3, and leaves the roaster at a temperature of 750 °F. The air for combustion enters at 77 °F and 50% relative humidity. The ore also enters at 77 °F. Assume that the mean heat capacity of the cinder is 0.18 Btu per (lb)(°F). Calculate the heat loss per hour from the roaster.

9.20 Pure pyrites (FeS_2) is fed to a Herreshof burner at about 25 °C, the resulting cinder being entirely Fe_2O_3. The air enters at 77 °F. The cinder leaves at 1000 °F and the gases (8.7% SO_2, 9.8% O_2, 81.5% N_2) leave at 800 °F.

Assuming that no SO_3 was formed in the combustion, calculate the heat radiated per ton of pyrites charged.

9.21 In the process of manufacturing nitric acid by oxidation of ammonia, the ammonia is burned to NO and H_2O at a temperature of 1000 °C. The NO is then oxidized, using the excess air from the first reaction to NO_2, and the NO_2 is absorbed in water to produce nitric acid and NO, by the reaction:

$$3NO_2 + H_2O = 2HNO_3 + NO.$$

A small fraction of the ammonia fed may also be thermally decomposed to nitrogen and hydrogen, the latter burning immediately to water.

In the absorption tower, water enters the top of the column, absorbs the NO_2, and leaves as 65% nitric at the bottom. The column operates at 100 psia and essentially at atmospheric temperature. Both sensible heat changes and radiant heat losses may therefore be neglected. The analysis of the converter gas is 6.12% NO, 11.4% O_2, 9.6% H_2O, 72.9% N_2. The gas leaving the absorption column contains oxides of nitrogen equivalent to 0.2% NO. Calculate:

a. fraction of the ammonia which decomposes in the reactor.
b. cooling requirement in the absorption column in units of Btu per ton of 65% acid.

9.22 One ton per hour of limestone is being calcined in a continuous rotary furnace. The analysis of the limestone is 85% $CaCO_3$, 14% $MgCO_3$, and 1% water. It is charged to the furnace at 75 °F. Calcination is complete and the product gases leave at 650 °F. Air from the surroundings at 100 °F leaks into the system so that the product gas is not pure CO_2, but only 92% CO_2. Calcined lime leaves at 1400 °F. Assuming no radiant heat losses, calculate the heat required to operate the process.

9.23 Carbon black is an important industrial product, large tonnage being compounded into rubber for tires. A large proportion of the industrial black is made by controlled combustion of fuel oils. The heat generated in the partial combustion is used to decompose the remainder of the fuel oil.

The gases and the black leave the furnace at 2500 °F and are quenched by spraying water, at 90 °F, directly into the gas stream. The carbon black is then removed by electrostatic precipitation and by scrubbing before the gas stream finally passes to the stack.

The fuel acts in all respects like hexadecane ($C_{16}H_{34}$). This fuel enters the furnace at 25 °C, as does air. The approximation of the latter as dry will introduce no appreciable error into the calculations.

In one such installation, the effluent gases analyze 28% H_2, 6% CO_2, 5% CO, and the remainder nitrogen and water vapor. The latter includes both the quench water and the water formed in the partial combustion. 60% of the carbon charged in the fuel is recovered as pure carbon black, with the remainder being burned to CO or CO_2. Calculate:

a. pounds of carbon black per pound of oil fired.
b. pounds of quench water required per pound of carbon black.
c. temperature of the products after quench.

9.24 A vertical lime kiln is charged with pure limestone ($CaCO_3$) and pure coke (carbon), both at 25 °C. Air, dry, at 25 °C, is blown in at the bottom and provides the necessary heat for decomposition of the carbonate by burning the carbon to CO_2. The lime (CaO) leaves the bottom at 950 °F, containing no carbon or unburned limestone. The kiln gases leave at 600 °F, containing no free oxygen. The molal ratio of $CaCO_3$:C in the charge is 1.5:1. Calculate:

a. the analysis of the gas leaving.
b. radiation losses.

9.25 In a contact process H_2SO_4 plant the gases fed to the converter analyze 10.5% SO_2, 10.0% O_2, 79.5% N_2, and enter at 250 °C. In the total two-converter system the SO_2 is substantially 100% converted to SO_3, 80% in the first converter, and heat losses from the system are negligible. The heat of reaction is used to raise the temperature of inlet gases to reaction temperature from the 250 °C inlet.

The temperature of gas leaving the last reactor is 475 °C; the gas is cooled to 375 °C in the first interchanger. Calculate the temperature of feed gas when it leaves this interchanger.

9.26 Benzene is burned completely, using 50% excess air, preheated to 150 °F. The benzene is charged at 25 °C. Calculate the theoretical flame temperature.

9.27 A producer gas analyzing 35% CO and 65% N_2 is burned with 25% excess air. The gas is fed at 25 °C. Calculate the theoretical flame temperature when:

a. air is at 25 °C.
b. air is at 500 °F.

9.28 Which gas will give the higher *maximum* theoretical flame temperature, Pyrofax (essentially propane, C_3H_8) or acetylene?

9.29 In a contact sulfuric plant, a mixture of gases containing 10% SO_2, 12% O_2, and 78% N_2 has been thoroughly cleaned and dried. The gas temperature is 90 °F. Before entering the catalyst chamber, gas must be brought to a temperature of 800 °F by passage through a heat exchanger, hot gases from a previous passage over the catalyst giving up the necessary amount of heat. The flow is 15,000 cu ft per hr, measured at inlet temperature and 30.4 in. Hg. Calculate the heat supplied per hour to produce the required temperature.

9.30 A plant produces nitric acid by oxidizing ammonia to nitric oxide, subsequently oxidizing this gas to the peroxide and absorbing the latter in water. The ammonia-air mixture must be fed to the converter at a temperature of 600 °C, and in the mole ratio of 1:6. However, the ammonia must not be preheated lest it decompose. Therefore, the air is preheated above 600 °C before mixing, so that it may contain enough enthalpy to bring the temperature of the mixture to 600 °C. Both ammonia and air enter at 25 °C. Calculate the temperature to which air must be preheated.

9.31 Formaldehyde is manufactured by the vapor phase oxidation of methanol with air on a silver catalyst. The primary reaction is

$$CH_3OH + \tfrac{1}{2}O_2 = CH_2O + H_2O.$$

However, a secondary reaction may occur consecutively, namely,

$$CH_2O + \tfrac{1}{2}O_2 = HCOOH.$$

The feed mixture, containing oxygen in the amount theoretically required, enters the catalyst chamber at 170° F. Analysis of the converted gas indicates that 60% of the methanol was oxidized. These gases leave the reaction zone at 1100 °F. It is estimated that not over 2% of the methanol reacting follows the second reaction. Calculate the heat lost (or gained) by the process per mole of methanol fed.

ADDITIONAL DATA Heat capacities, in Btu per (pound)(°F) may be assumed to have the following constant values: methanol vapor, 0.45; formaldehyde vapor, 0.55; formic acid vapor, 0.65.

9.32 Carbon disulfide can be manufactured by reacting methane with sulfur vapor to form 1 mole of carbon disulfide and 2 of hydrogen sulfide. The reaction takes place on a silica-gel catalyst and the gas must be fed at 700 °C to insure reaction. The reacted gas is passed to an interchanger as shown in the flow sheet and the recovered heat is used to vaporize the sulfur and to preheat the methane gas.

FLOW SHEET

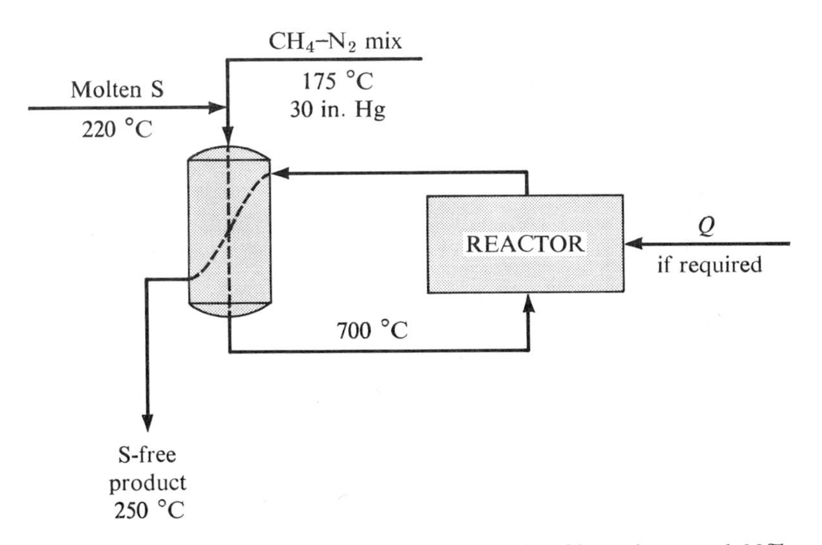

Preliminary runs indicate that, with a feed of 80% methane and 20% nitrogen, and a ratio of 2 moles of methane per mole of sulfur, the reaction will be 85% complete. Assume that there are no radiant heat losses either from the reactor or from the heat interchanger. Then calculate:

a. whether sufficient heat may be obtained in the interchanger to raise the reacted gases to 700 °C., *or*,
b. the amount of additional heat input needed.

ADDITIONAL DATA Heat capacities, in Btu per (pound-mole)(°F) are as follows: H_2S, 9.5; CS_2, 7.6; sulfur liquid, 7; sulfur vapor, 8.7.

Sulfur fed molten at 220 °C; natural gas, fed at 175 °C, 30 in. Hg; minimum product temperature is 250 °C, after the interchanger.

9.33 An ammonia converter is fed a stoichiometric mixture of nitrogen and hydrogen at 425 °C. Sixteen percent conversion is obtained. The resulting mixture has the ammonia removed by cooling and condensation; the remaining gas is then reheated to 425° and recycled.

Assume that the ammonia is condensed at 120 °F and that the recycle gas is saturated at this temperature. The process is conducted at 250 atm. Obtain the latent heat and vapor pressure of ammonia from the *Chemical Engineers' Handbook*. Calculate the net heat removed in the cooler, with no radiant heat loss.

9.34 The gases to an SO_2 converter analyze 8.7% SO_2, 9.8% O_2, and 81.5% N_2. They enter the converter at 425 °C and 80% of the SO_2 is converted to SO_3. The process is adiabatic. Before entering the second converter, they must be cooled to 400 °C. Calculate:

a. temperature leaving the first converter.
b. heat abstracted in the cooler per pound of sulfur.

9.35 A catalytic reactor converts naphthalene (N) to naphthaquinone (NQ) and phthalic anhydride (PA) by partial combustion processes as shown:

$$\underset{\text{N}}{C_{10}H_8} + \tfrac{3}{2}O_2 \rightarrow \underset{\text{NQ}}{C_{10}H_6O_2} + H_2O.$$

$$\underset{\text{N}}{C_{10}H_8} + \tfrac{9}{2}O_2 \rightarrow \underset{\text{PA}}{C_8H_4O_3} + 2CO_2 + 2H_2O.$$

All materials are in the gas phase. 40% of the N fed is converted to NQ, the remainder to PA. Two hundred and fifty percent excess air is used over that required for NQ.

The process* is conducted as follows: N is fed at 218 °C, its boiling point; air is fed at 30 °C; the products are removed at 240 °C, the boiling point of PA being 284.5 °C.

DATA AVAILABLE

Heat Capacities

N: solid; liquid: 0.402 at 87 °C; vapor: nothing available; PA, NQ: nothing available.

Latent Heats

A. *Fusion*

Naphthalene: 35.6 cal/gm at 80.2 °C.
PA, NQ: nothing available, melting points, respectively: 130 °C and 94 °C, estimated.

B. *Vaporization*

Naphthalene: 75.5 cal/gm at 218 °C.
PA: nothing available, boiling point, 284.5 °C.
NQ: nothing available, boiling point, 245 °C, estimated.

Vapor Pressures

t, N	t, PA	P, mm Hg
145.5	202	100
167.7	228	200
193.2	257	400
217.9	284.5	760

• • • • •

* These statements bear no necessary relation to fact.

Standard Heats

Of combustion to form CO_2(g), H_2O(l):

Naphthalene (s): 1232.5 kcal/mole.
NQ (s): 1100.8 kcal/mole.
PA (s): 783.4 kcal/mole.

Calculate the required heat input for 100 lb of naphthalene consumed.

9.36 Preliminary calculations are being conducted on a process involving the following reaction:

$$\underset{\text{Aniline}}{2C_6H_5NH_2(g)} \xrightarrow{\text{cat}} \underset{\text{Diphenylamine (DPA)}}{(C_6H_5)_2NH(g) + NH_3(g)}$$

The process* is intended to run as follows: charge vapor-phase aniline at its boiling point; remove the products at the boiling point of DPA; operate at atmospheric pressure. Only 10% of the aniline charged will be converted to DPA; the remainder will be recycled after DPA and ammonia are removed in an auxiliary operation. Calculate the heat that must be added or removed to maintain the steady state.

The following data are available:

Heat Capacities

NH_3 (g): $5.92 + 8.963 \times 10^{-3}T - 1.764 \times 10^{-6}T^2$.
DPA (s): 0.343 cal/(gm)(°C) at 26 °C.

DPA (l):

t °C	Sp Ht
54	0.437
56	0.441
66	0.480
53	freezing point

Aniline (l):

t °C	Sp Ht
0	0.478
50	0.521
100	0.547

Aniline (g): sp ht = 0.8 (estimated).

Latent Heats

DPA: fusion, 25.23 cal/gm at 53 °C.
Aniline: vaporization, 103.68 cal/gm at 183 °C.

• • • • •

* Duplication of conditions to be expected in an operating process is not intended.

Vapor Pressures

p, mm Hg	Aniline, t °C	DPA, t °C
20	82.0	175.2
100	119.9	222.8
200	140.1	247.5
400	161.9	274.1
760	184.4	302.0

Standard Heats

1. Formation, 25 °C: NH_3(g): −11.04 kcal/gm-mole.
2. Combustion, 25 °C : Aniline (l): −812 kcal/gm-mole.
   DPA (s): −1536.2 kcal/gm-mole.
   Products: CO_2(g), H_2O(l), N_2(g).

9.37 Phosphate rock is comprised of 75% $Ca_3(PO_4)_2$, 20% SiO_2, and 5% inerts. The phosphate reacts with sand (SiO_2) and coke (carbon) in an electric furnace. Phosphorus vapor, CO_2, and CO are the gaseous products. A slag is formed, comprised of $CaSiO_3$, the inerts, unburnt carbon, and unconsumed sand. Quantities charged are figured as follows:

a. sand should be added in sufficient quantity to provide a 5% excess over the stoichiometric supply for the phosphate.
b. coke is added in 30% excess over the quantity required if all were to burn to CO_2; CO_2:CO in the gaseous products is in the ratio of 9:1.
c. twenty per cent of the rock is unreacted and recycles to the furnace.

All feeds are at 1 atm and 25 °C. All products leave at 1 atm, the gas at 300 °C and the slag at 1600 °C. Calculate the net heat gain (or loss) per ton of phosphate rock charged.

ADDITIONAL DATA

	ΔH^0_{F298} kcal/mole	Boiling Point °C	Fusion °C	Approx. C_p, Pcu/(mole)(°C) T in °K*
CO_2	−94.052			$18.04 + 0.0000447T$
CO	−26.416			$6.89 + 0.001436T$
P	0 (solid)	280	44.1	Solid: 5.5(0–44 °C) Liquid: 2.8(44–280 °C) Gas: 2.0
SiO_2	−203.4		2230	$10.90 + 0.0055T$
$Ca_3(PO_4)_2$	−988.9		975	
Inerts			2000	0.25 Btu/(lb)(°F)
C	0		Decom-posed	$2.673 + 0.00262T$
$CaSiO_3$	−378.6		1540	$27.95 + 0.00206T$

* All C_p expressions have been truncated to simplify the calculations and may be used for phases and the temperature ranges involved in the problem.

Latent heat of vaporization of phosphorus—11.88 kcal per gm-mole.
Latent heat of fusion of phosphorus—0.615 kcal per gm-mole.

Molecular Weights

P — 31; Ca — 40; Si — 28; C — 12; O — 16.

▶ Mechanical Energy Balances

9.100 Two moles of an ideal gas, in a rigid container originally at a temperature of 70 °F, are heated to 150 °F. Heat losses from the system during the heating may be considered negligible. Assuming that the heat capacity may be taken as constant, and that an acceptable value of C_p is 7.5 (molar units), calculate:

a. Q,
b. ΔU,
c. W.

9.101 The situation is exactly as described in Problem 9.100 except that the gas is now contained in a cylinder fitted with a frictionless piston by which a force of one atmosphere is maintained constant during the process. Calculate the same quantities.

9.102 Oxygen gas is stored in a tank under a pressure of 100 psig at a temperature of 25 °C. The capacity of the tank is 1.5 cu ft. Gas is allowed to flow from the tank to the atmosphere until the tank pressure has dropped to one half its initial value. Assume reversibility, the ideal gas law and $C_p = 7$ (molar units). Calculate the mass of oxygen remaining in the tank when the expansion has taken place:

a. isothermally,
b. adiabatically.

9.103 SO_2 is expanded reversibly and isothermally from 250 °F and 5 atmospheres to a pressure such that, after cooling at constant volume to 32 °F, the final pressure is atmospheric. Assuming the gas to be ideal and that the heat capacity may be treated as a constant with $C_p = 10$, in molal units, calculate for the whole process, per mole of SO_2:

a. Q, c. ΔU,
b. W, d. ΔH.

9.104 Repeat Problem 9.103, using the following relationship for C_p, but still assuming that the SO_2 obeys the ideal gas law.

$$C_p = 6.945 + 10.01 \times 10^{-3}T - 3.415 \times 10^{-6}T^2,$$

where T is in degrees Kelvin.

9.105 The conditions under which a gas is stored are changed from 75 °F and 1 atm to 200 °F and 10 atm. Name at least two different reversible processes by which this batch operation might be accomplished. Then, assuming applicability of the ideal gas law and constant heat capacities, $C_p = 7$, $C_v = 5$, calculate per mole of gas for each process:

a. Q, c. ΔU,
b. W, d. ΔH.

9.106 Assuming the applicability of the ideal gas law and constant heat capacities, derive expressions for the slope of the lines representing, on a P-v diagram, each of the following nonflow processes:

a. isothermal,
b. adiabatic.

For the same initial conditions the line representing which process is indicated to lie above the other on the diagram.

9.107 A cylinder 6 in. in diameter is closed by a piston. With air in the cylinder and the piston position 1 ft from the closed end of the cylinder, the air temperature is 80 °F. The total weight of the piston and adjusting weights is 700 lb. Two processes will be considered.

a. the gas is heated at constant pressure until the volume occupied by the gas has increased by 75%.
b. the gas is expanded adiabatically by sudden removal of 80% of the total applied weight, but catches will stop the piston after the expansion has again increased the volume by 75%, if the expansion tends to progress beyond this point.

Calculate, and if necessary discuss, the work done by the system (i.e., the gas alone) in each case.

9.108 Water from a reservoir is discharged through a turbine, the discharge pipe delivering the effluent through a 24-in. diameter pipe open to the atmosphere 100 ft below the reservoir level. The process is isothermal, at 40 °F. The turbine delivers 300 hp. Frictional losses may be neglected. Calculate the rate of discharge in gallons per minute.

9.109 Water is taken from a lake through a 1-ft diameter line to a turbine and discharged from the turbine through an 18-in. diameter line. The total drop in level, from lake to end of the 18-in. line is 100 ft, with the turbine being located 50 ft below the level of the water in the lake. Assume that frictional effects are negligible. Calculate the horsepower theoretically delivered by the turbine when water is flowing isothermally at 40 °F and at a rate of 9500 gal per min.

9.110 Oil, whose specific gravity is $0.91^{25/4}$, flows by gravity into a tank, 3 ft in diameter, maintained under atmospheric pressure. A pump picks up the oil from the tank at such a rate that the level of oil in the tank remains constant. The oil is pumped at 80 °F, isothermally, through a 2.0-in. diameter iron pipe to the top of an absorption tower and is discharged freely from the end of the line, 75 ft above the level of the liquid in the tank. The tower is under 150 psig. The pump and motor have a combined efficiency of 78%, the motor being rated 10 hp. Assuming frictionless flow, calculate the rate of oil flow in gallons per minute.

9.111 A pump capable of delivering exactly 1 hp is placed in a line composed of clean iron pipe with 2.00-in. inside diameter. The line carries an incompressible fluid (ρ = 58 lb per cu ft) at a rate of 2.62 lb per sec in isothermal flow, and this line terminates in a spray nozzle discharging to the atmosphere. The supply of fluid comes from a large tank, also under atmospheric pressure, whose level is 10 ft above the nozzle. Assuming that friction in the line and the pump can be neglected, calculate the pressure which can be expected immediately behind the nozzle.

9.112 The situation is exactly as in Problem 9.109 except that the water temperature increases 1 °F within the system. The process is essentially adiabatic; i.e., $Q = 0$, frictional losses accounting for the rise in temperature. Calculate:

a. the frictional loss, in feet of fluid flowing.
b. the horsepower delivered by the turbine.

9.113 The situation is exactly as in Problem 9.111 except that: first, the temperature of the outlet fluid must be 30 °F higher than that at inlet, and, second, a heater is inserted in the line to assure that this condition may be met. The pressure behind the nozzle is to be set at 90 psig. The average heat capacity of the oil is 0.3 Btu per (lb)(°F) and may be assumed to be unaffected by pressure. Calculate:

a. the thermal energy requirement to be supplied in the heater.
b. the thermal energy requirement if the pressure in the tower were atmospheric.
c. the frictional loss in horsepower in the nozzle, assuming that the droplets that emerge as spray contain negligible kinetic energy.

9.114 A clean steel pipe line is 10 miles long. Oil is to be pumped isothermally at a rate of 60,000 barrels (1 barrel = 42 U.S. gal) per day throughout the year. The discharge end is 20 ft, and the inlet end 300 ft, above sea level. Both ends of the line may be represented as tanks under atmospheric pressure. Three pumping stations, each of 200 hp and each with overall efficiency of 68.5%, will be installed. Calculate:

a. required standard pipe size.
b. maximum amount by which capacity could be increased in summer in an emergency.

AUXILIARY DATA

Maximum summer temperature: 120 °F.
Minimum winter temperature: 21 °F.
Specific gravity of the oil can be assumed constant at 0.82.
Fittings in the line make the equivalent length of straight pipe 10% greater than the distance traveled.

t °F	Viscosity Centipoises
14	123
130	20.5
210	5.8

NOTE The frictional losses will vary with the change in viscosity of the oil, which is a function of temperature, as indicated above. See Figure 9.10 as a means of estimating the frictional loss.

9.115 A standard 1-in., schedule 40, brass line (I.D. = 1.049 in.) carries diphenyl at a temperature of 350 °F (sp gr = $0.992^{73/4}$) in a heating plant. The line extends from a tank with fixed liquid level 4 ft above ground to a boiler below the liquid level. A $\frac{1}{2}$-in. diameter orifice, placed in the line as a flow-measuring instrument, registers 6 in. on a mercury manometer. Assume (1) isothermal flow, (2) atmospheric pressure at each end of the system, and (3) frictional losses other than in the orifice, negligible with respect to other power requirements. Calculate the percentage of the power requirement represented by the orifice.

9.116 A standard orifice, Venturi, or nozzle is placed in a vertical section of a line carrying an incompressible fluid and connected to a manometer in the usual manner. Derive expressions showing the relation of the velocity through the orifice to the reading on the manometer for flow in either direction.

9.117 A conical nozzle is placed in the bottom of a vertical pipe whose inside diameter is 6.0 in. The nozzle diameter is 0.856 in. Water enters the side of the 6-in. pipe at a constant rate. After operating in this fashion for a

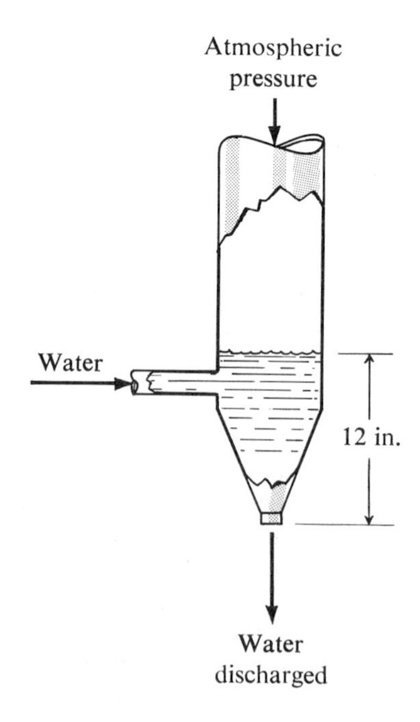

sufficient length of time, the level of water in the pipe stabilizes to a height of 1 ft above the discharge end of the nozzle. Flow is isothermal at 70 °F. The nozzle coefficient of discharge is known to be 0.98. Calculate:

a. the rate of flow in pounds per second.
b. the height of liquid in the pipe if the flow rate is 5.0 lb per min.

9.118 An orifice in a standard, schedule 40, 4-in. line (I.D. = 4.026 in.) is 1.00 in. in diameter and is sharp-edged. The line carries 50% glycerine (specific gravity of $1.126^{20/4}$) at a rate of 70 gal per min at an average temperature of 70 °F.

The substitution of a Venturi for the orifice is being considered. If installed, its throat diameter will also be 1.00 in. In either case a mercury manometer will be used. Calculate:

a. the manometer readings with either instrument.
b. the reduction in horsepower effected by the installation of the Venturi.

9.119 A crude petroleum, whose specific gravity is $0.90^{25/4}$, is flowing through a pipe whose inside diameter is 15.0 in. With a 6-in. diameter sharp-edged orifice in this line, with oil flowing at 80 °F, and with a mercury manometer indicating the pressure drop across the orifice, the manometer reading is 10 in. Calculate the manometer reading if the orifice is replaced by:

a. a nozzle with 6.00-in. throat diameter;
b. A Venturi with 6.00-in. throat diameter.

9.120 A light oil at 70 °F is flowing at a rate of 10,000 gal per hr through a 3-in., standard, schedule 40, iron pipe (I.D. = 3.068 in.). The specific gravity of the oil is $0.80^{20/4}$. An orifice or a Venturi is to be installed in the line, either one with throat diameter = 1.00 in. The more economical of the two is to be selected.

AUXILIARY DATA Installation will be operated 24 hr per day, 365 days per yr. Cost of orifice—$100 installed; cost of Venturi—$600 installed. Fixed charges—20% of total cost, annually. Cost of power—2¢ per kwhr. Efficiency of pump and motor—80%.

9.121 Air, at an average velocity of 20 ft per sec, enters the lower end of a standard 2-in. (I.D. = 2.067 in.) steel pipe, 100 ft long and inclined at an angle to 45 degrees to the horizontal. The entrance pressure is 10 lb per sq in., gauge, and the temperature is 20 °C. Under isothermal conditions, and assuming that the air will act as an ideal gas, calculate the exit pressure and velocity.

9.122 Air at 150 lb per sq in. gauge and 80 °F flows through a 1.5-in. orifice in a standard 3-in. (I.D. = 3.068 in.) line. The pressure drop over the orifice is measured and found to be 10 lb per sq. in. Calculate the flow rate, in pounds per minute, assuming that the air acts like an ideal gas.

9.123 In equipment similar to that described in Problem 9.122, the following changes in conditions are of interest:

a. the gas is changed from air to pure CO_2 without change in any other conditions.
b. the inlet air pressure is reduced to 100 psig but the pressure drop over the orifice is unchanged.
c. the inlet air temperature is increased to 120 °F.

In each of these individual situations calculate the percentage change in the mass rate of flow.

9.124 Compressed air is delivered to a long straight, horizontal section of clean iron pipe with inside diameter of 3.068 in. Conditions at inlet are 40 lb per sq in. gauge and 70 °F. Fifty pounds per minute will flow through 1000 ft of this line, the flow being essentially isothermal. Assume that the ideal gas law applies. Calculate the pressure at the end of the line.

9.125 Consider the following changes in the system described in Problem 9.124 and calculate the exit pressure in each case:

a. the equivalent length of the line is unchanged but the exit end is 500 ft above the entrance.
b. the flow rate is reduced to 50% of the previous value.
c. the gas is 21% CO_2 and the remainder N_2.
d. the length of the line is increased to 2000 ft with a 100 ft rise from entrance to exit.

▶ Unsteady State and Length-Dependent Energy Balances

9.200 A flat brass* rod, surrounded by air at the constant temperature of 25 °C, is heated at one end only. The rod is 5 ft long, $\frac{1}{8}$-in. thick, and 2 in. wide. The heat transfer coefficient between the rod and air is estimated to be 1.0 Btu hr^{-1} ft^{-2} $°F^{-1}$. The heated end is maintained at a temperature of 100 °C. When the steady state temperatures are attained, calculate:

a. temperature at the unheated end.
b. temperature at the midpoint.

9.201 The experiment outlined in Problem 9.200 is conducted without knowledge of the heat transfer coefficient. The following data are taken:

x ft	t °C
1.0	52.6
2.0	35.0
3.0	28.5
4.0	26.6

Calculate the heat transfer coefficient.

9.202 The same rod listed in Problem 9.200, and under the same conditions, is heated from both ends, each maintained at the same temperature. Calculate:

a. temperature at the midpoint.
b. temperatures at one-quarter and three-quarter points.

• • • • •

* Thermal conductivity relatively constant and equal to 60 Btu ft^{-1} hr^{-1} $°F^{-1}$.

9.203 The same situation listed in Problem 9.202 again exists, but this time with one end at 100 °C and the other end at 125 °C. Calculate:

a. temperature at midpoint.
b. point of minimum temperature.

9.204 The same rod heated from one end only, is now surrounded by a jacket through which 50 cu ft per hr of air flow countercurrent to the heat flow in the rod. The air flow rate is measured at inlet, where the temperature is 25 °C. The pressure will be atmospheric throughout and heat transfer coefficient is estimated to be 2.0 Btu hr^{-1} ft^{-2} $°F^{-1}$. Calculate:

a. temperatures at midpoint and at unheated end.
b. repeat the calculations if the rod is square but of the same cross section as the flat rod.
c. repeat the calculations if the rod is circular but of the same cross section as the rectangular rod.

9.205 Repeat the calculations required in Problem 9.204 if the same quantity of air enters cocurrently.

9.206 A very thin, trapezoidal, copper* fin extends from a flat wall whose temperature is constant at 100 °C into air whose temperature is constant at 95 °F. The fin is 3 in. at the base, 2 in. at the free end, and 3 in. in length. The heat transfer coefficient is estimated to be 40 Btu hr^{-1} ft^{-2} $°F^{-1}$. Calculate the temperature-length profile, assuming that heat flows only in the axial direction in the fin.

9.207 The situation described in Problem 9.206 is unchanged except that the fin is of annular cross section on a pipe of 1-in. outside diameter. The outside diameter of the fin is 3 in. Calculate:

a. temperature at the extremity of the fin.
b. temperature 1 in. from the tube.

9.208 A 50-gal, well-agitated, jacketed kettle has 25 sq ft of heating surface. The contents, for which specific heat = 1.0, are heated from 70 to 120 °F in 2 hr, using saturated steam at 240 °F in the jacket, the condensate leaving saturated. Calculate:

a. the additional time required to heat from 120 to 170 °F.
b. the heat transfer coefficient, in Btu hr^{-1} ft^{-2} $°F^{-1}$.

9.209 In a well-agitated tank, the exposed surface is 125 sq ft. The liquid loses (or gains) heat through this surface at a rate of 2.5 Btu hr^{-1} ft^{-2} $°F^{-1}$. The surroundings remain at 90 °F.

The tank contains 1000 gal of water at 75 °F. It is planned to heat this water using saturated steam at 250 °F in a steam coil whose heat transfer area is 10 sq ft. Condensate will leave saturated. From previous tests the heat transfer coefficient is known to be 200 Btu hr^{-1} ft^{-2} $°F^{-1}$. Neglecting the heat capacity of the tank walls, calculate the length of time to heat the water to 200 °F.

• • • • •

* Thermal conductivity may be taken to be 60 Btu ft^{-1} hr^{-1} $°F^{-1}$.

9.210 If, to the tank described in Problem 9.209, fresh water at 75 °F is added continuously at a rate of 5 gal per min, and heated water is withdrawn from the tank at the same rate, calculate:

a. the possibility of the water in the tank reaching a temperature of 200 °F, and if possible, the time to do so.
b. the temperature of the water in the tank at the end of 2 hr, if the steam supply is turned off at the end of the first hour.

9.211 A holding tank of 150 gal capacity is one-quarter full of 66 ° Bé sulfuric acid. The contents of the tank are kept well agitated and are heated by steam in a coil. The heat transfer coefficient is estimated to be 125 Btu $hr^{-1}\ ft^{-2}\ °F^{-1}$. The steam and condensate are both saturated at 212 °F. The acid in the tank is at 80 °F and fresh acid of the same concentration and a temperature of 50 °F is added at a constant rate of 20 gal per min, while acid is withdrawn at a constant rate of 10 gal per min. Calculate:

a. temperature of acid when the tank is full.
b. any time when a minimum or maximum temperature occurs.

9.212 Given a cylindrical tank, as shown, originally half full. Initial water temperature is t_0. Water at temperature, t_1, is added at rate $Ke^{-B\theta}$, where θ represents time and the constant B has units of reciprocal time. The constant K has units of pounds per unit time.

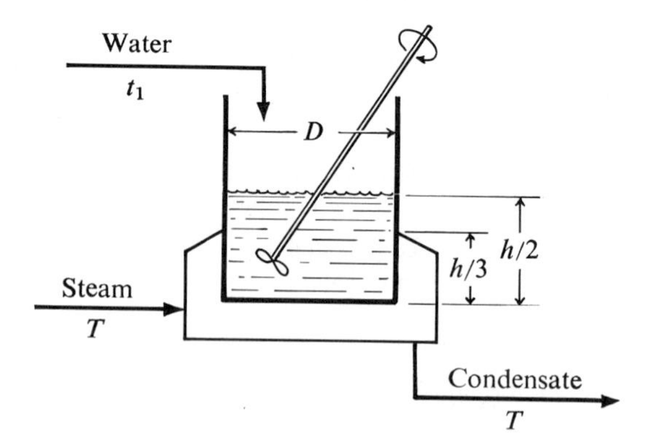

Steam is fed to the jacket at the same instant that water begins to be added to the tank. The heat transfer coefficient, U, is constant; heat capacity may be assumed to be constant.

Set up the ODE's required to calculate:

a. time to fill the tank; *solve* this equation.
b. temperature in the tank when full; do not solve but arrange in good ODE form, and indicate boundary conditions to be employed.

NOTE The datum temperature in part (b) is at the discretion of the engineer. Choose first a general datum, t_d. Subsequently establish the best datum to simplify the solution.

9.213 A perfectly stirred kettle containing 50 gal of a liquid with sp ht = 0.6 and sp gr = $0.9^{25/4}$, is heated electrically. The electrical input is sinusoidal and can be represented as

$$q = A + B \sin \omega\theta.$$

The inlet temperature is 20 °C. The constants have the following values:

$$A = 300 \text{ Btu},$$
$$B = 50 \text{ Btu},$$
$$\omega = (\pi/10) \text{ rad per min}.$$

Calculate:

a. temperature after 1 hr of heating.
b. temperature-time curve over the first 20 min.

9.214 Repeat Problem 9.213 but add fresh liquid, with the same characteristics as the original batch, at a rate of 10 gal per hr and remove liquid from the kettle at the same rate.

9.215 Repeat Problem 9.213 but let the sinusoidal input be voltage. Assume that the resistance of the heater is constant and equal to 45 ohms.

$$A = 440 \text{ volts},$$
$$B = 100 \text{ volts},$$
$$\omega = (\pi/10) \text{ rad per min}.$$

9.216 A steam heater is available with one hundred $\frac{3}{4}$-in. diameter tubes in parallel, each 14 ft long. The unit is built for use with low pressure steam, 5 psig maximum. Six hundred gallons per minute of aniline at 75 °F is to be heated. At this rate the heat transfer coefficient is estimated to be 250 Btu hr^{-1} ft^{-2} $°F^{-1}$.

Calculate the temperature to which the aniline could be heated under these circumstances.

DATA FOR ANILINE sp ht = 0.512; sp gr = $1.022^{25/4}$.

9.217 In a double pipe heat interchanger, 20 ft long, the cooling agent is water which enters the jacket at 90 °F at a rate of 20 gal per min. Kerosene, at 150 °F enters countercurrently in the inner pipe at a rate of 10 gal per min. The inner pipe has an inside diameter of 1.049 in. Kerosene has a specific gravity of $0.805^{60/60}$, a specific heat of 0.58, and a thermal conductivity of 0.09 Btu hr^{-1} ft^{-1} $°F^{-1}$; these figures may be assumed not to depend appreciably on temperature in the range encountered.

The kerosene leaves the interchanger at 120 °F. Calculate:

a. exit temperature of the water.
b. heat transfer coefficient, in usual units, assuming that it is constant throughout the interchanger.

9.218 The cooling effected in the interchanger described in Problem 9.217 is found to be insufficient. A second interchanger of identical design is to be added in series, so that the total length available will be 40 ft. Both fluids will flow countercurrently through the entire length, entering at the same temperatures listed in Problem 9.217. For the purpose of calculating the expected outlet temperatures, the heat transfer coefficient is estimated to be 300 Btu hr^{-1} ft^{-2} $°F^{-1}$. Calculate:

a. outlet temperature for kerosene.
b. outlet temperature for water.

9.219 From past experience your supervisor estimates that the heat transfer coefficient in an interchanger such as described in Problems 9.217 and 9.218 will rise 2 units per degree rise in kerosene temperature. For the 40-ft interchanger, he wants you to recalculate both outlet temperatures based on this assumption, using $U = 300$ at $T = 150$ °F.

SUPPLEMENT TO CHAPTER 9

A. Extracting Roots from a Polynomial

There are many cases in engineering requiring the extraction of a root of a polynomial. Our present interest is due to the practice of reporting heat capacities in polynomial form. Then, when temperature is the unknown, a solution of a polynomial is required. The degree of the polynomial is usually at least the third if heat capacities of gases are involved. It is, therefore, pertinent to have a good method for the extracting of roots from polynomials.

S9.1A Number of Real Roots

Although there are two roots for a quadratic equation, three for a cubic, etc., there is the possibility that some of the roots are not real. In the problems with which we are concerned in this book, the *roots which we seek* shall all be real. It is a matter of some concern to know the maximum number of real roots available, and their sign. To start, we know that complex roots occur in complementary pairs. A cubic must, therefore, have one real root. It need not necessarily be positive. If we are solving for temperature in the cubic equation representing an enthalpy balance we must have a positive real root or the equation is incorrect, indicating some error in drawing the balance. For the maximum number of real roots, we can resort to the sign test, which states:

a. The maximum number of real positive roots of the polynomial $f(x)$ is indicated by the number of sign changes in $f(x) = 0$; and
b. The maximum number of real negative roots of the polynomial $f(x)$ is indicated by the number of sign changes in $f(-x) = 0$.

As an illustration take the $f(T)$ in Example 9.8.

$$f(T) = aT^3 + bT^2 + cT + d = 0,$$

where

$$\begin{aligned} a &= 0.885 \times 10^{-6}, \\ b &= -0.881 \times 10^{-3}, \\ c &= -6.386, \\ d &= 2132.8 \end{aligned}$$

The number of sign changes is two (one between a and b and one between c and d). There are no more than two real positive roots. Then,

$$f(-T) = -aT^3 + bT^2 - cT + d,$$

and there is one sign change between b and c, so that there is no more than one negative root. The necessary condition that one positive root can exist is met.

S9.2A Location of Roots

A root exists at the point $x = z$, when $f(z) = 0$. Then, if we designate a value of x slightly below z as $z-$, and another slightly above z as $z+$, there must be a sign change between the values of $f(z-)$ and $f(z+)$. The general region in which a root shall lie can be ascertained by substituting a series of values of $x = z$ and locating the interval containing a root as that from $z-$ to $z+$ when $f(z+)$ is positive and $f(z-)$ is negative, or vice versa.

As illustration, use the $f(T)$ from Example 9.8. When $T = 320$, $f(320) = 20.3$, and when $T = 330$, $f(330) = -46.5$. Therefore a root exists between 320 and 330 °K. Of course, in this instance the previous trial-and-error iteration had provided this information.

Knowing the general location of a root, the engineer needs a method of narrowing this region until the root is located with any desired precision. The *Newton-Raphson* method performs admirably in this respect. It converges rapidly. It gives an improvement in precision of approximately one order of magnitude per iteration. It is a good method for a computer since the algorithm can be simply programmed. It is a good method even for a desk calculator, since the convergence is rapid. It is particularly good in this latter instance since the operator can relax the exact predicted trial value for the next iteration, using rounded values near the predicted value when such relaxed values will simplify the calculation in the next iteration. Naturally, this method is not the only one which is useful in extracting roots. It is chosen as one that is good under circumstances commonly encountered.

S9.3A Newton-Raphson Method

Take any $f(x)$ with root at z, so that $f(z) = 0$. Now choose a value $z^{(1)}$. It must follow that $z^{(1)} + h = z$. Therefore, $f(z^{(1)} + h) = 0$. Expanding the function about the point $z^{(1)}$ in Taylor series and truncating after the term linear in h,

$$f(z^{(1)} + h) = 0 \doteq f(z^{(1)}) + hf'(z^{(1)}).$$

It then follows that

$$h \doteq -\frac{f(z^{(1)})}{f'(z^{(1)})}.$$

However, since the series was truncated, this value of h is only an estimate and $z^{(1)} + h$ does not equal z, the root, but is merely a second approximation of this root. We could then write

$$z^{(2)} = z^{(1)} + h,$$

or in more general terms

$$z^{(i+1)} = z^{(i)} - \frac{f(z^{(i)})}{f'(z^{(i)})}. \tag{S9.1A}$$

Equation S9.1A is the algorithm for the calculation of successive iterations.

The iterative procedure is continued until $h^{(i)} \leq \varepsilon$, where ε is some maximum allowable error, established by the engineer. Since any value of ε can be chosen, the precision with which the root can be determined is essentially unlimited.

As an example, take the cubic equation in Example 9.8. Obtain the root in the region of 122 °F (50 °C) with $\varepsilon \leq 0.5$ °F. This value of ε will allow us to round accurately to 1 °F. The equation reads,

$$f(T) = 0.885\left(\frac{T}{100}\right)^3 - 0.0881\left(\frac{T}{10}\right)^2 - 6.386T + 2125 = 0.$$

Then,

$$f'(T) = 2.655\left(\frac{T^2}{10^6}\right) - 0.1762\left(\frac{T}{100}\right) - 6.386 = 0.$$

Since 122 °F = 50 °C = 323 °K, we shall choose 320 °K as $z^{(1)}$. We also note that $\varepsilon \leq 0.28$ in this scale. Then,

$$\begin{aligned} f(320) &= 28.0652, \\ f'(320) &= -6.678, \\ h(320) &= -4.4026. \end{aligned}$$

and $z^{(2)} = 320 - (-4.4) = 324.4$ °K.

Since $|z^{(2)} - z^{(1)}| > 0.28$ another iteration is required. We can let $z^{(2)}$ equal 324.4 as calculated, or we can relax this estimate to obtain a simpler number with which to deal. In this case, use 324 °K. Recalculation, as shown in the following table, indicates that $z^{(3)} = 324.60$ °K, so that $\varepsilon = 0.4$ and no further iteration is required.

i	$z^{(i)} = T_i$	$f(T_i)$	$f'(T_i)$	$h_i = \varepsilon$	$z^{(i+1)}$
1	320	28.0652	−6.678	+4.4	324.4
2	324	1.3529	−6.678	+0.202	324.6

The root of the equation is $T = 324.2$ °K $= 51.2$ °C $= 124$ °F.

S9.4A Graphical Interpretation of Newton-Raphson Method

The first iteration is illustrated graphically in Figure S9.1A. In that figure, the root, z, is located at the intersection of the curve with the x axis. The

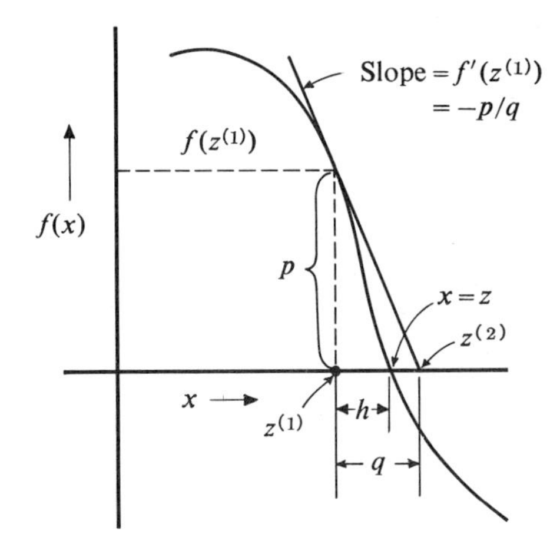

FIGURE S9.1A Graphical representation of first Newton-Raphson iteration.

choice of the point $x = z^{(1)}$ established the value of $f(z^{(1)}) = p$, and the slope of the curve, $f'(z^{(1)}) = -p/q$. Then the new estimate of z will be

$$z^{(2)} = z^{(1)} - \frac{f(z^{(1)})}{f'(z^{(1)})},$$

$$= z^{(1)} + \frac{p}{p/q} = z^{(1)} + q.$$

A characteristic second iteration is shown in Figure S9.2A.

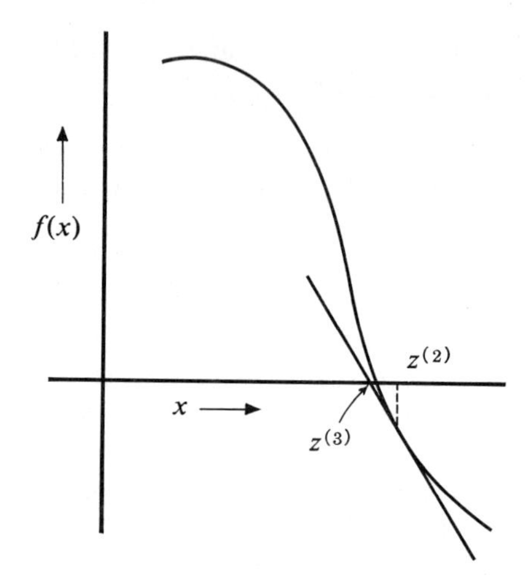

FIGURE S9.2A Graphical representation of second Newton-Raphson iteration.

LIST OF SYMBOLS

Latin letters

a, b, c, d, p, q	Arbitrary constants
$f(x)$	Any function of x
$f'(x)$	Derivative of $f(x)$
h	Defined by $z^{(i+1)} = z^{(i)} + h$
T	Absolute temperature
x	A variable
z	A root of $f(x) = 0$
$z^{(i)}$	The ith estimate of z; i = an integer

Greek letter

ε	A stated small increment

SUPPLEMENT* TO CHAPTER 9

B. Degrees of Freedom in Process Specification

The original discussion of degrees of freedom, in the Supplement to Chapter 4, contained restrictions specifically limited to material balances. We shall now remove these special restrictions so that the principles of degrees of freedom may be applied to energy balance situations. One modification will be the inclusion of pressure and temperature as important specifications. Another is the need for the inclusion of one or more energy streams.

S9.1B Variables

In *heat balances* (i.e., no important mechanical terms) the variables in each stream continue to be $(N_{sp} + 2)$, composed of $(N_{sp} - 1)$ to specify composition, one to specify rate (or quantity in a batch process), one to specify temperature, and one to specify pressure. A new variable is added, the transient energy, which in heat balances may be written as Q, the net heat flow. The number of variables in this restricted class of processes may then be written

$$N_v = N_S(N_{sp} + 2) + 1. \qquad \text{(S9.1B)}$$

When using Equation S9.1B it should be clearly understood that the one variable added for an energy stream must then represent a *net* heat flow. If there is interest in a heat supply *and* in a radiation loss (e.g., a distillation column) the net heat is the supply required by the process *plus* the heat loss. When there is interest in both of these streams the number of variables must be modified to read

$$N_v = N_S(N_{sp} + 2) + 2. \qquad \text{(S9.2B)}$$

In subsequent remarks and examples dealing with heat balances, Equation S9.1B is used exclusively, with proper care being taken to consider it as accounting for a net heat stream only.

When *mechanical balances* are encountered, it is no longer true that composition, temperature, pressure, and rate will completely specify the stream. At least two other quantities are required, definitions of the kinetic

• • • • •

* No problems are supplied with this supplement. As noted in the Supplement to Chapter 4, degree of freedom calculations may be performed on any of the problems pertinent to the general subject of the chapter. All problems in Chapter 9 are therefore suitable.

and potential energies contained in the streams. No additions are required for the specification of the Pv-work. Other forms of energy that may be present (e.g., electrical, atomic, chemical) will be ignored in subsequent remarks. If they do exist, there must be further modification of the number of variables to complete the specifications of these forms.

Velocity and vertical position will specify the kinetic and potential energies. There are, then, not $(N_{sp} + 2)$ variables per stream but rather $(N_{sp} + 4)$.

Further, one net energy stream is a poor accounting in mechanical processes. We would be better advised to include three, one for the heat flow, Q, one for the mechanical work, W, and one for the frictional loss, F. Then the total number of variables for this situation would be written

$$N_v = N_S(N_{sp} + 4) + 3. \qquad \text{(S9.3B)}$$

S9.2B Restrictions

The energy balance constitutes a restriction not counted in the case of material balances. There is, of course, only one possible energy balance, in contrast to the C material balances. In the case of processes without important mechanical effects, the energy balance becomes the heat balance.

Inherent Restrictions

The previous inherent restrictions which removed pressure and temperature from consideration are now inapplicable. It is true, with heat balances, that pressure may be classed as unimportant as long as enthalpies may be assumed to be independent of pressure. However, the phases in each stream must be specified, and the usual manner of doing so is to include information on pressure.

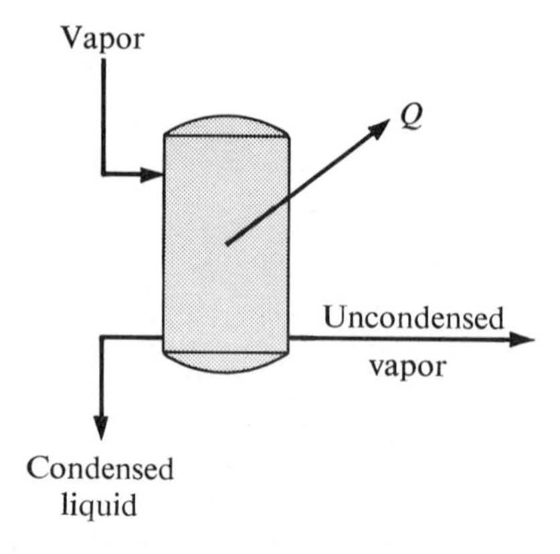

FIGURE S9.1B Diagram of partial condenser.

Certain new inherent restrictions may be added. For example, in the stream divider (Figure S4.2) we noted that the compositions of the two exit streams B and C must be identical, providing $(C - 1)$ restrictions. (Since the process is physical we can use N_{sp} and C interchangeably in this situation.) We can now specify that temperatures and pressures of these streams must also be identical, making the total number of inherent restrictions $(C + 1)$. We can also find cases where the compositions are *not* necessarily identical, but two streams leave a single unit and must therefore be identical in temperature and pressure. Such instances occur in any situation where two phases leave the same point in any element or unit. Examples are the partial condenser and the partial reboiler. These elements are represented in Figures S9.1B and S9.2B. A stage in a physical separation unit (e.g., plate in

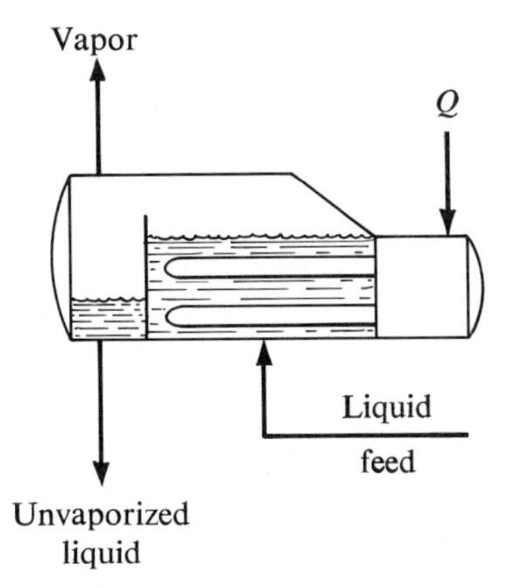

FIGURE S9.2B Diagram of partial reboiler.

a distillation column) is another example. The relationship between the compositions of the streams leaving such elements may be governed by phase equilibrium, in which case we shall have to include restrictions imposed on the compositions of the streams because of the establishment of physical (or chemical) equilibrium.

S9.3B Degrees of Freedom

Degrees of freedom will be calculated as before:

$$N_{df} = N_v - N_r.$$

When no new inherent restrictions have been introduced by the addition of energy considerations, we have added one new variable to N_v (i.e., the energy stream) and countered by removing it through the use of the energy balance as an additional restriction. Thus, without new inherent restrictions, the degrees of freedom are the same as those calculated previously plus 2 N_S, this latter addition to N_{df} resulting from retention of temperature and pressure as important variables.

Examples

We shall illustrate the calculation of degrees of freedom for a few of the elements considered in the Supplement to Chapter 4, looking critically at the additional $2N_S$ degrees of freedom. One new element, the partial condenser (or partial reboiler) will be added. Balances in situations where mechanical effects are important will also be considered.

Example S9.1B The Condenser

This element, as indicated in Figure S4.1, is a two-stream element. It is then reasonable, using a net Q, to state that its degrees of freedom are the same as calculated in the Supplement to Chapter 4, but with $2N_S = 4$ new degrees of freedom to account for temperatures and pressures of the feed and effluent streams. The distribution of the resulting $(C + 4)$ degrees of freedom might include $(C + 2)$ to specify the vapor feed, with the other two for the pressure of the condensate stream and its temperature. It is *possible* to specify the net heat flow rather than temperature but the latter is more likely to be used in design and certainly more likely to be measured in performance.

Example S9.2B The Stream Divider

In the Supplement to Chapter 4 the degrees of freedom for this element were $(C + 1)$. For the three streams involved (see Figure S4.2), we should expect to add six additional degrees of freedom for temperatures and pressures to obtain $(C + 7)$. Normally pressure and temperature identities are assumed to exist in the two outlet streams. We have then two new inherent restrictions, leaving $(C + 5)$ degrees of freedom. Again it would be common to use $(C + 2)$ of these to specify the feed completely. Two of the other three might well be assigned, one to the ratio of rates of the two streams and one to their common pressure. The remaining one would normally be used to specify heat losses expected from the element.

It is worthwhile to note that the identities of temperature *and* pressure assume that there are not separate valves in the two outlet lines, but that the divider is more in the nature of a vane. Two separate valves would eliminate the inherent restriction of equal pressures. The system would then become more complicated, for the losses due to friction in the valves would be related to the ratio of streams required. Let us suppose that this situation is the one which faces us. Then, if we use two degrees of freedom to specify these outlet pressures and one to specify the ratio of rates, we have allowed nothing for heat loss. Since the heat loss is due to geometry and the temperature of the system with relation to the surroundings, it cannot be varied independently and some information regarding it must be assigned to utilize a degree of freedom. In this instance we have *not*, then, properly utilized the degrees of freedom. The logical revision is to specify the heat losses and let the pressure of the inlet stream be unspecified. It will build up to the necessary value to overcome the frictional losses of the valves and to provide the outlet pressures specified. It is a quantity which should be calculated, not specified.

This example illustrates the danger of thinking in terms of complete freedom in the assignment of degrees of freedom. Knowledge of the equipment and how it operates is essential. The engineer must have this information. The study of the design and operation of equipment is an obvious necessity in his education.

Example S9.3B The Mixer

The degrees of freedom for material balance calculations only were found to be $C(N_S - 1)$, where $(N_S - 1)$ were the number of streams mixed. If we now add two degrees of freedom for the temperature and pressure of each stream we shall have $(CN_S + 2N_S - C) = (N_S - 1)(C + 2) + 2$. This number is only two greater than the number of degrees of freedom required to specify the feed streams. These two might be used to specify the outlet pressure and the heat loss, or outlet pressure and temperature.

Another, and probably more common, problem with mixtures would be the preparation of a given mixture from $(N_S - 1)$ streams of known composition. Counting $(C + 2)$ specifications for the mixture and $(N_S - 1)(C - 1)$ for the specification of the known compositions, there will be left $(3N_S - C - 3)$ as the number of additional specifications which might be made. We might decide to use two degrees of freedom to specify the temperatures and pressures of each of the feed streams. Then we shall have $N_S - (C + 1)$ unspecified items left. If this number is to be such that all the degrees of freedom are specified,

$$N_S - (C + 1) = 0,$$

and

$$N_S = C + 1.$$

Examination of mixing processes will indicate that this number of streams is quite logical. If the mixed product is, for example, three-component, then three streams will be mixed, in general, to give the fourth stream, the product.

We have, in the preceding analysis, overlooked one important point. The mixing process does not ordinarily include a heat stream. What we have specified requires a very specific Q to maintain all the specified temperatures. The heat leak will have to provide this Q. It might, then, have been more logical to specify that the heat leak is zero (or of known value) and omit the specification of the outlet temperature.

Example S9.4B The Distillation Column

The degrees of freedom for material balance calculations were found to be $2C$. The element contains three streams (see Figure S4.4). With temperature and pressure to be considered, the degrees of freedom will now number $2C + 6$ as long as one *net* heat stream, included in the variables, is canceled by the energy balance restriction. Then we might use these degrees of freedom to specify:

The feed stream completely	$C + 2$
Pressure of product and bottoms	2
Temperature of product	1
Subtotal	$C + 5$

If we assign C degrees of freedom to outlet compositions as before the remaining degree of freedom could be used for the temperature of the bottoms, in which case the net heat could be calculated, or for the net heat in which case the temperature of the bottoms could be calculated.

In this case we shall probably want to consider a definite amount of heat leak. Now we must either consider that this item is not a specification, being only part of the net heat stream, or we can treat the number of variables as constituted by Equation S9.2B, in which we have allowed for separate specification of the heat leak and the heat supply. Either method is satisfactory.

Example S9.5B The Partial Condenser or Partial Reboiler

These two elements are shown schematically in Figures S9.1B and S9.2B. In principle they are similar. A stream enters. In the element it is split and leaves as two streams, one liquid and one vapor. Vapor is fed to the condenser and liquid to the reboiler; otherwise they are identical.

$N_v = 3(C + 2) + 1.$

N_r:

Material balances	C
Enthalpy balance	1
t and p identity of outlet streams	2
Total	$C + 3$

$N_{df} = (3C + 7) - (C + 3) = 2C + 4.$

These degrees of freedom may be distributed in many ways, but a common one would be:

$C + 2$ to specify the feed stream.
1 to specify the Q-stream (or temperature of outlet streams).
1 for pressure of the outlet streams.

$C + 4$ to this point, leaving C to be specified.

If the compositions of the two streams are related through physical equilibrium, as is often assumed, there are no further degrees of freedom, as these relationships use C degrees of freedom. If not so related, the C remaining degrees of freedom can be used:

$C - 1$ to specify one stream's composition.

1 to specify one stream's rate.

The following two examples deal with problems solved in Chapter 9 as illustrations of the energy balances.

Example S9.6B Cooler

For this example refer to Example 9.10. The element is a simple cooler. Referring only to the gas to be cooled, there are two gas streams and a heat stream.

$$N_v = 2(C + 2) + 1 = 2C + 5.$$

N_r:			
	Material balances	C	
	Energy balance	1	$C + 1$

$$N_{df} = (2C + 5) - (C + 1) = C + 4.$$

These degrees of freedom may be distributed with $C + 2$ for the feed and one for the outlet pressure. The remaining one can specify either the outlet temperature or the Q-stream. Specification of either one allows calculation of the other. The two basic types of problems (i.e., temperature unknown or Q unknown) described in Chapter 9 are thus justified by the analysis of degrees of freedom. Again note that two Q-streams may be indicated, one for required input and one for heat loss. In such a case, $N_{df} = C + 5$, and heat loss may be independently specified.

Analysis of the cooling fluid will provide exactly the same result. If the Q-streams are unspecified, the difference in the quantities calculated on the two sides of the element is the heat leak from or to the unit.

Example S9.7B Gas Producer

Refer to Example 9.12. The feed streams are the producer gas and air and the outlet stream is the flue gas. There is a net Q-stream which must take into account the heat of reaction combined with heat gain or loss from external sources.

$$N_S = 3.$$

$$N_{sp} = 4(CO_2, CO, O, N).$$

Independent Reaction

$$CO + \tfrac{1}{2}O_2 = CO_2.$$

$$C = 3(\text{C, O, N}).$$

$$N_v = N_S(N_{sp} + 2) + 1 = 18 + 1 = 19.$$

N_r:

Material balances	3
Energy balance	1
Zero concentrations	5
Fixed ratios (O/N in air)	1
Total	10

$$N_{df} = 19 - 10 = 9.$$

These degrees of freedom might be assigned as follows: 6 to temperatures and pressures of all three streams, one to rate and one to the percentage of CO in the fuel, and one to rate of air flow. Then the composition of flue gas and the heat loss are obtained consecutively from the material and heat balances.

Alternately, three degrees of freedom are assignable to the pressures, two to temperatures of fuel and air, two to rate and composition of the fuel, one to rate of air, and one to heat loss. The composition and temperature of the flue gas are obtained from the material and heat balances, respectively.

A third option finds the following disposition of the degrees of freedom: six to temperatures and pressures, three to rate and composition of the flue gas. Rates of fuel and air are then obtainable from material balances and the heat loss is subsequently obtained from the heat balance.

Example S9.8B The Pump

In this type of unit the material streams are two in number. The number of species is identical to the number of components. Therefore, from Equation S9.3B,

$$N_v = 2(C + 4) + 3.$$

The restrictions are all embodied in the material and energy balances, numbering $C + 1$. The number of degrees of freedom will be $C + 10$.

Usually we should consider that the difference in height from center line of inlet to center line of outlet is zero, or negligible and that the process is adiabatic. These three items can be considered as additional restrictions or as use of the three of the degrees of freedom. By either procedure, there are $C + 7$ specifications which can be made.

The usual assignment of degrees of freedom would recognize $C + 2$ items for the feed stream. Adding information on the inlet and outlet line sizes will provide for the calculation of the two kinetic energies and will use two degrees of freedom. Frictional losses can be specified (possibly as a pump efficiency) for one more. Two are left. If they are used for outlet temperature and work input, the pressure developed by the pump may be calculated. Alternately, if both outlet temperature and pressure are specified, the work required may be calculated.

Example S9.9B Pumping System

Refer to the pumping problem of Example 9.2. There are only two material streams, a net heat input, and a work input.

$$N_v = 2(C + 4) + 3.$$

The restrictions are $C + 1$ for material and energy balances and two for the inherent restrictions of atmospheric pressure at the two ends of the system. The degrees of freedom then number $C + 8$. In this case we really should reduce this number by one since we are only interested in a *net* heat stream, the frictional losses not being separately considered.

These $C + 7$ degrees of freedom are assigned in the problem, $C + 1$ to the feed (pressure already having been accounted in the inherent restriction), one to the outlet temperature, two to the heights of the two ends of the system, two to the velocities at the two ends of the system and one to horsepower input. The problem is suitably specified.

If the situation had been one involving flow through a line, or flow through a piece of equipment such as an interchanger, frictional losses could be estimated from the geometry. However, a good deal of information about the *internal* parts of the system would be required to eliminate this one degree of freedom. For example, with a pipe line, we should need to know the lengths of pipe of each diameter, the character of all fittings, and any obstructions or instruments. There is no anomaly in this requirement of a great deal of information to calculate one item in the energy balance. In writing the energy balance, the *external* boundaries of the system only are required. The *internal* parts of the system are used to establish the composite of frictional losses which is then represented as a single energy stream leaving the whole system. Therefore, in dealing with this sort of problem, the engineer must clearly differentiate between the boundary conditions and the internal parts. He must not assign a numerical equivalent, as degrees of freedom, to each item of information.

Example S9.10B Pumping System, Incompletely Specified

The following problem, somewhat paraphrased, can be found in a prominent chemical engineering text.

A pump takes water from a river and discharges it at a level of 100 ft above the river, with a velocity of 100 ft per sec. The river flows at a rate of 2 miles per hr. If 5000 gal per min is pumped, calculate the horsepower requirement of the pump, neglecting frictional losses.

Streams = 2; Components = 1.

$N_v = N_S(C + 4) + 3 = 13$ (See Section S9.1B)

N_r:		
	Material balances	1
	Energy balance	1
	Total	2

$N_{df} = 13 - 2 = 11.$

Specifications in the problem are as follows:

The 100-ft difference in height	2
The two velocities	2
(The inlet velocity is zero, since there is no velocity component in the direction of pumping; i.e., vertically.)	
The two pressures, assumed to be atmospheric	2
The rate of flow	1
Zero friction	1
Total	8

Obviously three specifications have been omitted. Information is required with respect to temperatures of both streams and to heat loss in the system. If the temperatures are specified, the net work, $Q - W$ can be calculated. W itself cannot be calculated unless Q is specified.

LIST OF SYMBOLS

Latin letters

C	Number of components
N_{df}	Number of degrees of freedom
N_S	Number of streams
N_{sp}	Number of species
N_v	Number of variables
Q	Heat in transit, positive when supplied to the system
W	Mechanical work, negative when supplied to the system

10 *Simultaneous Material and Energy Balances*

In Chapters 4 and 9, material and energy balances were discussed separately. In Chapter 9, problems involving energy balances were solved successively, first the material balances to obtain missing quantities and second the energy balances to calculate temperatures of outlet streams or quantities of heat. The present chapter deals with situations where the successive approach will not produce the missing quantities. The two sets of balances must be treated simultaneously rather than successively. We shall first survey the conditions that lead to this type of situation in both physical and chemical processes. Second, we shall illustrate representative problems in this area. In no instance shall we consider the case of fissionable materials, in which simultaneous energy and material balances would be automatically required.*

10.1 Need for Simultaneous Solutions

We are concerned with the case where the data supplied may be insufficient for a solution of all of the material balances that can be drawn independently

• • • • •

* The discussion of the situations in which simultaneous solution of energy and material balances is required will depend to a large extent upon an understanding of the principles of degrees of freedom. This material appears in the Supplement to Chapter 4 and in Chapter 9. When the subject is introduced in Chapter 4, Supplement, it is stated that the reader need not cover this material in order to understand the remainder of the book. Despite the reliance that is about to be placed on degrees of freedom in this chapter, the remark still holds. The reasoning behind the need for, and recognition of, situations involving simultaneous solution of the material and energy balances is the only aspect of this subject demanding this specific knowledge. The solution of the problems can be readily handled without the knowledge of the reasons for their existence.

around a process. This situation usually results when part of the basic knowledge of the process displaces information on mass quantities by an equivalent amount of information regarding heat quantities. Then one or more of the essential known quantities relates to energies, or enthalpies, and necessary items required to solve the material balances have been sacrificed in order to provide this information. The material balances may not be solved uniquely and will contain $C + 1$ unknowns in the C equations. Further, the energy balance may not be solved by itself since it contains unknown quantities of mass. But the two together contain exactly enough information to allow a simultaneous solution.

The most common specification of an energy term involves some statement regarding heat gain (or loss) by (or from) the system. Such a statement is made, for example, when the approximation of adiabatic operation is assumed, or when it is possible to predict in advance the exact amount of heat added or removed from the system. Another instance would include specifications of stream temperatures. Every such specification will not, however, require simultaneous solution of material and energy balances.

Each situation can be analyzed from the point of view of degrees of freedom and only in this manner can the need for the simultaneous solution be satisfactorily rationalized and readily detected.

Suppose that some general process is analyzed. There are N_S streams with C components and N_{sp} species, and in the N_S streams there is a total of z zero concentrations. Then the number of variables, N_v will be $N_S(N_{sp} + 2) + 1$, including one net energy stream. The number of restrictions will be: C for material balances, one for the energy balance, and z for zero concentrations or a total of $C + z + 1$. The degrees of freedom, N_{df} are

$$(N_S N_{sp} + 2N_S + 1) - (C + z + 1) = N_S(N_{sp} + 2) - C - z.$$

If the same process is analyzed for the number of degrees of freedom available for material balances, including temperatures and pressures, the number of variables will be reduced by one, since the net energy stream will not be considered. The number of restrictions will also be reduced by one, since the energy balance will not be involved and will, consequently, not be counted. The number of degrees of freedom will be identical, not counting restrictions in the material balance case for the temperature and pressure specifications. We can now examine the assignment of this number of degrees of freedom to ascertain under what conditions it is possible to solve successively material balances followed by energy balances. When this successive calculation is not possible, we have a situation requiring simultaneous solution.

It might be presumed that any time one of the degrees of freedom is used to specify anything about the net Q-stream, one of the necessary degrees of freedom required for unique solution of the material balances had been sacrificed. That this is not the case can readily be seen from energy balance calculations in Chapter 9, where an invariable specification of Q was made in any instance where an outlet temperature was to be calculated. In all these instances the material balances could be completed prior to the attack upon the energy balance. A degree of freedom has been used to specify information about the net Q in this instance, but one of the outlet temperatures has not been specified. In material balances it is assumed that all temperatures are unimportant, which is equivalent to assuming that all are known. When Q is specified and one temperature is unknown, we have used one degree of freedom (for Q) and omitted one specification (a value of t) *not* required for material balances, and the material balances are capable of solution prior to the attack upon the energy balance.

When an energy balance is required to obtain an estimate of the net q-stream, the second major type of energy balance as defined in Chapter 9 (i.e., Q unknown), there are also no omissions of specifications needed in material balances. Thus, the material balances are again capable of solution prior to the attack upon the energy balance.

Then the next logical question is: how might degrees of freedom be assigned so that it is *not* possible to solve the material balances and the energy balances successively. The answer is, of course, quite obvious: when we use one degree of freedom to specify something about the net Q-stream and others to specify temperatures of *all* streams. Now we have not regained the degree of freedom needed to supply information concerning Q by the omission of a temperature specification unimportant to material balances. Some illustrations will be helpful. The first three will treat situations, and the last two will solve specific problems.

Example 10.1 Simple Continuous Still

Simple continuous still and continuous flash distillation units have the same basic characteristics. The assumption is normally made that the liquid and vapor leave in physical equilibrium.* The unit is shown in Figure 10.1. We wish to analyze the degrees of freedom to see when simultaneous heat and material balances may be required by the choice of specific items of temperature, pressure, mass (rates), and concentrations.

Since this system deals with physical changes, components and species are identical in number. There are three streams. Therefore the number of variables, N_v, is $3(C + 2) + 1$.

• • • • •

* This specification of equilibrium between two outlet streams requires the introduction of two new inherent restrictions due to this relationship. Since we have previously ignored this type of restriction, special attention will be given to it in the treatment of this situation.

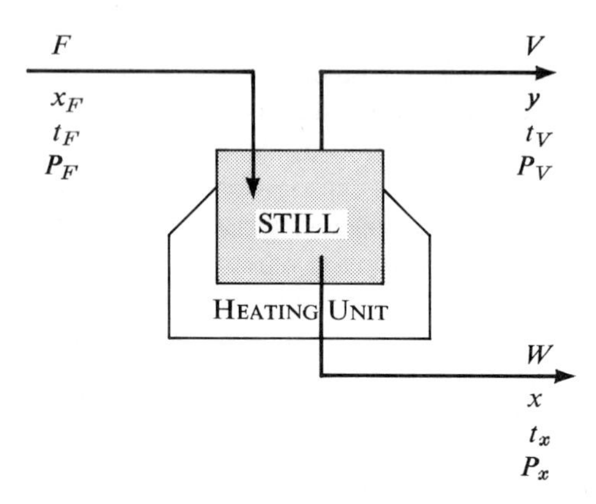

FIGURE 10.1 Flow sheet of simple continuous still.

The restrictions may be counted in the usual categories, material balances, an energy balance, and inherent restrictions. The balances number $C + 1$. The inherent restrictions are of two types. The first is due to the fact that both liquid and vapor leave the same element and, consequently, must leave at the same temperature and pressure. These restrictions therefore number two.

The number of restrictions that account for the equilibrium relationship between liquid and vapor are C in number. We argue this number in the following fashion. When the composition, temperature, and pressure of one stream is known, the composition of the other stream is known and the temperature and pressure identities have already been taken into account. There are fixed mass relationships between the two streams dependent upon the $(C - 1)$ compositional relationships. These composition and mass relationships total C inherent restrictions. [The normal argument of the number of restrictions due to physical equilibrium has not been presented in order to avoid a dependence on a background in physical chemistry. This argument may be found in Smith's *Design of Equilibrium Stage Processes* (New York: McGraw-Hill Book Co., 1963), p. 71. Smith shows that, with C components and P phases in equilibrium, the inherent restrictions number $C(P - 1)$. In our case, where $P = 2$, the restrictions must number C.]

The total number of restrictions in the present situation is, then, $2C + 3$. The degrees of freedom number $(3C + 7) - (2C + 3) = C + 4$.

For the purpose of discussion of the allocation of these degrees of freedom, simplify by choosing $C = 2$, the case of the binary mixture. Then $N_{df} = 6$. Start by assuming that the feed conditions are known, using $C + 2$, or 4, degrees of freedom and leaving two to be assigned. For one of these we might well be able to make a statement regarding the pressure in the still. The process probably operates at, or near, constant pressure and the choice of pressure will be at the discretion of the designer. Since both distillate and bottoms leave the same compartment, fixing one automatically fixes the other, as recognized in the restrictions.

There is one final degree of freedom to be assigned. Two logical selections might be: (a) the temperature of one (and therefore both) of the effluent streams; and (b) the net rate of heat supply to the process.

These situations require simultaneous solutions of material and energy (heat) balances. While they do not seem to fit the generality previously proposed, in fact, they do. But we must look at the unit as containing *two* energy streams to see the agreement. These two streams are the heat loss (or gain) from the unit and the heat requirement to keep the distillation operating at the required rate. By considering two streams, we have increased the degrees of freedom, since we have added a stream but no new restrictions. Now the same two specifications previously proposed would read: (a) the temperature of one (and therefore both) of the effluent streams *plus* negligible heat loss from the unit; or (b) the net rate of heat supplied to unit, where "net" indicates the difference between the two streams, implying that we must have information of some sort about each.

Now the situation is exactly as generalized in Section 10.1, and simultaneous solution of the heat and material balance equations will obviously be necessary.

Example 10.2 Continuous Rectification

Change the process in Example 10.1 only slightly. For the simple continuous still with distillate and bottoms in equilibrium substitute a continuous rectifying column in which the distillate and bottoms compositions are not in physical equilibrium. There are still three streams plus the heat input and C components in each stream. Therefore,

$$N_v = 3(C + 2) + 1.$$

The restrictions consist only of material and energy balances totaling $C + 1$ and leaving $2C + 6$ degrees of freedom. In each, $C + 2$ of the degrees of freedom will be assigned to the feed, leaving $C + 4$ for other assignment.

CASE 1 The temperatures and pressures of both effluent streams will be specified leaving C degrees of freedom to be used for composition specification. For a binary, $C = 2$ and the use of one each for compositions of distillate and bottoms would establish the composition of these two streams completely. The material balances may then be solved prior to investigation of the heat balance. The same compositional assignments will be recalled to have been made in the Supplement to Chapter 4.

CASE 2 Identical to Case 1 except that one of the outlet temperatures will not be specified and will be replaced by the specification of the net rate of heat input. It is clear that this change does not in any manner affect the items required for solution of the material balances prior to attack on the heat balance. The unspecified effluent temperature may be calculated by solution of the heat balance.

CASE 3 All temperatures and pressures, *plus* the net rate of heat input, will be specified. Including the feed specification, we have now used $C + 7$ of the degrees of freedom, leaving $C - 1$ for compositions of distillate and bottoms. For a binary, this number is *one*. The compositions of effluent streams are not now sufficiently specified to allow the solution of material balances alone. A simultaneous solution is required.

Example 10.3 Vertical Lime Kiln

A vertical lime kiln (see Figure 10.2) is fed a solid fuel and a limestone at the top. Air is blown in at the bottom. The oxygen in this stream is used to burn the fuel providing heat to decompose the limestone. The combustion gases, mixed with the CO_2 from the $CaCO_3$, leave at the top and the raw lime (CaO) leaves at the bottom. We shall not try to attack this problem in general terms but shall rather set up some specific conditions in order to avoid confusing generalities with respect to components, N_{sp}, and zero concentrations in the various streams.

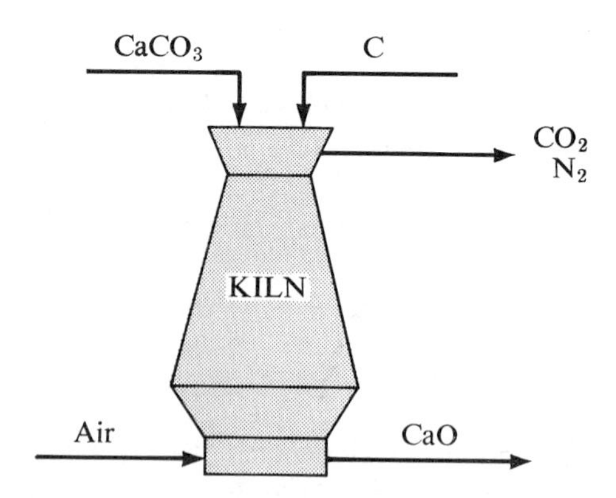

FIGURE 10.2 Flow sheet of vertical lime kiln.

The simplest possible situation would be a charge of pure $CaCO_3$ in one stream and pure carbon in the other, combined with complete decomposition of the carbonate and no excess air. These conditions are noted in Figure 10.2.

$$N_s = 5 + \text{(the heat stream)},$$
$$N_{sp} = 6(\text{CaO, C, O, N, } CO_2\text{, } CaCO_3).$$

Independent Reactions

$$CaCO_3 = CaO + CO_2,$$
$$C + O_2 = CO_2,$$
$$C = 4(CaO, C, O, N),$$
$$N_v = N_S(N_{sp} + 2) + 1 = 5(6 + 2) + 1 = 41.$$

Restrictions, N_r

Material balances	$C = 4$
Energy balance	1
Zero concentrations	23
Fixed ratios (O/N in air)	1
Total	29

$$N_{df} = 41 - 29 = 12.$$

In the assignment of these degrees of freedom we must bear in mind that the statement of the problem has eliminated the need for any statements of concentrations except in the outlet gas. Further, although two streams leave, there is no necessity for them to be at the same temperature and pressure, since they leave different parts of the apparatus. Pressure drop due to frictional flow could well cause a higher pressure at the bottom than at the top and the bed of solids can have temperature gradients due to different rates of the two reactions throughout the unit.

We shall start by assuming that all pressures are nearly atmospheric, the small differences which surely exist being unimportant to the enthalpies in comparison to temperature changes. We have used five degrees of freedom for those pressure specifications. Now assign inlet temperatures to three streams, using three more degrees of freedom. Name a temperature of the effluent CaO at which it is quite sure that all carbonate is decomposed. Nine of the degrees of freedom have now been named. The heat losses from a unit of this sort should be small compared to the heats of reactions and the enthalpies, so we shall assume that the process is adiabatic. There are only two unassigned degrees of freedom at this point, and no rates have yet been named. Choose the rate of one of the solid feeds, logically the carbonate since we are naturally interested in the capacity of the unit with respect to lime production. Now one degree of freedom remains. Choose it in one of two ways: (a) the rate of one of the other feeds, carbon or air; and (b) the remaining temperature, that of the outlet gas.

Two completely different types of solutions are generated. In the first, there is enough information to perform all the material balances solving for all quantities before the heat balance is undertaken. This heat balance will provide the temperature of the outlet gas. In the second, the amount of carbon required (and consequently the ratio of $CaCO_3$:C to be charged) will be calculated by simultaneous solution of the material and heat balances. The calculation for this second case appears in Example 10.5.

Example 10.4

▶ STATEMENT: A mixture consisting of 50% benzene by weight, 50% toluene by weight is to be distilled in a simple continuous unit (see Figure 10.1) operating without appreciable heat loss. The entire unit can be presumed to be at atmospheric pressure. The feed will be charged at a rate of 1000 lb per hr, and at 75 °F. The liquid and vapor streams from the unit may be assumed to be in physical equilibrium.

Calculate all missing temperatures, compositions, and heat inputs given any one of the following specifications of the last available degree of freedom:

a. the vapor composition is 65 mole % benzene.
b. the vapor temperature is 95 °F.
c. the net rate of heat input to the still can be as high as 150,000 Btu per hr.

AUXILIARY DATA (Obtained from handbooks.) The effluent liquid and vapor compositions are related by the expression

$$y = \frac{\alpha x}{1 + (\alpha - 1)x},$$

where x and y have the usual connotations, mole fractions of benzene in bottoms and in distillate, respectively, and where α is assumed to be sufficiently constant so that the value of 2.50 will be satisfactory under the specified conditions and atmospheric pressure. Heat capacities and latent heats of vaporization will be assumed to have the following constant values over all temperature ranges encountered:

Latent heats in calories per gram:
Benzene: 94.3;
Toluene: 86.5.
Specific heats: 0.45 for each, liquid state.

▶ SOLUTION:

PART (a) The basis of 1000 lb of feed seems as reasonable as any. We note that considerable care must be taken with respect to mass units since some of the data are in molal units (x and y) and some in pounds (feed rate, heat quantities).

1000 lb of feed
500 lb of benzene = 6.41 moles
500 lb of toluene = 5.43 moles

Total = 11.84 moles

Mole fraction of benzene in the feed = 0.541.

Material Balances (*in Molal Units*) where $D \equiv V$ in Figure 10.1.

$$\text{Total:} \quad D + W = 11.84 \quad \text{(a)}$$
$$\text{Benzene component:} \quad 0.65D + xW = 6.41 \quad \text{(b)}$$

Since $y = 0.65$, by the specification of this part of the problem, x may be determined by the equilibrium relationship which we now solve for x:

$$x = \frac{y}{2.5 - 1.5y} = 0.426.$$

Multiplying Equation (a) by 0.65 and substituting 0.426 for x in Equation (b):

$$\begin{aligned} 0.65D + 0.65W &= 7.70 \\ 0.65D + 0.426W &= 6.41 \\ \hline 0.224W &= 1.29 \\ W &= 5.76 \text{ moles} \\ D &= 11.84 - 5.76 = 6.08 \text{ moles.} \end{aligned}$$

The exit temperature at known pressure (atmospheric) and known composition ($x = 0.43$) can be determined from equilibrium data. The determination is not difficult but falls in the realm of physical chemistry, and outside of the scope of this text. It is found to be 94 °C.

The material balances are now complete. Enthalpies of all streams can be calculated, assuming additivity in the mixture for both sensible and latent heats. The heat balance will then provide information regarding the amount of net heat input required to operate the process.

Heat Balance

A datum plane is required. We choose 75 °F (23.9 °C) for the convenience of eliminating any enthalpy in the feed stream. Then the chosen phase should be liquid.

Input Enthalpies in Pcu

$$\begin{aligned} \text{In feed:} &\quad \text{zero (by choice of datum)} \\ \text{External heat supply:} &\quad \underline{Q} \\ \text{Total:} &\quad Q. \end{aligned}$$

Output Enthalpies in Pcu

The bottoms consists of 5.76 moles

$$\begin{aligned} \text{Benzene:} &\quad 2.45 \text{ moles} = 191 \text{ lb} \\ \text{Toluene:} &\quad 3.31 \text{ moles} = \underline{304 \text{ lb}} \\ &\quad \text{Total} \quad = 495 \text{ lb} \end{aligned}$$

Enthalpy in bottoms:

$$(495)(0.45)(94.0 - 23.9) = 15{,}615 \text{ Pcu } (1.561 \times 10^4).$$

The vapor consists of 6.08 moles, 505 lb.

$$\begin{aligned} \text{Benzene:} \quad (6.08)(0.65) &= 3.95 \text{ moles.} \\ \text{Toluene:} \quad 6.08 - 3.97 &= 2.13 \text{ moles.} \end{aligned}$$

To heat the 505 lb to boiling:

$$(505)(0.45)(94.0 - 23.9) = 1.593 \times 10^4 \text{ Pcu.}$$

To vaporize, using molal latent heats:

$$\begin{aligned} \text{Benzene:} \quad (3.95)(94.3)(78) &= 2.903 \times 10^4 \text{ Pcu.} \\ \text{Toluene:} \quad (2.13)(86.5)(92) &= 1.695 \times 10^4 \text{ Pcu.} \end{aligned}$$

$$\begin{aligned} \text{Total enthalpy out:} \quad & (1.561 + 1.593 + 2.903 + 1.695) \times 10^4 \\ &= 7.752 \times 10^4 \text{ Pcu,} \\ &= 13.95 \times 10^4 \text{ Btu.} \end{aligned}$$

Since the only input item is Q, $Q = 13.95 \times 10^4$ Btu.

PART (b) If the effluent temperature is specified, the composition boiling at this temperature is fixed and the problem resolves into the same type of solution just encountered in Part (a), reversing only the calculation of y (and x) and t.

PART (c) With the specifications given in this part we cannot establish y or x. Therefore we write the material balances (in moles), retaining these compositions as unknowns:

$$\begin{aligned} D + W &= 11.84, \\ yD + xW &= 6.41. \end{aligned}$$

Here we apparently have two equations and four unknowns; actually there are only three real unknowns since y and x are related through the equilibrium expression, a third mathematical relationship. Combining the two balances:

$$\begin{aligned} 6.41 &= yD + x(11.84 - D). \\ \therefore\ 6.41 &= (y - x)D + 11.84x. \end{aligned} \tag{c}$$

Now, using the same datum as chosen in Part (a), write the heat balance:

Input Enthalpies

Feed: zero.

External heat supply: 150,000 Btu $= 83.3 \times 10^3$ Pcu.

Output Enthalpies

Liquid: $0.45W'(t - 23.9)$ Pcu.

Vapor: $0.45D'(t - 23.9) + D'L_v$, where D' and W' are expressed in pounds and L_v is the latent heat of the mixture, in Pcu per pound.

Total: $450(t - 23.9) + D'L_v$, since $D' + W' = 1000$.

Heat Balance

$$450t - 11.84 \times 10^3 + D'L_v = 83.3 \times 10^3.$$

The equation appears to contain three unknowns but any chosen value of t establishes y and x with D' then obtainable from Equation (c). Then if the heat balance is written in the form

$$450t + D'L_v - 95.17 \times 10^3 = \Delta, \tag{d}$$

the equation is solved when the choice of t results in $\Delta = 0$.

TRIAL 1 $t^{(1)} = 95$ °C.

At this temperature and atmospheric pressure, $y = 0.63$. We calculate x as before,

$$x = \frac{0.63}{2.5 - 1.5(0.63)} = 0.405.$$

Then, from Equation (c)

$$0.225D + 11.84(0.40) = 6.41,$$
$$D = 7.15 \text{ moles.}$$

Benzene:	4.51 moles	= 376 lb
Toluene:	2.64 moles	= 243 lb
	Total	= 619 lb

Now substituting into Equation (d):

$$(450)(95) + (35{,}460 + 20{,}020) - 95{,}140 = \Delta,$$
$$\Delta = 2.19 \times 10^3.$$

TRIAL 2 $t^{(2)} = 92.5$ °C.

At this temperature, $y = 0.70$ and x can be calculated to be 0.482. Then, from Equation (c), $D = 3.20$ moles and the heat balance yields

$$\Delta = -29.5 \times 10^3.$$

TRIAL 3 $t^{(3)} = 94$ °C.

We know that this value is not correct since it is the temperature of Part (a), where the net heat input is only 140,000 Btu. But this trial value is between $t^{(1)}$ and $t^{(2)}$ one of which gave a negative Δ and the other of which gave a positive Δ. This intermediate value should complete the trials and allow a graphical interpolation to the t where $\Delta = 0$. At 94 °C, $y = 0.65$ and $x = 0.426$. From Equation (c) we find $D = 6.08$ moles and

$$\Delta = -6.7 \times 10^3 \text{ Pcu.}$$

Using the three values of Δ from trials 1, 2, and 3, we can plot the values of Δ *vs.* the trial temperatures, obtaining the answer where the line through these points crosses $\Delta = 0$. The answer is 94.8 °C. All other quantities can now be calculated.

Example 10.5

▶ STATEMENT: A vertical lime kiln is charged with pure limestone ($CaCO_3$) and pure coke (carbon), both at 25 °C. (See Figure 10.2.) Dry air at 25 °C is blown in at the bottom and provides the necessary heat for decomposition of the carbonate by burning the carbon to CO_2. Lime (CaO) leaves the bottom at 950 °F and contains no carbon or calcium carbonate. The kiln gases leave the top at 600 °F and consist only of CO_2 and N_2. Radiation losses are negligible. Calculate the ratio of $CaCO_3$:C to produce these conditions.

AUXILIARY DATA (Obtained from handbooks.)
Standard heats of formation at 25 °C, in kilocalories per gram-mole:

$CaCO_3$(s)	−289.5,
CaO(s)	−151.7,
CO_2(g)	−95.052,
C(s) and O_2(g)	0.

Molal heat capacities (deliberately chosen in different form from those in Table 7.1c):

$$\text{CaO:} \quad C_p = 10 + 4.84 \times 10^{-3}T - \frac{108{,}000}{T^2}.$$

$$\text{CO}_2\text{:} \quad C_p = 10.34 + 2.74 \times 10^{-3} - \frac{195{,}500}{T^2}.$$

$$\text{N}_2\text{:} \quad C_p = 6.5 + 1.00 \times 10^{-3}T.$$

▶ SOLUTION: All the temperatures are specified as well as the net Q. Thus, from the discussion in Example 10.3 we know that we cannot complete the material balances without aid from the energy balance. We shall set up the material balances with as few unknowns as possible.

BASIS Since most of the data are in moles, we choose a molal basis, 100 moles of $CaCO_3$ fed. We can write four material balances, since there are four components. Convenient components will be CaO, O, N, and C.

CaO-balance: the 100 moles of $CaCO_3$ will provide 100 moles of CaO as effluent.

C-balance: The charge of carbon is unknown. Assign to it the letter X. Then CO_2 comes from $CaCO_3$, in amount of 100 moles, and from the carbon charged all of which burns to CO_2, and therefore provides an amount X. CO_2 in effluent gas: $100 + X$ moles.

O_2-balance: All the oxygen charged appears in the CO_2. There is no excess. Therefore, the oxygen charge is X moles.

N_2-balance: The nitrogen charged with the oxygen in air is $3.76X$ moles.

Effluent gas

$$CO_2: \quad 100 + X \text{ moles};$$
$$N_2: \quad 3.76X \text{ moles}.$$

We can proceed no further with the material balances.

Heat Balance

(Datum at 25 °C and the states of aggregation chosen in writing the chemical reactions.)

Heats of Reaction

$$\begin{array}{ccccc} CaCO_3(s) & = & CaO(s) & + & CO_2(g) \\ -289.5 & & -151.7 & & -94.052 \end{array}$$

$$\Delta H^0_{298} = -151.7 - 94.052 + 289.5 = 43.75 \text{ kcal},$$
$$= 43{,}750 \text{ Pcu}.$$

$$\begin{array}{ccccc} C(s) & + & O_2(g) & = & CO_2(g). \\ 0 & & 0 & & -94.052 \end{array}$$

$$\Delta H^0_{298} = -94{,}052 \text{ kcal},$$
$$= -94{,}052 \text{ Pcu}.$$

Total heat input from reactions, in Pcu:

$$100(-43{,}750) + (X)(94{,}052).$$

Input Enthalpies

All materials charged are at the datum temperature, 25 °C, and at the states specified in the chemical equations used to calculate standard heats of reaction. Therefore, the total sensible heat input is *zero*.

Output Enthalpies

$$CaO: \quad 100\int_{298}^{783} C_{p_{CaO}}\,dT = 100\left(10T + 2.42 \times 10^{-3}T^2 + \frac{108{,}000}{T}\right)\Bigg]_{298}^{783}.$$

$$CO_2: \quad (100 + X)\int_{298}^{589} C_{p_{CO_2}}\,dT.$$

$$= (100 + X)\left(10.34T + 1.37 \times 10^{-3}T^2 + \frac{195{,}500}{T}\right)\Bigg]_{298}^{589}.$$

$$N_2: \quad 3.76X\int_{289}^{589} C_{p_{N_2}}\,dT = 3.76X(6.5T + 0.50 \times 10^{-3}T^2\Big]_{298}^{589}.$$

Total: $893{,}300 + 10{,}610X.$

Heat Balance

$$-4{,}375{,}000 + 94{,}052X = 893{,}300 + 10{,}610X,$$
$$X = 63.0 \text{ moles of carbon}.$$

Therefore, the required ratio is

$$\frac{CaCO_3}{C} = \frac{100}{63.0} = 1.58.$$

Example 10.6

For the same process as described in Example 10.5, change the specifications only to the following extent. Do not specify the outlet gas temperature, but specify the ratio of carbonate to carbon which will be charged. Suppose that this ratio is set at 1.70. Now the outlet gas temperature is unknown but all the material balances may be completed prior to the introduction of any calculations involving the heat balance.

We shall not perform the calculations. Although the operations are quite straightforward in the light of those appearing in Example 10.5, they are tedious since the unknown is a temperature. This type of problem has already been thoroughly covered in Chapter 9.

LIST OF SYMBOLS

Latin letters

C	Number of components
D	Number of moles of distillate
D'	Number of pounds of distillate
L_v	Latent heat of distillate mixture, per pound
N_{df}	Number of degrees of freedom
N_s	Number of streams
N_{sp}	Number of species
N_v	Number of variables
P	Number of phases
Q	Quantity of heat in transit
T	Absolute temperature
t	Temperature in general
W	Number of moles of residue from distillation
W'	Number of pounds of residue from distillation
X	An unknown quantity
x	Mole fraction of more volatile component in liquid
y	Mole fraction of more volatile component in vapor
z	Number of zero compositions

Greek letters

α	Relative volatility, used in equilibrium relationship between y and x
Δ	A symbol for quantity to be reduced to zero by iteration

PROBLEMS

It will be apparent that many of the problems in Chapter 9 can be modified so that the specifications would require the simultaneous solution of heat and material balances. The following problems are largely in this category. The processes operate at atmospheric pressure unless otherwise specified.

10.1 Hydrochloric acid at 25° Bé is to be produced by absorption of HCl in water. The gas fed to the tower is an air-HCl mixture containing 25% HCl. The exit gas will contain 0.2% HCl. The gas stream and the water both enter at 20 °C and the liquid is to be maintained at this temperature so that the effluent hydrochloric acid is also at 20 °C. The heat radiated from the tower is estimated to be 70,000 Btu per hr. Calculate the tons of hydrochloric acid which can be produced per day.

10.2 Pure pyrites (FeS_2) is fed to a Herreshof burner at 25 °C, the resulting cinder being entirely Fe_2O_3. The air also enters at 25 °C. No appreciable amount of SO_3 is formed in the combustion. The effluent gas contains 9.8% oxygen and leaves the burner at 800 °F. The cinder leaves at 1000 °F. The heat radiated from this unit is estimated to be 100,000 Btu per hr.

Calculate the rate at which the pyrites can be burned to maintain the specified temperatures.

10.3 Limestone analyzes 85% $CaCO_3$, 14% $MgCO_3$, and 1% H_2O. It is charged to a kiln at 75 °F. Heat is supplied externally so that the flue gases from the fuel do not mix with the carbon dioxide evolved. Calcination of the limestone is complete; the product gases leave at 650 °F and the calcined lime leaves at 1400 °F. The net heat that can be supplied to the process is 10,000,000 Btu per hr.

Calculate the rate of production of calcined product.

10.4 In a kiln calcining the same lime described in the preceding problem, fuel is burned so that the flue gases are in direct contact with the solid and mix with the carbon dioxide evolved from the limestone. Temperatures are as stated in the previous problem.

The fuel is oil consisting of 85.7% carbon and the remainder hydrogen, with a net heat of combustion equal to 18,000 Btu per lb at 25 °C. 15% excess air is used in the combustion, no carbon monoxide appears in the products, and a quarter of a ton of oil is used per hour. The kiln is well insulated and is estimated to radiate not more than 50,000 Btu per hr.

Calculate the rate at which the limestone may be calcined.

10.5 In a contact sulfuric process the gases fed to the converter analyze 10.5% SO_2, 10.0% O_2, 79.5% N_2. These gases enter a two-converter system in which the SO_2 is substantially converted to SO_3 with negligible heat losses from the system. In the first converter the gases enter at 425 °C and the outlet temperature is measured at 525 °C.

Calculate the percentage of SO_2 converted to SO_3 in the first converter.

10.6 Formaldehyde is manufactured by the vapor-phase oxidation of methanol with air on a silver catalyst. The primary reaction is

$$CH_3OH + \tfrac{1}{2}O_2 = CH_2O + H_2O.$$

A small portion of the formaldehyde may be subsequently oxidized to formic acid.

The feed mixture contains oxygen in the amount theoretically required for oxidation of the methanol to formaldehyde and the mixture enters the catalyst chamber at 170 °F. Analysis of the converted gas indicates that 60% of the methanol was oxidized. The exit gases leave the reaction zone at 1100 °F. The heat lost from the reactor is negligible.

Calculate the fraction of the formaldehyde oxidized to formic acid.

Additional data may be assumed to be the same as reported in Problem 9.31.

10.7 Carbon black is an important industrial product, large tonnage being compounded into rubber used for tires. A large proportion of the industrial carbon black is made by controlled combustion of fuel oils. The heat generated in partial combustion is used to decompose the remainder of the fuel oil. The gases leave the furnace at 2500 °F and are quenched to 500 °F by spraying water at 90 °F directly into the air stream. The carbon black is subsequently removed by electrostatic precipitation and by scrubbing.

The fuel oil acts in all respects like hexadecane ($C_{16}H_{34}$). This fuel enters the furnace at 25 °C as does air. The approximation of the latter as dry will introduce no appreciable error in the calculation.

Ninety per cent of the hydrogen formed in the decomposition of the hexadecane does not burn to water, and 50% of the carbon burned forms carbon monoxide with the remaining 50% burning completely to carbon dioxide. Sixty per cent of the carbon in the fuel is decomposed to carbon black. Assume that heat losses from the furnace are negligible. Calculate:

a. composition of the exit gas prior to quenching.
b. ratio of fuel oil to air.

10.8 A vertical lime kiln is charged with pure limestone ($CaCO_3$) and pure coke (carbon), both at 25 °C. Air, dry, is blown in at the bottom and provides the necessary heat for decomposition of the carbonate by burning the carbon to CO_2. The lime (CaO) leaves the bottom at 950 °F, containing no carbon or undecomposed limestone.

The kiln gas leaves at 600 °F and contains no oxygen. Radiation losses will be assumed to be zero. The air has been supplied at 25 °C but means are now at hand to increase this temperature to 250 °C.

For CaO, use:

$$C_p = 10 + 0.484 \times 10^{-2}T,$$
T in °K;
C_p in Pcu/(lb-mole)(°K).

Calculate:

a. the ratio of the $CaCO_3$:C in the charge.
b. the increase in the ratio by the increase of air temperature.

Appendix A

Table of Atomic Weights of the More Common Elements
(*Based on* Carbon^{12})

Element	Symbol	Atomic Weight
Aluminum	Al	26.9815
Antimony	Sb	121.75
Argon	Ar	39.948
Arsenic	As	74.9216
Barium	Ba	137.34
Beryllium	Be	9.0122
Bismuth	Bi	208.980
Boron	B	10.811
Bromine	Br	79.909
Cadmium	Cd	112.40
Calcium	Ca	40.08
Carbon	C	12.01115
Chlorine	Cl	35.453
Chromium	Cr	51.996
Cobalt	Co	58.9332
Copper	Cu	63.54
Fluorine	F	18.9984
Gold	Au	196.967
Helium	He	4.0026
Hydrogen	H	1.00797
Iodine	I	126.9044
Iron	Fe	55.847
Krypton	Kr	83.80
Lead	Pb	207.19
Lithium	Li	6.939
Magnesium	Mg	24.312
Manganese	Mn	54.9380
Mercury	Hg	200.59
Molybdenum	Mo	95.94
Neon	Ne	20.183
Nickel	Ni	58.71
Nitrogen	N	14.0067
Oxygen	O	15.9994
Palladium	Pd	106.4

(*continued*)

Table of Atomic Weights of the More Common Elements
(Based on Carbon^{12})

(continued)

Element	Symbol	Atomic Weight
Phosphorus	P	30.9738
Platinum	Pt	195.09
Potassium	K	39.102
Rhodium	Rh	102.905
Selenium	Se	78.96
Silicon	Si	28.086
Silver	Ag	107.870
Sodium	Na	22.9898
Strontium	Sr	87.62
Sulfur	S	32.064
Tantalum	Ta	180.948
Tin	Sn	118.69
Titanium	Ti	47.90
Tungsten	W	183.85
Uranium	U	238.03
Vanadium	V	50.942
Xenon	Xe	131.30
Zinc	Zn	65.37
Zirconium	Zr	91.22

Source: Selected from data appearing in *Handbook of Chemistry and Physics*, 44th ed. (Cleveland, Ohio: Chemical Rubber Publishing Company, 1962–63).

Appendix B

Selected Critical Constants and Values of a and b to Be Used with van der Waals' Equation*

Compound	T_c	P_c	a	b
Acetylene	309.1	61.7	1,129	0.8232
Ammonia	405.5	111.5	1,075	0.598
Benzene	561.6	47.7	4,820	1.935
n-Butane	426.0	36.0	3,675	1.944
Carbon dioxide	304.1	72.9	925	0.686
Carbon monoxide	134.4	34.6	381	0.639
Chlorine	417.	76.	1,668	0.90
Cyclohexane	554.1	40.6	5,513	2.242
Diethyl amine	496.4	36.58	4,923	2.229
Dimethyl amine	437.7	52.4	2,665	1.372
Ethane	305.2	48.8	1,391	1.028
Ethyl chloride	460.3	52.	2,970	1.455
Ethylene	282.8	50.7	1,150	0.9165
Helium	5.2	2.3	8.57	0.372
n-Heptane	540.0	26.8	7,931	3.311
n-Hexane	507.9	29.5	6,374	2.829
Hydrogen	33.2	12.8	62.8	0.426
Hydrogen chloride	324.5	81.6	942	0.654
Hydrogen sulfide	373.5	88.9	1,145	0.691
Isobutane	407.1	37.0	3,265	1.808
Isopentane	460.9	32.92	4,704	2.300
Methane	191.1	45.8	581.2	0.6855
Methyl chloride	416.2	65.8	1,920	1.041
Monoethyl amine	456.3	55.54	2,735	1.351
Monomethyl amine	430.0	73.6	1,832	0.960
Nitrogen	126.0	33.5	346	0.618
Oxygen	154.3	49.7	349.5	0.510
n-Pentane	470.3	33.0	4,886	2.342
Propane	369.9	42.01	2,374	1.446
Propylene	364.8	45.0	2,155	1.332
Sulfur dioxide	430.3	77.7	1,737	0.910
Sulfur trioxide	491.4	83.8	2,105	0.964
Water	647.3	218.2	1,400	0.488

Source: Selected from comprehensive data appearing in B. F. Dodge, *Chemical Engineering Thermodynamics* (New York: McGraw-Hill Book Company, 1944).

*T_c is in °K; P_c is in atm; a is in (atm)(ft^6)/(lb-mole)^2; b is in ft^3/lb-mole.

Appendix C

Properties of Saturated Steam

Abs Pressure lb/sq in. P	Temp °F t	Specific Volume cu ft/lb: Saturated Liquid v_f	Specific Volume cu ft/lb: Saturated Vapor v_g	Enthalpy Btu/lb: Saturated Liquid H_f	Enthalpy Btu/lb: Evaporation L_v	Enthalpy Btu/lb: Saturated Vapor H_g
1.0	101.74	0.01614	333.6	69.70	1036.3	1106.0
2.0	126.08	0.01623	173.73	93.99	1022.2	1116.2
3.0	141.48	0.01630	118.71	109.37	1013.2	1122.6
4.0	152.97	0.01636	90.63	120.86	1006.4	1127.3
5.0	162.24	0.01640	73.52	130.13	1001.0	1131.1
6.0	170.06	0.01645	61.98	137.96	996.2	1134.2
7.0	176.85	0.01649	53.64	144.76	992.1	1136.9
8.0	182.86	0.01653	47.34	150.79	988.5	1139.3
9.0	188.28	0.01656	42.40	156.22	985.2	1141.4
10	193.21	0.01659	38.42	161.17	982.1	1143.3
14.696	212.00	0.01672	26.80	180.07	970.3	1150.4
15	213.03	0.01672	26.29	181.11	969.7	1150.8
20	227.96	0.01583	20.089	196.16	960.1	1156.3
25	240.07	0.01692	16.303	208.42	952.1	1160.6
30	250.33	0.01701	13.746	218.82	945.3	1164.1
35	259.28	0.01708	11.898	227.91	939.2	1167.1
40	267.25	0.01715	10.498	236.03	933.7	1169.7
45	274.44	0.01721	9.401	243.36	928.6	1172.0
50	281.01	0.01727	8.515	250.09	924.0	1174.1
55	287.07	0.01732	7.787	256.30	919.6	1175.9
60	292.71	0.01738	7.175	262.09	915.5	1177.6
65	297.97	0.01743	6.655	267.50	911.6	1179.1
70	302.92	0.01748	6.206	272.61	907.9	1180.6
75	307.60	0.01753	5.816	277.43	904.5	1181.9
80	312.03	0.01757	5.472	282.02	901.1	1183.1
85	316.25	0.01761	5.168	286.39	897.8	1184.2
90	320.27	0.01766	4.896	290.56	894.7	1185.3
95	324.12	0.01770	4.652	294.56	891.7	1186.2
100	327.81	0.01774	4.432	298.40	888.8	1187.2
110	334.77	0.01782	4.049	305.66	883.2	1188.9

(*continued*)

Properties of Saturated Steam (continued)

		Specific Volume cu ft/lb		Enthalpy Btu/lb		
Abs Pressure lb/sq in. P	Temp °F t	Saturated Liquid v_f	Saturated Vapor v_g	Saturated Liquid H_f	Evaporation L_v	Saturated Vapor H_g
120	341.25	0.01789	3.728	312.44	877.9	1190.4
130	347.32	0.01796	3.455	318.81	872.9	1191.7
140	353.02	0.01802	3.220	324.82	868.2	1193.0
150	358.42	0.01809	3.015	330.51	863.6	1194.1
160	363.53	0.01815	2.834	335.93	859.2	1195.1
170	368.41	0.01822	2.675	341.09	854.9	1196.0
180	373.06	0.01827	2.532	346.03	850.8	1196.9
190	377.51	0.01833	2.404	350.79	846.8	1197.6
200	381.79	0.01839	2.288	355.36	843.0	1198.4
250	400.95	0.01865	1.8438	376.00	825.1	1201.1
300	417.33	0.01890	1.5433	393.84	809.0	1202.8
350	431.72	0.01913	1.3260	409.69	794.2	1203.9
400	444.59	0.0193	1.1613	424.0	780.5	1204.5
450	456.28	0.0195	1.0320	437.2	767.4	1204.6
500	467.01	0.0197	0.9278	449.4	755.0	1204.4
550	476.94	0.0199	0.8424	460.8	743.1	1203.9
600	486.21	0.0201	0.7698	471.6	731.6	1203.2
650	494.90	0.0203	0.7083	481.8	720.5	1202.3
700	503.10	0.0205	0.6554	491.5	709.7	1201.2
750	510.86	0.0207	0.6092	500.8	699.2	1200.0
800	518.23	0.0209	0.5687	509.7	688.9	1198.6
850	525.26	0.0210	0.5327	518.3	678.8	1197.1
900	531.98	0.0212	0.5006	526.6	668.8	1195.4
950	538.43	0.0214	0.4717	534.6	659.1	1193.7
1000	544.61	0.0216	0.4456	542.4	649.4	1191.8
1100	556.31	0.0220	0.4001	557.4	630.4	1187.8
1200	567.22	0.0223	0.3619	571.7	611.7	1183.4
1300	577.46	0.0227	0.3293	585.4	593.2	1178.6
1400	587.10	0.0231	0.3012	598.7	574.7	1173.4
1500	596.23	0.0235	0.2765	611.6	556.3	1167.9
2000	635.82	0.0257	0.1878	671.7	463.4	1135.1
2500	668.13	0.0287	0.1307	730.6	360.5	1091.1
3000	695.36	0.0346	0.0858	802.5	217.8	1020.3
3206.2	705.40	0.0503	0.0503	902.7	0	902.7

Source: Selected from data appearing in S. H. Keenan and F. G. Keyes, *Thermodynamic Properties of Steam* (New York: John Wiley & Sons, Inc., 1936).

Answers to Selected Problems

Chapter 2

2.1 a. 19.84; 11.02.
b. 1902; 1057.
c. 36,000; 20,000.
d. 7.35 in all sets of units.
e. 72.51.
f. 1.167×10^6; $(g_c/g)\ 1.167 \times 10^6$; 0.589.
g. 2373.5×10^3; 659.3; 2373.5×10^{10}; 2373.5×10^{10}.
h. 0.393×10^{-3}.
i. 282.9.

2.2 8.93×10^{-8}.
2.5 0.908; 0.946 liters/qt.
2.8 0.751; 0.751×10^4.
2.12 4.82×10^5; 0.65×10^6; 0.244; 0.646×10^4.
2.15 1.118×10^{-6}; 0.652×10^{-6}.
2.19 P.E. = $(g/g_c)\ 10^6$ = K.E.
2.20 6.59% CH_4; 92.29% C_2H_4; 26.7; 26,850.
2.25 0.231; 0.119; 7.5; 7.23.
2.26 °F = 3.427 °A − 169.6.
°C = 1.904 °A − 112.
2.30 1.337; 2.93.
2.32 0.1576; 3.65; 0.228; 0.00365.
2.35 0.0281; 497 °C.

S2.1A 10.46 ft of water.
S2.4A A. a. 0.503; b. 1.161.
B. a. 0.442; b. 1.161.
S2.7A 2.16 mm.

S2.10A a. 12.95 psia; b. 1.021 psi.
S2.16A a. 0.77 psia
b. 0.71 psia −1.76 mm Hg.

Chapter 3

3.2 9.2.
3.3 0.93.
3.6 40.74% O_2.
3.9 a. 1203.5; b. 50.67% B.
3.12 a. 44.3; b. 2620.
3.16 a. 15.41% CO_2; b. 13.18.
3.20 3.99.
3.23 4672 atm.
3.26 1.71 lb.
3.31 1152 mm Hg.
3.33 168 °C.
3.34 a. 0.651; b. 0.00651; c. 0.01751.
3.36 52.7 or 55.5.
3.40 81.1%.

Chapter 4

4.2 a. 72.8% CH_4; b. 46.6.
4.5 a. 0.464; b. 0.536; c. 3.26.
4.6 a. 2.14/1; b. 0.413.
4.9 a. 219.2, 288.4, 492.4; b. 780.8.
4.13 a. 9.6 acid, 40.4 water; b. 48.
4.15 a. 928.4, 7.35; b. 105.2.
4.19 377.
4.22 a. 6645; b. 38.2.
4.23 a. 123,000; b. 268,000; c. 270 lb; d. 4400 lb.

4.102 7.05, 4.67, 10.71.
4.104 9:1.
4.107 15.1*T*.
4.112 a. 9.50% CO_2 wet; 11.73% CO_2 dry.
b. 6.54% CO_2 wet; 7.5% CO_2 dry.
4.114 a. Yes; b. 30.1.
4.118 a. 19,450 air, 18,800 F.G. (STP); b. 39.1;
c. 3.92% H, 6.38% combined water, 9.7% ash.
4.120 a. $0.367/*T*; b. $2.93.

4.202 138.
4.205 a. 1.55; b. 38%; c. 95.4%.
4.210 2400 lb, 174 gal.
4.213 a. 9.58%; b. 0.0361%; c. 5610.
4.218 a. 820; b. 9500; c. 98.8%.
4.224 $C_{107}H_{141}$; 0.682.

4.227 46.7%.
4.229 a. 10.99% CO_2; b. 1.54%.
4.232 a. 0.0354; b. 32.8.
4.237 a. 86.3% C; b. N.
4.241 a. 49% FeO; b. 72.2%; c. 1.070T; d. 0.633T.
4.245 a. 28.7; b. 5.24, 2.96; c. 4.14, 18.96, 76.9.
4.251 a. 1.96; b. 8.5% C_2H_4; c. 20.76% C_2H_5OH.

Chapter 5

5.1 264.5 min.
5.2 a. 46.1; b. 63.2%.
5.3 2000.
5.6 20 min.
5.8 b. 16 min.
5.11 a. fill; b. ∞ sec.
5.15 960 lb.
5.17 A. 0.293; B. 0.245.

Chapter 7

7.1 b. 4.1365×10^{-2} cal/(sec)(cm)(°C).
c. 0.3 cal/(gm)(°C).
7.6 $C_{pm} = \int_{t1}^{t2} C_p\, dt/(t_2 - t_1)$
7.7 Acetanilid (1): 45.3 Btu/(lb mole)(°F).
Acrylic acid (1): 37.6 Btu/(lb mole)(°F).
7.9 b. 14.00 Btu/(lb mole)(°C).
c. 14.00 Btu/(lb mole)(°C).
d. 17.96 Btu/(lb mole)(°C).
7.10 b. 74.24 Btu.
7.14 67.7%.
7.17 21,330 Btu.
7.19 a. 177.3 Btu/lb.
b. 167.8 Btu/lb.
7.20 198 cal/gm.

Chapter 8

8.1 a. −92,934 Btu.
j. +7,555 Btu.
8.2 a. +11.6 kcal.
8.3 a. −118.36 cal.
8.6 −30 kcal.
8.7 b. −13,360 cal.
c. −13,355 cal.
d. −16,697 cal.
8.9 ΔH^0_{298} for Process (a) = −202.2 kcal; for Process (b) = −116.47 kcal.

8.10 c. -9.56 kcal.
8.12 $B = 94.052$; $C = -34.16$; $G = 70.96$ when ΔH_F^0 is in kcal.
8.15 a. -216.24 kcal.
b. -213.7 kcal.
8.18 b. -11.7 Btu/lb soln.
d. -88.2 Btu/lb soln.
8.19 1.934×10^7 Btu.

Chapter 9

9.2 13,540 lb/day.
9.3 a. 8620; b. 15,520; c. 249 Btu/lb.
9.4 46,000.
9.8 25.9 Btu.
9.11 5.35×10^6 Btu/day.
9.15 a. 97.5 Btu added.
9.17 234 °C.
9.20 3.85×10^6 Btu/ton.
9.23 c. 1800–1850 °F.
9.27 a. 1510 °C; b. 1590 °C. (Answers dependent on $\mathbf{C}_p$ data used.)
9.30 725 °C.
9.35 220,000 Pcu; sensitive to assumptions and approximations.

9.100 879 Btu.
9.102 a. 0.486 lb; b. 0.584 lb.
9.105 A. $Q = -2146.7$; $W = -2773.7$; $\Delta H = 875$; $\Delta U = 627$, all in Btu.
B. $Q = -1574.2$; $W = -2201.2$; $\Delta H = 875$; $\Delta U = 627$, all in Btu.
9.108 13,470.
9.109 234.9 hp.
9.111 88.6 psig.
9.114 16 in. STD (15 in. I.D.); 114% increase.
9.117 6.5 ft.
9.120 Orifice.

9.200 a. 75; b. 75.1.
9.202 a. 26.7 °C; 32.5 °C.
9.203 b. 2.59 ft.
9.208 a. 3.10 hr; b. 2.91.
9.212 a. $-B^{-1} \ln (1 - BW_0/K)$ is the form of general solution.
9.217 a. 97 °F; b. 310.
9.218 107.5 °F, 99.9 °F.

Index